# RETHINKING
# THE PRINCIPLES
# OF WAR

# RETHINKING THE PRINCIPLES OF WAR

Anthony D. Mc Ivor

EDITOR

Naval Institute Press
Annapolis, Maryland

Naval Institute Press
291 Wood Road
Annapolis, MD 21402

First printing in paperback, 2007
ISBN-10: 1-59114-482-5
ISBN-13: 978-1-59114-482-3

The Library of Congress has cataloged the hardcover edition as follows:

Rethinking the principles of war / Anthony Mc Ivor, editor.
    p. cm.
  Includes bibliographical references.
  ISBN 1-59114-481-7 (alk. paper)
  1. Military art and science. 2. War (Philosophy) 3. Military art and
  science—History—21st century. 4. United States—Military policy.
  5. Military doctrine—United States. 6. World politics—21st century.
  I. Mc Ivor, Anthony.
  U102.R48 2006
  355.0201—dc22

                                                                2005024654

Printed in the United States of America on acid-free paper

14 13 12 11 10 09 08 07    9 8 7 6 5 4 3 2
First printing

# CONTENTS

## Part 3
## OPERATIONAL ARTS: IRREGULAR WARFARE

## Part 4
## POST-CONFLICT AND STABILITY OPERATIONS

Part 5
## INTELLIGENCE—WINNING THE SILENT WARS

# FOREWORD

War is about more than combat, and combat is about more than shooting. That is, war entails the interrelated actions and support of diverse people, public and private organizations, and institutions. If these relationships—and their implications for interagency integration and multinational coalitions—were so obvious, then we would have less debate about what constitutes a principle of war and what they mean collectively. Furthermore, if this truism were more widely recognized, then our nation would have a less ad-hoc approach to responding to friction—the known unknowns of warfare. The hierarchical relationship of war to combat and to shooting is analogous to the relationship between strategy, operations, and tactics. This analogy indicates that some unifying theme or theory binds the activities.

Some theorists have argued that the unchanging nature of war, an intense form of human competition involving violence, profound risk, and mutual danger, leads to the principles of war. But the only principle that type of thinking leads to—quite rightly—is that war should be avoided except as a last resort, which is an essential element of just war theory. Put another way, the unchanging nature of war leads to an enduring just war theory, while the ever-changing character of war leads to changes in the underlying principles governing how it is actually conducted. That is why this anthology is important.

Still other strategists have argued that we have an unchanged unifying theory of war's conduct complete with guiding principles in Carl von Clausewitz's nineteenth-century treatise *On War*. That Clausewitz is more often quoted than read is partly the result of the early-nineteenth-century writing style that renders much of the work inaccessible to the modern reader. But more so because one can find within Clausewitz's work justification for any point on war, strategy, or the nature of combat that one chooses to make. This is not an attack on the enduring value of his work, but merely

reflects a growing recognition that many of the tenets espoused are less relevant in today's information age. The time is ripe, then, for a rethinking of the
principles of war. This anthology is a laudable first step in catalyzing a new
intellectualization of the changed competitive landscape confronting those
who would chose war in the twenty-first century.

This is not a blasphemous proposal. Intellectualizations on war frozen in
the nineteenth century offer a useful departure point for a new debate. War
cannot be divorced from the era in which it takes place. Too often the study
of war has been "didactic and normative," with experts seeking to ferret out
certain "immutable principles" to serve as "guides for the efficient conduct of
war in the future," according to British military historian Sir Michael
Howard. However, to "abstract war from the environment in which it is
fought . . . is to ignore a dimension essential to the understanding, not simply of the wars themselves but of the societies which fought them," Howard
says in his concise study *War in European History.*

The strategic situation regarding war is even more dynamic than the world
Clausewitz faced when he set out to write a compelling vision on the nature of
war. Today, the sets of rules that governed or regulated much of the dynamics
of military confrontation during the past fifty years have been swept aside by
the fall of Communism, have been rendered less relevant given the unmitigated pace of globalization, or have been overtaken by the information age.
We are experiencing a period of rapid change wherein many of the international bodies that guided much of the postwar political, economic, and military discourse, such as the United Nations, NATO, and even the International
Monetary Fund, are now under increasing pressure to transform. The era is
generating profound changes in terms of how militaries maintain both their
competency for the age and their relevancy for the times.

Clausewitz wrote *On War* in the aftermath of the Napoleonic Wars that
imposed wholesale changes in the established European order and fundamentally altered the continent. He sought to account for the enormous
changes in warfare wrought during that era. We must now undertake a similar venture to understand how war is changing and how that will influence
our future. The time is ripe for a reassessment of Clausewitz's nineteenth-
century dictums.

That is why an initiative like *Rethinking the Principles of War* is such an
important undertaking. As the conceptions of preemption versus prevention
become codified into the U.S. National Military Strategy, it is imperative that
we seek to generate a public debate on the emerging character of war. Open-

ness is important. It contributes to the richness/reach phenomenon that characterizes the information age, in which the dynamics of information sharing increase quality. So, the more this debate can be done in the open, the greater is the participation of interested parties, the greater the intellectual rigor of the overall enterprise.

There are clear distinctions to be made, however, between the *nature of war* and the *character of war*. This is not a semantic issue. Language is important in this context because it conveys culture. While the nature of war is fundamentally unchanging, the character of war is always in flux. It changes as societies and political entities evolve because it is a product of those separate interactions. Nation-states, and even substate and transnational actors, wage war in ways consistent with their culture, values, and resources. So, the way of war reflects the way a society creates, measures, and disposes of power and wealth.

On today's battlefield, we can witness new metrics being created that are the entry fee to the types of capabilities future forces must possess. These are access, speed, distribution, sensing, mobility, and networking. These are society's new metrics. They are scale free and valid at every level of warfare—tactical, operational, and strategic. How are these new metrics impinging on the traditional principles of war delineated by Clausewitz nearly two hundred years ago? That is what this anthology is all about.

Whether in force building or actual force operations, new characteristics are emerging that imbue those militaries that embrace them with new competitive advantages. In a sense, these can be termed the metrics of success for information age operations in rapidly changing times. These are the ability to

- create and preserve options
- develop high transaction rates
- instill high learning rates
- achieve overmatching complexity at scale

Because war takes place in the context of the societies so engaged, knowledge and cultural understanding are linchpins in this formulation. I am heartened that Maj. Gen. Bob Scales, U.S. Army (Ret.), has focused attention on these elements in his contribution. Changing culture is at the heart of transformation more so than any technological or organizational elements. As the pace of change accelerates in the information age, so must the transaction rates that create new capabilities from that "learning." Stagnation of institutional learning comes at the expense of future advantage.

Developing a culture committed to learning and understanding adversaries and how they adapt is critical to attacking the barriers that now prevent the U.S. military from enjoying the fruits of such concepts as the global scouts that Scales so forcefully articulates in his writing. Even while facing the daunting challenge of prosecuting the global war on terrorism, the U.S. military must never become too busy to learn, as the British Army belatedly recognized during the nineteenth century—and Scales rightly illuminates. Our adversaries are adapting and evolving at the speed of modern business; we cannot afford to be operating at the pace of bureaucratic doctrine. Faster institutional knowledge is critical in order to take advantage of what the new age offers.

With war increasingly less likely to pit nation-state against nation-state—the very foundation on which Clausewitz's principles are based—a new understanding of the dynamics taking place across the spectrum of conflict is required. Michael Vlahos perceptively explains that this development is more akin to medieval relationships and much less like those ushered in after the Treaty of Westphalia to which Western nations have long grown accustomed. New threats are emerging from groups and people who are "disconnected" from the wave of globalization and not connected to the "core" of emerging societies, according to Tom Barnett in *The Pentagon's New Map*. This holds enormous implications for the principles of war. Great power war has seemingly been taken "off the table," at least for the near future. The locus of violence is shifting dramatically downward to the level of the individual. This is a much more nuanced threat—defined by the vague, the inconsistent, and the irrational—that we are still groping to measure. As a result, we are discovering that forces must be rebalanced and realigned to fit this fundamentally different strategic context. The character of war is changing. If and when great power war reemerges, as some strategists argue it must, it will reappear in a new and shocking form to challenge our current conception of the character of war.

Long-held principles are chaffing against new realities on the battlefields. When this happens, rules change. But this rethinking of long-held principles should not simply end with the publication of this anthology. This body of work is focused at the operational level of warfare. There are new questions arising at the strategic level of war that are also worthy of extended thought and examination. Foremost is the need for a new intellectualization spanning all levels of war and all modes of competition. I have long proposed that network-centric warfare provides an emerging theory of war for that purpose. But what are the principles of war that would be similarly spanning?

As we have witnessed in both Afghanistan and Iraq, achieving political victory is far different from "winning" on the battlefield. In fact, success on the battlefield is only loosely coupled with achieving political aims. This gap yawns before us. The nation's military force must be an adaptive instrument of national power. It must provide *political* utility across a much more diverse and difficult range of scenarios and circumstances. This force must act as a flexible instrument of policy engagement, not simply provide a larger sheaf of thunderbolts.

In an era where national borders and the seizing of territory hold less and less relevance in the face of determined individuals equipped with weapons of mass destruction, there is a dire need for a new conceptualization to govern the creation of military power and to bolster the moral underpinnings for its application. Clinging to outdated vestiges of what victory once meant only brings risk. In his magisterial study *The Shield of Achilles,* Philip Bobbitt writes that with nation-states in decline and threats like those of 9/11 in the ascendancy, "there will be no final victory in such a war. Rather, victory will consist in having the resources and the ingenuity to avoid defeat." Is that good enough for a general principle, or is it merely as close as we can come to victory in a specific mode of warfare? For the general case, perhaps Bobbitt aims too low. Are victory and defeat merely events, or are they conditions? We must expend the intellectual capital now, not only to ensure our future, but also to create it.

VICE ADM. ARTHUR K. CEBROWSKI, U.S. Navy (Ret.)

---

Admiral Cebrowski served as the founding director of the Pentagon's Office of Force Transformation until February 2005. A former president of the Naval War College and a combat aviator in Vietnam, Cebrowski's thinking is widely considered to be the intellectual genesis of the concept of network-centric warfare.

# PREFACE

The *Rethinking the Principles of War* project began with a simple yet complex question. In late 2002, prompted by direct combat experience during Operation Enduring Freedom in Afghanistan, we asked whether the traditional principles of war had changed. That, in turn, led to other questions. Was the very concept of principles of war still relevant to warfare in the twenty-first century? What, then, to make of the list of principles embedded in our fundamental service doctrines since the 1920s? Moving beyond the legacy nine, it seemed to us that a thematic focus on the principles of war could open an opportunity to explore *the idea* of first principles.

In talking about first principles, we had in mind those elements, conceptual and physical, that animate the present national security environment and shape the outcomes of military action. While such principles are not deterministic, either individually or in tandem, in the past we ignored them only at great peril. Whether they still wield their original instructive or admonitory powers is an intriguing question. But it was not our immediate purpose. Instead, our objective was to examine the notion of principles, per se. The project—an extended conversation—would not seek to replace one set of principles, hostage to time and place, with another set equally constrained. We acknowledged that there would be no perfect, quick, or easy answers. As a start, we simply sought to elicit the right questions.

In the early days, the project benefited immeasurably from the foresight and encouragement of the Honorable Gordon R. England, secretary of the navy, and Vice Adm. Arthur K. Cebrowski, founding director of the Office of Force Transformation. As intellectual mentors firmly committed to a no-holds-barred approach to every relevant question, they made all the difference.

Our first concern was to create the widest range of participation. To ensure that, we enlisted several program partners, and with them we planned a variety of program components. We were fortunate indeed to have the collaboration of the Johns Hopkins University's Applied Physics Laboratory, the U.S. Naval Institute, and the Office of Force Transformation. The members of the *Rethinking* Steering Committee, Fred Rainbow, Duncan Brown, Eric Van Camp, John Laraia, Robert Tomes, and in particular Harlan Ullman, contributed wit and wisdom, time and again. We are much in their debt.

The result of this joint approach was a senior seminar series that attracted more than a thousand participants during 2004–5; a national essay contest that drew nearly nine hundred entries; and the publication of this anthology of commissioned essays. The defense community's response to these various undertakings indicated a sustained level of interest in rethinking the place of principles—as a part of broader reassessment of national defense. The project also benefited from strong currents of intellectual ferment cutting across the larger national security community.

The context for these activities was an unusually far-ranging and public examination of the policies and programs that underwrite the American way of war. Intelligence reform proposals and counterproposals dominated the headlines for months as the greatest restructuring of the intelligence community since 1947 got under way. The Pentagon's budget negotiations forecast dramatic cuts in traditional weapons programs in a move away from systems designed for state-on-state warfare. Reflection on the rules of military engagement itself emerged in appointment hearings on Capitol Hill, and one Judiciary Committee member was moved to suggest that "a new law of war" was needed to clarify U.S. support for the Geneva Convention.

All of this took place under the looming shadow of the twin processes of the BRAC and force restructuring, while the emerging agenda for the upcoming Quadrennial Defense Review generated wide interest. In the midst of this intense and profound reassessment, the craving for clear ideas and new thinking was reflected in a widely circulated *Inside the Pentagon* article that offered a reading list of book picks from prominent commanders and defense experts; earlier, the *New York Times* chief book critic offered a similar survey titled "How Books Have Shaped U.S. Policy." In short, an audience for *Rethinking* already existed.

The essays selected for this volume reflect the extraordinary range of issues on the table today: who to recruit, what to acquire, when to train, how to fight—and why. The contributors, distinguished scholars and warriors, have actively shaped various facets of the present policy environment. In turn, their views have been shaped by it. Whether as participants or observers, their presence on the field brings us important new vantage points and, occasionally, an unexpected lesson. But rarely are these given in concert. If we were to agree on just one thought, it might be this: we (friends and foes alike) are embarked on a vast and enduring process of accelerated change, about which we have much yet to learn and within which the future of warfare can yet only be very dimly perceived.

Looking ahead, we believe the *Rethinking* approach offers a mechanism for constructive engagement and discovery. The broadest possible engagement is crucial to meeting the kaleidoscope of irregular risks that characterizes our time. Most importantly, the project extends an invitation to think anew, across the traditional barriers of discipline, politics, and habit. The authors in this volume have accepted the challenge. In doing so, they have given us fresh perspectives and new points of departure.

*Rethinking* is not a prescription. It is a journey, an exploration. We are still asking questions. Perhaps this collection will prompt you to join the conversation and the search for answers. We hope to see or hear from you at some point along the way.

# RETHINKING
# THE PRINCIPLES
# OF WAR

# Get On with It

## James M. Dubik

**M**ost of the professional literature written about warfare in the post–Cold War and information age concerns the conduct of war: how war will be fought. The fundamental claim, correct, is that throughout the history of the military profession, major changes in either the strategic environment or in technology have altered the way war is fought.[1]

Today, for example, the authors of *The Future of War* report that the manner in which wars are waged is undergoing "a dramatic transformation."[2] They are not alone; the same outlook is central to the argument in *The New Face of War, Network Centric Warfare,* and *Fighting for the Future.*[3] Although each of these authors, and many other theorists and warriors, differ as to the precise ways in which they believe the conduct of war will change, all agree to two central claims. First, that war in the post–Cold War and information age will be or should be fought differently from that of the Cold War and industrial age. Second, that the Cold War and industrial age armed forces we had, and to some degree still have, are not completely the ones we need.

We seem to have an emerging consensus. No mere "tweaks" to the Cold War/industrial age paradigm are sufficient. Business as usual with a few adjustments here and there simply will not do. Only fundamental changes, a transformation in the way we approach the profession of arms, will suffice. So, how to get that done?

The profession of arms does not like to talk about theories. But the fact of the matter is that theories are practical mental constructs. They describe a portion of reality; they explain how that portion works; they serve to guide the actions of those operating in that portion; and they help forecast what

1

will happen next. The principles governing the profession of arms—its theory—are in the midst of profound change. And this change is far from over.

A brief look at scientific revolutions provides a glimpse of how fundamental is the continuing transformation in the profession of arms. From Aristotle, who held that heavenly bodies move by their own nature, to Galileo, who posited an empirical theory of motion; from Ptolemy's earth-centered explanation of the movement of heavenly bodies to the Copernican sun-centered explanation; and from Newtonian physics to that of Einstein—scientific revolutions are instructive with respect to military and national security transformation. Each revolution represented not just a slight adjustment to the prevailing theory; each represented a fundamentally new way of looking at the world. Therefore, each had immense practical consequences: what counted as evidence, as explanation; who got chairs of science at universities; where grant money or state sponsorship went, who was ultimately included in the professional community, etc.[4] The same is true for military and civilian security professionals: the post–Cold War strategic environment and emergence of the information age are not just "theoretical changes"; they have immense practical consequences for all associated with our nation's security.

Another interesting parallel arises when comparing scientific revolutions to the emergence of a new framework governing national security matters: when the current scientific view—whether that of Aristotle, Ptolemy, or Newton—faced difficulties, the adherents of that view tried to explain away the difficulties as "anomalies." Then, as the number of anomalies multiplied, they tried to make minor adjustments to the prevailing view—tweaking the accepted practices, policies, organizations, and programs. Ultimately, tweaking proved inadequate to explain the growing numbers of "anomalies" that they faced, and a new scientific theory emerged—a fundamental change of the way scientists looked at reality and conducted their business.

Literally, Galileo saw, interpreted, reached conclusions, and explained reality differently from Aristotle; Copernicus, differently from Ptolemy; and Einstein, differently from Newton. Fundamental changes in a governing scientific paradigm, says Thomas Kuhn in *The Structure of Scientific Revolutions,* causes "scientists to see the world . . . differently."[5]

A similar fundamental shift is required in the way national security professionals, military and civilian, must look at reality and conduct their business. Simply put, the national security "map" of reality has changed fundamentally—for military and civilian security specialists alike. The framework that guided it for the past fifty years has lost its explanatory power. We're on a new map sheet, so-to-speak. The key to understanding how much we have to change lies in realizing and accepting how radically different the new map sheet is.

As the 1990s progressed and the need for change became more and more apparent, the first focus for each of the American armed services—independently and collectively with Joint Forces Command—has been experimenting with new organizations, new headquarters, new operational procedures, and new weaponry. The American profession of arms generally recognizes the need for fundamental change in the way it conducts wars of all varieties—from conventional combat to irregular war to stability operations. The profession understands that application of the old Cold War and industrial age framework is unsatisfactory. The profession is arguing within itself as to what specific changes are necessary. Furthermore, each service—individually and together with the joint community—has made significant changes as a result of experimentation, study, and operational lessons learned. In the last decade, especially the last five years, the American armed forces have radically transformed their war-fighting capabilities and style.

Given that the future is still unfolding, only time will tell whether the profession "got it right," or perhaps more importantly, "didn't get it too badly wrong." Only in hindsight will we be able to judge fully and completely whether the American profession of arms could bring itself to the complete set of fundamental change called for by the new framework of the post–Cold War and information age.

A second aspect of the fundamental change in the conduct of warfare concerns the principles of war, principles that guide war fighting. These principles are as follows: maneuver (one side dislocates the other), offense (one side attacks the other), economy of force (minimize divergent activity), mass (focus on one point in space and time), objective (focus on one purpose), unity of command (have one commander), surprise (one side preempts the other), simplicity (focus on one idea), and security (one side forestalls the other).[6] In some senses, these principles are timeless. They can be found nascent in Sun Tzu's *Art of War* and explicitly stated in Clausewitz's *Principles of War,* as well as in "The Fundamental Principles of War" in Jomini's *The Art of War.*[7] The nine principles quoted above are sometimes described as the modern principles of war that were codified in the early-twentieth century.

The debate as to whether these principles have changed or should change continues. Leonhard's excellent book presents a powerful, complete, and cogent argument that the principles of war that govern the fighting of an information age force do change. The figure that follows depicts his view of the new principles of war for the information age.[8]

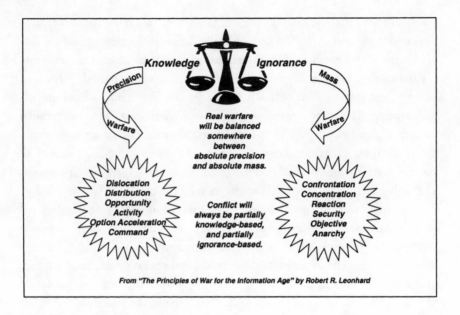

Precision Warfare

Knowledge    Ignorance

Mass Warfare

Real warfare
will be balanced
somewhere
between
absolute precision
and absolute mass.

Dislocation
Distribution
Opportunity
Activity
Option Acceleration
Command

Conflict will
always be partially
knowledge-based,
and partially
ignorance-based.

Confrontation
Concentration
Reaction
Security
Objective
Anarchy

From "The Principles of War for the Information Age" by Robert R. Leonhard

But others are unconvinced. They argue that the principles remain the same; only the application differs.

Again, only time will tell which side of this debate is correct, or whether some middle ground proves to be right. The important fact with respect to this volume of essays is that the American profession of arms recognizes that the new framework of the post–Cold War and information age requires a reexamination of the principles governing the conduct of war.

The operating forces that conduct war, as well as the principles that govern fighting, are getting ample professional attention. There is a separate set of forces, those that generate and sustain the operating force, however. These forces also need professional attention, because the ways in which a Cold War and industrial age operating force are generated and sustained will not produce and sustain a post–Cold War and information age operating force. We cannot just change the operating forces; we must also change the generating and sustaining forces.

The principles by which an operating force is generated and sustained, unfortunately, are largely unstudied. In fact, while we have a codified set of principles that govern the conduct of war—even if there remains debate as to their content and applicability in the post–Cold War and information age— no such set exists to describe how an operating force should be generated and sustained.

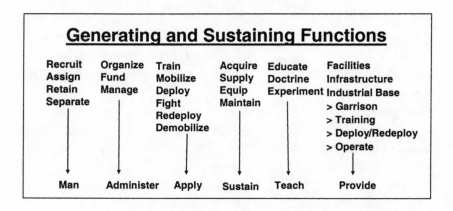

The functions of the generating and sustaining force, depicted in the figure that follows, are extensive.

These functions are not only extensive, but also complex. They are shared among several large institutions. not the least of which is Congress, the Department of Defense, the uniformed services and combatant commanders, as well as the American industrial base.

A further complicating factor is this: each of the services and the joint force portions of the generating and sustaining forces have guided the experimentation and change of the operational forces. No doubt, some operating forces, service and joint, participated in experiments, and leaders in service and joint operational forces help structure the decisions associated with their change. Equally without doubt, however, is that service and joint generating and sustaining forces have been primarily at the helm of operational change. Generating and sustaining forces have been able to stand back and provide some sense of objectivity and intellectual rigor.

Now it's their turn, for few of the generating and sustaining functions that were designed and continuously improved to serve industrial age armed forces created to succeed in the Cold War will function well in today's and tomorrow's environment. As hard as change has been and continues to be for the operating forces, it will be much more difficult for the generating and sustaining forces.

A comprehensive and cohesive experimental program for the generating and sustaining forces is as important as the service and joint programs associated with the operating forces. Such an experimental program, already under way in some sectors, will have to involve an even wider community than did the experiments associated with changing the operating forces. The appropriate start point should be an assumption that the generating and sustaining force that now exists—a generating and sustaining force designed and improved for an industrial age army to succeed in the Cold War environment—is not appropriate for an information age armed force that must succeed in the post–Cold War strategic environment.

We have fundamental issues to work out, difficult issues that get at the heart of transforming the ways in which we man, administer, apply, sustain, teach, and provide for our post–Cold War and information age operating forces. The issues raised in the following paragraphs are merely a sample of the hard questions the profession must ask itself—in conjunction with Congress, the Department of Defense, and the American industrial base.

The rules governing a volunteer armed force must adapt; what was satisfactory in the 1980s no longer works in many ways for the 2000s. How does the nation protect citizen-soldiers in the National Guard and U.S. Army Reserve so that they can serve their nation when called and continue to rise in their chosen civilian careers? How does the nation change our citizen-soldier program to account for the new requirements of the post–Cold War strategic environment? How should the leader-to-led ratio throughout the operating as well as generating and sustaining force adjust to account for the requirements of a more joint force? Can the armed forces really compete monetarily with the growing numbers of high-salary contractors? If not, what's an alternative approach to the retention of high-demand military skills? How does an information age armed force recruit and retain knowledge workers at every rank in an armed force used to "up or out" policies? Are the current percentages of enlisted, noncommissioned officers, warrant officers, and commissioned officers the right mix for the armed forces we are building? Which part of the generating and sustaining force should rightly remain uniformed, which government service, and which contract?

Distributive collaboration, virtual teaming, online simultaneous staffing are all becoming commonplace in the global, net-centric corporate world. These methodologies are also becoming common in the operating force. The hierarchical, sequential, bureaucratic methodology reigns supreme, however, in service and joint staffs, civilian secretariats, and Congress. How should an information age generating and sustaining force be administered? How can the generating and sustaining forces take advantage of collaborative tools to make corporate decisions faster and more effectively without the loss of unity of effort?

If a post–Cold War and information age armed force fights in a joint, interagency, and multinational (allies or coalition partners) way, how can it practice the application of force—military and nonmilitary, lethal and non-lethal, kinetic and non-kinetic? What are the changes required in America's interagency processes and what are the organizations necessary in light of the post–Cold War strategic environment and information age tools? How can we build depth into nondefense agencies so that they can plan and train with their defense partners? What are the right mobilization and demobilization policies applicable for the post–Cold War strategic environment? How do so many contractors and other civilians now common on the battlefield affect the laws of war? How should conventions, treaties, and laws accommodate the expanding understanding of "war"—especially war against non-nation-states—and the movement of some war into the space between conventional combat and crime?

How does an information age armed force reconcile the ease with which leaders can go online for peer collaboration and to get current "best practices" with the benefits that derive from an officer and sergeant education system? Expert knowledge is migrating from military schoolhouses and manual writers to the field, so how does an information age armed force bring the power of this near-real time, online learning into the classroom as many commanders have already done for home station training? How can peers in nondefense agencies keep up with the pace of this learning? What changes in the overall military and civilian professional education systems are needed to acknowledge that war is fought by a combination of military, interagency, and multinational partners? Should an information age force operating, generating, and sustaining in the post–Cold War strategic environment make a distinction between "skill improvement training" and "culture changing education"? If so, how? How does the set of defense and nondefense agencies needed to prosecute and succeed in war develop leaders who can adapt as fast as the pace of change in the information age? What new learning methods should the armed forces and their nondefense partners use to increase the experience base of junior leaders in less time? What common frameworks are required among military, nonmilitary governmental agencies, nongovernmental agencies, and multinational partners so that they can team faster?

Operational and garrison boundaries are blurring. Information age tools link deployed forces with those who stay behind; both support each other in essential ways. Traditional boundaries among training programs at home stations, service and joint training centers, and schoolhouses are also in flux. Our posts, bases, and stations must accommodate training for all components—active, guard, and reserve. They must also often serve as deployment and redeployment centers as well as mobilization and demobilization platforms.

Infrastructure that serves family support and readiness requirements is as important as that which serves training needs. Our forces are positioned around the globe based upon the requirements identified during the Cold War; how do we reposition for the future?

What does "industrial base" mean in a globalized world where civilian specifications are often more stringent than military and when nonmilitary supplies and civilian replacement parts are more available than military?

Many of these questions have been asked already. Some answers have been suggested, others have been implemented, and still others are being worked as pilot programs or experiments. There is no lack of activity, that's not the issue. Rather, the issue is this: what is the coherent approach to learn our way toward answers and solutions that go beyond tweaks of prevailing organizations and method and get to fundamental change? What set of principles should we use to generate and sustain a post–Cold War and information age force? How should we conduct the necessary experiments so that we can make interim decisions quickly and adapt from them—learning on the move?

The profession of arms is an action-oriented profession of practitioners. The temptation to act quickly, to do something, to change will be great. And act we must. The issues at stake in the generating and sustaining force, however, are serious, and the "players" are not all military. Furthermore, the changes we make will have a long-lasting impact across the nation, not just within the armed forces. Experimenting and making quick changes as we learn is even more difficult in our period of hyperchange, but no less necessary in this period where we are changing-as-we-fight.

The end of the Cold War and the emergence of the information age have shifted the very foundations of the profession of arms, and there's an argument to be made that this shift extends to the very foundation of the set of institutions associated with our national defense. Clearly, the framework that guided decisions and actions in the past is just that, past.

In the past decade plus, scores of books have been written, briefs given, and conferences conducted on the need for new approaches to security. Three recent ones come to mind—Thomas Barnett's *The Pentagon's New Map*, Thomas Friedman's *The World Is Flat*, and Samuel Huntington's *The Clash of Civilizations and the Remaking of World Order*—but there are many others of equal merit.[9] The changing conduct of war has received equal attention in the past decade. Among the many excellent sources, the recent work of Maj. Gen. Robert Scales Jr., U.S. Army (Ret.), the essays of Ralph Peters, the study of land warfare by Gen. Gordon Sullivan, U.S. Army (Ret.), and the examination of future war by defense scholar Bruce Berkowitz are all thought-provoking and valuable.[10]

Absent from this discussion, however, is an equally rigorous and thorough discussion of how the end of the Cold War and the emergence of the information age change the generating and sustaining forces. The absence of this discussion is certainly not complete; one can find pockets of discussion and initial change if one looks hard, but not to the same robust degree as is the case in both the areas of security strategy and operating forces.

This should not be much of a surprise. The generating and sustaining forces, their organization, and the processes they follow are among the most difficult to understand and analyze. Further, the generating and sustaining forces represent a fluid tangle of military and civilian institutions and bureaucracies—a military/industrial Gordian knot at the end of the transformation rope. Climbing the rope is hard enough, but seeing a knot at the end would discourage even Sisyphus. The profession of arms has an old saying, though: "The difficult gets done immediately; the impossible takes a little longer." The profession is tackling the difficult now; on the road ahead, it's faced with the impossible. The essays collected in the present volume indicate that some serious thinking is under way. Let's get on with it.

## Notes

1. There are many good books on this subject. For a quick summary, see Maj. Gen. Robert H. Scales Jr., *Future Warfare* (Carlisle Barracks, Pa.: U.S. Army War College Press, 1999), pp. 1–13. For longer accounts see in Martin Van Creveld, *Technology and War* (New York: The Free Press, 1989) or Christopher Bellamy, *The Evolution of Modern Land Warfare: Theory and Practice* (New York: Routledge, 1990).

2. George and Meredith Friedman, *The Future of War: Power, Technology, and American World Dominance in the 21st Century* (New York: Crown Publishers, Inc., 1996), p. ix.

3. Bruce Berkowitz, *The New Face of War: How War Will Be Fought in the 21st Century* (New York: The Free Press, 2003); David S. Alberts, John J. Garstka, and Frederick P. Stein, *Network Centric Warfare: Developing and Leveraging Information Superiority* (Washington D.C.: Department of Defense Cooperative Research Program, 1999); Ralph Peters, *Fighting for the Future* (Mechanicsburg, Pa.: Stackpole Books, 1999).

4. For a complete description see Thomas S. Kuhn, *The Structure of Scientific Revolutions* (Chicago: The University of Chicago Press, 1970), especially pp. 1–10, and 43–135.

5. Ibid; p. 111.

6. The principles and their short description are from Robert R. Leonhard, *The Principles of War for the Information Age* (Novato, Calif.: Presidio Press, 1998), pp. 8–11.

7. Sun Tzu, *The Art of War,* Ralph D. Sawyer trans. (Boulder, Colo.: Westview Press, 1994); Karl von Clausewitz, *Principles of War,* and Antoine Henri Jomini, *The Art of War,* both in *Roots of Strategy, Book 2* (Harrisburg, Pa.: Stackpole Books, 1987).

8. Leonard, *Principles of War for the Information Age,* p. 252

9. Thomas P. M. Barnett, *The Pentagon's New Map* (New York: Putnam and Sons, 2004); Thomas Friedman, *The World Is Flat* (New York: Farrar Strauss, 2005); Samuel P. Huntington, *The Clash of Civilizations and the Remaking of World Order* (New York: Simon and Schuster, 1996).

10. Scales, *Future Warfare;* Peters, *Fighting for Future;* Gen. Gordon R. Sullivan and Col. James M. Dubik, "Land Warfare in the 21st Century," in *Envisioning Future Warfare* (Fort Leavenworth, Kan.: CGSC College Press, 1995); and Bruce Berkowitz, *The New Face of War: How War Will Be Fought in the 21st Century* (New York: The Free Press, 2003).

## Part 1

# AN AMERICAN
# WAY OF WAR?

# ONE

# The American Way of War

## Critique and Implications

### COLIN S. GRAY

### The American Way of War:
### Critique and Implications

In the history of American strategy, the direction taken by the American conception of war made most American strategists, through most of the time span of American history, strategists of annihilation. At the beginning, when American military resources were still slight, America made a promising beginning in the nurture of strategists of attrition; but the wealth of the country and its adoption of unlimited aims in war cut that development short, until the strategy of annihilation became characteristically the American way in war.[1]

### What Is the Question?

As excerpted above, Russell F. Weigley's now-classic study, with its bold, assertive title *The American Way of War*, comprises an invaluable extended statement about American strategic and military culture. But, much has happened since Weigley wrote the words quoted above in the dying phase of America's protracted and ultimately futile adventure in Vietnam. All too plainly, the characteristically American way failed to deliver strategic and political success in Southeast Asia. Was that undeniable fact a result of endemic and enduring weakness in the American way of war; did it just reflect the country's way of war, its national style in warfare, at a particular time; or did the American way attempt mission impossible in Vietnam? How dynamic is the American way of war? Does it evolve to such a degree that it

13

is far from fixed by allegedly deep-rooted cultural influences? To press scepticism further, is it even sensible to talk of *the* American way of war? That it is a familiar concept and that Professor Weigley wrote a well-regarded book about it certainly confer some legitimacy to the idea. Nonetheless, many an unsound idea has survived because of the familiarity granted by repetition, because of the blessings of ill-applied scholarship, and sometimes because of official adoption.

I must first declare the attitude and purpose of this essay. Notwithstanding the scepticism of the previous paragraph, this discussion will argue that there is important merit in the concept of an American way of war. More to the point, perhaps, it will suggest that recognition of that characteristic "way" is of the most signal importance for the current debate over military transformation and, indeed, over U.S. defense strategy as a whole. The principal reason why this should be so is all too easy to identify. There may be a characteristically American way of war that must shape, perhaps in some ways directly, the process of military transformation. Moreover, that American way will find expression in a style of military and strategic behavior that cannot really be transformed. This is all highly debatable.

It is useful to postulate two clear opposing positions. On the one hand, it may be argued that there has been, and remains, a dominant American way of war. Weigley's 1973 book made this case, while later scholars have claimed that America's public, strategic, and military cultures have a persisting, even a vital, influence. On the other hand, there are theorists who maintain that the American way of war shifts in response to stimuli from contextual changes. Eliot A. Cohen has offered an explicit statement of this view in an essay entitled "Kosovo and the New American Way of War." He claimed, reasonably enough, that "any way of war has its strengths and its weaknesses; all ultimately succumb to altered political circumstances and changing techniques and technology."[2]

This opening section leads with a challenge: "What is the question?" I suggest that the question is not, at least not quite, which of the two polar positions is correct. Rephrased: Does America have a timeless way of war, a way that is all but culturally mandated? Or, does America have a way of war that evolves, even alters radically, as the various contexts of conflict shift? As ideal types, those polar opposites are useful. They do serve to anchor the necessary debate on a spectrum, recognition of which has the most significant implications for policy and defense planning. The question(s), to repeat, is not which polar position is correct, but rather which is more nearly correct, in what ways, and why.

In case the salience of this apparently academic discourse has eluded some readers, I must explain that our subject is nothing less than the mutability and adaptability of the American way of war. To pose the question bluntly, can the U.S. armed forces do what they promise, and what they say they need to do, and transform themselves into a more agile and adaptable instrument of policy? Scarcely less important, can those forces lock into what they claim will be a permanent process of adaptive transformation? Or, must they be fatally constrained by the influence of cultural and other structural, all but permanent, factors? The prophets of a "new American way of war" must not be permitted to forget that America, as with any country, will adapt, execute, and exploit a new way of war in a distinctively American way. Today's transformation plans talk bravely of commitment to a revolutionary change in military culture, but we will discover, perhaps rediscover, that the cultural contributors to a characteristically American way of war are not so easily to be overturned for an apparently better fit with new circumstances.[3] Cultural assumptions certainly are not immutable, but neither are they to be replaced wholesale either by administrative fiat or by an act of will.

I have saved the most troubling question to last. Specifically, even if the U.S. armed forces are transformed to be more adaptable to meet the challenges posed by a deeply uncertain strategic context with its new risks, as well as old risks, how relevant will that transformation be to the effectiveness of the American way of war? A persisting problem with the American way of war has been not so much how well Americans fight, but rather how well or poorly that combat and sacrifice have served the country's political goals. The American defense debate is severely hampered by the popular conflation of war with warfare and, by extension, the confusion of principles of war with principles of warfare.[4]

## The Way of War Hypothesis and the Problem with Very Big Ideas

The American defense community, at least in its strategic intellectual dimension, shares some characteristics with the fashion industry. Expert defense professionals quite literally follow the fashion in ideas. They do this both to maintain their status as experts, and, more prosaically, in order to keep their funding. The bigger the idea, the greater its conceptual reach and hence its organizing potency, and hence the more compelling the felt need to jump aboard the intellectual bandwagon. There are a few, a very few, truly original minds in the ranks of defense professionals, but even their notable and distinctive contributions must be appreciated in light of the fact that there are really no new ideas bearing on war and warfare. If Thucydides, Sun Tzu, Clausewitz, and Rear Adm. J. C. Wylie did not say it, it probably is not worth saying.[5]

This apparent side trip from the main thrust of the essay is intended to highlight the necessity for historical perspective. Such a perspective is not merely a desirable extra; it is literally essential. The ever quotable Max Boot hit the mark squarely when he wrote, "[t]he past is an uncertain guide to the future, but it is the only one we have."[6] The U.S. defense community is vulnerable to capture by the buzzword of the week, the concept of the year, the big new-sounding idea that appears to promise rich rewards. The reason is because the typical essential vacuity or, at best, true familiarity of these notions is hard to recognize by intellectual consumers who are largely unprotected by historical knowledge.

A defense community that is historically challenged is ever liable to seduction by its hopes at the expense of the lessons of its experience, as well as the experience of everyone else. Naturally, if those lessons are not harvested, or simply are forgotten by institutions that have no memory, it is impossible to learn from experience. The pertinence of this line of thought to the story arc of this essay should be plain enough. The debate over an American way of war, and particularly over the extent to which it can be transformed, needs the debaters to be able and willing to confront the challenge of historical continuity as well as discontinuity. If historical perspective is missing, the debate is unlikely to be an intellectually high calorie contest, or one loaded with nuggets of sound analysis and advice for policy, defense strategy, and force planning.

As an optimistic, future-oriented superpower, the United States is almost uniquely prone to fall for the attractive fallacy that its future, in this case its strategic future, is a blank sheet to be filled at the national discretion. All too obviously, the idea of transformation has enormous cultural appeal, an attraction matched fully by its potential to disappoint. If transformation is approached as a bold mission in a way largely innocent of the complexity of strategy and the continuities in what, for want of a better term, we must label culture, then exceptional perspicacity is not required to foresee its failure.[7]

Not for nothing did Sun Tzu insist upon self-knowledge as a vital element in the intelligence mosaic.[8]

It is probably important to reemphasize that this essay addresses the idea of an American way of war from two interrelated, but distinctive, perspectives. First, we must consider the question of whether or not the United States will be able to transform its armed forces so that they should be capable of conducting future warfare effectively. While, second, and of no lesser significance, we need to inquire whether or not an evolving, even a radically transformed, American way of war is likely to meet the Clausewitzian test of serving faithfully as a potent tool of policy.[9]

Unlike much of the faddish jargon, easy familiarity with which marks out the defense experts, the very big concept of a national way of war is not short of substance or policy relevance. That claim, however, admittedly is at least somewhat contentious. Obviously, if one believes that the country can revolutionize at will the way in which it prepares for and wages war, it must make scant sense to talk about a national way of war. Such a way would refer simply to the current military style implicit and explicit in policy guidance, defense plans, and military capabilities. To be meaningful, the high concept of an American way of war has to refer to important enduring preferences, habits of mind, and modes of behavior. Properly regarded, a way of war will reflect the persisting influence of what British historian Jeremy Black has insightfully termed cultural assumptions.[10]

The U.S. Army may call heroically for a cultural revolution in its ranks as essential to the success of its long, indeed permanent, process of transformation. But, it is not self-evident that revolutions in culture can be achieved by effort, no matter how sincere and determined. After all, cultural assumptions are the products of historical experience, more precisely of how that experience is interpreted, national (or other) ideology, and a sense of identity, inter alia. It may be hard for Americans to accept the proposition that their future strategic performance is both enabled and constrained by what they are as Americans, by their cultural software. It would be a serious mistake to believe that America's multidimensional potency as the sole contemporary superpower allows it to remake itself strategically, and to behave adaptably in any way that new or old security challenges appear to require.

Somewhat belatedly, perhaps, I must sound the tocsin over the perils of Very Big Ideas. Invariably, indeed of necessity, they oversimplify in ways that can obscure or bury important matters that do not fit the grand scheme on offer. The inventor, or rediscoverer, of a master narrative tends to transition rapidly and painlessly from the role of innovative theorist to combative prophet of a new belief. If the American way of war seems unduly grand, too compound in its inclusivity, and overly vague, what are we to make of Victor Davis Hanson's far more grandiloquent concept of a "Western way of war"?[11]

That Western way is contrasted, of course, with an Eastern or oriental way. Fortunately, I am not obliged in this essay to describe or pursue the debate over Hanson's genuinely exciting thesis. I mention it solely in order to illustrate the claim that Very Big Ideas are prone to be very seriously flawed.[12] Especially is this the case when the theorist overreaches empirically. Such overreach is virtually certain when he or she advances what amounts to a master theory that reveals all. There will always be inconvenient historical items that do not fit the master narrative. This is not to condemn all grand theories, at least not quite. But it is to say that we pay a price in richness of detail ignored, and generally of contrary phenomena discounted, for the benefit of the great explanation. To quote Professor Black again:

> It is important to be wary about meta-narratives (overarching interpretations), and to be cautious about paradigms, mono-causal explanations and much of the explanatory culture of long-term military history. Instead, it is important to emphasize the diversity of military practice, through both time and space, and to be hesitant in adducing characteristics and explanations for military capability and change.[13]

Amen to that. However, we must take note of the historian's professional bias against, and resistance to, large ideas that, in the interest of big picture clarity, sacrifice richness of detail and some candidate evidence that does not slot in easily. Not a few historians can be described fairly as being conceptually challenged. Each profession has its skill biases. We social scientists, for example, are apt to theorize from a dangerously thin empirical base. As my one-time colleague Herman Kahn was fond of remarking, "When you are concentrating on the forest, occasionally you walk into a tree." This essay contends that despite the perils of the meta-narrative, the concept of an American way of war is both empirically sustainable and of high relevance for the country's future strategic performance. However, before proceeding to justify and explain that position, I should make clear why this discussion is so scornful of fashionable jargon, buzzwords, and concepts *du jour.*

Instead of breeding contempt, conceptual familiarity promotes acceptance, and confers legitimacy and eventually authority. As practical people doing their best to cope with the current demands of policy, planning, and execution, defense professionals tend not to be friendly to scholarly deconstruction of the big idea of the moment. Such activity is generally regarded as unhelpful and at best irrelevant to real-world concerns. But, for good or ill, concepts matter. Some Very Big Ideas can feed very big expectations, which are in sore need of critical scrutiny.

On a personal note, when I pointed recently to the essential vacuity of some concepts fashionable in the Department of Defense, I was advised that the words in question were the current "terms of art" and, therefore, were beyond useful criticism. As some readers may have guessed already, this essay is heading toward a meeting—I will refrain from calling it a confrontation—between two unusually masterful concepts: an American way of war and military transformation. At present, although there is much reference to an, or the, American way of war, the idea is not attracting the close attention that it merits. Of course, if you believe that the American armed forces and, by extension, presumably, their way(s) of war can be so altered as to warrant a claim for transformation, then the very integrity of a culturally shaped way of war must be in question. I deal with this central matter later in the essay.

Meta-narratives generally convey some valuable insights. However, the number and value of those insights vary widely. If we consider some examples from the sales catalogue of bigger concepts over the past fifteen years, by and large we discover that each idea was, and is, not without value. We discover also that that value was only to be expected, given the longevity of the concept at issue. Select your favorites. I mention with some affection: competitive strategies; RMA, of course; effects-based operations (EBO); asymmetric threats; network-centric warfare (NCW); and fourth generation warfare (4GW). The difficulty with each of these popular, indeed in some cases official, notions is not so much their empirical overreach, but rather their essential banality.

Is not strategy, singular or plural, necessarily intended to be competitive? Can anyone seriously challenge the proposition that periodically, if irregularly, the character of warfare changes radically? How could one conduct operations other than for the purpose of securing some particular desired effects? Belligerents invariably are more or less asymmetrical in strategically significant ways. To strive to offset one's disadvantages by seeking leverage from threats that the enemy is not well able to counter is really what the conduct of war, inalienably "a duel on a larger scale," is all about.[14]

NCW is a good idea, and it always was. If we can afford its literal technical realization, and if the people at the sharp end of our spear are not overwhelmed with data, NCW is obviously desirable. As for 4GW, it transpires on closer inspection to be both a statement of the glaringly obvious—that much of contemporary warfare is irregular in kind—and a perilous venture into the highly improbable.[15] The latter criticism refers to the fact that 4GW vitally shortchanges the strong probability of a return of great-power conflict.[16] Far from acting as the key which unlocks deep understanding of the past several centuries of strategic history, 4GW instead offers fairly uncontentious, if empirically unduly exclusive, history at the price of neglecting contrary evidence and possibilities.

Is the concept of an American, or indeed any society's, way of war also guilty of the overreach, the undue historical selectivity, perhaps even the essential vacuity, and the banality of obviousness, which bedevil the ambitious ideas just discussed very briefly?

## Critique and Reply

Because this analysis is friendly to the concept of an American way of war, it is especially necessary that the main problems with the hypothesis be recognized. Four difficulties in particular need to be noted and dealt with if we are to employ the way of war hypothesis to useful effect.

First, the concept of a way of war can be charged with being unduly vague. What is cultural and what is not? There is the peril of circularity. If, necessarily, enculturated Americans must wage war in an American way, the concept of an American way of war is in danger of melting down into a truism. The American way is what Americans think and do. Do Americans, despite their culture, sometimes behave in an un-American way, out of strategic and military character as it were? Directly put, how do we verify the American way of war? What is the evidence for American culture at work? What other explanations of thought and behavior compete for primacy? These methodological issues are not trivial; they bear down menacingly upon the very integrity of the concept of a national way of war.[17] Surely, much of strategic and military thought and behavior is universal and is essentially common to all societies. When the United States takes military action, how can we distinguish between the cultural influence of an American way of war and the universal logic of America's strategic and military contexts?

The difficulties just cited cannot be idly dismissed and neither can they be resolved with complete satisfaction. Culture is a vague concept. Its presence is as certain as its relative influence is uncertain. Those who share my conviction that the way of war hypothesis is useful, with its cultural underpinning, should neither panic nor have to resort to scholastic contortions in an attempt to rescue the concept from the charges of vagueness and unverifiability. Instead, we should simply maintain that the idea is, in general, valid and useful; it is empirically plausible even though it is not definitively verifiable. With some good reason critics argue that American strategic and military behavior can as well be explained according to the realist as to the cultural paradigm.[18] In other words, Americans are apt to behave as any people would, were they in the same strategic situation. In my opinion it is not helpful to contrast realist with cultural approaches to explanation, because every supposedly realist agent, every Strategic Person, cannot help but be enculturated by some particular brand of cultural programming. There are no "de-cultured" or otherwise acultural Strategic People.

Second, acceptance of the concept of a national way of war can hardly help but encourage the underrecognition, even the discounting, of evidence contrary to the main body of the thesis. In his admirable book *The Savage Wars of Peace,* Max Boot affixes the claim for "Another American Way of War" to his preface.[19] As with all meta-narratives or master explanations, the concept of an American way of war necessarily oversimplifies. Theory does that. In pointing up the most significant, it is obliged not to muddy the water with a host of qualifications. A powerful explanatory tool, which is what good theory should be, need not be capable of explaining everything. It follows that although the way of war thesis can be charged with an undue inclusivity and, as a consequence, with theoretical overreach, these should not be regarded as fatal flaws. Social science deals with human behavior, not the laws of nature. Because there will be exceptions to the strategic and military behavior we identify as the American way of war, we should not conclude that the thesis is thereby invalidated.

Third, the way of war hypothesis may be charged with encouraging an excessive expectation of continuity in a national style in strategic and military behavior. Three closely related sins tend to flow from this error: the possibility of radical change is discounted; adversaries are stereotyped; and strategic thinking becomes deterministic. These strategic intellectual pathologies can have dire consequences. We may be surprised to discover that some societies are capable of executing RMAs that require thoroughgoing change in their ways of war. Countries that we thought we understood strategically quite suddenly may be revealed no longer to fit our fixed opinion of their strategic and military cultures. Allies and, more especially, adversaries that we have neatly categorized strategically with our settled notion of their ways of war may jump out of the boxes that the thesis has provided and behave in utterly unexpected ways.

In short, the way of war hypothesis can be charged with discouraging the necessary alertness to strategic change. It cannot be denied that the way of war hypothesis can desensitize us to change. If we have a settled view of how the Chinese, the Indians, or the Russians "do it," we are likely to be slow to recognize evidence of a noteworthy shift in their ways of war. That granted, those who find this compound criticism persuasive need to ask themselves how the benefit of appreciating differences compares with the cost of the risk of slipping unbeknownst into a deterministic frame of mind.[20] All theories have the vices of their virtues. The thesis discussed here postulates that societies develop characteristic ways of war that persist and warrant description as cultural. That is an important and, in the opinion of this theorist, both a generally valid and a useful idea. But it is also an idea that has the potential to imprison the imagination and as a consequence lead us to fail to recognize developments that do not fit our paradigm of, say, "the Chinese way."

Fourth and finally, the way of war hypothesis can be found guilty of licensing an undisciplined search for the alien, the eccentric, the bizarre, and the purportedly characteristically quirky.[21] The hypothesis is apt to privilege identification of that which is different and distinctive, whether or not that distinctiveness is strategically significant. After all, notwithstanding cultural diversity, there is a fairly common body of military science, parallel discovery of ideas and technologies is entirely usual, and those ideas and technologies that lack for some national parents are certain to spread by the multifarious processes of diffusion.[22] Making due allowance for differences of geostrategic context, the world's military establishments resemble each other quite closely in most essentials. What is more, the identical claim can be made for the world's insurgents. Up to a point, a belligerent has cultural discretion to "do it my way," but there is a universal lore of warfare, regular and irregular, which cannot safely be treated with disdain for reason of cultural distaste.

This fourth charge against the hypothesis is superficially plausible. If we look for the uniquely characteristic, we are certain to find it. There is no doubt that the idea of a national way of war all but instructs us to locate such a way; it feeds our expectations. Recall the general truth about intelligence assessment: we tend to find what we expect to find. Despite the tendency of the way of war thesis to overstate distinctiveness, its application, even when highly arguable (e.g., with reference to the purposes for which, and the way in which, forces will be used), can be crucially important. We must grant the peril of the pathology of an undue focus on the odd and generally unusual. But, nonetheless, it remains an empirically verifiable fact of immense importance that different security communities are always liable to apply new strategic ideas and technologies for reasons and in ways that are far from common.

## The American Way of War

The way of war hypothesis has five claims at its core. It insists that

1. There is a distinctively American approach to war and warfare.
2. This distinctively American approach is so rooted in the nation's historical experience, and the beliefs that Americans hold about that experience, including myths and legends, that it merits ascription as cultural.
3. This postulated American way of war, though cultural, rests upon significant and persisting material realities.
4. Americans behave in new strategic contexts, and with new material assets, in a fashion shaped, at least influenced, by their culture as reflected in the national way of war.
5. The American way of war is always subject to some revision, at least temporarily in practice, in the face of enemy challenge at every level: political, grand strategic, military strategic, operational, and tactical. Not even the United States is able to wage war, as it were, autonomously, with the enemy of the day consigned expediently to the role of helpless victim or target set.[23]

It is scarcely necessary to emphasize the practical implications of the way of war hypothesis. Above all else, it claims that Americans are not, at least are not likely to prove to be, highly adaptable to new strategic circumstances, except in ways that privilege an American way in war and warfare. To put more bite into the claim, the hypothesis implies that the American way may well prove effectively resistant to attempts to effect major change in strategic and military behavior. Given the contemporary commitment to military transformation, the party of revolution in the Department of Defense is likely to find that it has underestimated the potency of unhelpful cultural factors. Moreover, when the way of war hypothesis is considered, as it should be, at its several levels of relevance, Americans are going to discover that they are the heirs both to a fairly distinctive way *with* war, and to a no-less-characteristic way *in* warfare. Even if the U.S. armed forces were progressively transformed along the ambitious lines specified, for example, in the army's *Campaign Plan* of 17 April 2004, and the *2004 Transformation Roadmap,* it is improbable that the transformed military instrument will be employed by policy in a manner that is transformed also.[24]

The paragraph immediately above risks overstating the negative influence of culture. We should not ignore the possibility that some of the attitudes, habits, and practices that will seem atavistic and reactionary to zealous reformers express a body of cultural lore that has been tried and tested over many a year and through many a conflict. It can be helpful to have a cultural brake on reform, lest the intended change is ill considered. On a more neutral note, recognition of merit in the way of war hypothesis should oblige American transformers to appreciate that they have no choice other than to pursue transformation in an American manner.

No matter how bold and apparently even revolutionary some of the steps toward transformation may be, they must be taken by Americans in an American way. Culture is inescapable. Military theorists cannot function beyond culture. They are enculturated Americans. For the moment, at least, I will ignore the argument which suggests that a wholly voluntary professional military establishment is, or can be, significantly transnational in its skills and attitudes. Similarly, I do not believe it would be profitable for this inquiry to challenge the apparent truism that societies make war according to their nature. Nonetheless, I commend to readers the apposite words of Gene Hackman's Capt. Frank Ramsey, U.S. Navy, in the movie (and book) *Crimson Tide:* "We are here to preserve democracy, . . . not to practice it."[25]

Writing in the early 1970s, and very much in the shadow cast by Vietnam, Russell F. Weigley did not hesitate to affirm the existence and persistence of an American way of war. If we fast-forward a decade, Samuel P. Huntington, one of America's most distinquished political scientists, indeed the author of the classic treatment of the country's civil-military relations, carried the same message as had Weigley.[26] In lectures delivered in the fall of 1985, Professor Huntington made a powerful case for the relevance of the way of war hypothesis. It is regrettable that these lectures on "American Military Strategy" are not known as widely as they merit, for he is worth quoting at some length. "My basic message is that American strategy and the process by which it is made must reflect the nature of American society. Earlier I criticized those who urged us to adopt a strategy that was at variance with the inherent character of American society."[27]

Professor Huntington was writing in the context of the debate of the early 1980s sparked by the military reform movement of those years. His words resonate for today.

> The principal prescription of the "lean, mean and deft" advocates for reform of the military establishment have much the same unreality as the "all or nothing" prescriptions do for politics. The U.S. military establishment is a product of and reflects American geography, culture, society, economy, and history . . . one should not be swept off one's feet by the romantic illusion that Americans can be taught to fight wars the way Germans, Israelis, and even British do. That would be both ahistorical and unscientific.
>
> American strategy, in short, must be appropriate to our history and institutions, both political and military. It must not only be responsive to national needs but also reflect our national strengths and weaknesses. It is the beginning of wisdom to recognize both.
>
> The United States is a big, lumbering, pluralistic, affluent, liberal, democratic, individualistic, materialistic (if not hedonistic), technologically supremely sophisticated society. Our military strategy should and, indeed, must be built upon these facts. The way we fight necessarily will reflect the way we live.[28]

Huntington proceeded to advise that it should be the American way: to plan to win quickly; to assume the offensive; to exploit technology; to "fight wars in a big way" (as a big country); and to use its armed forces "to achieve military objectives."[29] No doubt this listing reflected in some measure the ethos of its time of composition. In particular, it expressed the rejection of the military's appalling Vietnam experience and offered a foretaste of Secretary of Defense Caspar Weinberger's famous, or notorious, "Six Tests" for the employment of U.S. combat forces abroad.[30] After we have allowed for the transient influences particular to the early and mid-1980s, some substantial nuggets of vital American self-knowledge remain. And those nuggets could hardly be more relevant to the contemporary defense debate. To state the matter as plainly as possible: it is possible, even probable, that some key elements in the vision of a transformed American military establishment will prove to be culturally infeasible. Why? Because America is what it is, and that existential reality, though somewhat dynamic, must find some expression in the country's way of war.

Two caveats in particular need to be stamped in red ink on every manuscript that presents a generally favorable view of the concept of a national way of war. First, it is necessary to beware of caricature; second, it is mandatory to be alert to evolution and change. Those warnings are, of course, much easier to issue than to heed. As we noted earlier, good theory pares away the inessentials. But some of those inessentials could modify what otherwise is an unduly clear, if elegant, picture. When Max Boot writes about "Another American Way of War," one pertaining to irregular conflicts, should we permit our dominant model to be enriched, corrected, or somewhat contradicted and confused by his story?

With respect to the second caveat, American society is not stationary in its attitudes. Culture is constantly in motion, albeit much more slowly than opinions. If those scholars are correct who argue that there has been a revolution in attitudes toward the military (RAM), surely we should expect a potentially revolutionary change in the national way of war that Americans will accept.[31] Since strategic and military culture must be influenced by public culture, an evolving American society should be evolving also in its way of war. If this line of thought has much merit, it may be that some of the more serious concerns expressed here will prove to be largely groundless. Needless to add, perhaps, this assumes that the alleged RAM and the transformed armed forces will be conducive to strategic and political effectiveness.

Two relationships, not one, are the subject of this discussion. The first is that between American society and its way of war. The second, even more important, is between America's way of war and the political, strategic, and military demands from abroad that the country decides it must meet, should the option to fight be discretionary. It is certainly true to claim that in a democracy, a way of war must first succeed at home if it is to succeed abroad. But it does not follow that success at home must translate into success abroad. There can be a sharp contrast between internal and external integrity. Americans may be comfortable with a particular style of combat and an attitude to the use of force, but will they work where and when it matters against foreigners?

The American way of war has been discussed throughout this essay, but by and large it has remained unspecified. This apparent neglect is explained by the fact that my primary mission has been to consider the value of the concept, and to warn of its pitfalls, rather than to identify the ever-arguable content of the American way. No longer. What follows is a personal characterization of the traditional, indeed cultural, American way of war. I specify twelve features, many of which are challengeable anecdotally with exceptions, but all of which are sound enough as generalizations. Readers are reminded that whereas a single exception must invalidate a scientific law (e.g., an apple that declines to obey the law of gravity), social scientific lore is far more tolerant of deviant cases. Rather than argue for each of my twelve chosen features, I will restrict myself simply to explanation. It should be understood that no authoritative listing exists. Indeed, there could hardly be such, given that the notion of a national way of war is what we social scientists term an "essentially contested concept."

Please note that the items pertain both to war as a whole and to its military conduct, warfare. At least three of the twelve features comprise a vital part of the case which holds that the United States tends to confuse Principles of Warfare with Principles of War. If the country appreciated and generally adhered to a well-drafted and culturally embedded set of Principles of War, principles that truly were Clausewitzian (and Sun Tzuan and Thucydidean), its strategic and political performance in conflict after conflict should be considerably improved. But, strategically for good and ill, Americans are what they are. If, as I claim, Americans persist in failing to reap desired political rewards from their military effort, even when the effort is largely successful, there are cultural, really structural, reasons why that should be so. Most probably, Americans can only remake their strategic performance if they first remake their society, and that is a task beyond the ability even of the most optimistic agents of transformation.

### Characteristics of the American Way of War

| | |
|---|---|
| 1. Apolitical | 7. Firepower focused |
| 2. Astrategic | 8. Large-scale |
| 3. Ahistorical | 9. Profoundly regular |
| 4. Problem-solving, optimistic | 10. Impatient |
| 5. Culturally ignorant | 11. Logistically excellent |
| 6. Technologically dependent | 12. Sensitive to casualties |

## 1. Apolitical

Americans are wont to regard war and peace as sharply distinctive conditions. The U.S. military has a long history of waging war for the goal of victory, paying scant regard to the consequences of the course of its operations for the character of the peace that will follow. Civilian policymakers have been the ones primarily at fault. In war after war they have tended to neglect the Clausewitzian dictum that war is about, and only about, its political purposes. Characteristically, U.S. military efforts have not been suitably cashed in for political advantage.[32]

## 2. Astrategic

Strategy is, or should be, the bridge that connects military power with policy. When Americans wage war as a largely autonomous activity, leaving worry about peace and its politics to some later day, the strategy bridge has broken down. The conduct of war cannot be self-validating. For a leading, truly awful, example of this malady, we must cite Vietnam. The United States sought to apply its newfound theory of limited war in an ill-crafted effort to employ modulated, on-off-on coercion by air bombardment to influence Hanoi in favor of negotiations.[33] To resort to Clausewitzian terms, while war has its policy logic, it also has its own "grammar."[34] It is prudent to take notice again of these words of wisdom from Professor Huntington: "Military forces are not primarily instruments of communication to convey signals to an enemy; they are instead instruments of coercion to compel him to alter his behavior."[35]

## 3. Ahistorical

America is a future-oriented, still somewhat "new" country, one that has a founding ideology of faith in, hope for, and commitment to, human betterment. It is only to be expected, therefore, that Americans should be less than highly respectful of what they might otherwise be inclined to allow history to teach them. A defense community led by the historically disrespectful and ill educated is all but condemned to find itself surprised by events for which some historical understanding could have prepared them. History cannot repeat itself, of course, but as naval historian Geoffrey Till has aptly observed, "The chief utility of history for the analysis of present and future lies in its ability, not to point out lessons, but to isolate things that need thinking about. . . . History provides insights and questions, not answers."[36]

## 4. Problem-Solving, Optimistic

Holding to an optimistic public culture characterized by the belief that problems can be solved, the American way in war is not easily discouraged or deflected once it is exercised with serious intent to succeed. That is to say, not when it is made manifest in such anti-strategic sins against sound statecraft as with the "drive-by" cruise missile attacks of the late 1990s. The problem-solving faith, the penchant for the "engineering fix," has the inevitable consequence of leading U.S. policy, including its use of armed force, to attempt the impossible.[37] Conditions are often misread as problems. Conditions have to be endured, perhaps ameliorated, and generally tolerated, whereas problems, by definition, can be solved.

### 5. Culturally Ignorant

Belatedly, it has become fashionable to berate the cultural insensitivity that continues to hamper American strategic performance.[38] Bear in mind American public ideology, with its emphasis on political and moral uniqueness, manifest destiny, divine mission even, married to the multidimensional sense of national greatness. Such self-evaluation has not inclined Americans to be respectful of the beliefs, habits, and behaviors of other cultures. This has been, and continues to be, especially unfortunate in the inexorably competitive field of warfare. From the Indian Wars on the internal frontier to Iraq and Afghanistan today, the American way of war has suffered from the self-inflicted damage caused by a failure to understand the enemy of the day. For a state that now accepts, indeed insists upon, a global mandate to act as sheriff, this lack of cultural empathy, including a lack of sufficiently critical self-knowledge, is most serious.[39]

### 6. Technologically Dependent

The exploitation of machinery is the American way of war. One may claim that airpower is virtually synonymous with the American way of war and that its employment as the leading military instrument of choice has become routine. So at least it appeared, in the 1990s, in the warm afterglow of airpower's triumph in the First Gulf War.[40] America is the land of technological marvels and of extraordinary technology dependency. It was so from early in the nineteenth century, when a shortage of skilled craftsmen—they had tended to remain in Europe—obliged Americans to invent and use machines as substitutes for human skill and muscle. Necessity bred preference, and the choice of mechanical solutions assumed a cultural significance that has endured. American soldiers say that the human being matters most, but in practice the American way of war, past, present, and prospectively future, is quintessentially and uniquely technologically dependent. The U.S. Army's transformation plans are awash with prudent words on the many dimensions of future conflict, but at its core lies a drive to acquire an all but unaffordable (at \$92 billion plus) Future Combat System, consisting of a network of fifty-three vital technologies, nearly all of which are technically unproven at this writing.

## 7. Firepower Focused

Gen. William C. Westmoreland, as commander of the Military Advisory Command Vietnam (MACV), once famously and characteristically told a press conference that the answer for counterinsurgency was "firepower."[41] It has long been the American way in warfare to send metal in harm's way in place of vulnerable flesh. This admirable expression of the country's machine mindedness undoubtedly is the single most characteristic feature of American war making at the sharp end. Needless to say, perhaps, a devotion to firepower, while highly desirable in itself, cannot help but encourage the U.S. armed forces to rely on it even when other modes of military behavior would be more suitable. In irregular conflicts in particular, heavy and sometimes seemingly indiscriminate, certainly disproportionate, resorting to firepower solutions readily becomes self-defeating. It is difficult to avoid concluding that the principal effect of the military transformation now under way is likely to be an ever-improving ability to "service targets." Instead of being considered in its cultural context, the enemy instead is reduced to the dehumanized status of the object of U.S. firepower. At its nadir, this characteristic was demonstrated in action in Vietnam with the prevalence of the U.S. artillery's practice of conducting unaimed "harassment and interdiction" fire.[42]

## 8. Large-scale

As a superpower, the United States tends to excel at enterprises on a scale that matches its total assets. Professor Huntington believes, at least he believed in 1985, that "the United States is a big country, and we should fight wars in a big way."[43] More controversially, he claimed that "bigness not brains is our advantage, and we should exploit it."[44] No doubt those words will irritate and anger many readers. However, there is an important self-awareness in Huntington's point. As a large rich country, for the better part of two hundred years the United States has waged its many wars, regular and irregular, domestic and foreign, as one would expect of a society that is amply endowed materially. Poor societies are obliged to wage war frugally. They have no choice other than to attempt to fight smarter than rich enemies. The United States has been blessed with wealth in all its forms. Inevitably, the U.S. armed forces, once mobilized and equipped, have fought a rich person's war. They could hardly do otherwise.

From the time of the Civil War onward, foreign observers have been astonished by the material generosity with which American troops have been supplied and equipped. Strategic necessity is the mother of military invention, and since the 1860s at least, Americans have had little need to invent clever work-arounds for material lack. It is not self-evident that the United States is able to wage war in a materially minimalist fashion, any more than today's volunteer soldiers and their families back home would tolerate campaign conditions of unnecessary discomfort. The American army at war is American society at war. This is not a problem; it is a condition.

### 9. Profoundly Regular

Few, if any, armies have been equally competent in the conduct of regular and irregular warfare. The U.S. Army is no exception to that rule. Both the U.S. Army and the U.S. Marine Corps have registered some occasional success in irregular warfare, while individual Americans have proved themselves adept at the conduct of guerrilla warfare.[45] As institutions, however, the U.S. armed forces have not been friendly either to irregular warfare or to those in its ranks who were would-be practitioners and advocates of what was regarded as the sideshow of counterinsurgency.[46] American soldiers, if I may resort to generic usage, overwhelmingly have been very regular in their view of, approach to, and skill in warfare. They have always prepared nearly exclusively for "real war," which is to say combat against a tolerably symmetrical, regular enemy.

Irregular warfare, or low-intensity conflict (LIC) as the 1960s term-of-art called it all too vaguely, has been regarded as a lesser but included class of challenge. In other words, a good regular army has been assumed to be capable of turning its strengths to meet irregular enemies, whereas the reverse would not be true. The United States has a vast storehouse of firsthand historical experience that should educate its soldiers in the need to recognize that regular and irregular warfare are significantly different. Thus far, the necessary education has failed to adhere intellectually and doctrinally, but perhaps times are changing. Anyone in need of persuasion as to the extent of the regularity of the mindset dominant in America's military institutions need look no further than to the distinctly checkered history of the country's Special Operations Forces.[47]

### 10. Impatient

America is an exceptionally ideological society and, to date at least, it has distinguished clearly between conditions of peace and war. Americans have approached warfare as a regrettable occasional evil that has to be concluded as decisively and rapidly as possible. That partially moral perspective has not always sat well with the requirements of a politically effective use of force. For example, an important reason why MACV was not impressed by the promise of dedicated techniques of counterinsurgency in Vietnam was the undeniable fact that such a style of warfare would take far too long to show major results. Furthermore, America's regular military minds, and the domestic public, have been schooled to expect military action to produce conclusive results. At Khe Sanh in 1968, for a case in point, MACV was searching for an ever-elusive decisive victory. As a consequence it was outgeneraled, lured into remote terrain. Today, cultural bias toward swift action for swift victory is amplified by a mass media that is all too ready to report a lack of visible progress as evidence of stalemate and error.

### 11. Logistically Excellent

It may be a cliché, but the maxim that "geography is destiny" is supported abundantly in American strategic history. The whole of American history is a testimony to the need to conquer distance. With few exceptions, Americans at war have been exceptionally able logisticians. With a continental-size interior and an effectively insular geostrategic location, such ability has been mandatory if the country was to wage war at all, let alone wage it effectively.

Recalling the point that virtues also have vices, it can be argued that America not infrequently has waged war more logistically than strategically, which is not to deny that in practice the two almost merge, so interdependent are they.[48] The efficient support of the sharp end of American war making can have, and has had, the downside of encouraging a tooth-to-tail ratio almost absurdly weighted in favor of the latter. A significant reason why firepower has been, and remains, the long suit in the American way of war is because there has been an acute shortage of soldiers in the combat arms, the infantry in particular. A large logistical footprint, and none come larger than the American, requires a great deal of guarding, helps isolate American troops from local people and their culture, and generally tends to grow, as it were, organically in what has been called, pejoratively, the "logistical snowball."[49]

Given that logistics is the science of supply and movement, America's logistical excellence, with its upside and its downside, has rested of necessity upon mastery of "the commons." Borrowing from Alfred Thayer Mahan, who wrote of the sea as a "wide common," Barry Posen has explained recently how and why the United States is master not only of the wide common of the high seas of Mahan's time of writing but also of the new commons of the air, space, and cyberspace.[50] Should this mastery cease to be assured, the country would have difficulty waging war against all except Mexicans and Canadians.

## 12. Sensitive to Casualties

In common with the Roman Empire, the American guardian of world order is much averse to suffering a high rate of military casualties, and for at least one of the same reasons. Both superstates had and have armies that are small, too small in the opinion of many, relative to their responsibilities. Moreover, well-trained professional soldiers, volunteers all, are expensive to raise, train, and retain, and are difficult to replace. Beyond the issue of cost-effectiveness, however, lies the claim that American society has become so sensitive to casualties that the domestic context for U.S. military action is no longer tolerant of bloody ventures in muscular imperial governance. The most careful recent sociological research suggests that this popular notion about the American way of war, that it must seek to avoid American casualties at almost any price, has been exaggerated.[51]

Nonetheless, exaggerated or not, it is a fact that the United States has been perfecting a way in warfare that is expected, even required, to result in very few casualties for the home team. U.S. commanders certainly have operated since the Cold War under strict orders to avoid losses. The familiar emphasis upon force protection as "job one," virtually regardless of the consequences for the success of the mission, is a potent expression of this cultural norm.

9/11 went some way toward reversing the apparent trend favoring, even demanding, avoidance of friendly casualties. Culture, after all, does change with context. The *National Defense Strategy* document of March 2005 opens with the uncompromising declamation, "America is a nation at war."[52] For so long as Americans believe this to be true, the social context for military behavior should be far more permissive of casualties than was the case in the 1990s. Both history and common sense tell us that Americans will tolerate casualties, even high casualties, if they are convinced that the political stakes are sufficiently serious, and that the government is trying hard to win. It must be noted, though, that Americans have come to expect an exceedingly low casualty rate because that has been their recent experience. That expectation has been fed by events, by the evolution of a high-technology way in warfare that exposes relatively few American soldiers to mortal danger, and by the low quality of recent enemies.

As noted already, when the context allows, it is U.S. style to employ machines rather than people and to rely heavily on firepower to substitute for a more personal, and dangerous, mode of combat. A network-centric army, if able to afford the equipment, carries the promise of being supported by even more real-time on-call firepower than is available today.

It is one thing to identify and analyze a somewhat arguable body of attributes labelled "the American way of war"; it is quite another to pose and answer the classic question of the strategist, "So what?" It is to the strategist's question that the essay now turns for the concluding discussion.

## The American Way of War Meets Transformation and the Twenty-first Century

This extensive trek into the underappreciated and contestable terrain contoured by the way of war hypothesis leads this strategic theorist to make six concluding points in response to the "so what?" question cited above.

First, the American way of war is primarily a way of warfare. In short, that American "way" suffers from an acute political deficit. Antulio J. Echevarria has made this point in uncompromising fashion, accusing the country of adhering to a theory of battle in the mistaken belief that it is a theory of war.[53] The American way, in effect, is to treat warfare as a near autonomous activity, all but separate from its political purposes and consequences. One might characterize the traditional assumption as one which holds that "if we win the fighting, the politics will take care of themselves as a necessary benign consequence." In America today, Clausewitz is widely respected, even revered, but the practice of the American way of war tends to advertise the necessity for his major text to be read with greater understanding, especially in Washington. The burden of error naturally is borne more heavily by American civilian policy makers than by military professionals.

Second, and logically as an evil partner to the political deficit just outlined, the American way of war persists in suffering from a severe strategy deficit also. This essay has explained that strategy is the bridge connecting policy purpose with military power. All too often in American strategic history, that strategy bridge has been either missing or in need of urgent repair. As the most wealthy and generally materially best-endowed society on earth, the United States has not been in the habit of needing to make the difficult choices that strategy requires. With war and politics viewed in the United

States as substantially distinct fields of activity, it is scarcely surprising that a strategy deficit has plagued the American way of war.

Third, it is meaningful to identify a characteristic American way of war. I assert as much despite the great width of the spectrum of actual and potential warfare, from terrorism at one extreme and effectively total war at the other. I hold to this assertion notwithstanding the variety of conflicts in which the United States has engaged. As was explained in the preceding section, a study of American strategic history reveals persisting traits that collectively warrant the label "the American way of war." Of course, the United States, in common with other countries—though probably less often because of the depth of its resources—sometimes is obliged to behave out of preferred character. More accurately stated, perhaps, the country can find itself in a context where it needs to behave contrary to the precepts of its dominant way of war.

By and large, a society, any society, will not excel in the performance of unfamiliar and profoundly unwelcome strategic missions. Vietnam is the most obvious example of this phenomenon. MACV did not understand the problems it faced, so, all too understandably, it waged the kind of war that it did comprehend. This is not to deny that a society and its armed forces can change and raise their game to play an unaccustomed role successfully. But that behavior against the strategic cultural grain will be only temporary, almost certain not to lead to a permanent shift in the national way in war. Rather, the atypical experience will be followed by a post-mortem that concludes "never again"—until the next time, I should add.

Fourth, we need to beware of what can be called "paradigmitis." We strategic theorists are overly fond of inventing paradigms, while our official clients are wont to be unduly uncritical of impressive sounding Very Big Ideas, especially when they are presented with the bells and whistles of Power-Point razzle-dazzle. Americans enculturated with many if not all of the characteristics that result collectively in the way of war described above are probably uniquely vulnerable to the siren call of flashy strategic novelty. The future does not belong to irregular belligerency, call it 4GW or any other grand label you favor. Rather, the future will see the United States obliged to worry about, and prepare for, warfare that is both regular and irregular. A culturally ahistorical, even anti-historical, society like America is always in danger of conceptual capture by some bright and shiny notion, which typically is a rediscovery of the long familiar.

Fifth, the small library of official documents that express the commitment of the official defense community to military transformation betrays no real appreciation of the scope and depth of the cultural impediments to change, let alone revolution. There is an important sense in which the benefits of the officially directed transformation must be undermined, even rendered null and void, if the principal weaknesses in the American way of war are not recognized and frankly addressed. Specifically, the persistence of the political and strategy deficits in the American way that I have highlighted have to reduce the significance of the incredibly expensive transformation.

Those readers strongly critical of my argument might assert that the military profession now recognizes the necessity for what could amount to revolutionary change in American military and strategic culture. My reply to such a claim would be to say, first, that bold, good intentions are easy to state but not so easy to implement. Second, American transformers should not harbor the fallacy that the strategic and military culture of their society can be "fixed," or radically altered, by acts of will. There are deep and, in a historical sense, valid reasons why the American way of war is what it is, admittedly for both good and ill. This essay recommends strongly that the intellectual parents of transformation hasten to review their vision in the light of the cultural argument advanced here. If I were a betting theorist, I would wager that despite their best endeavors, the prophets and principal executives of transformation will find their great task shaped, reshaped, somewhat hindered, and possibly frustrated in its potential for new strategic advantage by an inconveniently persisting, cultural American way in warfare.

Sixth, and really to underline the immediately previous point, this essay concludes that the way of war hypothesis is extremely useful. This claim is made despite the fact that the thesis cannot help but simplify an extremely complex and varied American strategic historical experience. We noted earlier that the American defense community appears to have woken up to its lack of cultural grasp concerning actual and potential enemies. My argument here is that this defense community needs, no less, to wake up to the cultural programming that has produced a national style worthy of the title the American way of war. The transformation drive should be interrogated with reference to its relevance to each of the features of the traditional American way. Does it, can it, will it address the pathologies among those features? If the answer is by and large in the negative, it follows that the intended transformation must have disappointing results. At a minimum, the transformation effort needs to be broadened to accommodate honestly its *American* cultural context.

# Notes

1. Russell F. Weigley, *The American Way of War: A History of United States Military Strategy and Policy* (New York: Macmillan, 1973), p. xxii.

2. Eliot A. Cohen, "Kosovo and the New American Way of War," in Andrew J. Bacevich and Cohen, eds., *War Over Kosovo: Politics and Strategy in a Global Age* (New York: Columbia University Press, 2001), p. 61. Cohen begins his essay with this claim: "The Kosovo war marks a departure from the traditional American way of war" (p. 38). See also Max Boot, "The New American Way of War," *Foreign Affairs* 82 (2003): pp. 41–58.

3. For example, "The Army will transform its culture, capabilities, and processes as an integral component of Defense transformation," U.S. Army, *Army Campaign Plan* (Washington, D.C.: Office of the Deputy Chief of Staff, U.S. Army, 12 April 2004), p.10.

4. I develop this argument in an essay titled "From Principles of Warfare to Principles of War: A Clausewitzian Solution," unpub. paper, University of Reading, UK, January 2005.

5. Robert B. Strassler, ed., *The Landmark Thucydides: A Comprehensive Guide to the Peloponnesian War,* Richard Crawley trans., rev. ed. (New York: Free Press, 1996); Sun Tzu, *The Art of War,* Ralph D. Sawyer trans. (Boulder, Colo.: Westview Press, 1994); Carl von Clausewitz, *On War,* Michael Howard and Peter Paret trans. (Princeton: Princeton University Press, 1976); J. C. Wylie, *Military Strategy: A General Theory of Power Control* (Annapolis, Md.: Naval Institute Press, 1989).

6. Max Boot, *The Savage Wars of Peace: Small Wars and the Rise of American Power* (New York: Basic Books, 2002), p. 336.

7. On the complexity of strategy see Colin S. Gray, *Modern Strategy* (Oxford: Oxford University Press, 1990), ch.1; *Strategy for Chaos: Revolutions in Military Affairs and the Evidence of History* (London: Frank Cass, 2002), chs. 4–5.

8. Tzu, *Art of War,* p. 179.

9. Clausewitz, *On War,* esp. pp. 605–10.

10. Jeremy Black, *Rethinking Military History* (London: Routledge, 2004), pp. 13–22.

11. Victor Davis Hanson, *The Western Way of War: Infantry Battle in Classical Greece* (London: Hodder and Stoughton, 1989); *Why the West Has Won: Carnage and Culture from Salamis to Vietnam* (London: Faber and Faber, 2001).

12. For views critical of Hanson, see John Lynn, *Battle: A History of Combat and Culture* (Boulder, Colo.: Westview Press, 2003); and Black, *Rethinking Military History,* pp. 1–3.

13. Black, *Rethinking Military History,* p. 1.

14. Clausewitz, *On War,* p. 75.

15. See Thomas X. Hammes, *The Sling and the Stone: On War in the 21st Century* (St. Paul, Minn.: Zenith Press, 2004).

16. I defend this view in my book *Another Bloody Century: Future Warfare* (London: Weidenfeld and Nicolson, 2005).

17. See Alastair Iain Johnston, *Cultural Realism: Strategic Culture and Grand Strategy in Chinese History* (Ithaca, N.Y.: Cornell University Press, 1995), ch. 1; Colin S. Gray, *Nuclear Strategy and National Style* (Lanham, Md.: Hamilton Press, 1986), ch. 2; Gray, *Modern Strategy*, ch. 5.

18. Michael C. Desch, "Culture Clash: Assessing the Importance of Ideas in Security Studies," *International Security* 23 (1980, pp. 141–70; John Glenn, Darryl Howlett, and Stuart Poore, eds., *Neorealism Versus Strategic Culture* (Aldershot, UK: Ashgate, 2004).

19. Boot, *Savage Wars*, pp. xiii–xx.

20. See Ken Booth, *Strategy and Ethnocentrism* (London: Croom, Helm, 1979).

21. See Gerald Segal, "Strategy and 'Ethnic Chic,'" *International Affairs* 60 (1983–84), pp. 15–30.

22. Emily O. Goldman and Leslie C. Eliason, eds., *The Diffusion of Military Technology and Ideas* (Stanford, Calif.: Stanford University Press, 2003).

23. Frederick W. Kagan, 'War and Aftermath,' *Policy Review* 120 (2003), pp. 3–27, points up this danger.

24. U.S. Army, *2004 Army Transformation Roadmap* (Washington, D.C.: Office of the Deputy Chief of Staff, U.S. Army Operations, Army Transformation Office, July 2004); U.S. Army, *Army Campaign Plan*.

25. Richard P. Henrick, *Crimson Tide* (New York: Avon Books, 1995), p. 97.

26. Samuel P. Huntington, *The Soldier and the State: The Theory and Politics of Civil-Military Relations* (New York: Vintage Books, 1964).

27. Samuel P. Huntington, *American Military Strategy*, Policy Paper 28 (Berkeley: Institute of International Studies, University of California, Berkeley, 1986), p. 33.

28. Ibid., p.13

29. Ibid., pp. 14–17.

30. The 'Weinberger Doctrine' is reprinted conveniently and discussed informatively in Michael I. Handel, *Masters of War: Classical Strategic Thought*, 3rd edition. (London: Frank Cass, 2001), pp. 307–25.

31. James C Kurth, "Clausewitz and the Two Contemporary Military Revolutions: RMA and RAM," in Bradford A. Lee and Karl F. Walling, eds., *Strategic Logic and Political Rationality: Essays in Honor of Michael I. Handel* (London: Frank Cass, 2003), ch. 12; Jeremy Black, *War and the New Disorder in the 21st Century* (New York: Continuum, 2004), pp. 13–17; Colin McInnes, *Spectator-Sport War: the West and Contemporary Conflict* (Boulder, Colo.: Lynne Reinner Publishers, 2002), ch. 4.

32. See Colin S. Gray, *Transformation and Strategic Surprise* (Carlisle, Pa.: Strategic Studies Institute, U.S. Army War College, 2005).

33. See Stephen Peter Rosen, "Vietnam and the American Theory of Limited War," *International Security* 7 (1982), pp. 83–113; Mark Clodfelter, *The Limits of Air Power: The American Bombing of North Vietnam* (New York: Free Press, 1989); and Robert A. Pape, *Bombing to Win: Air Power and Coercion in War* (Ithaca, N.Y.: Cornell University Press, 1996), ch. 6.

34. Clausewitz, *On War*, p. 605. The intellectual godfather of coercive diplomacy, American-style, was Thomas C. Schelling. See his timely influential tour de force, *Arms and Influence* (New Haven, Conn.: Yale University Press, 1966). Another very American period piece that had real-world consequences was Herman Kahn, *On Escalation: Metaphors and Scenarios* (New York: Prager, 1965).

35. Huntington, *American Military Strategy*, p. 16.

36. Geoffrey Till, *Maritime Strategy and the Nuclear Age* (London: Macmillan, 1982), pp. 224–25.

37. See Stanley Hoffman, *Gulliver's Troubles: On the Setting of American Foreign Policy* (New York: McGraw-Hill, 1968).

38. Robert H. Scales Jr., "Culture-Centric Warfare," U.S. Naval Institute *Proceedings*, 130 (2004), pp. 32–36, is a hopeful sign of the times.

39. Colin S. Gray, *The Sheriff: America's Defense of the New World Order* (Lexington, Ky.: University Press of Kentucky, 2004).

40. Eliot A. Cohen: "The Mystique of U.S. Air Power," *Foreign Affairs* 73 (1994), pp. 109–24; "The Meaning and Future of Air Power," *Orbis* 39 (1995), pp. 189–200.

41. Quoted in Boot, *Savage Wars*, p. 294.

42. Ibid., p. 301. In 1966, for example, two-thirds of the American shells expended were fired at no known targets.

43. Huntington, *American Military Strategy*, p. 15.

44. Ibid., p. 16.

45. See Sam C. Sarkesian, *America's Forgotten Wars: The Counterrevolutionary Past and Lessons for the Future* (Westport, Conn.: Greenwood Press, 1984); John M. Collins, *America's Small Wars: Lessons for the Future* (Washington, D.C.: Brassey's, 1991); Anthony James Joes, *America and Guerrilla Warfare* (Lexington, Ky.: University Press of Kentucky, 2000); and Boot, *Savage Wars*.

46. See Andrew F. Krepinevich Jr., *The Army and Vietnam* (Baltimore, Md.: Johns Hopkins University Press, 1986).

47. Alfred H. Paddock Jr., *U.S. Army Special Warfare, Its Origins: Psychological and Unconventional Warfare, 1941–1952* (Washington, D.C.: National Defense University Press, 1982); Susan L. Marquis, *Unconventional Warfare: Rebuilding U.S. Special Operations Forces* (Washington, D.C.: Brookings Institution Press, 1997).

48. See Herman Hattaway and Archer Jones, *How the North Won: A Military History of the Civil War* (Urbana: University of Illinois Press, 1983), p. 720, for the primacy of logistics; and Thomas M. Kane, *Military Logistics and Strategic Performance*

(London: Frank Cass, 2001), for a superior explanation of logistics as the enabler of strategy.

49. Henry E. Eccles, *Logistics in the National Defense* (Harrisburg, Pa.: Stackpole, 1959).

50. Alfred Thayer Mahan, *The Influence of Sea Power upon History, 1660–1783* (London: Methuen, 1965), p. 25; Barry R. Posen, "Command of the Commons: Military Foundation of U.S. Hegemony," *International Security* 28 (2003), pp. 5–46.

51. Peter D. Feaver and Richard H. Kohn, eds., *Soldiers and Civilians: The Civil-Military Gap and American National Security* (Cambridge, Mass.: MIT Press, 2001).

52. Donald H. Rumsfeld, *The National Defense Strategy of the United States of America* (Washington, D.C.: U.S. Department of Defense, 1 March 2005), p. 1.

53. Antulio J. Echevarria II, *Towards an American Way of War* (Carlisle, Pa.: Strategic Studies Institute, U.S. Army War College, March 2004).

TWO

# The Second Learning Revolution

ROBERT H. SCALES

The Vietnam experience revealed serious faults in the way the American military prepared for war. In that conflict soldiers learned the lesson that superior technology alone was not sufficient to ensure victory. The enemy offset materiel inferiority with, often, exceptional skill at arms. Decades of real war experience gave the North Vietnamese an experiential advantage that American training and educational institutions could not easily overcome. The first training revolution actually began during the war with the U.S. Navy's Top Gun program followed soon thereafter by the U.S. Air Force's Red Flag exercises. The successes of both in restoring the American advantage in air-to-air competitiveness set the stage for similarly imaginative successes in preparing warriors to fight on the ground.

The air services sought to make better fighter pilots, but the U.S. Army sought to make better combat battalions and brigades. This focus on the operational rather than the tactical level of war was thought necessary for two reasons. First, the bitter experience of tactical warfare in Vietnam, particularly during the latter stages of the war, soured many senior leaders on the wisdom of transforming the army's training system at the squad and platoon levels. Second, the Israeli experience in the 1993 Yom Kippur War convinced the army that its doctrinal salvation rested with the ability to defeat the massive Soviet armored formations as they approached the inter-German border. Thus, the emphasis from the beginning was to become the finest and most formidable operational maneuver force in the world by combining brigade maneuver formations with deep fires provided by air force and army aviation as well as deep fires from cannon and rocket artillery.

The first training revolution continued in earnest after Vietnam with the creation of a system of force-on-force free-play exercises that were scored realistically and fought against a world-class adversary imbued with a serious desire to win. The army's laboratory for creating the revolution was the National Training Center in the California desert. When the U.S. Army and Marine Corps deployed to Desert Storm, both services had embedded the spirit of the combat training centers into their collective cultures. The results on the ground spoke volumes about the need for realistic training followed by forthright assessments during post-exercise "after action reviews." By the beginning of the kinetic phase of Operation Iraqi Freedom, the army had gotten it exactly right. Large army and marine armored formations supported by massive aerial strike forces were able to execute the first truly joint operational takedown in history thanks mostly to the skills learned by the army, marine corps, and air force in the deserts of California.

Subsequent events in Afghanistan and Iraq, however, suggest that the enemy understood and accepted America's superior ability to win on the sea and in the air. The enemy ceded to the ground services their ability to win the joint operational battle and sought, instead, to win at the tactical level of war. The enemy's logic was simple, its intent both logical and diabolical. It has learned that the surest way to gain advantage is to negate American big war technologies by moving the fight into complex terrain such as jungles, mountains and, most recently, cities. Prosecution of war against such an enemy has devolved increasingly into a series of tactical engagements fought principally at the squad and platoon levels. Joint warfare and the participation of other elements of military power are increasingly being applied at lower and lower levels to the extent that functions formerly considered the purview of senior commander are being taken up by combat leaders of much lower rank and experience. The challenge today is to create a second training and educational revolution that prepares our military leaders to fight in this new age of warfare. As the center of focus for fighting tomorrow's conflicts shifts downward, so too must the systems that teach soldiers and marines how to fight.

Learning science has also evolved to a point where the distinction between training and education has become blurred. The two are often now combined in several important aspects. Training prepares a young soldier to deal with expected situations on the battlefield; education prepares him to deal with uncertainty. On the modern battlefield a soldier knows that to survive he will have to be capable of fighting his weapons and following the orders of his leaders. But he will also be expected to demonstrate resourcefulness, initiative, creativity, and inventiveness demanded by a battlefield in which confronting the unexpected and new is considered to be routine. Tactical proficiency must be matched with a soldier's ability to speak the language and understand the culture of the society he is seeking to protect. Likewise, a soldier studying in a classroom will have access to virtual and synthetic environments that immerse him in a simulated battle that closely resembles real war. Thus, the nature of modern war is compressing both ends of the learning spectrum toward the middle.

More than ever war is a thinking man's game. Wars today must be fought with intellect as well as technology. Reflective senior officers returning from Iraq and Afghanistan are telling us that wars are won as much by creating alliances, leveraging nonmilitary advantages, reading intentions, building trust, converting opinions, and managing perceptions, all tasks that demand an exceptional ability to understand people—their culture and their motivation. While wars have become more complex, responsibility for those who fight them has increasingly slipped down the chain of command. Sergeants today make strategic decisions that previously were the preserve of colonels and generals. Yet sergeants and junior officers are the very ones who have very little time or maturity to prepare them to make strategic decisions. Even the time available to perform strategic tasks is diminishing as back-to-back deployments erode the time available to learn, particularly with the traditional institutional means. In a word, today's military has become so overstretched that it may become too busy to learn at a time when the value of learning has never been greater.

A second revolution is made possible in large measure by advances in learning science. When Gen. Paul Gorman led the effort to change the way the military learned how to fight in the late seventies, he had few technological tools available for assessing, measuring, and propagating the learning that occurred at the nation's training centers. Then, the World Wide Web was just an experiment and the science of personal and group testing was in its infancy. The centers were the finest live training experiences in the world, but the science of virtual and constructive training had yet to mature sufficiently to be applied realistically and on a large scale. Today, the military can draw on the experiences of academic, commercial, and corporate learning institutions to greatly improve how soldiers learn, and to propagate opportunities to a much larger field of learners.

# Creating Exceptional Individuals and Small Units

So, what must the military do to begin a "second training revolution"? This revolution must focus on creating extraordinarily proficient small units. No revolution can occur in learning unless supporting systems are changed to optimize opportunities for all soldiers to learn any time, any place. The military personnel systems of all services must be changed to accommodate these realities. The system must give small units very long periods to become proficient at combat tasks. Military learning must shift from an institutional to a soldier-based system that rewards individual performance rather than institutional efficiency. Soldiers must be given time and support to study and to improve their fighting skills continuously, over a lifetime rather than episodically at a time convenient to the bureaucratic needs of service personnel systems. Leaders must be taught at a much earlier age to lead indirectly, to think in time and to "see" a battlefield that is dispersed, complex, hidden, and ambiguous.

Make no mistake; the quality of performance among today's close combat soldiers is very high. Take a look at any news report or photograph of tactical engagements in this war and you will notice enemy soldiers running about shooting wildly. American soldiers move in tightly formed groups and even in the tensest moments carry their rifles with fingers outside the trigger wells. These images prove the value of rigorous training, and no one respects and appreciates first-rate training more than close combat soldiers. They consistently rate good training as more highly prized than pay and benefits because they, more than anyone, understand that first-rate preparation for war is the best life insurance one can buy.

However, past performance in combat provides no guarantees for the future. The compartmentalized, isolated, and unforgiving nature of war in the future—particularly when fought on the urban battlefield—will demand a new set of very demanding collective close combat skills. Battles will be isolated, compartmentalized affairs where small units will have to perform as self-contained, autonomous entities that may be forced to perform a multitude of complex tasks without external help. Thus, soldiers and marines will be proficient in a multitude of tasks formerly performed by such supporting units as intelligence, medicine, fire support, and communications.

In Vietnam, two-thirds of all small unit combat deaths occurred during the first two months in the field. Losses occurred in part because the training system of that era mass-produced close combat soldiers with too little time to properly prepare them for the incredibly complex and difficult task of close-in killing. Small units must be able in the future to undergo far more rigorous and definitive pre-combat conditioning. No unit should be allowed to go into a shooting situation until both leaders and followers have experienced this bloodless battle first.

Soldiers will also have to possess the flexibility and skill to transform themselves instantaneously from close combat specialists to providers of humanitarian assistance and social services. Often they will be obliged to shift between the two opposite roles several times during a deployment. Soldiers and marines such as these cannot be mass-produced. Training regimens for tasks such as these may take years rather than months. Think of tomorrow's close combat soldier transitioning from apprentice to skilled close combat journeyman under the tutelage of his "master craftsman" squad leader. Given that image it may be in the best interest of the U.S. Marine Corps to take a close look at the custom of keeping young marines in the ranks only through a few deployments before they are mustered out. It might be more productive for the corps to foster a culture wherein marines are kept in the force much longer.

The isolation inherent in urban fighting puts even greater demands on small units and will demand a degree of cohesion never before seen in the American military. A soldier's bond to his buddy often lasts long after the danger has passed, sometimes for a lifetime. Not much is known about how to generate this primal bonding agent, nor are commanders terribly skillful at creating conditions for effective bonding to occur. The one ingredient necessary for creating a closely bonded unit is time. Like a good wine, the aging of a good unit cannot be hurried. Platoons need at least a year to develop full body and character. The U.S. Army's effort to create individual soldier stability is admirable, but keeping a soldier stable is meaningless if he goes into combat a stranger within his unit. Perhaps the definition of "stability" might be recast to embrace the centrality of small unit stability, specifically close combat squads and platoons.

The challenge for the future is to develop training systems and methods that will allow small units to regain the advantage. The Cold War military learning system was predicated on progressing individuals and units to the point where they were able to perform tasks sufficiently well to "meet the standard." A standards-based system was revolutionary for the time because it induced individual and collective accountability and demanded that everyone perform to a level that could be measured. Thus, large organizations could be collectively categorized as to their level of training proficiency with some degree of reliability. In today's learning environment, merely meeting the standard is no longer sufficient to meet the demands of modern battle. Thus, a revised measurement is necessary to determine the proficiencies of smaller units, squads, and platoons. Likewise, meeting the standard should be replaced with a new set of performance indicators that raise the bar by creating an open-ended set of performance criteria. The "standard" is now a limiting factor in shaping the performance of highly capable and well-bonded units. Units today really don't know how good they can actually be. We should develop criteria for performance that demands and rewards performance for units and individuals without limiting criteria.

Training and education today are episodic. Just before they are deployed, units are at their peak. Soldier assignments, both internally and externally, have been stabilized and turbulence is at a minimum. Distractions are few. Leaders concentrate learning solely on those tasks to be accomplished in battle. Most deployments are preceded by a combat training center rotation where leaders get the opportunity to practice in the reality of rigorous simulated combat. But all too often, even in combat these skills deteriorate very quickly. The deterioration accelerates after soldiers return to their home station, when units dissolve and the learning focus is diffused by the distractions imposed by routine daily garrison life.

Unfortunately, the enemy rarely has downtime. The tempo within the theater of war continues while fighting skills atrophy at home. A first priority for the second revolution must be to keep the combat proficiency curve from dipping below the minimum necessary for a unit to reenter the combat zone without extensive train up or last-minute reshuffling of personnel. Likewise, individual learning in the form of officer or noncommissioned officer (NCO) learning must be constant and not subject to the same pendulum swings that diminish field proficiency. Keeping units and individuals within the zone of proficiency will demand a new set of attitudes and policies. Training evaluations will have to be random, persistent, and totally objective. Unit commanders will have to be held accountable for the learning of their units both in garrison and in the field. Such a reform will require constant reinforcement and validation, particularly at the small unit and individual levels.

## Cultural Preparation for Battle

The remarkable humanity of the American soldier occasionally gets him killed. Many of our past enemies have noted the relative naïveté of young soldiers new to close combat. Thanks to the oceans that surround us we are a relatively well-protected culture that rarely has faced massive invasions or traumatic intrusions into our homeland. This explains why so many soldiers recall that their first instinct in a firefight is disbelief that someone unknown really wants to kill him. The positive side of this social affectation is a congenital predisposition for American soldiers to befriend strangers, even enemies. German and even Japanese veterans often remark that they were astounded at how quickly American soldiers sought to bond with them and to forgive their aggressions once the battle ended. Children in particular were often the objects of this innate propensity to make friends.

Unfortunately, the gulf between West and East has never been greater than among American soldiers and Iraqis. In this war, the American soldier's proclivity to connect with alien societies is blocked by a barrier of cultural differences between American and Islamic societies. Few soldiers speak Arabic. Very few American soldiers have spent any time in Arab countries or even in the presence of Middle Eastern peoples. Close combat forces cannot again be sent into a tactical environment where they are forced to fight as complete strangers to the land and its inhabitants. This is a new style of war in which the strategic center of gravity is nested in the will of the Iraqi people. Soldiers cannot hope to fight such a war without possessing an intimate knowledge of how the enemy thinks and acts.

Every young soldier should receive cultural and language instruction. The purpose would not be to make every soldier a linguist but to make every soldier a diplomat in uniform, equipped with just enough sensitivity and linguistic skills to understand and converse with the indigenous citizen on the street. The mission of soldier acculturation is too important to be relegated to last-minute briefings prior to deployment. Acculturation policy should be devised, monitored, and assessed as a joint responsibility.

The military spends millions to create urban combat sites designed to train soldiers how to kill an enemy in cities. But perhaps equally useful would be urban sites optimized to teach small units how to coexist with and cultivate trust and understanding among indigenous peoples inside foreign urban settings. Such centers would immerse young soldiers within a simulated Middle Eastern city, perhaps near a mosque or busy marketplace, where they would be confronted with various crises precipitated by expatriate role-players who would seek to agitate and incite a local mob to violence. Interagency and international presence would be as evident in these centers as the services and joint agencies with, perhaps, a State Department, CIA, or allied observer controller calling the shots during an exercise.

## Improving Combat Units

During the Cold War, units fought in comfortably dense aggregations. Future battlefields will be far less comfortable. As the battlefield expands and becomes more uncertain and lethal, it also becomes lonelier and enormously frightening for those obliged to fight close. If the recent past is prologue, American campaigns will continue to be fought in unfamiliar weather and horrifically desolate terrain. To meet the challenges of tomorrow's more challenging battlefield, greater attention must be given to the psychological and physical preparation of close combat soldiers, particularly if they are to perform to the exceptional degree demanded of them. Modern science offers some promising solutions. Soldiers can now be better tuned psychologically to endure the stresses of close combat. Written instruments, tests, assessments, role-playing exercises, and careful vetting will reduce the percentage of soldiers who suffer from stress disorders after coming off the firing line.

The biological sciences offer promise that older, more mature soldiers will be able to endure the physical stresses of close combat for longer periods. This is important because experience strongly supports the conclusion that older men make better close combat soldiers. They are more stable in crisis situations, are less likely to be killed or wounded, and are far more effective in performing the essential tasks that attend close-in killing.

Junior leaders today are asked to make judgments and command decisions that in previous wars were reserved for far more senior officers. A corporal standing guard in Baghdad or Falluja can commit an act that might well affect the strategic outcome of an entire campaign. Sergeants in Afghanistan made decisions about where to deliver precision munitions that had enormous consequences for the overall success of the strategic mission there. Yet the intellectual preparation of these very junior leaders is no more advanced today than it was during the Cold War. Thankfully, the native creativity, innovativeness, and initiative exhibited by these very junior soldiers in close combat belie their relative lack of formal intellectual preparation. However, it seems clear that they could do even better if learning institutions took greater care to educate them earlier and with greater rigor.

During the Cold War, force-on-force engagements at combat training centers made better battalions and brigades. But the new era of warfare fought principally by smaller units demands that the same principles and techniques used to make operational units better must now be applied to make tactical units better as well. Squads and platoons must receive the same level of objective scrutiny. They must be exposed to rigorous and accountable engagements against an opposing force that closely replicates the skills and dedication of an Islamic terrorist insurgent. Future combat leaders should be required to demonstrate the same level of proficiency in realistic and stressful simulated combat environments that battalion and brigade commanders experience at combat training centers.

Today's close combat soldiers need more time to develop to peak fighting efficiency than their industrial age antecedents. Years, not months, are required to produce a close combat soldier with the requisite skills and attributes to do the increasingly more difficult and dangerous tasks that await him in the future. At least a year together is necessary for small units to develop the collective skills necessary to fight as teams. An infantry squad is the same size as a football, soccer, or rugby team. For a moment replace the "Willie and Joe" stereotype with an image of a National Football League team. Blocking and tackling are not enough to win the Super Bowl. Instead, a pro player must undergo a scientific regimen of physical conditioning. He does "two a days" during summer camp and watches the films and studies the coach's plays intently at night. He always has to fight for his position on the team because there is always the eager and hungry rookie looking to take his spot on the starting roster.

The quarterback knows that he cannot perform unless he has a team that follows his lead. A lack of confidence in the team leads to disaster. A collection of pros playing as individuals, rather than a team, will always lose. This is the image that we must build for our close combat soldier of the future. Not all need apply and very few should expect to join. Any shortcoming in performance should threaten a soldier's place on the team. Finally, every general manager knows that winning teams are purchased at a premium. Those who are willing (and likely) to die for our country should be held to no lesser standard, and those who pay for having such men on the roster should be willing to pay for the privilege of their presence.

## Identifying and Preparing
## Tactical Combat Commanders

Leaders ought to be trained for certainty and educated for uncertainty. In the industrial age, junior officers were expected to lead their men on the battlefield directly by touch and verbal commands. They were trained to follow instructions from their immediate commanders and to react and conform to the enemy. The image of sergeants and captains acting alone in the Afghanistan wilderness, innovating on the fly with instruments of strategic killing power, reaffirms the truth that today's tactical leaders must acquire the skills and wisdom to lead indirectly—skills formerly reserved for officers of a much higher grade and maturity. They must be able to act alone in ambiguous and uncertain circumstances, lead soldiers they cannot touch, and think so as to anticipate the enemy's actions—they must be tactically proactive rather than reactive.

Senior leaders must have the tools and the authority to identify and develop those who have the operational capacity and need, plus the intellectual "right stuff," to rise. History teaches that great combat commanders have one trait in common. They possess a unique, intuitive sense of the battlefield. They have the ability to think in time, to sense events they cannot see, to orchestrate disparate actions such that the symphony of war is played out in exquisite harmony. Perhaps no more than one in a hundred among many superbly qualified commanders has this unique gift. Often those with the operational right stuff are found only by accident. Commanders at the National Training Center often observe that it is the most unlikely commanders who perform well in the heart of battle. Perhaps they lack a certain pedigree, are rough around the edges, are even profane, but they know how to fight.

In the past, the only sure venue for exposing the naturals was in battle. Soldiers' lives had to be expended to find commanders with the right stuff. But today, learning science offers the ability to identify those who can make decisions intuitively in the heat of battle. The Germans called this gift *Fingerspitzengefuhl,* literally fingertip sense, or instinct, intuition. Many managers can make the right decision if given enough time, advice, and data, but only combat leaders can make the right decision at the right time in a crisis when the fog of war is greatest: when they are tired, fearful, and isolated.

Learning science offers the opportunity to find the naturals without bloodshed. The services must exploit this science by conducting research in cognition, problem solving, and rapid decision making in uncertain, stressful environments like combat. Leaders must be exposed during peacetime to realistic simulations that replicate conditions of uncertainty, fear, and ambiguity such that those who demonstrate *Fingerspitzengefuhl* are identified early, perhaps as early as commissioning. Those with the right stuff should be cultivated and exercised continuously to sharpen their decision-making prowess before they lead soldiers into real combat.

Military intellectual institutions must conduct research into a greatly expanded effort to understand the cognitive decision-making process. As much attention should be given to understanding how culture-centric systems interpret and use data as to how net-centric systems collect data. We need to better understand what information really is necessary for making decisions. Important in this effort is an understanding of how different commanders use information. Cognitive systems capable of customizing the decision-making process will emerge from that understanding. Perhaps soon commanders will be offered exercises and decision aids that will optimize their ability to make the right decisions in the midst of the mountain of information that will invariably descend on them in the heat of battle.

The requirement to better anticipate and shape performance in battle is made all the more challenging by today's conflict environment. Good commanders know how to lead in combat. Great commanders possess the unique intuitive sense of how to transition very quickly from active, kinetic warfare distinguished by fire and maneuver to a more subtle kind of cultural warfare distinguished by the ability to win the war of will and perception. Rare are the leaders who can make the transition between these two disparate universes and lead and fight competently in both.

## Acculturate Every Soldier to Prospective Theaters of War

One division commander in Iraq told me that his greatest worry was that his soldiers comprised "an army of strangers in the midst of strangers." During the early months of occupation, cultural isolation in Iraq created a tragic barrier separating Iraqis of good will from the inherent goodness that American soldiers demonstrated so effectively during previous periods of occupation in such places as Korea, Japan, and Germany. This cultural wall must be torn down. Lives depend on it.

Cultural and language instruction should be required for every young soldier. Today's e-learning technologies will permit such a program to be distributed over the Web. Soldiers should be able to achieve proficiency at home and demonstrate their knowledge using assessment tools administered by the Department of Defense or the Joint Chiefs of Staff before any soldier deploys overseas.

To assist in the acculturation process the Department of Defense should be required to build databases that contain the religious and cultural norms for world populations—to identify the interests of the major parties, the cultural taboos—so that soldiers can download the information quickly and use it profitably in the field.

## Institutional Learning: Creating a Habit of Lifelong Learning

This new era of war requires soldiers equipped with a fingertip sense of the battlefield and an exceptional degree of cultural awareness, as well as an intuitive sense for the nature and character of war. Where can they learn these skills? The military turns to universities for access to cutting-edge technology and resources for studying the nature of war. Unfortunately, however, American universities today have become less able to provide such wisdom. The study of violence is too removed from the sensitivities of contemporary academia to warrant their serious attention. Thus, very few civilian universities today offer comprehensive curricula in military history or war studies. Think tanks traditionally focus on short-term issues related to weapons procurement or on political and diplomatic policies. The only place where the art of war can be studied competently is within those institutions that retain responsibility for fighting wars—military schools and colleges.

Unfortunately, higher-level military schools and colleges fail to meet the learning needs of the services. Very few military leaders are fortunate to be selected to attend institutions that teach war. Those selected are chosen based solely on job performance rather than for the excellence of their intellect. Personnel policies affecting the purpose of senior military education have transformed these institutions partly into meeting places intended to achieve interservice, interagency, and international comity. The price for socialization has been a diminishment in the depth and rigor of war studies within these institutions. Thus, the central elements necessary to gain a deeper understanding of the nature and character of war—military history (primarily), along with war games and military psychology and leadership—often are slighted in an effort to teach every subject to every conceivable constituency to the lowest common denominator.

If the services are to get the greatest benefit from existing war studies programs, the system of professional education must be reformed to accommodate two imperatives. First, every military leader, particularly those whose job it is to practice the art of war, must be given every opportunity to study war. Learning must be a lifelong process. Every soldier, regardless of grade or specialty, should be given unfettered and continuous access to the best and most inclusive programs of war studies. Every soldier who takes advantage of the opportunity to learn must receive recognition and professional reward for the quality of that learning. Contemporary distance learning technology allows the learning process to be amplified and proliferated such that every soldier can learn to his or her capacity and motivation.

Distance learning technology permits military students to learn in groups, virtual seminars, even when on the job in some distant theater of war. The task of learning should therefore maximize the sharing and distribution of learning. Our officers and NCOs understand this phenomenon. The remarkable success of Web sites like companycommander.com and platoon leader.com testify to the need of our young leaders to learn by sharing. Soldiers should have the opportunity to learn continuously. Scholars have long known that learning is lifelong, not episodic. Therefore, soldiers should become members of a Web-based community of learners from the moment they join the service.

Second, those who demonstrate exceptional brilliance and whose capacity for higher level strategic leadership is exemplary should be afforded a unique opportunity to expand their knowledge to a degree unprecedented in the past. In this scheme the traditional staff and war colleges would focus attention exclusively on a constituency selected principally on intellectual merit. Every officer would be given the privilege of competing for a seat in these selective courses in residence. The courses would be dedicated exclusively to the study of war. The opportunities for attendance would be limited. The pedagogical model for the school would be based on the very successful advanced seminars already extant at all service schools (known within the U.S. Army as the School of Advanced Military Studies at the intermediate level and the Advanced Strategic Art Program at the senior level).

The military has too few learning resources to train and educate its leaders adequately. The commodity in shortest supply is time. Soldiers are often too busy to learn, and for that reason learning has taken second place to action in today's operationally focused force. The new learning environment should be centered on the student rather than the institution. Every learning opportunity should be crafted to ensure that the right methods, both pedagogical and methodological, are used to give the military learner just what's needed when it's needed using a suitable blend of on-site and Web-based delivery. Every concession must be made to lessen the burden of learning. First preference should be given to learning at home over the Web. Schools should be held responsible for monitoring and assessing the quality of the student's achievement while minimizing time students spend away in some distant classroom.

Immediately after commissioning, an officer would become part of a joint seminar composed of a dozen or so peers from the same and other services who share a common specialization. Operators would be paired with operators, communicators with like officers in another service. Senior educators from middle- and higher-level service schools would moderate educational seminars. Students' unit commanders would actively serve as their mentors, responsible for counseling and evaluating their progress. The learning interaction between the commander and his young officers will give the commander a unique opportunity to thoroughly articulate his intent (or how he practices command in combat) and how he solves practical military problems.

The opportunity for commanders to invest themselves personally in the learning processes of their junior leaders will most certainly strengthen the bonds of trust and confidence that come from honest debate and shared understanding. The program would be history-based and thoroughly joint in composition. Young officers would begin early to become indirect leaders by demonstrating their ability to grasp higher-order concepts and by learning the essential nature and character of war. These lifelong seminars would be rich in Web-based simulations and gaming, which would be evaluated and critiqued by monitors, mentors, and peers. Students would be required to produce original research papers sometime during their course of study in order to gain academic credit leading toward a master's degree from an affiliated, accredited institution of higher learning.

Those who demonstrate extraordinary talent would earn the privilege of pursuing a master's or a doctorate degree. Officers would be required to take individualized specialty courses online before assuming specific duties within a unit. For example, an officer would have to demonstrate an acceptable level of knowledge of personnel management just before assignment to a position as personnel officer within a battalion or squadron. No officer would be allowed to proceed to hands-on training at site-based training centers until he had passed the Web-based, foundation-setting portion of his training. The limited time available for site-based training and education must be devoted strictly to the practical side of learning with time spent in the field, in regional training centers, or in collective training exercises.

A Web-based method of learning would allow the creation of a lifelong electronic learning portfolio for every student. The portfolio would include an objective accounting of the student's performance in every course, seminar, simulation, field exercise, and game. The portfolio would be managed strictly within the "academy," independent of service personnel bureaucracies and with access limited to promotion boards and assignment officers. The commander-mentor and academic monitor would share responsibility for maintaining the portfolio. The monitor would be free to comment on how well the student learns the material and grasps the content; the commander on how well the student applies the knowledge.

Of course, the services might exploit each officer's portfolio as dictated by individual service cultures. However, in keeping with established precedents, the chairman of the Joint Chiefs of Staff would be authorized to use the portfolio to set assignment and promotion guidelines based on an officer's educational preparation for selection or promotion into joint assignments.

## Unit-Based Learning

During the last decade corporations have learned the value of educating their employees. Increasingly, some of the best managed companies have created chief learning officers and have given managers the responsibility to ensure that their subordinates are properly prepared intellectually to transition to new levels of responsibility. The military can learn from this example. Soldiers do best what their commanders demand from them. Commanders focus energy on what their commanders deem to be most important. In the past, responsibility for learning has been relegated to military learning institutions. If we are to create a body of leaders in the future capable of fighting asymmetric wars, responsibility for learning must be shifted to those most responsible for success—unit commanders.

Unit-based leader development must be perceived as a condition for unit readiness overall. A cycled rotation system will permit enough scheduled downtime to allow commanders to establish and actively superintend a disciplined study program for junior officers. Nonetheless, a method for monitoring time for professional development must be established by a disinterested authority, again divorced from service personnel systems.

The level of responsibility for critical decision making in the services continues to drift downward. Today, sergeants make strategic decisions that were reserved just a decade ago for officers of very senior grade. In Afghanistan, Special Forces sergeants succeeded in defeating the Taliban by establishing trust and mutual effort between the Northern Alliance and U.S. forces. Sergeants called in those precision strikes from strategic bombers that proved so successful in breaking the back of Taliban resistance. Thus, NCOs must be educated as well as trained for this new style of war. All NCOs should be given cultural and language training; those with the greatest promise should be offered the opportunity to pursue the study of war and alien cultures either in advanced military or in civilian educational institutions.

## Nine Initiatives

A second learning revolution is necessary to meet the needs of a military facing a new and unforeseen destiny. The ambiguity, uncertainty, and complexity of modern war demand that a new generation of soldiers and leaders be brought into a new culture of open inquiry, universal access, and heightened regard for both combat acumen and intellectual excellence. A credible start can be made if the services commit to

- Creating an attributable, measurable service-wide system for training small units. Consider establishing a consortium of small unit or even individual equivalents to Top Gun or Red Flag.
- Replacing the Cold War "task, condition, standard" learning paradigm with one that includes universal, open-ended goals, with each level of achievement measured objectively and recognized with a sliding scale of rewards for exceptional excellence that increases arithmetically and without limit.
- Reforming service-specific personnel programs such that individual accountability for excellence is reflected in promotions and selection for greater opportunities. Progress will be measured and learning rewarded by learning institutions not by personnel officers.
- Reforming the learning system such that lifelong learning is a requirement for personal progress. All soldiers would be given complete access to learning opportunities, but all would also be help accountable for their progress.
- Formulating an objective, realistic system for identifying small-unit leaders who have the special qualities that will permit them to perform with distinction in combat.
- Demanding a comprehensive, uniform, and accountable standard for cultural immersion of soldiers and their leaders.
- Creating a training culture that reflects the need for close combat proficiency as well as standards for performing well in operations that

require cultural sensitivity and the ability to work effectively with civilian, interagency, and international agencies and partners.

- Placing greater emphasis on NCOs. Translate the same open-ended standards expected of junior and mid-level officers at least to the degree that they can become proficient indirect leaders before they are put into circumstances that require such skills.
- Creating a truly universal and accessible training and learning resource library that clearly lays out expectations for excellence and includes all of the learning tools necessary for small-unit leaders to build their own effective training systems and programs without having to rely on higher-level learning institutions.

To be sure, a learning reformation such as has been suggested here will require a commitment perhaps as encompassing as the Goldwater Nichols Act (1986) or perhaps as culturally transformational as the Caldwell reforms or the Root reforms of the late-nineteenth and early-twentieth centuries. But changes as sweeping as these are essential if the services are to overcome the inadequacies of the current system and move into an era where the fighting and cognitive abilities of our military keep pace with those of the enemy.

# THREE

# Principles of War or Principles of Battle?

Antulio J. Echevarria II

In the aftermath of successful military operations in Afghanistan, a number of political figures and defense analysts proclaimed the arrival of a "new" American way of war, one purportedly based on the novel application of knowledge, speed, and precision. The initial phases of Operation Iraqi Freedom, which featured the fall of Baghdad in record time and with remarkably low casualties on both sides, seemed to validate this view. As subsequent stages of the campaign unfolded, however, it became clear that announcements proclaiming a new way of war were premature.[1] The so-called new American way of war is not yet a way of war, but—much as the old one—still a way of *battle*.[2] A way of war implies a war focus, which in turn necessitates a holistic view of conflict, one that grasps how—in the atmosphere of violence that is war—political, social, economic, and military activities may contribute to, or detract from, the accomplishment of preferred ends. Under this view, the purpose of conflict is to achieve a policy aim, normally a better peace, beyond the implied task of defeating an opponent. In other words, a war focus considers the defeat of the enemy as a means to an end; the way in which, and the extent to which, an adversary is to be overcome must, therefore, facilitate achieving the desired political ends.

A battle focus, in contrast, concentrates primarily on defeating an opponent militarily. It seeks tactics and stratagems that will destroy in the most efficient way an opponent's physical and psychological capability to resist. Overcoming an adversary—whether accomplished tactically, operationally, or strategically, or all together—thus becomes an end in itself. With that, the task of war is considered done. Under this view, the attainment of larger policy objectives, while clearly important, does not fall within the scope of war, even though their realization might require the use of military force. Put simply, the difference between a war focus and a battle focus is whether one approaches conflict, and thus so arranges one's doctrine and organizations, to achieve policy success or military victory. To be sure, as Clausewitz pointed out, political and military ends need not be at odds; policy goals can have a "warlike" quality that makes them almost indistinguishable from military aims.[3] Even in such cases, however, policy goals tend not to be realized until well after military ones, that is, with the completion of political, social, and economic reconstruction, or the exploitation of resources acquired through conquest.

Throughout history, military writers in the West concerned themselves more with how to achieve military victory than policy success. The former, after all, seems a natural prerequisite to the latter. It is not surprising, therefore, that the so-called principles of war really evolved as principles of battle, regardless of their label, or whether they were referred to as "truths, axioms, guides, rules, laws, fundamentals, maxims, or lessons."[4] To be sure, while the principles associated with war varied greatly over time, they still pertained almost exclusively to the task of overcoming an opponent. The chief exception came in the early 1990s when the U.S. military introduced principles for "operations other than war" (objective, unity of effort, legitimacy, perseverance, restraint, and security) into its doctrine, thereby implying a difference between war operations and other military operations.[5] The U.S. military has since recognized the fallacy of that distinction, however, and removed it, combining the "traditional" principles of war and the principles for operations other than war into a single set of "principles for joint operations." Nonetheless, whether the U.S. military calls them principles for joint operations or something else, they are little more than principles of battle, that is, so far as they qualify as principles at all.

All of this begs the question: If the U.S. military had genuine principles of war, rather than principles of battle, what might they look like? This essay demonstrates how the principles of war from Machiavelli to the U.S. military's current principles for joint operations have consistently maintained a battle focus, and then proposes two principles of war for consideration by the joint community.

## From There to Here

A survey of military history reveals that it was the quest for military victory, rather than the desire for policy success, that drove the development of principles of war in the West. The principles were deliberately conceived to help commanders achieve victory on land, at sea, and more recently, in the air. Machiavelli's *Art of War,* for instance, which some scholars regard as the first truly modern treatise on warfare, is essentially a commander's how-to manual derived from the theories and practices of the ancient Greeks and Romans.[6] It is, in fact, almost exclusively about winning battles rather than using military force to accomplish political objectives. Machiavelli's discussion of various "rules that generals should observe"—which is drawn from the historical examples of Frontinus, Caesar, and Vespasian—is a case in point. It focuses solely on how generals should prepare themselves and their forces to fight effectively.[7] The *Art of War* also discusses such topics as battle formations, commands, provisioning the army, choosing campsites, suppressing mutiny, and military discipline. Of course, considering Machiavelli's *Art of War* together with his other major works, *The Prince* and *The Discourses,* provides a more comprehensive view of his ideas. Nonetheless, the essence of those ideas is remarkably consistent and, for the West, remarkably representative. For Machiavelli, military victory advances military power, which in turn advances political power. The aim of conflict is thus power, and it is usually achieved by overpowering, outwitting, or otherwise disposing of one's rivals.

In a similar manner, Clausewitz's initial treatment of principles of war, as detailed in his essay *The Most Important Principles on the Conduct of War to Complement My Instruction to His Royal Highness the Crown Prince,* essentially focuses on how to defeat an opponent in battle.[8] Also, in his undated prefatory note to *On War,* he listed a number of "statements" (*Sätze*) that he felt were demonstrably true, and thus can be considered principles

> that defense is the stronger form of fighting with a negative purpose, and attack the weaker form with a positive purpose; that major successes help bring about minor ones; that strategic effects can, therefore, be referred to certain centers of gravity; that a demonstration is a weaker use of force than a real attack, and therefore demands special conditions; that victory consists not only in the occupation of the battlefield, but in the destruction of the enemy's physical and psychic forces, and this is usually only attained in the pursuit after the battle is won; that success is always greatest at the point where a battle has been fought and won, and that the changing over from one line and direction to another can only, therefore, be regarded as a necessary evil; that a turning movement can only be justified by general superiority or by the superiority of

our lines of communication and retreat to those of the enemy; that flank-positions are also governed by the same considerations; that every attack weakens as it advances.[9]

Moreover, in *On War* itself, he enumerated and discussed at length several fundamentals of strategy, all of which function as principles.[10] Indeed, without a body of principles, Clausewitz could not have begun to construct a theory of war, the focus of his life's work. "If concepts combine of their own accord to form that nucleus of truth we call a principle, it is the task of the theorist to make this clear."[11] Principles, which he defined as deductions reflecting the "spirit and sense" of a law, formed the foundation for his overall theory of war. He also carefully distinguished principles from laws, rules, and prescriptions and methods. Laws, which explain the cause-and-effect relationship between events, reflect universal and absolute truths (such as the laws of physics). Rules were inferences based upon experience (such as infantry should form squares when opposed by cavalry). Prescriptions and methods were merely the regulations and standard operating procedures by which an army conducted its business.[12] Throughout his work, Clausewitz stressed, though somewhat vaguely, that one must grasp the "meaning" of principles rather than treat them as "rigid rules."[13] Presumably, he meant that they should be applied with a certain mental flexibility. Of course, this begs the question as to how much flexibility one should allow; embracing too much flexibility deprives a principle of its character as a principle.

In any case, all of Clausewitz's principles essentially concern the attainment of military victory. Many of them actually resemble the U.S. military's current principles of operations as defined by Joint Publication 3-0, *Doctrine for Joint Operations,* revised first draft (RFD) (see Appendix 1).[14] Although Clausewitz repeatedly warned against taking any fundamentals as absolutes, he clearly believed that those he discussed qualified as principles. His understanding of the term *principles*—as general but not absolute truths—accords with the modern sense of the term. One could well argue, therefore, that Clausewitz was indeed the father of the modern principles of war, despite the prevailing conventional wisdom to the contrary.[15]

### Clausewitz's Principles of War
### From *The Most Important Principles on the Conduct of War:*

- Use *all* forces with the utmost energy.
- Concentrate as much power as possible where the decisive blow is to be struck.

- Act quickly, waste no time.
- Exploit all successes with the utmost energy.

**From *On War,* Book III, "On Strategy in General":**
- Chap. 6, "Boldness" ~ *Offensive*
- Chap. 8, "Superiority of Numbers" ~ *Mass*
- Chap. 11, "Concentration of Forces in Time" ~ *Mass*
- Chap. 12, "Unification of Forces in Time" ~ *Mass*
- Chap. 14, "Economy of Force" ~ *Economy of Force*
- Chap. 9, "Surprise" ~ *Surprise*
- Chap. 10, "Cunning" ~ *Simplicity, Security, & Surprise*

As the various principles of war evolved from the late-nineteenth through the twentieth centuries, they remained decidedly battle oriented in focus. The British army first codified an official list of principles in 1920 (the U.S. Army followed in 1921), largely because of the influence and energy of J. F. C. Fuller. Fuller took the position that martial principles were "eternal, universal, and fundamental," though the principles he developed continued to vary in number from one manuscript to another. He initially identified eight so-called strategic principles—objective, offensive, mass, economy of force, movement, surprise, security, and cooperation—and three tactical principles—demoralization, endurance, and shock.[16] The strategic principles later became his and the British army's principles of war, and their similarity to the U.S. military's current set of principles is as substantial as it is obvious.[17] They are thus often referred to as the "traditional" or industrial-age principles of war. In content, these principles owe much to those of Clausewitz; indeed, they hardly advance beyond his.[18]

Even recent efforts to revise the traditional principles of war in light of the challenges of the information age have really only produced new principles of battle. For instance, Robert Leonhard's *Principles of War for the Information Age* is refreshingly original; the principles he proposes differ markedly from the traditional ones in both content and form.[19] His approach seeks to establish an almost yin and yang balance between opposites, such as command and anarchy. Leonhard maintains that the U.S. military should organize the principles of war into three categories. The first category is principles of aggression, which includes (1) dislocation and confrontation and (2) distribution and concentration. The second grouping is principles of interaction, which consists of (3) opportunity and reaction and (4) activity and security. The third category is that of control, which includes (5) option acceleration and objective and (6) command and anarchy. All of these principles depend on one "independent" principle—knowledge and ignorance; the extent to which one is able to exercise command or must endure anarchy, for instance, depends on how much knowledge one has.

Under Leonhard's system, military leaders would *not* have to memorize "eternal truths" for later application; instead, they would learn to weigh the pros and cons of opposing concerns, which are themselves a function of how much one is willing to pay in time, money, personnel, and other resources for knowledge. Knowledge is, of course, *not* an independent variable, as Leonhard maintains; neither is ignorance. The fact that one must expend a certain amount of resources for knowledge means that knowledge is *dependent* on them, and on one's willingness to pay for them, as well as on one's basic ability to interpret raw data and turn that into genuine knowledge. Also, in an environment such as war, knowledge often has to be fought for; it is, therefore, just as dependent on the other principles of war that Leonhard describes as they are on it. They in fact share a symbiotic relationship.

Despite the revolutionary nature of Leonhard's principles, moreover, they still pertain more to battle than to war. As he readily admits, it is as if "we believe that *war* is nothing more than a *battle* that has gotten out of hand."[20] Its faults notwithstanding, Leonhard's approach deserves thoughtful consideration. It offers much of value to a military profession still attempting to transition from industrial-age to information-age war, even if the focus of that approach remains on fighting.

Similarly, the principles ascribed to the so-called *new* American way of war—speed, knowledge, jointness, and precision—are also essentially principles of fighting, or battle (see Figure 2).[21] Speed and knowledge can apply to the strategic, operational, and tactical levels of warfare. However, they concentrate exclusively on defeating an enemy rather than on "winning the peace" and securing one's policy objectives.

### Principles of the *New* American "Way of War"
- Speed—ability to act quickly, to get inside the enemy's decision cycle.
- Knowledge—better intelligence and the ability to act on it rapidly.
- Jointness—services integrated to maximize power and lethality.
- Precision—ability to deliver devastating damage more accurately.

As explained earlier, a war focus involves much more than simply applying a strategic perspective to the conduct of war. To be sure, the battle focus that characterizes both the traditional and the newer principles of war is an important one. After all, defeating an opponent in battle is a difficult task. A holistic view of war without a commensurate eye for defeating an opponent would have little point. The battle focus is thus not wrong; it just does not go far enough, especially given that, if the last ten to fifteen years are any guide, twenty-first century conflicts, particularly those involving the global war on terror, may entail increased requirements for political, economic, and social reconstruction.

## U.S. Principles of Joint Operations (from JP 3-0 RFD)

JP 3-0 RFD lists the twelve principles (of war) now recognized by the U.S. military. The manual combines nine traditional principles—objective, offensive, mass, economy of force, maneuver, unity of command, security, surprise, and simplicity—with three others originally developed for operations other than war—restraint, perseverance, and legitimacy. As previously mentioned, the new set is now referred to as the principles of joint operations (which are reproduced in the appendix for the reader's convenience).[22] The shift in title is a fortuitous one, since none of the twelve principles recognized by joint doctrine can be considered a genuine principle of war. Moreover, JP 3-0 RFD does not define the term "principle." Consequently, it tends to confuse assertions, cautions, preconditions, or desirable conditions with principles.

### *Objective*

Objective can apply to the strategic, operational, and tactical levels of war. Strategies are developed (or should be) with certain end states or objectives in mind; operations and battles are conducted to achieve those ends. However, the fact that the principle of the objective applies to every level of war does not make it a principle of war, unless objective emphasizes the need for ways and means to translate military victory into policy success. Unfortunately, JP 3-0 RFD addresses objective (which it never actually defines) only as it pertains to the military dimension of conflict: "the purpose of objective is to direct every *military* operation toward a clearly defined, decisive, and attainable objective."[23] The roles that political, economic, and informational objectives might play in achieving the overall aim, and the need to coordinate them with the military effort, are not mentioned, despite the fact that joint doctrine continues to stress the critical nature of interagency coordination. In any case, the real crux of the issue is not remaining steadfastly fixed on accomplishing the objective or being flexible enough to adapt quickly when it changes, as the verbiage implies. Instead, it is knowing *if, when,* and *how* the objective should change. Objective, as it is described in JP 3-0 RFD, is thus not a true principle of war, though if modified it could provide the basis for one.

## *Offensive*

Like objective, the offensive can also apply to every level of war. However, the offensive is an unsubstantiated assertion about a presumed advantage, not a principle. It presupposes that the act of being on the offensive gives the attacker certain coveted advantages, such as the opportunity to strike first and the ability to seize and retain the initiative. However, these advantages are more apparent than real. In martial arts, for example, striking first is not always wise; a skilled opponent can take advantage of an attacker's momentum to throw him off balance and render him vulnerable, thereby turning a presumed advantage into a disadvantage. Similarly, the term *initiative* is often confused with having the upper hand. Simply put, it means having the power to force one's opponent to react, and perhaps to establish a desirable tempo. However, when the attacker's actions play into the hands of the defender, as they often do in martial arts, it is the defender who actually has the upper hand, even though he is merely reacting to the blows of the attacker and the tempo set by him. The offensive (like the defensive), therefore, is just a descriptor, and a potentially misleading one at that: a combatant is said to be either on the offensive or on the defensive, but neither is necessarily synonymous with having the advantage. A combatant can, of course, be on the offensive strategically, while also being on the defensive operationally or tactically, a fact that further undermines the value of both terms as descriptors. Moreover, asserting, as JP 3-0 RFD does, that the purpose of the offensive is "to seize, retain, and exploit the initiative" tends to make the offensive an end in itself, and overlooks the influence that political leaders should have in determining what the purpose of a particular offensive or defensive action should be.[24] The offensive is thus not a principle, a principle of operations, or a principle of war.

## *Mass*

Due to the destructive power of modern chemical, biological, radiological, nuclear, or high-explosive (CBRNE) weapons, the U.S. military modified the principle of mass from its original sense—namely, a concentration of forces to maximize combat power at the point of decision—to the idea that mass really meant the "massing of effects" not forces. Concentrating one's forces tends to turn them into a single lucrative target, which makes their destruction by airpower or nuclear, chemical, or other weapons that much easier. Hence, the purpose behind the idea of massing the effects of one's weapons, rather than the weapons themselves, was to accommodate the need for greater dispersion. However, a closer examination of modern conflict reveals that concentration of forces still occurs, it just does so on a larger geographic scale. In Operation Iraqi Freedom, for example, five (eventually six) of twelve carrier battle groups deployed to the Persian Gulf and the Mediterranean Sea—and thus were not available for use elsewhere. Such deployment does indeed represent a concentration of forces when viewed globally. Just as in classical antiquity, the ranges of weapons tend to limit the extent to which they can be dispersed, and thus require them to be concentrated to some degree. We need to become more accustomed to viewing conflict on a larger—indeed, a global—scale. Put differently, massing effects has always required, and still does, considerable concentration of forces. Regardless of the scale involved, mass applies strictly to the act of overcoming an opponent. It is thus a principle of battle, not of war.

### Economy of Force

Economy of force is merely the complement of mass. By exercising economy of force in certain locations, one can achieve mass elsewhere. In a sense, Operation Iraqi Freedom saw an attempt to supplant mass with economy of force. That attempt succeeded well enough in the initial phases of the conflict, but it failed completely when military operations shifted from major combat operations to providing security for reconstruction efforts. Mass, albeit broadly applied, thus proved more appropriate than economy of force for maintaining control over key people and places, particularly as the conflict morphed into an insurgency. Nonetheless, like mass, economy of force is a principle of battle rather than of war.

### Maneuver

The underlying purpose of maneuver—which is to put one's opponent at a positional disadvantage—can apply to diplomacy as well as to combat. However, it has rarely been viewed in the former sense, despite the fact that the idea of "political maneuvering" is not an unfamiliar one. Machiavelli, for instance, underscored its importance in *The Prince*. Maneuver can apply to any level of war, but most of the literature on the topic addresses only tactical and operational maneuver. Its essential components have been fire and movement, though some debates among military professionals in the 1980s perceived a dichotomy between firepower and maneuver, a dichotomy that eventually proved false in a dispute full of misunderstandings. In recent years, the concept of "operational maneuver over strategic distances" emerged to describe how increases in the speed and range of military systems now permit coordinated fire and movement to occur on a much greater scale geographically. However, even operational maneuver over strategic distances overlooks the importance of integrating political, informational, economic, and military maneuver into a synchronized concept of "global maneuver."[25] Thus, the principle of maneuver has traditionally functioned as a principle of battle rather than of war, though it has obvious potential for application beyond mere physical fighting.

### Unity of Command

Unity of command essentially means unity of effort. While it has been a military shibboleth for some time, at the level of policy formulation it runs counter to such fundamental safeguards as checks and balances and balance of power. The classic example of its violation in a military sense is that of Republican Rome's practice of assigning consuls to command the Roman army on alternating days. The Carthaginian general Hannibal Barca took advantage of this practice at the Battle of Cannae in 216 BC when he baited the impetuous consul C. Terrentius Varro and his sixty-five thousand legionnaires into a battle of encirclement and annihilation; historians surmise that Varro's alternate, the cool-headed L. Aemilius Paullus, would not have taken the bait. The principle underscores the need to be of one mind in military endeavors, and it applies equally well to the strategic, operational, and tactical levels of war. The value of coordinating the efforts of allies and coalition partners is obvious. The need for coordination is just as apparent for directing the elements of national power. However, in this case the value is more theoretical than practical, since political, military, economic, and informational power tend to operate at different rates of speed. Some types of military action can have an impact long before economic sanctions or embargoes take effect. With regard to policy formulation, moreover, unity of effort is hardly a healthy concept for a democracy unless the singleness of purpose it represents aims in the direction of preserving constitutional liberties, democratic values, and individual freedoms. Otherwise, unity of effort opens the door to autocracy. For all of these reasons, therefore, it is a principle more appropriate for battle than war.

### Security

Security assumes a new meaning when considered within the context of current reconstruction efforts. Traditionally, security applied only to friendly forces, and it essentially became synonymous with the term *force protection,* in effect a counterpoint to surprise. However, Operation Iraqi Freedom illustrated the importance of security for reconstruction efforts—vital activities when regime change and the establishment of a viable democratic government are the objectives of the conflict. Thus, the principles of mass and security go hand in hand, especially with regard to the conduct of security and stability operations during the so-called post-conflict phase of hostilities. They remain principles of operations rather than of war, but their value in the whole process of turning military successes into political ones has been underappreciated. In many ways, both are prerequisites to the realization of policy aims.

### Surprise

Although surprise can occur on any level of war, it amounts to a principle of battle because it deals almost exclusively with overcoming an adversary. Essentially, surprise is the use of unexpected ways or means to gain an advantage by causing the psychological and perhaps physical dislocation of an opponent.[26] The purpose of surprise is thus to render the opponent easier to defeat by diminishing his psychological and physical capacity to resist. Deception plays a key part in achieving surprise. The ways and means employed to achieve it, of course, need not be military in nature. However, surprise hardly applies to stability and support operations, or to reconstruction efforts—the critical end game of war where policy aims tend to be realized. A vital advantage in defeating an opponent, surprise is therefore a principle of battle.

### Simplicity

The principle of simplicity was originally intended to help mitigate the effects of fear, friction, chance, uncertainty, and exertion, all of which tend to produce high levels of stress among combatants. That, in turn, makes the activities associated with warfare more difficult to accomplish. As Clausewitz said, "in war everything is simple, but even the simplest thing is difficult."[27]

Stress affects perceptions, memory, reasoning, and physical strength and stamina. The assumptions underpinning simplicity as a traditional principle of war are, first, that the simpler a plan is, the greater its chances of being understood and executed properly. Second, the fewer moving parts a system has, the less complex it is, and the less likely it is to malfunction. However, these assumptions are not always valid. Hannibal's plan at Cannae was more complex than Varro's, but the Carthaginian's scheme succeeded whereas the Roman's did not. A javelin has fewer moving parts than an assault rifle, but under most circumstances, troops will opt for the latter. To be sure, a holistic view of war is more complex than one centered simply on battle. However, it is absolutely essential if the criterion for success is—as it seems to be now— the accomplishment of policy aims. Simplicity is less a principle than an admonition, therefore. It should serve merely as a warning to avoid extraneous considerations in doctrine, planning, and execution, without eschewing the importance of acknowledging the complexity of the physical world.

### Restraint, Perseverance, and Legitimacy

Restraint, perseverance, and legitimacy were originally introduced into joint doctrine as principles for operations other than war, but they are not principles, that is, they are not general rules requiring judgment in application. Restraint— or the judicious use of force—and perseverance—or physical and psychological capabilities to persist in the execution of protracted operations—really amount to little more than common-sense advice. Restraint is a function of a combatant's perception of how much force is necessary to accomplish a particular mission, the discipline needed to refrain from exceeding that level of force, and the ability to control the application of force itself. One does not apply perseverance in the way that one applies the principle of mass; rather, one endeavors to persevere. Whether or not one succeeds depends on one's physical and psychological capacities relative to an opponent's. Restraint and perseverance thus fall into the category of cautions rather than principles.

Legitimacy—the perceived legality and morality of war aims and operations conducted in pursuit of them—has become more important in the globalized environment of the twenty-first century where political statements and physical actions can be captured and retransmitted almost instantly. Still, it is not a principle, but a subjective condition that can become an advantage with constant cultivation. In any struggle, both sides will endeavor to claim legitimacy for their own aims and actions. Global communications make all but the most concealed actions subject to public scrutiny. While possessing legitimacy is clearly not necessarily a prerequisite to waging war successfully, combatants whose aims and actions are perceived as legitimate have a better chance of garnering the support of the international community. Such support can contribute a great deal toward achieving a successful outcome in modern conflict. To be sure, achieving and maintaining legitimacy requires an active information campaign, purposeful planning, and conscientious execution. However, it is neither a principle of war nor of battle.

In sum, the U.S. military's principles of operations—at least those that are in fact principles at all—are merely principles of battle. They focus almost exclusively on the goal of defeating an opponent rather than on using force to achieve policy aims.

## On Principles of War

To address the challenges of the current strategic environment more fully— particularly if one of the chief policy aims of the United States' global war on terror remains, as it likely will for some time, the elimination of the conditions that spawn terrorism—the U.S. military has two choices. It can continue to debate the merits of its principles of operations, which may or may not result in their revision, reaffirmation, or even reinvention. Or, it can develop a set of genuine principles of war and use them as a guide for understanding conflict in the twenty-first century holistically. Based on that enhanced perspective, it might then generate a more fruitful debate regarding the appropriateness of its traditional principles of operations. The two principles discussed below offer a starting point for a holistic approach. Here, the term *principles* means guidelines for thinking.

### 1. Understand the Nature of War

Understanding the nature of war is important because the nature of a thing tends to determine how it can be used. The nature of a hammer makes it difficult, if not impossible, to employ a hammer as a screwdriver, for instance. But, what *exactly* is the nature of war? Defense intellectuals and soldiers often refer to it, but rarely define it. Unfortunately, much of the literature dealing with the nature of war actually fails to differentiate between *war* as a state of conflict between two or more parties and *warfare* as a process.[28]

However, most of those who do use the term *nature of war* properly claim that their understanding of it either comes from or agrees with Clausewitz's, though some give him credit for the initial spark but challenge what they understand as his view. Indeed, one can find as many misunderstandings of Clausewitz's view of the nature of war as of his other ideas. The most popular misconception is that the Prussian's concept of the nature of war was essentially political, as exemplified by the oft-quoted phrase "War is the mere continuation of political intercourse by other means."[29] This misconception is reflected in the bulk of the modern literature on war, including the memoirs of prominent military commanders, and was reinforced by the post-Vietnam renaissance of Clausewitz in military and scholarly literature.[30] The underlying rationale for this view is, since Clausewitz claimed that war is essentially a political act, policy (or politics) must form the central element of war's nature. Still others, such as the late historian Russell Weigley, have maintained that politics tend to become an instrument of war rather than the other way around: "War once begun has always tended to generate a politics of its own: to create its own momentum, to render obsolete the political purposes for which it was undertaken, and to erect its own political purposes."[31] Yet, as Clausewitz explained, this phenomenon merely reflects the fact that political and military aims sometimes coincide to the point where they become nearly indistinguishable.[32]

Another view holds that war's nature resembles Clausewitz's concept of absolute war, by which Clausewitz meant something akin to pure war, that is, continuous and all consuming. Clausewitz was quick to point out that while absolute war is possible to conceive of intellectually, in reality it is a physical absurdity because the inevitable friction and other constraints, to include political ones, imposed by the physical world will simply not permit war to reach such an extreme. However, Clausewitz's warning did not preclude attempts to equate absolute war with such concepts as total war, though one might well question to what extent total wars have actually been total.[33] Still others wrongly believe that Clausewitz maintained that the nature of war—consisting of such imponderables as violence, friction, chance, and uncertainty—was essentially unchanging; only its character, or the way it is waged, changes over time.[34] Yet, others contend that war itself has no specific nature; instead, it takes on the forms and purposes that societies and cultures give it.[35]

Since these misunderstandings claim to begin with Clausewitz, so must we. Actually, the most important aspect of Clausewitz's view of the nature of war has gone unnoticed, namely, that war has a dual nature, not in the bipolar sense where wars can be limited or unlimited, but in the sense that derives from German philosophical traditions in which phenomena are considered to have objective and subjective natures.[36] The objective nature consists of all aspects of a phenomenon that are universally valid; the latter concerns those that are true only for a specific time and space. The objective nature of war thus includes those elements—violence, friction, chance, uncertainty, fear, and exertion, for example—which all wars have in common, no matter where or when they are fought. These elements are universally present, though the degree to which they are present and the influence they exert may vary. Conflicts can range in kind from an unrestricted war to a war of observation (peacekeeping), for instance, but each will have all of these elements present in one degree or another.

By contrast, the subjective nature of war encompasses those elements—military forces and their doctrines and weapons, for example—that vary from war to war, and thus make all wars unique. The objective-subjective construct, therefore, enabled Clausewitz to explain why all wars are, at once, the same and yet different. Maritime conflicts, for instance, clearly differ from wars on land, but they are nonetheless similar with respect to their objective characteristics. Even the same conflict can change its subjective aspects over time as different combatants enter or leave the contest, or introduce new weapons, tactics, and techniques into the fray, as occurred in the Thirty Years War. Interestingly, its objective characteristics can change as a result of subjective changes.

Under Clausewitz's concept, the objective and subjective natures of war interact continuously. New weapons or methods can increase or diminish the degree of violence or uncertainty, but probably never eliminate them entirely. Similarly, a war's political motives can cause the combatants to use, or refrain from using, certain types of weapons or tactics, as in the Cold War where both the United States and the Soviet Union essentially established a number of treaties designed to prevent escalation to nuclear war. Thus, the objective and subjective natures of war remain inextricably linked. In Clausewitz's words, war's objective and subjective natures make it "more than a simple chameleon" that only changes its color as its environment changes.[37] A chameleon can change its color, but its internal organs remain exactly the same. War's internal tendencies, on the other hand, can change in intensity, proportion, and relative significance as the external features themselves transform. The objective nature of war, therefore, cannot be separated from its subjective nature, and vice versa.

Under Clausewitz's concept, then, the nature of war is not immutable; its objective qualities are always present, but they vary constantly in intensity and significance.

Understanding the nature of war is thus the first and most important principle because it will (or should) preclude attempts to use war to achieve something that it cannot, at least not on a consistent basis. Knowledge of the nature of war should, for example, preempt the development of rigid, one-sided theories or "McNamaraesque" formulae that purport to measure progress toward victory. The nature of war is too dynamic for that. It should show that, because of the interdependent relationship between the subjective and objective natures of war, we can choose weapons and ways of fighting that might well reduce the elements of violence, chance, friction, and physical exertion, for example. However, because our actions take place in the physical world and not in a vacuum and because our opponent can take preemptive action or introduce effective countermeasures, the results can never be certain. An element of risk is therefore unavoidable. Knowledge of the nature of war is thus important because it informs the choices that political leaders make when deciding to resort to war to achieve certain aims, or when making decisions about force structure; it also informs the choices that commanders make when equipping and training their troops for war. As the preceding paragraphs illustrate, our understanding of the nature of war is fundamentally flawed. Establishing "knowledge of the nature of war" as the first principle of war may help remedy that.

## 2. Understand the Nature of the War at Hand

Understanding the nature of *the* war one is engaged in, or about to be engaged in, is just as important as comprehending the general nature of war. Indeed, understanding the latter without linking it to, and thus developing an appreciation for, the former can turn the activity of thinking about war into a largely academic exercise. Guerrilla wars and conventional wars share the same basic objective nature, for instance, but the element of violence differs in intensity and significance. In a guerrilla war, the level of violence is generally lower than in a conventional conflict, but its political significance is nonetheless high. In a conventional conflict, the level of violence may be higher, but the percentage of it that is politically significant is lower. The subjective natures of each kind of war differ in obvious respects, particularly as regards the tactics and types of forces employed. Of course, two wars of the same kind can in fact differ markedly from one another in both objective and subjective natures, depending on the belligerents involved, the objectives at stake, and the weapons at hand, etc. Hence, we must refrain from using that knowledge as a template to dictate courses of action. In other words, coming to grips with the nature of *the* war at hand helps generate questions that can lead to a better grasp of the potential power and apparent limitations of that particular form of war, especially in terms of the political aims at stake and the strengths and weaknesses of our opponent's ways and means in relation to our own.

As stated earlier, all belligerents influence how specific objective and subjective elements might combine, or not, in a particular conflict. The Cold War was a case in which both sides determined that the violence of a nuclear exchange was to be avoided, as the vast destruction likely to be wrought by such an escalation would be self-defeating. Thus, both sides worked toward limiting the objective element of violence, which in turn gave the Cold War a unique subjective nature, since the emphasis shifted away from conventional conflict toward insurgencies and counterinsurgencies. In the war on terror, however, Al Qaeda and its ilk seek to intensify the element of violence by acquiring weapons of mass destruction (and by other means), while the United States and its allies and strategic partners endeavor to prevent that escalation from happening. In another example, the subjective nature of the war in Iraq has shifted from a conventional conflict to an insurgency. In this case, one side was able to exert more influence on the subjective nature of the war than the other, though that will not necessarily thwart the latter's ability to accomplish its policy aims. Nonetheless, appreciating when a conflict's subjective nature has changed (or even anticipating the various ways in which it might change and preparing for them) is critical for accomplishing one's aims, or recognizing that they must change.

Comprehending the nature of the particular war one is engaged in, or about to be engaged in, is thus the second principle of war. This understanding should play a key role in shaping the discussion of detailed policy aims and the strategies—to include the ways and means—appropriate for accomplishing them. A key question strategists might ask themselves under this principle, for example, is whether the nature of the war in question would require reconstruction of some sort, whether post-conflict or otherwise, for policy success. For the nature of the war on terror, the answer would appear to be yes.

It would not be necessary to be able to predict what kind of war lies round the corner; indeed, under this principle, that particular hobby would be exposed for the fallacy it is. Anticipating differs qualitatively from predicting. The former permits hedging one's bets, while the latter renounces it as unnecessary. Hedging one's bets, of course, entails building a force capable of fighting effectively across the spectrum of operations. Predicting, on the other hand, appears to be more cost-effective, but actually requires accepting higher risks across the majority of that spectrum. If military forces are going to be used with any frequency to accomplish strategic aims, then strategists must learn to anticipate rather than predict. Indeed, by assuming they can predict the nature of the next war, or the next shift in the nature of the war at hand, they demonstrate that they fail to understand the nature of war in general.

## Conclusion

One could well argue that the U.S. military ought not to dictate how the other elements of national power are to be used in conflict, since that knowledge lies outside of its expertise. As a profession, however, the U.S. military has the duty to think beyond battle to understand war in its entirety, in all its various dimensions and stages. Only in this way can it be sure that it fully understands the requirements and limitations of military force in contemporary conflict, and thus offer the best advice. No other profession or occupation in the defense community has the experiential base and abstract knowledge of warfare to think about war holistically, at least not in a systematic way. Developing genuine principles of war is a vital step in that process. Of course, the greatest risk in proposing any principles of war, or of battle, is precisely what drove J. F. C. Fuller to walk away from them in the end, namely, that strategists and other military thinkers will take them as more prescriptive than they were meant to be. Principles must be regarded guidelines that facilitate or inspire thought, not formulae that guarantee victory. That, at least, is one principle that we should never change.

## Notes

1. In fact, one of the proponents of the "new" way of war now admits its principles were inadequate. Compare Max Boot, "The New American Way of War," *Foreign Affairs*, Vol. 82, No. 4, (July/August 2003): 41–58 and Max Boot, "The Struggle to Transform the Military," *Foreign Affairs*, Vol. 84, No. 2 (March/April 2005): 103–18.

2. This argument is developed further in Antulio J. Echevarria II, *Toward an American Way of War* (Carlisle, Pa.: U.S. Army War College, Strategic Studies Institute, March 2004).

3. Carl von Clausewitz, *Vom Kriege. Hinterlassenes Werk des Generals Carl von Clausewitz*, 19th Ed., ed. by Werner Hahlweg (Bonn: Ferd. Dümmlers, 1991), Book I, Chap. 1, pp. 211–12.

4. John I. Alger, *The Quest for Victory: The History of the Principles of War* (Westport, Conn.: Greenwood, 1982), 8–9.

5. Russell W. Glenn, "No More Principles of War?" *Parameters* (Spring 1998): 48–66, describes the change.

6. Niccoló Machiavelli, *The Art of War*, Trans. Ellis Farneworth, Rev. and introduced by Neal Wood (New York: Da Capo, 1990).

7. Machiavelli, *Art of War*, 122–25.

8. Carl von Clausewitz, *Die wichtigsten Grundsätze des Kriegführens zur Erganzung meines Unterrichts bei Sr. Königlichen Hoheit dem Kronprinzen*. Reprinted in Clausewitz, *Vom Kriege*.

9. Clausewitz, *Vom Kriege*, 71.

10. Clausewitz, *Vom Kriege*, Book III.

11. Clausewitz, *Vom Kriege*, VIII, 1.

12. Clausewitz, *Vom Kriege*, II, 4. Clausewitz's definitions of law, principle, and rule parallel those used by Immanuel Kant in *Critique of Judgment*.

13. Clausewitz, *Vom Kriege*, II, 5, "Kritik."

14. Department of Defense, *Doctrine for Joint Operations*, Joint Publication 3-0 Revised First Draft (Washington, D.C.: Department of Defense, September 15, 2004). The previous version of JP 3-0, dated February 1, 1995, referred to them, inaccurately, as principles of war. The revision of JP 3-0 is scheduled to be completed in December 2005, with the U.S. Army's FM 3-0: *Operations* following shortly thereafter; the two manuals are to be harmonized in terms of content and structure. See Program Directive, dated March 25, 2004, Subject: Revision of JP 3-0.

15. For the conventional wisdom, see Alger, *Quest for Victory*, 186.

16. Alger, *Quest for Victory*, 115–16.

17. Fuller's eight strategic principles essentially became the British army's official "Principles of War" in 1920 and, in terms of basic content, have survived, albeit intermittently and in modified form, to the present. Alger, *Quest for Victory*, 122–25.

18. Ironically, Fuller's influence not only contributed to the publication of the first official list of such principles, but it also played a part in their later abandonment, albeit temporarily, as criticism of his book *Foundations of the Science of War* and his political views mounted, and caused him to fall from public favor. Alger, *Quest for Victory*, 125–26.

19. Robert Leonhard, *Principles of War for the Information Age* (Novato, Calif.: Presidio, 1989).

20. Leonhard, *Principles of War*, 9–10.

21. Summary of Lessons Learned, Prepared Testimony by Secretary of Defense Donald H. Rumsfeld and General Tommy R. Franks, presented to the Senate Armed Services Committee, July 9, 2003. See also the remarks by Vice President Dick Cheney, "A New American Way of War," to the Heritage Foundation, May 1, 2003, which ascribes many of the same characteristics to a new style of American warfare.

22. JP 3-0 RFD, Appendix A.

23. JP 3-0 RFD, A-1. Emphasis added.

24. Ibid.

25. Antulio Echevarria, "Interdependent Maneuver," *Joint Force Quarterly.* (Autumn 2000): 11–19.

26. In this sense, surprise is indeed a "subset of dislocation," as Leonhard points out. Leonhard, *Principles of War,* 193.

27. Clausewitz, *Vom Kriege,* 261.

28. William E. Odom, *America's Military Revolution: Strategy and Structure after the Cold War* (Washington, D.C.: American University Press, 1993) purports to discuss the changing nature of war but really only addresses changes in weaponry; Martin van Creveld, "Through a Glass Darkly: Some Reflections on the Future of War," *Naval War College Review* LIII, No. 4, (Autumn 2000): 25–44, in effect, only includes changing forms of war when discussing changes in the nature of war; and the Cantigny Conference Series, *The Changing Nature of Conflict* (Chicago: Robert R. McCormick Tribune Foundation, 1995), 32, actually only discusses changes in war's root causes.

29. The German is: "Der Krieg ist eine bloße Fortsetzung der Politik mit anderen Mitteln." *Vom Kriege,* 210.

30. See Bernard Brodie, "A Guide to the Reading of On War," in Carl von Clausewitz, *On War,* ed. and trans. Michael Howard and Peter Paret (Princeton: Princeton University Press, 1976), 641–46; and Eliot A. Cohen, *Citizens and Soldiers: The Dilemmas of Military Service* (Ithaca: Cornell University Press, 1985), 22–24, which maintains that "politics pervades war and the preparation for it"; and Eliot Cohen, "Strategy: Causes, Conduct, and Termination of War," in *Security Studies for the 21ˢᵗ Century,* ed. by Richard H. Shultz Jr., Roy Godson, George H. Quester (Washington, D.C.: Brassey's, 1997), 364–66, which uses "politics" as the defining element in war; and Colin S. Gray, *Modern Strategy* (Oxford: Oxford University Press, 1999), who places it at the center of his concept of strategy. See also Colin Powell, *My American Journey: An Autobiography* (New York: Random House, 1995); and Wesley K. Clark, *Waging Modern War: Bosnia, Kosovo, and the Future of Combat* (New York: Public Affairs, 2001). On the Clausewitzian renaissance, see Christopher Bassford, *Clausewitz in English: The Reception of Clausewitz in Britain and America, 1815–1945* (Oxford: Oxford University Press, 1994); Stuart Kinross, *Clausewitz and the American Military Tradition,* PhD Dissertation, Kings College, 2002, chapters 1 and 3; and for an example, see Harry G. Summers Jr., *On Strategy: A Critical Analysis of the Vietnam War* (Novato, Calif.: Presidio, 1995).

31. Russell Weigley, "The Political and Strategic Dimensions of Military Effectiveness," in *Military Effectiveness,* ed. Williamson Murray and Allan R. Millett (Boston: Allen and Unwin, 1988), vol. 3, *The Second World War,* 341.

32. Clausewitz, *Vom Kriege,* Book I, Chap. 1, pp. 211–2.

33. See Col. Joseph I. Greene, "Foreword" to Karl von Clausewitz, *On War,* trans. O. J. Matthijs Jolles (New York: Modern Library, 1943), xiii.

34. Colin S. Gray, "How Has War Changed Since the End of the Cold War?" *Parameters* Vol. 35, No. 1 (Spring 2005): 14–26; Colonel James K. Greer, "Operational Art for the Objective Force," *Military Review* (September–October 2002): 22–29, esp. 24; G. F. Freudenberg, "A Conversation with General Clausewitz," *Military Review* Vol. 57, No. 10, (October 1977): 68–71.

35. John Keegan, *A History of Warfare* (New York: Alfred A. Knopf, 1994), 24, 46. See also Martin van Creveld, *The Transformation of War* (New York: Free Press, 1991), 33–62, 124–26; and "*The Transformation of War* Revisited," *Small Wars and Insurgencies* 13, No. 2, (Summer 2002): 3–15.

36. Peter Paret, *Clausewitz and the State: The Man, His Theories, and His Times* (Princeton: Princeton University Press, 1985), 154, noted that Clausewitz used the same objective-subjective construct in an earlier essay entitled "Strategie und Taktik" (1804). This construct was not uncommon in Clausewitz's day, particularly in philosophical discourse. It was used in the works of Immanuel Kant, for example, from whom Clausewitz borrowed several analytical tools; for example, compare Clausewitz, *Vom Kriege*, Book I, Chapter 3, "Military Genius," to Kant's description of genius in the *Critique of Judgment,* Sections 46–50. Under Kant's system, *objective* laws were those recognized as valid for the will of every rational being. *Subjective* laws were those regarded as valid only for an individual; see Book I, Chapter I, Section 1 and the Table of Categories in the *Critique of Pure Practical Reason.* Some otherwise excellent analyses of Clausewitz's thought do not mention his use of this construct, such as Raymond Aron, *Clausewitz: Philosopher of War,* trans. Booker & Stone (Englewood Cliffs, N.J.: Prentice-Hall, 1985).

37. Clauswitz, *Vom Kriege*, Book I, Chap. 1, p. 213.

FOUR

# On War

## *Enduring Principles or Profound Changes?*

### Harlan Ullman

Has the nature of "war" changed and, if so, how, why and where? If war has changed in some fundamental way, what does that imply for the so-called principles of war that, for most of history, many have regarded as inviolable? Or, are these changes really accommodations and adjustments to the realities of the day rather than profound transformations in how states use force and violence for their own purposes and to their own ends? And, to what degree has "transformation" and the revolution in military affairs anticipated and reacted to these questions? From this inquiry, recommendations and prescriptions follow for how the United States should harness its intellectual assets in better understanding the nature of war in contemporary and longer-term time frames. Whether anyone in authority is interested and listening is perhaps a more relevant question, assuming that the answers to the others convey a sense of urgency in assessing contemporary wisdom.

The thesis of this chapter is purposely and doubly provocative. First, it argues that "war," as traditionally defined and understood, has indeed been transformed in at least three fundamental ways. Distinctions between "winners" and "losers" in war, at least regarding advanced states that possess thermonuclear (and possibly biological) weapons that could destroy much of the adversary's existing society even after its own had been devastated, have been rendered largely meaningless. In cases where powerful states such as the United States attack and defeat lesser adversaries such as Iraq, physical risk to the homeland, to paraphrase Clausewitz, has shifted from military retaliation and the threat of mass destruction to other means (discussed below).

A second revolution has affected weaponry. For all of history—certainly until the early 1990s—nearly every rock, stick, arrow, spear, projectile, bullet, and missile hurled, thrown, or launched missed. Today, the opposite is true. Not only will weapons hit on the first shot; a high probability is that the target will be killed or destroyed on the first shot. This revolution in lethality and accuracy marks a profound difference from the past.

Finally, in the post–9/11 world and America's declaration of war against global terror, "war" is no longer waged between organized and legitimate states with, more or less, a set of agreed-upon rules and laws. America's pre-occupation with the global war on terror has elevated the adversary to an equivalent status of state without the need to possess similar (or any) levels of authority, legitimacy, or definition. These three changes, indeed revolutions, will have profound implications for how "war" is waged and how force and violence are used by states and non-state actors to achieve policy ends and ambitions that may be driven by other than the traditional concept of national interests.

The second provocative assertion is made only as general background, namely, that the United States may be at greater risk today than at any time in its history since the Civil War. This argument was advanced in my last book, *Finishing Business: Ten Steps to Defeating Global Terror* (Naval Institute Press, 2004). In summary, this argument had three parts.

First, Americans do not fully comprehend the extent of the danger posed by our adversaries who are out to grab power using a perverted interpretation of Islam as ideological cover and grounds for legitimacy. These ambitions are collaterally supported by "terrorists," who like the anarchists of the late-nineteenth century, are prepared to die in the process of waging jihad. Second, our society is permanently vulnerable to disruption and it is the threat of disruption that our enemies will exploit. Third, it is arguable whether a system of government designed by the best minds of the eighteenth century can deal with the rigors of the twenty-first century when the danger is not a state or states and cannot be defeated by conventional military or means that have worked in the past. That our enemies have enveloped themselves in a protective religion raises profound constitutional challenges and questions as to how we can combat that combination of radical Islam and unyielding political ambition in which terror is a crucial tactic without shattering our notions of governance, freedom, and liberty. Indeed, checks and balances central to American governance have become the targets for our adversaries to exploit.

While this argument is not central to this chapter, it is to the future. Put differently, our adversaries may be at about the same point of maturity as Lenin and the Bolsheviks were a century ago and Hitler and the Nazis were in the decade after World War I. No one thought Russia could be turned into the Soviet Union or the Kaiser's Germany into the Third Reich. They were. Those are not guarantees or predictions that Islamic jihadists will seize power in the Middle East. It does mean that this possibility must be taken seriously now if it is to be prevented in the future.

## Back to the Beginning— First and Unchanging Principles

War, as Clausewitz and Sun Tzu tell us, is a violent clash of wills between states waged by more or less organized forces and armies. A (or the) principal instrument of war is, of course, military force supported by all or many of the tools and resources of the state. War was waged against and between well-defined states, giving it a certain legitimacy and enabling rules and laws of war to govern (some of) its conduct. A principal object of war was to defeat the enemy usually by destroying the will to resist. Clausewitz argued that this was achieved by eliminating the means to fight, which meant defeating or destroying the enemy's military forces or creating the perception that resistance was useless.

The notion of enduring principles of war was established by and generally reflected the notions of the greatest and even lesser philosophers and practitioners of war. Among these general principles were the key aim of disarming the enemy as a necessary condition for victory; that wars were fought between and among recognized states; that war aims ranged from all-out or unconditional victory to more limited outcomes, such as seizing territory or gaining a concession; and that knowledge and intelligence about the adversary were crucial if not always attainable commodities.

Not all wars were "winnable," and many were fought to a draw. But one fact remained inviolable at least until 1945: the defeated power had no means to destroy the enemy once its army was neutralized and its ability to resist was eliminated. True, "Pyrrhic" victories occurred, such as Carthage's costly defeat of Rome that led to its ultimate defeat. In 1945, the invention of nuclear and later thermonuclear weapons forever altered that principle. No longer did winning and losing have the same traditional meanings if society could be destroyed in the "victorious" states regardless who "won" or who "lost."

Had there been a war between the Soviet Union and the United States or China, it could have been nuclear and thus thermonuclear. A thermonuclear or H-bomb is about one thousand times as powerful as the weapons that destroyed Hiroshima and Nagasaki. In the nuclear age, as Chairman Mao Tse-tung understood, "a few atom bombs were enough." But the "defeated" nation still had sufficient nuclear strength to launch an attack on the other, with the power to destroy the so-called winner many times over. Hence, the notion of deterrence based on the threat or ability to obliterate society, called mass or assured destruction, fundamentally changed prior enduring notions about winners and losers. What is interesting is whether the principles of deterrence that emerged during the nuclear years have endurance and indeed relevance to the twenty-first century and the threat of stateless groups using terror as a tool and mass disruption rather than destruction of society as the lever to achieve political ends. And, should biological agents be developed to the point where they could threaten the existence of societies, that too could have profound effects on the strategic balance and how wars against terror and rising political movements will be aged.

As a result, with the end of the Cold War and the events of September 11, 2001, that destroyed New York's Twin Towers and part of the Pentagon in Washington, D.C., the question of what constituted war was no longer fully answerable by traditional logic and experience.

Clearly, terror was as old as war. Terrorist organizations threatened many states and orders. In Europe, toward the end of the nineteenth century and the beginning of the twentieth, terrorists and anarchists tried to collapse the political order then in place. Terrorists targeted heads of state. World War I was precipitated by the assassination of Archduke Franz Ferdinand of Austro-Hungary. Yet, that war and the others of the last century were fought among states. Even Vietnam, which began as an insurgency, ended with the North Vietnamese Army, organized in traditional formations, overrunning the South.

With the declaration of a global war on terror by President George W. Bush, and an assault against the ill-defined "terrorists" and terror networks that lacked borders and boundaries, in which military force is only one of many tools, it is fair to ask whether the changing conditions of international politics have indeed changed how "war" is viewed, waged, and defined and whether this is a temporary phenomenon or will have enduring qualities. Adversaries no longer need borders. In a sense, the Internet and instant global communications provide a technological alternative for groups seeking to achieve their own political agenda. Against that background there has been another inescapable technical reality with the most profound consequences for war and how it is conducted.

For at least through the late 1980s, except in close-in, hand-to-hand combat, precision and accuracy were of exceedingly low order. As noted, virtually all weapons fired missed. In World War II, for example, it took many dozens of rounds to kill a tank and huge numbers of bombs to destroy a major complex. Those statistics no longer apply in what can still be described as "conventional" war, and for the United States, the opposite effect applies. Combined with the notion of "effects-based targeting" or operations in which a target is broken down to the smallest vulnerability to determine the minimum effort required to neutralize it, a real revolution has occurred. The caveat applies to conventional war. What, for example, is occurring in Iraq today or in many other regions where insurgencies or the more violent end of criminal activity constitutes the threat is not as easily affected by this revolution in accuracy, precision, and targeting.

All of this raises the most profound questions about war, its nature, and how it might be fought in the future. On the one hand, "conventional" war could break out on the Korean peninsula, between India and Pakistan, or, more remotely, in the Taiwan Straits between the People's Republic of China and Taiwan. War in the Middle East, something that has happened every decade since World War II ended, can never be assumed away. On the other hand, as criminal activity and terror take on increasingly more visible and politically relevant roles, and proponents of violence are less constrained by lack of access to technologies (particularly informational ones that can do great electronic damage), traditional notions of war may be less or no longer applicable. Instead, law enforcement, intelligence, and nonmilitary tools may prove to be the more appropriate means to address these challenges.

To understand, even partially, the evolutionary and revolutionary aspects of these factors, it is crucial to begin with the background and forces that led to the so-called revolution in military affairs, how the United States viewed that revolution (which became one of "transformation") in terms of future direction, and what all of this may mean for the international order about what is to be done.

## Transformation and the RMA—Looking Back

Whether one agrees or not with the magnitude of changes in the nature of war noted above, their implications have not yet held full sway in how the

United States shaped or reshaped its military forces. The end of winners and losers in the thermonuclear age led to efforts to exploit nonnuclear forces and supporting technologies. The revolution in accuracy and lethality was not fully understood during the period when the revolution of military affairs defined conventional wisdom. Nor was it fully understood in the period of transformation that followed in the sense that force structure was not dramatically revised to fit this new reality. And the end of state versus state conflict as the dominating factor of "war" has failed to register in transformation to the extent that military forces are still largely configured to fight major wars against like and symmetrical threats.

Much has been written about the revolution in military affairs. Interestingly, the term came from the Soviet Union and is often confused with the "Military Technical Revolution." Hence, a bit of history is important to review in order to understand the relevance of the term, what it is taken to mean, and how it has affected and shaped both military thinking and military forces, qualitatively and quantitatively.

During the Cold War, one of the presumptions was that the United States generally maintained a large technical superiority over the Soviet Union and China. American weapons were usually regarded as superior once America put its mind to it. Of course, that was not always true. During the early part of the Korean War, the Soviet T-34 tank was virtually invulnerable to the U.S. 2.36-inch "bazooka," the RPG of its day. When the Russians sent Sputnik into orbit in 1957, the Americans mistakenly believed they were losing the missile race. And Americans often ridiculed Soviet fighter aircraft for relying on cathode ray tube and non–solid state technology, dismissing the reliability of those old tubes and the fact that they were not vulnerable to electromagnetic pulse energy from nuclear explosions that would "cook" solid state Western systems, making them useless while having no effect on "ancient" electron tubes.

From President Dwight Eisenhower in the 1950s to President Ronald Reagan just prior to the end of the Soviet Union, the United States consciously attempted to overcome Soviet numerical superiority by exploiting technological qualitative advantage. When Jimmy Carter was president (from 1977 to 1981), his Secretary of Defense, Harold Brown, explicitly made exploiting technology as a means of overcoming Soviet quantitative strength part of the strategy. Clearly, the Soviet Union recognized this.

The great fear on the part of the United States and its European allies was that because of the huge numerical advantage of the Red Army, the alliance was at a decided disadvantage in Europe should conventional war break out. As the nuclear balance became perceived as more equal, reliance on nuclear deterrence appeared more problematic. Hence, the United States spent considerable resources investigating how "smart" or revolutionary technology could defeat the Red Army on other than a "man for man" basis. Of course, the truth was that with Turkey's huge army, NATO had more troops under arms than did the Soviet Union. However, the constraints inherent in that alliance meant it was virtually impossible to rationalize its forces to achieve maximum military power. Technology had to be the solution to end this impasse.

The United States, through the improvement of weapons accuracy provided by such so-called smart technology as lasers, radar, and the global positioning system, or GPS, began developing the capability for "deep attack." Deep attack meant that Soviet forces could be placed at risk far from the front lines. The aim was to destroy as many of Soviet forces as possible before they could join battle. By the time the Reagan administration took office in 1981, deep attack had evolved into the "air-land battle" in which the combination of ground and air forces with these new technologies would be able to destroy much larger Soviet formations and thus win the first battle in any war in Europe.

The Soviets in many ways followed and read American military publications and doctrine more closely than did the Americans. Aware of American plans and American technology, the Soviets began writing about the "reconnaissance-strike complex" and the ability to attack deeply into enemy territory with very accurate and lethal weaponry. The term of art to describe these changes was the "Military Technical Revolution" (MTR), and it meant how technology was changing the nature and the conduct of war. The extent of the revolutionary nature of the MTR, or the Revolution in Military Affairs (RMA), the phrase the United States borrowed from its adversary, would be fully felt as the Soviet Union was disintegrating. Andrew Marshall, a well-known American strategist who has headed the Pentagon Office of Net Assessment since 1973, had become the father of the RMA and made sure it was widely understood in U.S. military circles. By 1990, the U.S. military had in fact experienced at least a partial RMA. With Saddam Hussein's invasion of Kuwait in August 1990 and the deployment of American and Coalition forces to the Persian Gulf, the impact of the RMA would become clear the following January when Operation Desert Shield turned into Operation Desert Storm and the hundred-hour war that liberated Kuwait.

As Coalition forces stormed across and around the so-called Saddam defense line on the Kuwaiti border with Iraq, the overwhelming technical, operational, and training advantages translated into one of the most one-sided battles in history. Iraqi Soviet T-62 and T-72 tanks, once thought to be unstoppable, were sitting ducks. Many were destroyed on the first shot at extended ranges before they could engage Coalition forces. Not a single U.S. M-1 Abrams tank was lost in action. The same results applied to air forces. There were no real battles at sea since Iraq did not have a navy.

Thus, at that stage, the RMA for the United States had translated into forces that were still based on largely Cold War formations but had been made far more lethal, with highly effective and accurate smart munitions, sophisticated command, control, and communications systems, and advanced reconnaissance from space, from the air in the form of airborne early warning aircraft (AWACS) and even with a few unmanned drones. Rather than a revolution, even though the combat results were revolutionary, the changes had been evolutionary.

From 1993–2001, during the Clinton administration, the RMA was still very much an important cornerstone for defense planning. Each of the services was required to assess its spending priorities in terms of how each fit the RMA and demonstrate how capabilities were improving along those lines. However, with the Soviet Union gone and, outside the Korean peninsula, no large-scale war in which America might fight in sight, it was difficult to find a metric to measure how well or badly U.S. forces were assimilating the RMA.

## Transformation

In 2000, candidate George Bush promised to "transform" the U.S. military for the twenty-first century. This campaign promise was a terrific sound bite. But putting flesh and specific changes on it would prove difficult. How and why, for example, would anyone wish to transform the best military in the world? To what would it be transformed and why? With Bush's election, the Clinton focus on the RMA was replaced by "transformation."

When he was chosen as Secretary of Defense, Donald Rumsfeld brought many distinguishing features with him. He had served once before under President Gerald Ford as defense secretary from 1975–76, the youngest ever. Now, Mr. Bush would make him the oldest to hold that job. Rumsfeld had chaired several panels on the ballistic missile threat and on space, so he was familiar with many defense issues. He had also been a very successful businessman and had made a run for president in 1988.

Rumsfeld entered office with the mandate to "transform" defense. Senior military officers were very skeptical at first. How do you make the best military in history better or stronger? they asked. Rumsfeld's first answers were not helpful.

The secretary convened a number of outside panels to help him define transformation, what it meant and how it would be executed. But translating a campaign slogan into practical terms and then implementing that program would prove difficult. Initially, transformation begot a series of other slogans for military forces that would be "more agile, lethal, and flexible." For senior and junior officers alike, the quest for those qualities did not always translate into directions and guidance that were helpful in changing the structure and capability of the forces. For the first months of the Bush administration, there was great frustration in the Pentagon over transformation and huge doubts about the ability of the secretary to make good on Bush's promise.

In one regard, the attacks of September 11 did change everything, at least as far as transformation was concerned. The attacks would transform Rumsfeld from the media's nominee as the first senior Bush appointee likely to be fired—Treasury Secretary Paul O'Neill would have that honor—to its darling. As the dust settled, the administration determined that, if the ruling Taliban in Afghanistan would not turn over Al Qaeda leader Osama bin Laden, presumed perpetrator of the attacks, the United States would intervene and take out Taliban rule. Given the unsuccessful occupation of the Soviet Union in Afghanistan two decades earlier, many observers believed the United States would be caught in a similar trap.

By force of personality, Rumsfeld cajoled, bullied, and persuaded the uniform military that a highly innovative and fundamentally different mix of forces and capabilities would remove the Taliban. Relying on the numerical advantage of the indigenous Northern Alliance and local Afghans opposed to the Taliban, and overwhelming precision weapons technology, the result was a strategy and plan that combined special operations troops from the U.S. services, CIA paramilitary operating in Afghanistan with warlords and the Northern Alliance, and selected airpower from land and sea bases to carry the day. The image of an American Special Forces soldier on horseback, equipped with GPS and calling in B-52 strikes to fire precision weapons on Taliban positions, was the most graphic representation of transformation in action.

Once the Taliban were defeated, the Bush administration turned to Iraq. Again, Rumsfeld believed that the conventional wisdom calling for overwhelming force in the form of half a million or so troops as were deployed in the First Gulf War was unnecessary. Systematically, he repeated the process of Afghanistan to convince or to direct the senior military to recognize that

America's overwhelming capability meant more could be done with fewer troops. Instead of the half-million troops deployed in the First Gulf War, the decision was made to go with a force of less than half that size. Orders were to advance as rapidly as possible to Baghdad to collapse the Hussein regime with a strategy of "shock and awe." Shock and awe turned out to be sound bites that quickly dissipated as the bombing of Baghdad was wrongly equated by the media as attacking civilians and nonmilitary targets. In fact, effects-based targeting and rules of engagement specifically made minimizing collateral damage high-priority instructions.

The original plan called for deploying the 4th Infantry Division through Turkey into western Iraq as part of a pincer movement to support the two-pronged assault from the south spearheaded by the U.S. Army and Marine Corps. Unfortunately, at the last minute, the Turkish government refused permission and so the attack had to be made with a smaller force, one that lacked the power of an additional—and indeed the army's most technologically capable—division advancing from the west.

In one of the most remarkable military campaigns in history, Coalition forces covered about four hundred miles in three weeks to capture Baghdad, routing an army of some four hundred thousand in the process. This was the best illustration of what transformation meant: flexible, swift, and lethal forces, operating with a different mindset rather than revolutionary new technologies, to defeat an adversary overwhelmingly with minimal rather than maximum force.

Sadly, that the war went so well had little positive impact on the peace. The reverse turned out to be the case. The Bush administration's plans for the postwar period simply failed. On the basis of the last war, the administration expected huge refugee problems, catastrophic damage to oil facilities, and (through use of weapons of mass destruction) the threat of famine and lack of water. Instead, the problems turned out to be political, social, and economic: to rebuild the country and provide enough proof of better lives so as to rally public support.

An insurgency broke out and threatened to expand to civil war. Unemployment and underemployment remained high. Had the 4th Infantry Division been permitted to stage through Turkey, perhaps some of these problems could have been averted because the United States would have had tens of thousands more "boots" on the ground. That did not happen. The situation in Iraq seems to have gone from worse to worst as the insurgency has caught fire and so frighteningly resembles civil war.

## Transformation—Looking Forward

So where is this headed? Transformation has evolved into a doctrine and intellectual understanding of the need for permanent change. There is a Leninist or *Trotskyist* quality, ironically, to advocating the need for permanent revolution. But this is the mind-set that has been created within the American defense establishment. Transformation was originally meant to cover only the central rationale and the legal basis of the Department of Defense—to be ready to conduct sustained operations incident to combat. In other words, since 1947 when the original national security act was passed by Congress and signed into law by President Harry Truman, the primary mission of the department was fighting and winning major or regional wars. September 11 and the threat of terror now challenge that premise and the centrality of war fighting as the single most and dominant priority for the military.

A further reality is in the form of a contradiction. The U.S. Department of Defense has two conflicting principal priorities. The first is transformation; the second is the successful "stabilization, reconstruction, and democratization" of Iraq. It is exceedingly difficult to transform in the middle of a war when virtually every army division, for example, is either deployed to, returning from, or preparing for duty in Iraq. And since the thrust of transformation is to enhance combat capability, the reality that postwar Iraq requires other skills compounds the difficulty in transformation.

A brief reading of the law, in this case the National Security Act first passed in 1947, reveals that the principal mission of the military services is to be prepared for "the prompt conduct of sustained operations incident to combat." Nowhere are the missions currently being undertaken in the global war on terror, Iraq, and Afghanistan explicitly authorized. A major and fair critique of the war in Iraq was that while the United States military performed brilliantly in the three weeks it took to defeat the enemy army and unseat Saddam Hussein, the department was not prepared for the aftermath. This is one area that needs transformation: what should the department be doing that it is not? At this writing, the quadrennial defense review is under way and will examine answers to this question.

Beyond this profoundly important matter of how far, if it all, the Defense Department should go to incorporate tasks that require military forces in operations short of conventional war, there are huge problems of people, money, and organization. The fact is that the All Volunteer Force based on mobilizing reserve and National Guard forces to fight a major war has been all but broken. A new model is needed. That the Pentagon requested an additional $80 billion supplemental in January 2005 above the $420 billion budget it already submitted is indicative of the monetary shortfall. With growing debt and deficits, these large increases will soon prove impossible to grant. Something fundamental will have to give. Finally, the organization and force structure of the services and the Unified Command Plan around which the combatant commands are structured must be changed to fund the new realities. These are each hugely difficult challenges and tests.

## War, Peace, and the Future

American movie mogul Sam Goldwyn was fond of saying many years ago, "Making predictions, especially about the future, ain't easy." So where are war, violence, and conflict headed? Clausewitz asserted that "war was an admixture of policy with other means." But, in an age where terror, surely for the United States, has taken on such a powerful and different political context, should there be different assessments about the nature of war and the role and use of force and violence to achieve political ends?

On the positive side of the ledger, while Korea, the Taiwan Straits, South Asia, and the Middle East have some possibility of provoking conventional war, the probabilities are not perceptively greater for such conflicts to erupt than in the past. Perhaps the war in Iraq may even have the effect of deterring war in other parts of the Persian Gulf, such as Iran or Syria, as the United States is fixated there.

On the other side of the ledger, terrorism and violence, as in the Sudan and Russia, are not waning. Regarding the former, to many states, there is nothing new about terror. It is something that has been part of the dangers, and September 11 really served to make America more sensitive to the threat that other countries long endured.

What else is changing? With America's declaration of a global war on terror, although its Congress has not legally passed such a declaration, the nature of the threat has shifted from a predominantly military one, at least regarding the most powerful weapons that could be brought to bear, to a very diaphanous and ill-defined danger. That means that determining the nature of the threat in quantitative or even qualitative terms is perhaps impossible and certainly illusive. It was relatively clear what the military strength and therefore potential of Nazi Germany and Soviet Russia were. And any ambiguities were neatly bypassed by shifting from focus on capabilities to intentions, the latter also being hard to define. However, because of the ideological or political consensus of the time, few would argue that the Americans could not assert understanding of Soviet intentions as a means of rallying and receiving general public support. Today, there is no consensus or understanding of what the scope of the threat of terror is, exactly, or what the real intentions of terrorists are beyond hating Americans and wishing to kill them.

Meanwhile, the strategic Cold War paradigm of massive destruction has fundamentally changed. In its place is the danger of mass disruption. Certainly a weapon of mass destruction can do a great deal of damage, but society is not at risk of total destruction. Instead, the aim of extremists is to use terror or its threat to disrupt, to make life as difficult and frightening as possible, whether through taking Russian schoolchildren hostage or threatening attacks to ensure that threat warnings are raised in the United States at great expense. The snipers that terrorized the Washington-Baltimore area in late 2003, and the fire in a New York subway that shut down that system for weeks, show how potentially vulnerable society is to disruption. And even in war-ravaged Iraq, the infrastructure—from electrical grids to oil production—is regularly subject to disruption by insurgents.

Al Qaeda and myriad other terrorist groups understand the leverage of disruption. They cannot destroy the United States or any other country even if they get their hands on a nuclear weapon. What became obvious post–September 11 was the capacity and leverage of disrupting American and other societies. A trip to any airport vividly shows the disruption of air travel caused by security procedures. Not long ago, Los Angeles airport was shut down because a tiny pocket battery caught fire and the fear was that this had been a failed terror attack. The March 11, 2004, bombings in Madrid can be seen as helping to change that regime and influence the election that threw Prime Minister José María Aznar out of office. There were other factors to be sure. However, disruption is the new aim of those seeking to use terror or its threat to achieve political purpose.

War therefore is no longer the dominant factor it once was. The rules and general agreement such as treatment of prisoners of war, banning certain weapons, and permitting international supervision of prisoners does not have the same standing. Today, the United States is dealing with captured enemy combatants as it did sixty years ago during World War II by subjecting them to military tribunals rather than trials and charges on the basis of due process. Then, in one case, German spies landed by submarine on the East Coast and were quickly captured, tried, convicted, and hanged. While there is no record of any executions, indefinite arrest and incarceration and denial of due process have stirred up a hornet's nest of protests. And, as "terror" and the war against it dominate the American agenda, no new rules, tools, or even metrics to determine how well or badly that war is going have been created. But not every nation agrees with this or with America's preoccupation with the global war on terror.

Meanwhile, because of the general effects of globalization and the revolutions in technology and information that connect much of the world instantaneously from bank accounts to bad music on the Internet, what happens in one place can and does have an impact on others. Terrorist threats directed against international airline travel are the best example. Flights are cancelled, schedules are changed, and disruption is immediately felt beyond the nation that happens to own the affected airline or aircraft.

About the danger, there is no real consensus. Many European and Asian states see terror as part of doing business, acts that are as old as history. Americans tend to see terror as directed against them on the grounds that terrorists hate things American, especially democracy and freedom, and therefore, motivated by hatred, will mindlessly kill. Sadly, that view is as wide of the mark as was the view that Communism was monolithic.

What should we conclude from this? First, our intellectual focus should be on the means to advance and support our interests rather than on war and its deterrence and conduct. Certainly, decades hence, China may find itself again arrayed against Japan, India, or Russia in a military confrontation in which war is a central part of policy, whether avoiding it or threatening it as a last resort. Part of this new focus should be directed at removing the basic causes that promote terror and violence for political means that threaten to disrupt stability and international comity and trade. Hence, loose informal or formal structures that share information among like-minded states that will help reduce the grounds for violence and force used in these ways are essential.

A hundred years ago, the century of relative peace and stability in Europe following Napoleon's defeat at Waterloo in 1815 and the Concert of Vienna that redefined the political order was coming to an end. That was not the case in Asia where there had been the Sino-Japanese war in 1894–95, the Boxer rebellions in China in 1900, and the 1904–5 Russo-Japanese conflict. That era exploded in World War I. From World War I came the seeds of World War II and then the Cold War. Today, the Cold War is long gone. Indeed, the role of war as has been traditionally defined and practiced may be in temporary suspended animation. Stateless or borderless theologies may be surfacing as the contemporary dangers.

What is needed is both an intellectual and a practical framework to capture these challenges and propose solutions. In fact, unless we mobilize our intellectual resources in the form of people, we will never win the struggle against determined adversaries. Better international relationships, alliances, and partnerships are critical to this end. The key is understanding. Americans tend to be culturally ignorant or indifferent. That must change.

Perhaps the study of war might be improved by future study of how all of the tools of statecraft can be put to use to build better and more pluralistic worlds for the diverse community of nations that must live and deal with that world. And perhaps we need to reshape the process by which we teach and educate those who serve in national security. How to do this forms the conclusions of this chapter.

We need to broaden national defense to incorporate the framework and foundation of national security. And we must focus education on "learning" across the broad commons that form the universe for assuring that security.

National Security can no longer be defined by national defense. We had that luxury during the Cold War. Now the aperture must be greatly expanded. That expansion applies well beyond the Department of Defense, the State Department, and the CIA—the traditional centerpieces for national security. Today, Treasury, Transportation, Homeland Security, Commerce, Health and Human Services, Agriculture, and Interior all have large national security roles. If we cannot keep our borders secure but sufficiently porous to permit legitimate flows of people and immigrants; if we cannot protect our food supplies and infrastructure; if we cannot turn off money laundering; if we cannot educate our citizens as to the dangers and opportunities of this environment; and if we cannot do thousands of other things, the nation will never be safer or more secure. Hence, virtually all of government is required in this task.

To accomplish this, a revolution in education is needed. Education should be transformed into a process of learning measured by what someone has learned, not degrees and courses taken. The National Defense University should be expanded into a National Security University, with students coming from all arms of government including Congress and its supporting agencies. Indeed, to drive this point home, the president of this new NSU should not be only of great stature. That person should also be appointed to sit on the National Security Council and be responsible for national security "learning" across government—actions that will bring the requisite authority and responsibility to make these reforms actually happen—something rare in government.

Thought should be given to expanding the service academies to national security colleges with larger enrollments. Graduates would be required to spend a requisite number of years in approved national security positions throughout government. For those not entering military service, reserve commissions would be awarded and those individuals would spend a certain period in the reserve. A "core" course in national security would be offered at these colleges so that all graduates would have the foundation for the assignments and careers on which they were embarked. And the focus would be on learning.

America has often won its wars by outspending its adversaries. The fact is that we probably had more ships during World War II than Hitler had torpedoes. Unfortunately, against today's adversaries, these old means no longer work when there are no armies, air forces, or navies to destroy. We must engage our intellects if we are to win. Learning and education are the best and possibly only ways that will work.

# FIVE

# Speed the Kill

## *Updating the American Way of War*

RALPH PETERS

In the wake of recent conflicts, any review of the "principles of war" is an invitation to play on words and declare that we have too damned many principles of every kind and too little grasp of war. Faced with enemies whose approach to conflict is free of our cherished principles, military or moral, our quest for rules that dominate, regulate, or mitigate warfare and its consequences borders on naïveté.

While the essence of war—a fundamental human endeavor—does not change, the modalities of making war do. We currently fight enemies who care nothing either for treaties and laws designed for conflicts between Western states or for principles deduced from the experience of war between formal armies. We, on the other hand, are "principled" half to death, arguing over a checklist of campaign rules inherited from Napoleon's disciples, while concerned about hurting our enemy's feelings. One variety of principles muddles the other, with a determination to wage moral war in constant tension with battlefield necessity.

We worry about the immorality of the process, failing to see that the greatest immorality is to be defeated by the foes of civilization. In wars waged against implacable murderers, the end does, indeed, justify the means.

We have lost a sense of warfare's immutable reality. Even the experiences of Afghanistan and Iraq have been insufficient to fully convince civilian decision makers that, in the words of that vicious man and splendid soldier Nathan Bedford Forrest, "War means fighting, and fighting means killing."

The principles of war, in the sense that doctrinal manuals propound them, do not change entirely. But the relative importance of each principle

95

shifts, from conflict to conflict, even from mission to mission. The initial campaign in Afghanistan—a distinct anomaly—was embraced as proof that massed land forces had become irrelevant, given the effectiveness of postmodern weaponry when backed by our military infrastructure and culture. Ground forces could be miniaturized like microchips. An eyeblink later, the occupation of Iraq demonstrated to those willing to look reality in the face that mass matters, after all; that there is still no substitute for adequate numbers of trained troops on the ground.

We'll sort this out. Reality has a way of forcing even the staunchest ideologue to grapple with facts eventually. The savagery of our enemies will lead us to respond with fewer moral reservations and greater effectiveness, once we realize that we have no other rational choice. The classic principles of war will suffer from fewer applications of the moral brakes. From recent wars in which we not only worried about our own casualties, but also agonized over the enemy's losses, we shall return to the true American way of war: Ruthless, overwhelming, and decisive.

Meanwhile, other developments extraneous to war's traditional parameters have driven an even greater shift in—and forged an addition to—the hierarchy of warfare's commonly agreed principles. In our age of instant communications and a metastasizing global media, the dominant principle on future battlefields will be speed—if we mean to win.

We are already incomparably swift in many respects. Strategically, we can shift fires and forces with remarkable speed. Operationally, our pace is relentless (unless interrupted by civilian interference). Even at the tactical level, mounted combat is often an affair of moments. From the strategic sphere down to the level of armored engagements, we're history's masters of the quick draw.

The problem is that our enemies know it. And the wiser among them refuse to play by our rules. Within the lifetimes of many still alive, the battlefield was literally a field—or a vast expanse of rural or semirural space. Armies avoided cities whenever they could. But in inter-civilization wars of belief in a madly urbanizing world, the prize is no longer a few hundred yards of mud or a few hundred miles of sand. The new "field" of battle is the city. Afghanistan, a literally medieval country beyond the bounds of its tattered cities, was a throwback to a lost world—yet even there the grand prize was Kabul. Even in states largely canopied with jungle or covered with colossal expanses of sand, possession of the land itself means nothing unless you master the cities.

To an extent, this was always so, but the historical goal was to conquer the cities by defeating armies in the field, not to fight in the streets (unless the inhabitants resisted to the last). Today, however, we face enemies who *choose* to fight in cities and care little for the empty spaces over which armies move efficiently and operate most effectively. (For Arabs, with their desert heritage and no legacies of statehood, cities long have had a totemic power: no Arab writer complains of the Mongol conquest of Mesopotamia, but of the destruction of Baghdad; where vast tracts of land had little worth, the city's value was magnified, and the military history of the Middle East is a litany of the defense or conquest of cities—Damascus, Jerusalem, Cairo, Esfahan, Mosul, Antioch, Aleppo, Edessa, Samarkand, Bukhara, Heart, and Constantinople, to name a very few).

Our military long resisted the obvious need to prepare to fight in cities. Finally, in the early 1990s, the U.S. Marine Corps took up the challenge and thought hard about urban combat. They worked on doctrine and, vitally, trained for street fights. Reluctantly at first, the U.S. Army followed suit (after several years of pretending that the Battle of Mogadishu was a unique event that would never be repeated). One army general, now retired and a vocal advocate of the preeminence of urban operations, told me ten years ago that, "The U.S. Army would never be stupid enough to fight in a city," as if the choice of environments were ours alone.

But rational analysis of a changing world won out. By the time they were ordered to march on Baghdad, the American marines and soldiers had made remarkable progress in preparing, mentally and practically, for urban operations.

Still, there was one fateful, if not fatal, flaw in the assumptions the services made: Influenced by the devastating casualty lists and poor exchange ratios of historical city fights, our doctrine was cautious and methodical, intended to protect both our own forces and, whenever possible, the urban infrastructure. We failed to understand how the macro-environment of battle had changed: The fight wasn't just against the shooting enemy any longer, but also against an enemy at which no soldier or marine could return fire—a hostile global media.

The reasons for media hostility are too complex to be examined in detail here; suffice it to say it is a phenomenon of herd behavior and mass hysteria triggered by a profound sense of failure and frustration in a jealous world. For the purposes of this limited essay, the important thing isn't the root causes, but the operative fact that much of the global media is instinctively hostile to our purposes and to our military. Nor is the situation likely to remedy itself fully within our lifetimes—if it ever does. We enter battle in a spiteful world.

Militarily, we conceptualize well about space, whether the clean geometries of air and sea, in which conflict is largely an engineering problem, or in organized land warfare, which becomes profoundly human as soon as the mechanical and electronic systems give way to the soldier engaged in direct-fire combat. The factor we neglected when we approached urban combat was *time*. Unthinkingly, we assumed that we would be the masters of that dimension.

The reality proved otherwise. The most instructive contemporary example is the First Battle of Falluja, in the spring of 2004. Approximately two thousand U.S. Marines were given the mission of taking back the city from a motley network of terrorists and insurgents. Those marines executed the mission by the book. By that standard, they performed extraordinarily well. Each day, they bit deeper into a city whose peacetime population approached three hundred thousand. Marine casualties were blessedly light, compared to every historical norm. The terrorists and insurgents were hard-pressed and losing. Some estimates put the marines within two days of collapsing the organized resistance.

Instead, the marines were defeated—not by the armed opponents firing at them, but by the media. Al-Jazeera, an Arab broadcast network based in Qatar, presents itself as a news outlet, but it is far more of an anti-American, anti-Israeli, anti-Western propaganda organ, an intellectually slovenly combination of outdated pan-Arabism, irresponsible religion, and the personal demons of its staff. Yet, "Freedom of the media!" was the mindless, wildly inappropriate chant from our own press and civilian administrators. Journalists were primarily interested in defending other journalists, no matter their misdeeds or the consequences for our national and Coalition efforts. The moral equivalent would have been a public defense of Islamist terrorists by our own military.

Our reward was that al-Jazeera daily broadcast blatant lies about events in Falluja. While the U.S. Marines were fighting under severe restrictions regarding which targets were legitimate, which munitions could be employed, and even when it was "legal" to return fire, al-Jazeera's newsreaders and commentators insisted that the U.S. military was slaughtering innocent women and children by the hundreds each day, that they were purposely targeting hospitals, schools, and mosques. The reports were pure fabrications, but they were delivered with a perfect-pitch understanding of al-Jazeera's audience, of the region's prejudices and atavistic fears. Watching al-Jazeera in Iraq during the battle, I was amazed by the breathtaking extent of the network's mendacity. But the lies worked because the viewers wanted to believe them. (And because we didn't have a clue how to challenge them effectively.)

Nor were the only hostile viewers those in the Arab world. The BBC and other European networks took up the spirit of al-Jazeera's campaign. The British and Italian governments came under extreme domestic pressure to "stop the slaughter" that was supposedly under way. Reportedly, the manufactured outrage led Prime Minister Tony Blair of the United Kingdom to telephone President George W. Bush with a combination plea and ultimatum. The political pressure upon Blair—already under fire for his courage in supporting an unpopular war—had simply become too great.

On the road to victory, the U.S. Marines traveled too slowly—and they lost.

After personally ordering the marines into Falluja in the wake of the murder and dismemberment of four American contractors, President Bush folded his cards. The offensive was called off. And the terrorists and insurgents were able to declare their first military victory since the fall of Baghdad a year before.

That victory was amplified by the lamentable error committed by American leaders when they trumpeted that our troops were going to take Falluja and rid it of our enemies. We ourselves set the public standard for success or failure. At that point, the terrorists only needed to retain control of a single ruined basement to claim a win. Thanks to al-Jazeera, they retained far more than that. Over the next six months, Falluja became a terrorist city-state. Briefly the world capital of terror, the city became a magnet for international terrorist recruits. Everybody loves a winner.

Inevitably, the U.S. military had to go back into Falluja. When our forces did so in November 2004, it was clear that commanders and planners had learned their lessons well: numbers mattered, mass was back. Crucially, they also recognized that time was *not* on their side, that they had to finish the main operation quickly, even if that meant higher casualties and greater destruction. Before committing almost fifteen thousand troops, most of whom were American soldiers and marines (with a few Iraqi battalions on training wheels), our forces prepared the battlefield through more than a month of special-operations raids and actions, the targeting of key defenses and support sites from the air, a blockade, and a detailed analysis of the enemy's defenses. The decision makers understood that the formal attack would act as a media magnet, that it had to overpower the enemy defenses and culminate in a victory so swift that the global media could not intercede effectively to protect the terrorists.

The Second Battle of Falluja was conducted vigorously and relentlessly. The combat power applied was irresistible. The major fighting was over in less than a week from the hour the leading elements of the assault force crossed the line of departure. Yet, even that swift operation may have neared the limits of the time window permissible for operations in a small city such as Falluja. Our counter-media efforts in this particular case were aided by a series of terrorist errors over the preceding months. A succession of video-taped beheadings and expanded savagery against Iraqi civilians backfired badly. A novelty at first, the taped executions repulsed regional audiences as they became routine, smacking of renegade cult behavior. Al-Jazeera made an operational error by broadcasting the new tapes, one after the other, long after they had lost audience approval. In a rare misjudgment, the network didn't grasp the extent to which the beheadings and other torments dismayed both religious and secular Arabs.

Giddy with past successes, the al-Jazeera staff judged the beheadings to be great television, riveting entertainment bound to galvanize resistance to American goals. They badly misread viewer psychology. The ritual murders sponsored by terrorist leader Abu Musab al-Zarqawi embarrassed Muslims, making the faith appear backward and bloodthirsty. The spectacle of the beheadings undercut popular support for the terrorists—without turning the mood pro-American, the atrocities moved the "Arab street" toward apathy.

Thus, our forces had greater leeway to fight boldly and decisively in Falluja the second time around: Zarqawi's fanatical terrorists were no longer viewed as a home team worth cheering on. As a result, the far greater destruction in the Second Battle of Falluja had less of an effect on public opinion than the "atrocities" invented by al-Jazeera during the first fight. Al-Jazeera and its media allies attempted to stir up passions—but the popular response was lethargic. And the U.S. military did not give the media time to regain its command of global sentiment.

Nonetheless, the rules are different now, and we can't go back to the old ones. We can no longer be as methodical as we would like to be (or remotely as pristine as civilian leaders would prefer). We have to speed the kill.

The implications of the emergence of speed as the dominant principle of postmodern war range from the rifle squad to the Oval Office. Planners can no longer use traditional methods to calculate how long a given operation should take, but must ask themselves, "How much time do we have? How long before global pressure on the president to call a halt becomes irresistible?" Once unleashed, our combat operations must be so furious and decisive that the key objectives are gained before civilian leaders can entertain second thoughts.

Even under the antique conditions of Desert Storm, every hour mattered. We failed to kill enough of the right Iraqis during the too-brief period of combat. A premature end to hostilities left a dictator in power and his tools of oppression essentially intact. What began as a war ended as a diplomatic circus. (Diplomats, too, are often the enemies of military effectiveness, although their power decreases while that of the media increases.) Then we watched as the enemy we had spared slaughtered Iraq's Kurds and Shias.

If there is any issue over which flag officers should be prepared to resign, it is the nonsense of proportional response and limited operations. Every operation, whether a special-operations raid or a full-scale invasion, should be decisive on its own terms—otherwise, why is it worth American lives? Only decisive operations deter potential enemies or validate alliances worth having. And, in the twenty-first century, "decisive" increasingly means swift.

There isn't a minute to spare as we pursue the complete destruction of our enemies, or the false mercies of civilian leaders will ensure that our enemies live to fight another day. As combat operations drag along, the situation reports passed up the chain of command by the military have less power to influence decision making, while the media's influence grows with every twenty-four-hour news cycle. Even the firmest presidential decision can be undercut by the specter of increasing diplomatic and political costs. No matter who is president, military commanders cannot count on open-ended support. Presidential decisions to engage the enemy must be protected by the achievement of rapid results.

In the 1980s, our military spoke of "operating within the enemy's decision cycle." We fairly well mastered that. Then the game changed. Now our forces have to "operate within the media cycle."

Of course, the rules and time frames for postmodern warfare aren't fixed—there are no tables to which we can turn (but we may assume that the general time windows acceptable for tactical combat will continue to shrink). The time tolerance for military operations will be subject to considerations of scope, purpose, provocation, perception, location, and friendly and enemy force composition—and the best-conceived plan can be undone by a single news cameraman who gets a "lucky" shot that mobilizes world opinion against us, as happened at the end of the Second Battle of Falluja, when the remarkable success of our forces (with friendly casualties a tenth of historical norms) was downplayed by the media in favor of a clip of a marine shooting a wounded Iraqi; we were fortunate that the incident wasn't filmed on the first day of the operation.

There will be subconscious equations and unique dynamics at work in every engagement, battle, or campaign. Different operations will be allotted different amounts of time in the popular psyche. The mission to subdue a hostile village in country X may have to be executed and concluded within twenty-four hours, while the need to reduce resistance in a city of several million would be given a longer dispensation (a U.S. Army division commander told me that, in Iraq, he realized that many smaller missions had to be initiated and concluded between midnight and dawn). Real casualty levels will matter, of course, but the *perception* of success or failure, of civilian losses and "wanton" destruction, increasingly will have less to do with battlefield reality and more to do with media representations.

*The media create parallel realities that may have little or nothing to do with the empirical reality facing our soldiers and marines.* We are dealing with competing worlds—something that might have been drawn from the science fiction of the past—and until our forces achieve a clear, irreversible, and undeniable outcome, the media-manufactured "reality" is more powerful strategically than the concrete reality on the ground. The media generate their own narrative of war—often wildly different from the facts. No military briefings, no matter how detailed and forthcoming, can compete with the version of events the media choose to portray. The tale told over the airwaves becomes the global truth (the impact of print media, although important, is distinctly secondary in this battle for days, hours, and minutes).

Consider Operation Iraqi Freedom. Conditioned to a 24/7 news cycle—life on methamphetamines—reporters and pundits were declaring disaster after less than four days. Every time a unit paused, we heard that the war had bogged down. The impatience of broadcast journalists who needed vivid stories to file before their deadlines was palpable. Halfway to Baghdad, some pundits were already declaring a new Vietnam because there were skirmishes—not battles—in some of the cities and towns behind our lines. *They didn't understand the complexity of war—and they didn't want to understand.* The reporters wanted a story, not explanations or knowledge.

Baghdad fell in less than three weeks, an astonishing achievement by any military or historical standards (his intelligence services had briefed Russia's president that it would take the U.S. military six months of hard fighting to get to the Iraqi capital). The media's response was a discordant combination of surprise that we'd won, despite their confidence in the Iraqi regime's capabilities to embarrass us, and a sense that it was about time we'd gotten a move on.

From local gunfights to entire "shooting" wars, the time allowed us to win has compressed dramatically and will continue to shrink. Even the occupation of Iraq was daily declared a failure because that country did not turn into Disneyworld overnight (as I write, Iraq's successful and inspiring first free national election has chastened the prophets of failure, but they'll be back).

Part of this is that expectations have changed profoundly and, sometimes, nonsensically. Even in our own country, some audience segments—untouched by war themselves—expect things to wrap up in time for the weekend football game and to be as neatly scripted as a television sitcom. The ersatz reality of the entertainment media now dictates the public's perception of appropriate time lines for actual warfare. And it isn't a matter of education or sophistication. When it comes to battle, red-state Americans are far less susceptible to unreasonable impatience than blue-state intellectuals who should "know better." Even setting aside the condescending hatred the urban intelligentsia feels toward the Bush administration, it's fascinating that those with the highest levels of formal education seem less equipped to judge events astutely than do those of lesser privilege—but, then, Nascar Universe is far closer to basic human reality than is the world of *The New York Review of Books*. Yet, the urban intelligentsia dominates the media—which has an immediate effect upon an insular Washington governing elite largely untethered to the reality of America beyond the Beltway.

This means that any military plan unwilling to sacrifice some degree of security and precision for speed risks confounding the president with media-induced cries for an early end to the operation. Even the most determined president will be able to provide only top cover for so long in the wars of the future. We must be prepared to win fast—and a swift victory that generates great destruction is nonetheless a victory, while a tidy campaign that ends in premature termination is a defeat, no matter now graceful the military ballet has been.

Victory is forgiven. After the cries of outrage settle—which they do with remarkable speed—the reality we have created on the ground prevails. By contrast, failure means that you have to fight again—if you're lucky. If you're unlucky, the consequences can be immeasurably bad and you won't get a second chance.

With refreshing accuracy, some voices have begun to speak of the current series of conflicts as a contest of wills. That means we must have the will to win, no matter the cost. And we have to do it fast.

Speed is the foremost principle for the twenty-first-century American way of war.

This requires a rejection of the accumulated nonsense of half a century of theorizing about war by civilians and their military accomplices. The increase and power of the defense industry far beyond anything President Dwight Eisenhower could have imagined when he warned the nation, in 1960, of the threat from the "defense-industrial complex" has created another alternative reality in the minds of decision makers: The mirage of bloodless war—or, at least, of war as a cheap date.

Although our experience in Iraq has begun to provide hard, countervailing evidence to the promises of contractors and the dreams of desk-bound theoreticians, we remain a long way from grasping the need for the resolute application of force in abundant mass in order both to defeat our enemies and to *convince* them that they have been hopelessly defeated (the latter is at least as important as the former). Since the Vietnam War—surely the most overinterpreted and misinterpreted conflict in history—political leaders have been only too willing to be seduced by promises that buying the next wave of high-technology weapons will lead to conflicts low in flesh-and-blood costs and of minimal destruction. American technological superiority became a disadvantage by creating an utterly false perception of what war is and can be.

Consider the absurd execution of "shock and awe" against the Saddam Hussein regime at the beginning of Operation Iraqi Freedom. A truly shocking and awesome air campaign—one not afraid to inflict real pain and widespread destruction on the enemy—might well have demoralized the opposing leadership in the crucial opening days of the war, while creating a useful panic among the civilian population. But the actual air effort was little more than a sound-and-light show over Baghdad. We wanted to win a war without breaking windows. Our cautious, pinprick air attacks had the opposite effect of what we intended: Instead of crushing the enemy's will, it convinced Saddam Hussein that he could survive our technological war making. Our mannerly, precision-guided approach *strengthened* the enemy's confidence.

We had warned the world of the mighty things the U.S. Air Force would do when the war began—then we made a little noise in the night and the planes flew home before dawn. Shock-and-awe seemed more like minor vandalism than a serious act of war. In the enemy's eyes, we looked weak and, for all our technology, amateurish.

In their self-deluding calculations, the designers of the air campaign against the Baghdad regime failed to take the enemy into account. Everything was supposed to break our way—a dangerous assumption in any conflict. Intelligence was supposed to locate Saddam Hussein and his key subordinates for surgical decapitation strikes. According to the theory, any ranking survivors would then be so disheartened that our troops could take a Sunday drive to Baghdad and round them up like sheep.

Of course, hundreds of billions of dollars of strategic intelligence hardware could not track the Iraqi leadership. Nor did our display of genuinely astonishing precision much impress our enemies, since we struck nothing that mattered. And even had we killed a substantial number of Iraq's key leaders, the others had no incentive to surrender. We wanted to persuade an enemy to give up based upon a brief display of our capabilities, but that enemy saw the war as a zero-sum game; surrender meant the loss of everything—wealth, power, freedom, and even life.

As noted above, the insipid nature of our initial strikes on Baghdad convinced the enemy's leaders that they had a fighting chance. So the war had to be won the old-fashioned way, with soldiers and marines fighting northward along lines of communication that have felt the weight of Macedonian sandals and Persian slippers, Arabian hooves and, at last, the grinding treads of Abrams tanks.

Thanks to the skills and moral superiority of U.S. Marines and soldiers, we won a splendid conventional victory. Yet, even then our restrained approach to punishing our enemies meant that the key enemy population— Iraq's Sunni Muslims—didn't *feel* defeated. The war hardly touched the now-infamous "Sunni triangle." Instead, the city and village populations that later hosted the insurgency barely felt our presence—many did not see a U.S. military vehicle for months in the wake of Baghdad's fall.

The sorry tale of botched occupation planning (slight as the planning was) and willful ignorance on the part of the Office of the Secretary of Defense (OSD) is best reserved for another essay; suffice it to note that the lack of sufficient troops to flood the Sunni triangle with soldiers and marines, along with a series of other flawed decisions, turned a lightning victory into a grinding occupation.

We learned—or should have learned—that in addition to the new requirement for decisive speed, we had to revive the time-honored concept of decisive force, of mass deployed without stinting. For political reasons, OSD decision makers wanted to limit the size of the force allotted to the campaign, to limit costs of all kinds, to limit destruction—even to avoid annoying our enemies too much in the process of defeating them. The result was an enemy who felt tricked and betrayed, not legitimately and hopelessly beaten.

A generation and more of contractor fables and bureaucratic wishful thinking about the power of technology got the psychological math profoundly wrong. By raising false expectations of low casualties and minimal destruction, precision weaponry made a decisive and lasting victory harder, not easier. This doesn't mean that such weapons have no utility—they do, indeed—but that they are far from dominating every dimension of warfare. Precision weapons are useful tools, not an autonomous means of achieving victory.

When human passions are aroused, as they are bound to be in the postmodern conflicts confronting us, it is insufficient to slap the enemy on the wrist. Our campaigns must be not only swift but also devastating. Above all, we must consider the psychology of our enemies and stop indulging ourselves in seductive fantasies of wars without consequences.

Had the situation been reversed, had the Iraqis been able to employ precision weapons to eliminate our most important operational or even our key national leaders, would we have surrendered? Or just quit and gone home? In this post-9/11 world? We never asked ourselves *why* the enemy would want to surrender. Yet, we ourselves would not have done so. Had the Iraqis been able to assassinate George W. Bush, we truly would have gone on a crusade. And had a precision strike killed Saddam Hussein on the first night of the war, he would have gone down as a martyr and a hero in the Arab world. How much better to have dragged him out of a hole in the ground, hair matted and eyes bewildered. We were lucky we missed.

All this means that the issue of speed cannot be divorced from sufficiency of effects. Not only must we win fast, we must also inflict a paralyzing sense of defeat on our enemies. Precision weaponry can support a comprehensive effort—as it did during the Second Battle of Falluja—but by itself it has insufficient physical and psychological impact to bring a campaign to a speedy, *enduring* conclusion. With precision weapons, we have elevated useful means to virtual ends in themselves. Such weapons may help the overall force achieve its objectives more promptly and at lower cost, but they are not an adequate replacement for that overall force. Nor is precision targeting a functional substitute for extensive graphic destruction in creating a psychology of defeat among our enemies.

It isn't enough to win on points. Opponents have to accept in the depth of their souls that they have lost and have no hope of recovery except under the terms we deign to set. We imagine that, through careful targeting, we can make our enemies love us. But the victor's first requirement is to be feared. Mercy can follow victory, but rarely should it precede it.

We must do what it takes to win—a very traditional concern—in a hypervelocity, twenty-first-century mode of warfare. Although the circumstances varied politically and practically, every significant conflict we entered between 1945 and 2001 had indecisive results because of our reluctance to employ the means necessary to gain an incontestable victory. From Korea, through Vietnam and Desert Storm, to Somalia and the Balkans, our national leadership's fundamental question seemed to be not "How can we win?" but "How cheaply can we get off, how little can we do, how can we get a gentleman's C in the school of strategy?"

The relevant maxim on war is cast in iron: Those unwilling to pay the butcher's bill up front will pay it with compound interest in the end. We fought bloody, indecisive wars because we did not fight to win from the start.

It's easy enough to write about the need for swift, decisive operations, but far harder for planners and operators to meet the challenge on the ground. There are always obstacles and restrictions, practical and political. As an urban battlefield, Falluja was little more than a playpen compared to the developing-world megacities in which we may be forced to engage in full-scale combat—or play cat and mouse with future terrorists and insurgents. The challenges before us are among the most perplexing in military history: How can we conduct swift, decisive operations in a complex city whose population numbers in the millions? How fast can we go? What traditional practices and inhibitions must be discarded? What does the urban combined-arms team of the future look like? Do inherited assumptions about fire, maneuver, and protection prevent us from grasping obvious new organizational, behavioral, and material requirements? How much can we afford to destroy (greater speed means greater destruction, at least for now)? How many civilian casualties dare we inflict for the greater good? Can the media's power ever be turned to our advantage, or even be forced to report with a semblance of honesty? Should we treat propagandists as combatants? Is the new strategic center of gravity the fickleness of civilian leaders under media pressure? Is our military's first responsibility to the National Command Authority just to win fast, no matter the cost?

Far more questions remain to be asked and answered, but on a practical level, our military certainly needs to think far more rigorously and critically about that new third party to every conflict, the media. Instead of being a subparagraph in a minor appendix to an unread annex, media management should be included in the mission statement and operations paragraph of every order generated above the battalion level (perhaps, even lower). No matter how well we fight, the global media will defeat us if we do not plan thoroughly and imaginatively—and operate within the media cycle. The length of that cycle will vary from conflict to conflict and between levels of operations, but it will rarely favor us.

If we are to achieve victory on the battlefields of the future, our military will have to learn to speed the kill and accept the consequences.

# SIX

# The U.S. Air Force and the American Way of War

### Grant T. Hammond

## Introduction

The U.S. Air Force (USAF) is the world's only "full service" air force in that it carries out all its roles and missions—strategic bombing, close air support, air mobility, aerial refueling, and intelligence, surveillance, and reconnaissance from both air and space—and it exercises global awareness, reach, and power.[1] While other services have air and space assets, it is the USAF that is the primary custodian of the third dimension. More than any other service or air and space force, it exercises exploitation and control of all that is up, over, and above. It is able to bring a variety of kinds of presence and power to bear nearly anywhere over the globe. Its ability to do so is unique, and the distinctive character of the power that can be deployed and employed in the third dimension makes airpower one word, not two. In this sense airpower means air and space power, the arena and capabilities of the third dimension.

In the last decade, in addition to providing air superiority and space superiority, the USAF has also taken on the role of providing information superiority. This has come about because of its investment in and global deployment of air and space assets for command, control, communication, and computing, and intelligence, surveillance, and reconnaissance (C4ISR) as well as its lead role in space where it operates most of the U.S. government's space capabilities. Because of its large and important space presence,

The views expressed here are those of the author and not necessarily those of the U.S. Air Force, the Department of Defense, or the U.S. government.

the USAF is the primary military custodian of space assets, even though most military communications utilize commercial satellites.

In the last fifteen years, airpower has achieved stunning military success, if not political victory, in the First Gulf War, Bosnia, Kosovo, Afghanistan, and Iraq. It has accomplished historically unique rapid, relatively cheap, low casualty, high technology success of stunning dimension. But in an era characterized as one of continuing conflict—a constant contest rather than episodic conquest—airpower may not be as effective. The strategic conditions of airpower's recent success are changing. Threats are increasingly non-state, involving weapons of mass destruction (WMD) on one end of the spectrum of conflict and unconventional warfare (terrorism and insurgency) on the other. Adversaries may use chemical and biological munitions, not conventional uniformed military forces, or asymmetric capabilities such as a suicide bomber—a poor man's cruise missile that is subsonic, with a terrain-following guidance system and a relatively small warhead that is difficult to intercept. Airpower and the U.S. military in general have succeeded admirably in interstate conventional warfare, but the verdict has yet to be rendered as to their defensive, as opposed to preemptive, capability in a different strategic landscape.

## The Principal Attributes of Airpower

Airpower provides the nation with global awareness, reach, and power. The awareness is essentially data collected from an array of air-breathing and space-based sensors of all kinds. These provide overhead imagery consisting of optical photographs, infrared images, radar, and multispectral imaging, as well as communications intercepts and other sources of information. It is that which can be gained from the perspective of altitude and overflight, be it at various altitudes, near space, or low, medium, or geosynchronous earth orbit. Second, it provides global reach. It can do so at great ranges and high speeds, reaching most places on earth in a matter of hours and days as opposed to weeks and months for most surface forces. It can do so from the continental United States (CONUS) or in an expeditionary mode deployed abroad. Last, it exercises global power and has achieved and demonstrated what amounts to hegemony in air and space in what has been called the "commons" of the atmosphere and near space. While there are air and space capabilities in other services, including satellites, air-breathing assets (both manned and unmanned), and missiles, the great bulk of air and space assets belong to the USAF. These capabilities give the United States the quality and quantity of capabilities that no other state possesses.

Space is an increasingly important part of the third dimension. What can be done from the perspective of and presence in space is growing in importance. The use of pagers, ATMs, and self-serve gas pumps that accept credit cards, plus the availability of satellite-transmitted information of all kinds are examples of routine dependencies on space. So too is the military's use of space for C4ISR. Though the threshold of weaponizing space has not been crossed formally yet, it was militarized long ago. The U.S. military has played a vital role in the development of missiles, satellites, and many space missions. Half of the shuttle cargoes have had military payloads and most astronauts have come from the ranks of the military. The United States is and, by virtue of geography, always will be, a maritime nation. But it has transformed itself into an air and space nation as well. While most of the cargo that leaves or reaches American shores comes and goes by ship, many of the people entering or leaving the country do so by plane. The nearly twenty-six thousand commercial airline flights carrying more than 1.7 million passengers a day in the United States have replaced reliance on inland waterways as the means of fast transit in the early days of the republic.[2] Increasingly, communications for business transactions—inventory controls and stock quotes, financial transactions, and global communications—flow through space. The accuracy of location in both time and space afforded by the global positioning system (GPS) has revolutionized both military operations and many civilian ones from farming to trucking. Important information about weather, military maneuvers, pollution, individuals lost on land, sea, or air; product locations in shipment; migrations, etc. flows from space. Our everyday reliance on it grows exponentially. Protecting this increased reliance on space-based assets will be a growing security concern in the future. If "the way we make war reflects the way we make wealth," as the Tofflers asserted over a decade ago, then space will become an arena of conflict because of the increasing reliance on it as a wealth-producing asset and significant military capability.[3] Given the reliance of the United States on space, it will have more to lose than others when that comes to pass. Space is increasingly both an advantage and a growing vulnerability as it is difficult and expensive to protect space-based assets from both kinetic and non-kinetic attack.

The capability to utilize air and space in war resonates well with the American public. If war is to be fought, most Americans would prefer it to be fought the way airpower does it—from a distance; relying on high doses of firepower, not manpower; using America's strength in high technology to fight a short, relatively bloodless war. Airpower can exercise greater strategic leverage at less risk to life than most forms of surface power. That it can do so to the degree it can is due largely to superior military technology, particularly the capability to use stealth and precision strike. What took flights of a thousand B-17s dropping in excess of nine thousand bombs to accomplish in World War II was achieved in one pass of a single F-117 in the First Gulf War.[4] For the U.S. 8th Air Force in World War II, the odds of destroying a single 60-by-100-foot building were very poor.[5] Today, a B-2 using Mark-82 five-hundred-pound joint direct attack munitions (JDAMs) could hit eighty separate targets of that size on a single sortie.

The First Gulf War demonstrated the devastating capabilities and strategic effects of airpower. Many authors have commented on this capability. One put it this way:

> What are the chief lessons of our experience with strategic use of airpower in the last war? . . .
>
> 1. One lesson is that the time we were given to make our preparations was an absolutely essential factor in our final success. . . .
> 2. Air power in this war developed a strategy and tactic of its own, peculiar to the third dimension. . . .
> 3. The first and absolute requirement of strategic air power in this war was control of the air in order to carry out sustained operations without prohibitive losses. . . .
> 4. We profited by the mistakes of our enemies. . . . To rely on the probability of similar mistakes by our unknown enemies of the future would be folly. The circumstances of timing, peculiar to this last war, and which worked out to our advantage, will not be repeated. This must not be forgotten.
> 5. Strategic air power could not have won this war alone without surface forces. Air power, however, was the spark to success. . . .
>
> Another war, however distant in the future, would probably be decided by some form of air power before the surface forces were able to make contact with the enemy in major battles. That is the supreme military lesson of our period in history.[6]

This is a reasonably succinct summary of the role of airpower in 1991 and a projection into the future. But it was not written about the First Gulf War. It was written by Gen Carl "Tooey" Spaatz in 1946, in a *Foreign Affairs* article titled "Strategic Air Power: Fulfillment of a Concept" as his assessment of World War II. Whether in a retaliatory or a preemptive mode, from CONUS or abroad, airpower is likely to be a "first responder."

It was not until the post–Gulf War experience of Operation Northern Watch and Operation Southern Watch in Iraq that the USAF began to reconfigure itself as an expeditionary force, harkening back to its tradition from World War II. Chief of Staff Gen. John Jumper's insistence and perseverance in this matter after the experience of Operation Allied Force (OAF) in Kosovo led to the ability to execute Operation Enduring Freedom (OEF) in Afghanistan and the initial phases of Operation Iraqi Freedom (OIF) in the following years. The ability to strike from CONUS and forward bases and to sustain operations over extended periods gave airpower persistence it had not had before. The combination of these capabilities and superior technology has made it highly unlikely that the USAF will be directly confronted in traditional aerial combat by enemy airpower for some time to come.

But airpower is not the only ingredient necessary for military success. The American public still defines war as "boots on the ground." Anything else may be an intervention, but not "real" war. That said, if airpower alone could prove sufficient, the public would certainly support it. The old debate about whether airpower is "decisive" is largely irrelevant. Airpower is necessary, if not sufficient. If by decisive one means important, it is. If one uses decisive to mean conclusive, it may not be. But U.S. airpower in the 1990s reached the apogee of success in state-centric war.[7] Airpower has become the preferred American way of war. It fulfilled Douhet's vision of more than eighty-five years ago and reiterated Spaatz's claim from World War II. Whether it can be used to the same effect against globally distributed insurgencies is not clear. Regardless, air and space assets are integral to the way the U.S. military fights.

## The American Way of War

Much ink has been spilled on the notion of an American way of war. Just what it is, or was, depends on one's interpretation. The idea itself was popularized with the publication of Russell Weigley's famous book by that name in 1973.[8] A military history of the United States, Weigley's *American Way of War* related a history of winning by overwhelming the enemy—a combination of sheer weight of numbers and a strategy of attrition and annihilation, as exemplified by Ulysses S. Grant in the Civil War, by John J. Pershing in

World War I, by the massive destructive power of strategic bombing in World War II, and—with declining success—massive commitments of troops and munitions in Korea and Vietnam. But that American way of war has evolved significantly. It has embraced airpower as its principal component in the process.

Many other interpreters of U.S. military history purport to have found the essence of a new American way of war. Former Vice Chief of Staff, Adm. William Owens, U.S. Navy (Ret.), saw an opportunity in *Lifting the Fog of War* to rely on new computer and communications technologies.[9] This initiative has been taken up and refined by former Director of Force Transformation, Adm. Arthur K. Cebrowski, U.S. Navy (Ret.), the "father" of "network centric warfare" (NCW). Cebrowski saw the American way of war as characterized by networking, shared situational awareness, speed, and precision, and he saw the U.S. military in need of "transformation."[10] NCW is now labeled "net enabled operations" as its utility and value transcend war and apply to humanitarian relief and other military operations. With information as the linchpin of *Joint Vision 2010, Joint Vision 2020,* the Joint Operational Concept (JOPC), and "Transformation" itself, there has been a tendency to focus everything through the "war in the information age" lens.[11]

Michael Ignatieff wrote of Kosovo as *Virtual War.*[12] To Samuel Huntingdon first, and Thomas Barnett later, war was to become *The Clash of Civilizations and the Remaking of World Order* on *The Pentagon's New Map.*[13] Max Boot observed OIF and declared it "The New American Way of War," which "seeks a quick victory with minimal casualties on both sides. Its hallmarks are speed, maneuver, flexibility, and surprise. It is heavily reliant upon precision firepower, special forces, and psychological operations. And it strives to integrate naval, air and land power into a seamless whole."[14] For others, "jointness" among the services has been the watchword since passage of the Goldwater-Nichols Act of 1986. Its effect on the way the U.S. military does business has been profound, if incomplete.

A number of others concentrate on the contributions of airpower. Former Chief of Staff, Gen. Ronald Fogelman, characterized the "new American Way of War" in a seminal speech in 1996.[15] He declared that it was based on a shift from "annihilation and attrition warfare" and force-on-force to one that sought to attack an enemy's "strategic and tactical centers of gravity" directly. The targets were "the leadership elite; command and control; internal security mechanisms; war production capability; and one, some or all branches of its armed forces—in short, an enemy's ability to effectively wage war."[16] After the experience of OEF in Afghanistan, Fareed Zakaria declared, simply, "Face the Facts: Bombing Works."[17] Furthermore, he pointed out that the information advantage that could be exploited by airpower was the key to military victory. Thomas E. Ricks, writing in the *Washington Post*, labeled the experiences in Afghanistan "the new American way of war, built around weapons operating at extremely long ranges, hitting targets with unprecedented precision, and relying as never before on gigabytes of targeting information gathered on the ground, in the air and from space."[18] Michael Kelly called OEF the demonstration of "The Air-Power Revolution," declaring, "Historians and military analysts have long stressed the limitations of air power. Their arguments are no longer tenable."[19]

Alas, things may not be so rosy. First, this record was achieved against the small, weak or incompetent. From Panama to Operation Iraqi Freedom in Iraq, the United States has not faced a truly formidable foe. Second, America may have become too good at the American way of war. Few adversaries will now go "head to head" with U.S. forces because they know they will be defeated. They will try other means. Just as the Romans learned that their legions could be defeated by barbarians in the Teutoburg Forest, so too might modern militaries incur defeat in a new way of war that the adversary prefers.[20] The problem for the future is that the preferred American way of war has become so successful nobody wants to play by those rules, with those kinds of forces, in that kind of battle space. Because potential adversaries can't win militarily in fighting that way, they may become more adept at what the Chinese call "unrestricted warfare" across the spectrum of conflict.[21] Moreover, this can be conducted by myriad new actors with different ends (purposes), ways (tactics, techniques, and procedures), different means (methods), and different motives. Our preferred manner of waging war may not be theirs.

Does this mean the passing of airpower and space power? No. Airpower is still essential and increasingly important as sensors, if not shooters, against terrorists and other non-state actors. It can also be useful against what infrastructure they have or against those who harbor and assist them. And since traditional state-based warfare will likely be around for some time to come, it is absolutely essential against a near peer or peer competitor state engaged in more traditional warfare over territory, people, or resources. Struggles between states frequently are about influence: access to, or control of, people, places, or things. Airpower and space power still play a vital role in the establishment of these conditions, though perhaps not in all cases to the same degree, as recently demonstrated in OIF.

Airpower is a vital national resource, a powerful tool, and a preferred manner of warfighting. But it cannot do everything and it should not be asked to accomplish things (like stopping ethnic cleansing in Kosovo) it is unable to do. Airpower misapplied, or poorly resourced, cannot perform miracles. The ability of the USAF to find, fix, track and target, engage, and assess (F2T2EA) from air and space is not, and never will be, perfect. But it contributes greatly to the ability of the United States to attempt to achieve information superiority, the centerpiece of American military strategy. That said, there is a need for greater investment in human intelligence, particularly about intentions as well as capabilities. Air and space assets may not provide this but they may help to either target or confirm other intelligence.

The USAF is being asked to do essentially three missions—air superiority, space superiority, and information superiority—with the budget share it used to have devoted to only one of these. If air-launched precision guided munitions (PGMs) are to be fully utilized, they require precision guided intelligence and precision effects based assessment to accomplish their tasks. It is the information portion of air and space missions that is increasingly important. And airpower is being asked to perform the stunning military successes of the recent past in a vastly different strategic setting of non-state actors, globally distributed insurgencies, and near instantaneous global news coverage where the PGMs that count most may be precision guided messages. This changing strategic setting and rapidly accelerating technologies (biotechnology, nanotechnology, information technology, and directed energy—both lasers and high-powered microwaves—among others) are creating a grave new world. Do these realities call for a revised set of the principles of war?

## The Principles of War

The so-called principles of war were never universally agreed upon. Each country adopted notions of the principles of war best suited to its geography, political form, circumstance and national history.[22] They reveal some interesting insights in comparison.

Those items in the chart in italics represent the only commonality across all countries. Those items in boldface are unique. What are we to make of these similarities and differences?

First, it would appear that all are agreed on the concept of mass, initiative, and surprise. Mass, however, is today more likely to refer to information,

| United States | Britain | USSR | France | People's Republic of China |
|---|---|---|---|---|
| Objective | Selection and maintenance of aim | | | Selection and maintenance of aim |
| Offensive action | Offensive action | | | Offensive action |
| *Mass* | *Concentration of force* | *Massing and correlation of force* | *Concentration of effort* | *Concentration of force* |
| Economy of force | Economy of force | Economy, sufficiency of force | | |
| *Maneuver* | *Flexibility* | *Initiative* | *Liberty of action* | *Initiative and flexibility* |
| Unity of command | Cooperation | | | Coordination |
| Security | Security | | | Security |
| *Surprise* | *Surprise* | *Surprise* | *Surprise* | *Surprise* |
| Simplicity | | | | |
| | Maintenance of morale | | | Morale, political mobilization |
| | | Mobility and tempo | | Tempo |
| | | Simultaneous attack on all levels | | |
| | | Preservation of combat effectiveness, interworking and coordination | | Freedom of action |

fires, and effects than manpower or weaponry per se. That said, the latter can be used to generate the former and for this reason, if no other, China presents a serious challenge. There is a certain quality in quantity. All value initiative and flexibility, or maneuver, although they are permitted in varying degrees by their respective doctrine. These are valued capabilities, if a limited reality, within different militaries. Last, all value surprise. This reflects the always sought after and rarely achieved accomplishment of the quick victory reflected in Douhet's pursuit of the knockout blow, Mahan's concept of the decisive fleet engagement, or the Clausewitzian decisive battle which ends (wins) the war.[23] Implicit in this notion is a quick victory and a short war, realities, as Shakespeare said, are "consummations devoutly to be wished," but alas, rarely achieved.[24] All the other principles would appear to be embraced by three or more of the five great powers of World War II and have endured into the contemporary era.

It would appear that the United States alone values, or says it values, simplicity. This no doubt stems from a former reliance on a conscripted military and the desire to implement the keep it simple, stupid (KISS) principle. The increasing professionalism of the All Volunteer Force (AVF) and its reliance on ever more sophisticated "smart" weaponry and rapid global communications suggest that such a notion is a quaint anachronism. Anyone with any knowledge of the application of force by the U.S. military in the last fifteen years would agree that the concept may have been straightforward and therefore perhaps "simple," but the implementation could hardly be considered so. That requires timing coordinated to seconds and movements choreographed across multiple services from around the globe in coalition actions of immense size and complexity. Any one component of the military service or a single phase of joint operations, let alone the full panoply of power at the discretion of U.S. and coalition military forces, can hardly be considered simple.

Other unique principles of war are the concept of political mobilization of the Chinese, a reflection no doubt of Mao Tse-tung's belief in the superiority of man over technology and the necessity to have an indoctrinated and committed military. In an era of nationalism and ideology, neither of which is dead in the twenty-first century, and in combination with the reality of globalization and global communication of images, ideas, and intentions, this would appear to be an important addition. Two Russian concepts from the USSR are noticeably absent in other formulations and also would seem to be of increasing utility. The concepts of tempo and the simultaneous attack on all levels are valuable principles to which the U.S. military is increasingly devoted whether it has proclaimed them as principles of war or not. The notions of tempo and timing lie at the heart of the notion of the observe, orient, decide, act (OODA) loop, reluctantly accepted but now a universally adopted concept promoted by USAF Col. John Boyd.[25] Getting inside the adversary's decision cycle by controlling the tempo of operations and creating difficulty for him to keep pace effectively represents the application of information superiority to combat operations. Simultaneous attack on all levels is USAF Col. John Warden's concept of parallel war and the ability to attack all five of his famous ring set simultaneously, ranging from fielded forces to leadership, and including infrastructure, command and control, and logistics.[26] These are accurate descriptions of how the U.S. military fights today and thus suggest a revision of the American principles of war.

Technology has also intruded to reshape the arena of conflict in powerful ways; that reshaped arena in turn has reshaped the nature of war and conflict. And technology has done so not only in terms of munitions and weaponry but also in the equally important terms of information and communication.[27] The rapid spread of advances in the latter half of the twentieth century—television; satellite imagery and communications; the fax machine; digital processing of voice, data, and visual images; the appearance of the Internet and the World Wide Web; small powerful computers and personal digital assistants (PDAs); Webcams, sensors of all kinds; the ubiquitous nature of cell phones; increasing reliance on GPS; wireless communications; the expansion of journalists to include "bloggers" and anyone with a picture phone—have radically transformed the arena in which we live, work, and fight. In such a world, secrecy is giving way to ever greater levels of transparency.[28] It is not likely that in an age of exponential growth in technology that these trends will slow. While tactical and strategic surprise may still be possible, they are becoming more and more difficult to achieve and duplicate. That said, the omnipotent vision of information superiority that lies at the heart of the *JV2020* and network-centric operations in anything other than a relative and episodic sense is a long way off. The human dimension of the problem, particularly intentions and the frailties and foibles of these imperfect animals, will no doubt continue to confound technology.

## Principles of War for the Twenty-first Century

So, what should the principles of war for the twenty-first century be? Indeed, is formulating such a list of universally applicable principles across the spectrum of conflict even possible? What is equally relevant to threats ranging from terrorist acts to massive state-centric warfare, utilizing weapons and tactics ranging from the conventional arms to WMD to information operations, and a panoply of participants that includes everyone from Tom Friedman's vision of "Super Empowered Angry Men" to states or fantasy ideological movements?[29]

To formulate a list of principles to guide in preparing for and continuing to wage conflict against enemies foreign and domestic is not only possible, it is essential. Without some guide to help navigate the extremes of the threats it may face, the United States will be ill prepared to survive and prosper as a nation. This list is not traditional and may be somewhat controversial, but it does contain essential guidance for survival in a complex, chaotic global environment in which we have been charged to cope with "the unseen, the uncertain, the unexpected, and the unknown."[30] The fact that this may be difficult, even impossible, in no way diminishes the importance of the effort. Doing so effectively is what providing for security is all about.

The principles of war (conflict)—for it is now a matter of continuous conflict across the whole spectrum of human endeavor—for the twenty-first century are radically different from those of the past. War and peace are not opposites. Nor are they interludes, one substituting for the other. Security has expanded to be not only national and international but also corporate, municipal, and personal. And it may no longer be principally dependent on the uniformed military to provide it. Conflict and even war may no longer be solely a matter of conquest of territory or people. They are increasingly a continuing contest that involves competition in activities ranging from scientific research and basic education to military capabilities, from economic health and prosperity to skill in public diplomacy. Providing security is in essence about the ability both to shape as well as to be shaped by the environment in agile and adaptive ways, being able to co-evolve with a complex, chaotic, constantly changing global environment. While the United States may be the dominant actor in many ways, Americans are not the sole inhabitants of the planet, and the rest of the world neither is, nor necessarily wants to be, just like us. American religious preferences, version of democracy, and economic system may or may not be desirable, appropriate, or effective in other settings for other people. And while having a strategic environment and global reality that is if not supportive, at least not harmful to the American way of life is entirely appropriate as a national goal, an international reality reflecting that may or may not come to pass.

What are the principles of war suitable for the United States in the twenty-first century?

*Focus:* This is a broader restatement of the objective. While one must have an objective, exactly what it is and how it should be achieved may change during the course of conflict. While that may be mission creep to some, it may also be an exercise in prudence in not overreaching or applying a variety of capabilities to modulate desired effects. Although one's instrumental goals and the ways and means to accomplish them may shift over time or in certain places in a particular conflict, one's focus should remain constant. Knowing what must be accomplished and why is prerequisite to the use of force and a successful military strategy.

*Versatility and Adaptation:* Given accelerating change in nearly all areas of human activity and a constantly evolving world of differential capabilities and desires, one must be ready and willing to adapt to these evolving circumstances. Maladaptation virtually guarantees failure.[31] In order to respond quickly and well to changing circumstances, forces must be versatile and able to perform multiple missions. Being able to adapt as necessary to a changing strategic and tactical environment is a critical capability for the nation, and its military, to survive and prosper.

*Initiative and Shaping:* Military forces must be ready and willing to seize the initiative at all times, to act rather than merely be acted upon and react. They must be able to shape as well as be shaped by the circumstances they confront. Shaping the nature of the contest, its dimensions, requirements, and implications are critical to success. This requires close coordination and integration of other elements of national power and attention to how others—friend, foe, and bystander—perceive the conflict and its course. Losing the initiative may be disastrous.

*Surprise:* Those in service to the state must be wary of surprise by their adversaries and constantly consider the means by which they could achieve surprise against their adversaries. Though increasingly difficult to achieve given growing global information sources, the successful results of surprise have manifold benefits and must be a part of any national military strategy. Being able to achieve surprise against an opponent requires detailed information about both the opponent and his circumstances, which can be exploited to advantage. To accomplish it against an opponent, one must know his values, fears, and purpose, not just an opponent's order of battle.

*Sustainment:* This refers not only to the logistical capability to sustain forces in the field in military requirements but also the ability to sustain the effort required in the contest politically, economically, and psychologically for as long as may be required to achieve success. Public opinion is the center of gravity for democracies and protracted war a potent weapon against them. Not all contests—the Cold War, Vietnam, and Iraq(?)—are short. Even short wars have dramatic OPTEMPO—that is, the pace of operations in terms of equipment usage—and PERSTEMPO—that is, the measure of time an individual spends away from home station—implications and must be sustained at high levels. Dedication to and sustaining of one's national security strategy is essential for success.

*Precision Effects:* Precision has substituted for mass. It is no longer a matter of attrition and annihilation to achieve the desired effect; rather, it is precision and timing. This does not vitiate mass but refines it. Achieving effects-based operations in the twenty-first century may be as dependent on a mass of visual images on the Internet as on a precise message delivered to the right person at the right time. The target is always the opponent's perception. Change his mind and one need not destroy all his fielded forces to attain military victory. Strategic effect may be as dependent on speed and timing as on massed fires or putting large numbers of people in harm's way to confront an enemy.

Collectively, these principles of war for the twenty-first century broaden the horizon of both conflict and the tools with which to address it. Following these, however, may not remedy problems in the application of military force or the wider concern for an effective national security strategy. Several years ago, Colin Gray wrote a very perceptive piece about the American approach to strategy; it was at the heart of the American way of war during the Cold War. It is just as valid now, and has been recently demonstrated. Its eight characteristics were

1. An indifference to history
2. The engineering style and technical fix
3. Impatience
4. Blindness to cultural differences
5. Continental *Weltanschauung*, maritime situation
6. Indifference to strategy
7. The resort to force, belated but massive
8. The evasion of politics[32]

Unless these tendencies are adjusted to the realities of the twenty-first century and the U.S. position in it, variations in the principles of war may account for little. It is the thinking, as well as the technology, that is important. Gray's comments suggest a greater concern for history, culture, strategy, and politics; more patience and less engineering; a more timely and measured resort to force; and a broader concern for conflict.

## Airpower in the Twenty-first Century

These new principles of war for a new century do not change airpower and military force. They do, however, suggest changing how to think about them. If one truly wants to utilize force quickly and effectively, at low cost in human lives, for strategic effect, the utilization of airpower broadly defined is essential. The American way of war can no longer consist of outproducing adversaries. It must involve outwitting them. To do so, the principles suggested above—focus, versatility and adaptation, initiative and shaping, surprise, sustainment, and precision effects—are useful principles in wedding strategic thinking and operational art. Strategy is more than targeting, and it is not only kinetic. Human conflict is about more than uniformed troops, weaponry, and munitions and their deployment and employment. It is about human motivation and moral causes and consequences.[33] Airpower, as with other forms of military power, is a means to a broader end. Using it in light of the principles above gives one the latitude to employ airpower more fully and creatively. It is no longer a matter of shortening the "kill chain," but of creating an effects chain.

Airpower—air, space, and information superiority—is an essential part of national security strategy, whether in a deterrent, defensive, or offensive application. The awareness, reach, and power that it brings are necessary if not sufficient ingredients for America to survive and prosper in the twenty-first century. Broadening how these can be utilized and applied through the principles suggested permits one to make better decisions about force development and employment. The technologies of the next quarter century are likely to provide warfighting assets that are increasingly remote, robotic, cheap, small, and swift. Investing in those that permit application of the principles above will help shape necessary decisions in a resource-constrained environment.

The human part of the equation, not just the technical, is critical if we are to achieve the effects-based operations we intend. In the words of the late USAF Col. John Boyd, "Machines don't fight wars. People do, and they use their minds."[34] Hence his dictum: "People first; ideas second; things third." Unfortunately, the U.S. military tends to reverse the priorities. Oddly enough, that is in part because of a concern for people. Rather than masses of manpower to accomplish a military mission, the United States seeks to put fewer people in harm's way, to substitute technology and firepower for manpower. The emphasis on technological solutions for problems is a peculiarly American trait. Nowhere is that more apparent in the high technology investments made in air, space, and information power. Airpower is the preferred means for the United States to utilize military power because it can deliver bytes, bombs, bullets, bread, or bodies to the combat theater, from a distance, at high speed, and with great accuracy. In so doing, it has a high ratio of potential strategic success to lives risked, both friendly and hostile. If one has to fight a war, this seems to be the preferred way to do so. Demonstrating the continued capacity to do so is an investment in forestalling war rather than fighting one.

## Notes

1. This conceptual framework comes from the AF 2025 Study done by Air University published in 1996 available at http://www.au.af.mil/au/2025/ and accessed on 10 February 2005.

2. Data derived from the U.S. Bureau of Transportation statistics website for November 2004 located at http://www.bts.gov/press_releases/2005/bts008_05/ html/bts008_05.html and accessed on 10 February 2005.

3. Alvin and Heidi Toffler, *War and Anti-War: Survival at the Dawn of the Twenty First Century* (New York, Little, Brown & Co., 1993), pp. 3, 57–63.

4. Brig Gen David A. Deptula, "Effect-Based Operations: Change in the Nature of War" (Aerospace Education Foundation, 2001), 31 p, published as part of the Defense and Airpower Series. It is available at www.aef.org/pub/psbook.pdf.

5. The calculations are that in WW II, to have a 90 per cent probability of hitting a 60 x 100 foot building would require 9,070 B-17s, with a CEP of 3,300 feet to accomplish the task. Briefing chart, AF/XOXW, Fall 1990.

6. Gen Carl Spaatz, "Strategic Air Power: Fulfillment of a Concept," *Foreign Affairs,* April 1946, pp. 394–96.

7. See among other works, Benjamin S. Lambeth, *The Transformation of American Air Power* (New York: Cornell University Press, 2000), and Stephen Budiansky, *Air Power: The Men, Machines, and Ideas That Revolutionized War from Kitty Hawk to Gulf War II* (New York: Viking, 2004.)

8. Russell F. Weigley, *The American Way of War: A History of United States Military Strategy and Policy* (New York: Macmillan, 1973). It became a standard text in U.S. military history and went through numerous editions.

9. William Owens, *Lifting the Fog of War* (New York: Farrar, Straus, Giroux, 2000).

10. Arthur K. Cebrowski and Thomas P. M. Barnett, "The American Way of War," *U. S. Naval Institute Proceedings,* January 2003, pp. 42–43.

11. *Joint Vision 2010,* Washington, Office of the CJCS, n.d., available at http://www.dtic.mil/jv2010/jv2010.pdf; *Joint Vision 2020: America's Military Preparing for Tomorrow,* Washington, D.C.: JCS, n.d., available at http://www.dtic.mil/ jointvision/jv2020a.pdf; JOPC is the abbreviation for the Joint Operational Concept and it lies at the center of the U. S. approach to the vision of a network-enabled operations among joint forces in which they utilize collaborative shared situational awareness and service specific competencies to achieve the rapid effects-based operations.

12. Michael Ignatieff, *Virtual War: Kosovo and Beyond* (New York: Metropolitan Books, 2000).

13. Samuel P. Huntington, *The Clash of Civilizations and the Remaking of World Order* (New York: Simon & Schuster, 1996); Thomas P. M. Barnett, *The Pentagon's New Map* (New York: Putnam, 2004).

14. Max Boot, "The New American Way of War," *Foreign Affairs,* July–August, Vol. 82, No. 4, pp. 41–58.

15. Gen. Ronald R. Fogleman, "Airpower and the American Way of War," speech presented to the Air Force Association's Air Warfare Symposium, Orlando, Fla., February 15, 1996.

16. Ibid.

17. Fareed Zakaria, "Face the Facts: Bombing Works," *Newsweek,* December 3, 2001, p. 53.

18. Thomas E. Ricks, article in the *Washington Post,* cited by Michael Kelly in "The New American Way of War," *The Atlantic Monthly,* June 2002, p. 17.

19. Michael Kelly, "The New American Way of War," *The Atlantic Monthly,* June 2002, p. 17; Michael Kelly, "The Air-Power Revolution," *The Atlantic Monthly,* April 2002, p. 20.

20. The Battle of Teutoburg Forest took place in 9 A.D. and saw the defeat of three Roman Legions under the command of Publius Quinctilius Varus by Germans under the command of Arminius. These legions represented 10% of Rome's trained manpower and were the finest soldiers of their era defeated by the "barbarians." Though the Germans did not follow up their victory with an invasion of Gaul, it did prevent Roman plans for a province between the Rhine and Elbe Rivers. It was one of the most stunning defeats of a superior force in European history.

21. See Liang Qiao and Xiangsui Wang, *Unrestricted Warfare* (Beijing: Chinese People's Liberation Army, 1999). Translated into English with an introduction by Al Santoli, *Unrestricted Warfare: China's Master Plan to Destroy America,* Panama City, Panama: Pan America Publishers, 2002. This gives a full range of ways to combat the United States that include traditional, irregular, disruptive and catastrophic strategies and tactics.

22. This somewhat dated listing of the Principles of War is taken from AFSC Pub 1, 1997, Figure 1-1, available at www.au.af.mil/au/awc/awcgate/prinwar.htm. It

was adapted from JT Pub 1, FM 100-1, AFM 1-1, FMFM 6-4, *Military Review,* May 1995, and Soviet Battlefield Development Plan.

23. Thomas Hughes, "The Cult of the Quick," *Airpower Journal,* Volume XV, No. 4, Winter, 2001, pp. 57–68, offers cautions against the embrace of speed as a key to success.

24. William Shakespeare, *Hamlet,* Act III, Scene 1, line 56ff.

25. For more on Boyd and his ideas, see, Grant T. Hammond, *The Mind of War: John Boyd and American Security* (Washington, D.C.: Smithsonian Institution Press, 2001, paperback edition, 2004) and Frans Osinga, *Science, Strategy and War: The Strategic Theory of John Boyd* (Delft, The Netherlands: Eburon Academic Publications, 2005).

26. Col. John Warden, USAF (Ret.), was the architect of the first Air Campaign in the Gulf War known as "Instant Thunder." His ideas on "parallel war," "strategic paralysis," and his famous "five rings" model of targeting (from inside-out: Leadership, Key Production, Infrastructure, Population, Fielded Forces) were important but both Boyd and Warden are still not formally embraced by the U.S. Air Force for their contributions to modern strategy and air warfare. For more on Warden and his ideas see, John Andreas Olsen, *Strategic Air Power in Desert Storm* (London: Frank Cass, 2003).

27. For more on the impact of improved communications in an earlier era, see Daniel R. Headrick, *The Invisible Weapon: Telecommunications and International Politics, 1851–1945* (New York: Oxford University Press, 1991).

28. On this issue, see, Lt. Col. Beth M. Kaspar, *The End of Secrecy? Military Competitiveness in an Age of Transparency,* Center for Strategy and Technology Occasional Paper No. 23, August 2001, available at http://csat.au.af.mil.

29. Thomas L. Friedman, *The Lexus and the Olive Tree: Understanding Globalization* (New York: Anchor Books, 2000, pp. 398 ff).

30. Secretary of Defense Donald H. Rumsfeld, "Transforming the Military," *Foreign Affairs,* May–June, 2002, Vol. 81, Issue 3, p. 23.

31. The late Carl Builder of RAND was tireless in his efforts to convince the U.S. military of the need to avoid this and become more agile and adaptive in co-evolution with the changing world about them. He was urging "transformation" before it came in vogue.

32. This list is excerpted from Colin Gray, "Strategy in the Nuclear Age: The United States, 1945–1991," in Williamson Murray, MacGregor Knox, and Alvin Bernstein, *The Making of Modern Strategy: Rulers, States and Wars* (Cambridge: Cambridge University Press, 1996), pp. 592–98.

33. For more on this line of reasoning and the conventional and unconventional images of war, see Grant T. Hammond, "The Paradoxes of War," *Joint Force Quarterly,* Number 4, Spring, 1994, pp. 7–16.

34. Col. John R. Boyd, USAF (Ret.), at http://www.d-n-i.net/ and accessed on 11 February 2005.

# SEVEN

# Pitfalls and Prospects

## The Misuses and Uses of Military History and Classical Military Theory in the "Transformation" Era

### JON SUMIDA

> The information you have is not what you want.
> The information you want is not what you need
> The information you need is not what you can obtain.
> The information you can obtain costs more than you want to pay.
> Finagle's Laws of Information[1]

In the nineteenth century, Antoine Henri Jomini, Carl von Clausewitz, and Alfred Thayer Mahan wrote history for military professionals, and laid the foundations of modern strategic and operational theory through derivation of principles or insight from the analysis of the past. While the immediate inspiration for the work of Jomini and Clausewitz was the great wars of the French Revolution and Empire, much of the demand for their writing and that of others was generated by industrialization and associated changes in political and social organization, which transformed the means and methods of warfare. In the absence of a general conflict between great powers using the latest weapons and techniques, discourse about the requirements of future war drew heavily upon contemplation of history and its putative lessons. From the late-nineteenth century, the work of individual authors writing in what was in essence a private capacity was augmented by the formation of service historical sections that produced staff histories of past conflicts.

---

This chapter was written while the author served as Major General Matthew C. Horner Chair of Military Theory at the Marine Corps University, Quantico, Virginia, 2004–6. Support for this position was provided by Thomas A. Saunders III, Mary Jordan Horner Saunders, and the Marine Corps University Foundation.

By the beginning of 1914, military history and theory was a major intellectual force with respect to war preparation. Clausewitz had emerged as the dominant thinker about war on land. Mahan had published his last book, but his earlier works of history and theory were still highly influential. Julian Corbett, Mahan's heir presumptive, had written the bulk of his major historical work as well as his one volume of theory; he enjoyed the confidence of high-placed British government officials and senior naval officers; and he had begun his career as a writer of official staff history. At the International Congress of Historical Studies in 1913, Corbett observed that in the past, history had been brought into contempt through its use as nothing more than a polemical device to "prove anything." The advent of methodological sophistication and rigor, however, meant that "officers no longer look upon history as a kind of dust heap from which a convenient brick may be extracted to hurl at their opponents," but "as a mine of experience where alone the gold is to be found, from which right doctrine—the soul of warfare—can be built up."[2]

Just a few months after the publication of these remarks, the peace of Europe was broken by the outbreak of a general war that would last for four years. Although Jean de Bloch, a Polish banker, had described important military aspects of this conflict with remarkable prescience at the turn of the century, the bulk of the military historical and theoretical canon provided little guidance for the protracted stalemate and unprecedented loss of life and consumption of materiel that characterized World War I. Skepticism about, if not disillusionment with, the practical utility of military history and theory followed in train. Serious deliberation about strategy and policy in the years following World War I was unlike that of the prewar period because of the ready availability of firsthand experience with the latest weapons in a major conflict, whose claim to relevance trumped those of broad historical investigation and comprehensive theorizing. Within a generation, World War II provided a new source of direct experience with a great war, revitalizing empirical impulses that were already strong. The advent of the atomic bomb, moreover, introduced a weapon of unprecedented destructive power, for which history could seemingly provide little guidance with respect to use or influence.

In the late-twentieth and early-twenty-first century, however, circumstances came into being that favored renewed interest in the practical utility of history with respect to the consideration of national security questions. By this time, the United States had not engaged a peer competitor in all-out war for a half century. Over these years, rapid technological change generated by competition with the Soviet Union not only produced enormous general improvements in weaponry, but advances in information technology also opened the door to fundamental alterations in force structure and operational technique. These conditions—namely, lack of war experience and radical military change—seem to have encouraged demand for military history and history-derived theory as in the nineteenth century. During the decade of the 1990s, interest in the relationship between technological innovation and military improvement prompted the emergence of discussion about what was first called a "military technical revolution" (MTR), then "revolution in military affairs" (RMA), and finally "transformation."

In the United States, the driving force behind the use of military history to examine MTRs, RMAs, and transformation was the U.S. Government's Office of Net Assessment, which sponsored a host of conferences, seminars, and publications. Its director, Andrew Marshall, had a keen interest in the strategic implications of the relationship of weapons procurement and innovative military doctrine. He was convinced, in addition, that study of certain major historical cases would prompt insights, and perhaps teach lessons, that could be applied to the present. Marshall's choice of subjects was a reflection of his belief that large strategic gain could be achieved through radical advances in military technique based upon the intelligent investment of resources in technological innovation. Strategic advantage based upon operational virtuosity was not, moreover, sought for its own sake. Failure to innovate appropriately, Marshall thought, would expose America to catastrophic defeat in the manner of France in 1940, or shattering setback as at Pearl Harbor in 1941. In addition, developing and maintaining state-of-the-art military capability would justify continued heavy government support of armaments research in the face of post–Cold War domestic imperatives to reduce spending on defense.

In effect, the Office of Net Assessment's concerns set the agenda for historical inquiry with respect to policy discourse. In general, the subject was the causes and consequences of radical military change. This perspective encompassed a broad range of subjects including the rise of state-funded and state-controlled standing armies in early modern Western Europe (the so-called Military Revolution), the advent of the mass conscript army during the French Revolution, and the effects of various inventions such as the breech-loading rifle, steam-propelled warships, internal-combustion driven land armored vehicles, and airplanes.

The Office of Net Assessment's policy focus was not on mere reaction or adaptation to change, however; rather, it was directed exploitation in peacetime of new or prospective technical and associated doctrinal breakthroughs in order to achieve decisive strategic advantage. This sharpened the focus of inquiry upon three major examples in the twentieth century: the British introduction of the all-big-gun battleship, the American and Japanese development of the aircraft carrier, and the German invention of combined-arms maneuver warfare. Engaging these topics properly, however, was easier said than done.

The basic problem of investigating the named case studies of directed technological innovation, or any other similar subject, is the centrality of policy motive. That is to say, the critically important issue is not just what was done but why it was done. And after ascertaining why, it is no less essential to determine exactly how. Answering these two questions is vital because it is otherwise impossible to assess either the process of implementing the policy or its degree of success or failure. Did the outcome, even if strategically productive, match expectation? Or put it another way, was what was achieved what was intended? Or to reframe the question in yet other terms, to what degree was success advertent or accidental? Affirmation of a positive connection between intent and outcome could legitimately be used to support directed innovation by the United States in the twentieth-first century. Negative or otherwise problematical conclusions with respect to objectives and consequences might serve as a warning of the perils of such action.

Why and how questions are largely about internal policy and administration. The archival record of these processes is usually less than complete and otherwise misleading. Where the subject matter involved either secret research and development or highly classified intelligence data, it is very often close to nonexistent. Policy failure has almost always prompted the destruction of implicating memoranda and the production of misleading internal reports or staff histories in order to deflect blame from prominent individuals. Many important papers were destroyed by war action, or discarded in the course of routine administrative housekeeping. In addition to gaps and misinformation, a great deal of documentary material is simply difficult to comprehend. Papers on engineering, budgets, procurement contracting, and military exercises are obvious sources of significant data that cannot, however, be extracted without specialized knowledge, and a willingness to read and analyze hundreds if not thousands of pages of technical or statistical writing. The reconstruction of policy from fragmentary information is possible in some cases, but it is a difficult and complicated task that takes a great deal of time and effort.

Contemporary reporting in popular newspapers and journals, doctrinal or technical discussions in professional journals, and statistical compilations in military annuals are voluminous and relatively easy to read but are, for what should be obvious reasons, unsatisfactory sources of information about internal policy decisions. Memoirs offer the lure of inside knowledge by major players but are likely to be tainted by self-interest, and restricted by secrecy regulations. Studies of directed military technological innovation based upon the open sources just described are untrustworthy, lacking the data inputs essential to sound analysis. Studies of directed military technological innovation based upon a combination of open sources augmented by superficial, selective, or otherwise non-rigorous study of archival sources have little more than the appearance of sound scholarship without its substance. Studies of directed military technological innovation based upon careful consideration of the available unpublished and published data require arduous and protracted labor, and are, as a consequence, rare. The very high cost of state-of-the-art archival research, moreover, seems to have discouraged government-sponsored projects requiring such activity.

The existing historical literature is thus highly uneven in quality, with very few works having the requisite conceptual rigor and research base to provide a sound starting point for serious discussion. Given the hostility of university history departments to military history in general, and policy history in particular, there is virtually no prospect of the supply of appropriate historical work from civilian sources increasing to meet demand. As in the case of advanced weaponry, therefore, the highly specialized history needed for policy discourse cannot be bought off the shelf—if it is to be had at all in quantity, large sums will be have to be spent to increase the number of producers as well as purchase their output. But in addition to being expensive to the point of impracticality, the manufacture of such history by the government would be subject to powerful politicizing forces that would almost certainly preclude disinterested inquiry, which would vitiate the integrity and therefore the value of the entire enterprise.

In light of the foregoing, it should be clear that attempts to theorize about the dynamics of directed technological innovation for strategic gain lack—and will continue to lack—a solid foundation of historical fact and interpretation. In the absence of viable contemporary theory, the work of the classical strategic theorists as an alternative may seem attractive, based on the presumption that Clausewitz, Mahan, and Corbett had discovered fundamental principles of strategy and operations that can be applied appropriately to a wide range of phenomena. Such a course is highly problematical because all three to a greater or lesser degree believed that their major theoretical work did not address issues relating to the effects of changing technology on warfare.

In *On War*, Clausewitz factored out materiel in general from his consideration of armed conflict, and by so doing avoided the subject of technological innovation and strategy altogether. Mahan wrote the "Influence of Sea Power" series of histories in large part to counter the belief that technological change had fundamentally altered the nature of war. Corbett's first major historical monograph recognized the importance of naval technological innovation in the sixteenth century, but his theoretical writing was directed at the study of strategic and operational decision making in war. In addition, in the essay of 1914 quoted previously, Corbett observed that "applying historical conclusions to the solution of modern technical problems" was beyond "the province" of "professed historians," and "must be left to experts in the Services."[3]

For those undeterred by the inadvisability of applying writing in ways that violate the known basic attitudes of the author, it is also worth observing that consensus with respect to the meaning of major strategic propositions associated with Clausewitz, Mahan, and Corbett does not exist. This is not to say that sound interpretation is unavailable, but that classical strategic theory is still contested ground among serious scholars, with signs that debate will continue and probably even intensify. Would-be appliers of classical strategic theory to present policy problems cannot assume that propositions such as "war is an extension of politics by other means" or terms such as "center of gravity" have simple and singular meanings that can sanctify arguments about a contemporary problem. Or to put it another way, knowledge of the author's specific intent is the prerequisite to intelligent deployment of classical strategic propositions in discussions of current strategy and policy, and acquiring that knowledge, given the difficulty of the original texts and the plethora of conflicting opinions about them, is in itself a challenging task.

The foregoing problems do not matter if the primary function of military history and theory is to serve as marketing tools for already agreed-upon policy initiatives. If this is the case, the actual quality of the historical and theoretical raw material is of secondary importance—any historical reference or theoretical formulation will do so long as it is consistent with preconceived conclusions and is plausible enough to defeat rejoinder. Self-conscious mendacity of this kind is probably uncommon among reputable participants in serious discussions about policy and strategy. Unfortunately its effects can be replicated with remarkable verisimilitude by good faith but ill-advised attempts to make history and theory do what they cannot do given the short-comings and difficulties of the existing literature. Misuse of history and theory in this case may not be premeditated, but the integrity of policy and strategy discourse will suffer nonetheless. This is not to say that history and theory are irrelevant to current discussions, but that they must be used with due appreciation of what a historical literature can and cannot accomplish.

Sound historical studies of directed technological innovation for strategic gain are too few in number to support conclusive argument with respect to either the advisability or inadvisability of pursuing such a course. On the other hand, the virtue of work that is known to be well conceptualized and researched is that it provides a standard by which other studies may be evaluated. If it is accepted, for example, that a properly constituted examination of the development of a major advanced weapons system should include consideration of strategy, operations, tactics, logistics, intelligence, finance, legislative politics, engineering, interdepartmental and interservice rivalry, production capability, competing or alternative technologies, the career prospects and ambitions of significant junior and senior officers, and other variables large and small, inquiry that does not do these things, or lacks convincing proof, can at the very least be viewed with skepticism. This means asking—and answering—questions such as: Have all significant policy vectors been identified? Is the evidence sufficient to support the analysis? Is resort to speculation in the absence of evidence justified and bounded by knowledge that can be verified? Are the descriptions of human and institutional behavior plausible? Has due allowance been made for the play of contingent circumstances major and minor? And perhaps above all, what has been assumed to be true but may not be? Knowledge of good work, in other words, facilitates critical discrimination. This may not provide answers, but it will counter the pervasive tendency to support policy propositions with histor-ical case studies that are either inapplicable or oversimplified to the point of giv-ing highly misleading direction.

The more general phenomenon of military change and its implications offers a more productive venue for historical and theoretical treatment than directed technological innovation for major strategic gain. Here the central issue is not the motives for complex executive action that was carried out largely in secret, but the more easily identifiable play of large impersonal forces and their consequences. The emergence of professional armies and navies over the course of the sixteenth and seventeenth centuries, for example, has attracted the attention of many scholars, including some of the world's leading historians. The resulting historical literature on the so-called Military Revolution has set high standards for the study of the interaction of the political, social, cultural, economic, financial, technological, and administrative factors that motivate and restrict the actions of national governments. The emergence of a body of writing of comparable size and quality on the industrialization of war from the late-nineteenth to the mid-twentieth centuries seems to be well under way.

Such work may contribute to the consideration of current strategy and policy because 'transformation' is likely to be less about the intelligent exploitation of specific advances in weapons technology and associated military doctrine and much more about the necessarily wider range of responses that will be required to deal with fundamental shifts in political, economic, social, and cultural as well as technological and military circumstances. In addition, discussion that relates current national security policy and historical topics may be usefully broadened to include phenomena other than military change, such as imperialism, religious conflict, irregular war, military occupations, man-made environmental shifts, and epidemiology, to name only a selection of potentially productive issues. There are, however, serious dangers in pursuing the course just described.

The great temptation is to use historical case studies as sources of ready-made lessons. To do so is to take a situation in the past that appears to resemble one in the present or one supposed to come into existence in the future, evaluate outcomes, and then, reasoning by analogy, draw conclusions about right or wrong action in the present. The basic flaw of this approach is that the past cannot be identical to what comes afterward, and that the degree to which this is the case is not easy to establish. Even where the salient characteristics of the past seem to coincide with the present, differences in personality, institutional dynamics, and a host of other factors that must be presumed to have been in play, but which cannot be known with precision, almost certainly matter. In the absence of clear knowledge about all significant points of difference as well as similarity between the past and present cases, it is impossible to judge the aptness of the comparison. Moreover, in complex situations, the interactions of multiple variables large and small can have significant effects, making the drawing of conclusions about the validity of any one particular decision inherently problematical.

What a historical case study can do, even when differences with the present are large, is prompt an insight that, when combined with other information, can serve as a point of departure for creative problem solving. Historical thinking does this by generating a sense of perspective—that is, an outlook capable of taking into account the antecedents and contemporary circumstances that conditioned, and distant effects that proceeded from, events. Good histories try to evaluate past events with both the advantage of hindsight and due appreciation of the limitations of human and institutional function in the face of complex, dynamic, and difficult circumstances. They accomplish this by always asking and sometimes answering questions. What was the long-term as well as immediate provenance of the policy problem? What were the long-term as well as immediate effects of policy action? What factors shaped behavior or influenced outcomes that were unappreciated or unknown to the leading actors? Were there good, if not compelling reasons, for leading actors to make what in the event turned out to be the wrong decision? Was major policy success the product of a combination of faulty reasoning and good fortune rather than sound planning and execution? To what degree did policy success depend upon the assertion of will in the face of daunting indicators of policy failure? Or to what degree was policy debacle the product of obstinate refusal to come to terms with reality?

Perspective on past happenings is valuable because it encourages, if not mandates, comparable reflection about activity in the present. Historical perspective with respect to ongoing affairs is of course impossible because immediate and long-term outcomes have yet to transpire. In addition, all major elements of causation may not be identified, and their relationship to policy dynamics established. That being said, possession of historical and general perspective about past events can promote an appreciation of the susceptibility of even well-conceived and well-executed policy to misadventure; of the predilection of perception and judgment of events as they are happening to be wildly in error; of the propensity of emotion to distort judgment in either positive or negative directions; of the possibility that even great policy success will birth dire consequences; of the prevalence of problems that defy any solution; of the probability that opponents will do the unexpected and inconvenient; and of the certainty that bold action in the face of highly unfavorable circumstances may on occasion be the only viable course of action. Such understandings do not prescribe courses of action, but they delineate a mindset that could be described as experienced, at least shrewd, or even wise.

The most appropriate use of history thus may be as an effective precep-
tor of militarized executive temperament, not as a fund of models to be emu-
lated. History properly applied to present requirements, in other words, is
about how to study and what to learn rather than how to act. This was
Clausewitz's opinion. He was convinced that military history could not be
codified into any system of lessons or principles that could be applied with
confidence to decision making in war. For this reason, Clausewitz was also
certain that productive study of the past could not be about the rightness or
wrongness of command as measured by obedience or non-obedience to any
fixed standards of conduct. Instead, he argued that the objective of historical
study was contemplation of dilemma—that is, not what made decision making right or wrong, but what made it difficult, and, in addition, what human
qualities were required to contend with, if not overcome, difficulty. Clausewitz, in short, wanted to comprehend process rather than evaluate outcomes,
and to develop the character of actors, not specify acts. Although he recognized the existence and even validity of principles of war as concepts, he
allowed them only a limited role in his vision of education, and none with
respect to real action in war.

Mahan shared Clausewitz's views on the importance of addressing command character, but without articulating methods of doing so comparable to
the clear instructions offered by his Prussian predecessor. His deployment of
principles of war in the consideration of historical problems thus made him
appear to be more doctrinaire than was actually the case. Mahan did not
believe that knowledge of history or principles was directly convertible into
effective policy or strategy. Indeed, insofar as the applicability of history and
theory to the present and future were concerned, he was manifestly skeptical.
Mahan believed that even sound historical analysis could not dictate policy in
the present nor provide accurate prognostication about the future because
human affairs were complicated and outcomes dependent upon complex
interactions and contingent forces. On the other hand, he made a good case
for the value of using historical understanding as the basis of considering a
range of possible courses of action. These could be contradictory, or even
mutually exclusive, and based upon premises that were also incompatible.
The objective of such heuristic activity was obviously not prescriptive, but
educative. What Mahan wanted historical and theoretical contemplation to
achieve was not a specific decision, but the expansion of the imagination as a
precursor to intelligent formulation of decision.

Corbett's fine histories of naval warfare in the age of sail gave him a reputation that enabled him to write on contemporary naval matters with a measure of authority. This made him useful to the service chief of the Royal Navy, Adm. Sir John Fisher, who appears to have given Corbett classified knowledge of Royal Navy operational and strategic thinking. Probably at the prompting of Fisher, Corbett published *Some Principles of Maritime Strategy* in 1911. In this, his one work of theory, Corbett denied that the executive ability that was a major concern of both Clausewitz and Mahan was susceptible to theoretical treatment, and focused his effort on the formulation of propositions that addressed what he believed were the particular strategic circumstances of Great Britain in the early-twentieth century. The result was a strategic synthesis that called for the coordinated action of the army and navy in a global maritime conflict over trade and colonies.

Corbett's elegant vision, however, was to have little relevance to the protracted war of industrial attrition in France and battle fleet standoff in the North Sea that erupted only three years after its publication. Although widely read in the late-twentieth and early-twenty-first century as a founding treatise of "jointness," *Some Principles of Maritime Strategy* was largely ignored in its own day, and indeed for long afterward. Corbett's career as a contributor to practical discussion of strategy and policy may thus serve as a cautionary tale—in spite of profound historical knowledge, a sophisticated theoretical intelligence, privileged access to information, and the confidence of the navy's chief executive, there is no sign of significant influence on the serious deliberations of the Admiralty.

The virtues and limitations of the historical and theoretical texts aside, the utility of history and theory with respect to present-day strategy and policy is much affected by the character of the fields that produce current historical and theoretical writing, namely, history and political science. Both approaches have drawbacks. Historians, with cause, regard political scientific use of historical information to be compromised by generalization from insufficient data. Political scientists, also with justice, consider historians to be so preoccupied with detail and disengaged from the present as to be incapable of applying their own findings to current problems. Or to put it another way, historians suspect that a political scientist will use one fact to write a book, two to create an institute, and three to found a religion, while political scientists are certain that a historian requires one hundred facts to write a sentence, two hundred to write a paragraph, and a thousand to write advertisement copy for the local dry cleaner.

Crude and unfair (but admittedly fun) as such generalization might be, the differences in the methodologies of history and political science *have* produced an imbalance in the present-day generation of relevant scholarly literature that disfavors history. Diligent research in archives is both expensive and time-consuming—historians as a general rule will thus require more time to produce a book or article on their subject than a political scientist working with secondary sources. In addition, the primary mission of most political scientists is to examine contemporary issues even when using historical materials, while this is not the case for historians, who for sound reasons study the past for its own sake. And while political science considers national security issues a centrally important topic in their field, the reverse is the case in history, because nearly all major civilian universities have marginalized the examination of military subjects in general, and strategy and policy in particular. Thus, good histories of strategy and policy topics that are informed by awareness of current affairs are relatively few. And by virtual default, therefore, the task of meeting high demand for relevant historical case study has fallen to political scientists.

The shortcomings of many if not most of these exercises in "applied history"—which, to be fair, often compromise the work of genuine historians as well—are several. First, the use of primary sources is limited, and what are exploited the product of opportunistic rather than systematic archival study. Second, published histories are used with little discernment—that is, information and concepts are drawn from defective as well as sound studies with little or no evidence that the author is aware of the differences in the quality of his or her sources. Third, a collateral effect of incomplete and ill-informed survey of the primary and secondary sources is to exacerbate any predilection that may exist to present only that evidence which supports predetermined conclusions. Fourth, statistics are deployed without sufficient care for the provenance of the numbers or their actual significance. Fifth, ignorance of relevant historical context, which is almost unavoidable if the author lacks trained expertise in the period in question, promotes ahistorical analysis—that is, characterization of the motives of individuals, groups, and institutions, or the dynamics of political, economic, social, and cultural processes, in ways appropriate to the present but not to the past period in question.

Given the qualitative unevenness of current production of relevant scholarship, the military and civilian consumers of history and theory are confronted by the problem of having to sift good from bad work as it is published in order to provide a sound point of departure for productive discussion. The institutional instruments available to perform this task are, however, weak. Reviewers of books in service journals too frequently do not possess the expertise to provide informed critical evaluation, while significant articles escape formal assessment and even notice altogether. Evaluation of scholarly literature is not a major function of service military history offices, which lack both the human and the fiscal resources required to perform the job properly. War College faculties, which in theory have the right personnel, are for the most part given insufficient time to keep up their own scholarship, let alone tracking the work of others. And the small number of historians of policy and strategy in civilian universities means that the amount of appropriate professional review capability available to civilian historical journals is inadequate.

Significant changes in the behavior of military intellectual institutions or the supply of civilian historians specializing in policy and strategy are unlikely to be forthcoming in the near term. The main burden of rigorous discrimination, therefore, will have to be carried by ad hoc consultations with individuals and groups of individuals in conferences. More conversation between historians and political scientists would probably be helpful, particularly with regard to epistemological issues—that is, questions concerned with how we know what we know, the boundaries of knowledge, and the difficulties of converting knowledge into action. And here as elsewhere, the key issue will be the exhibition of as much concern for the integrity of the process of inquiry as the applicability of the outcomes.

In the nineteenth and early-twentieth centuries, military history and theory were valued as major contributors to the formulation of standards of strategic and operational conduct in war—that is, doctrine. In the late-twentieth and early-twenty-first centuries, however, the demand for military history and theory is to a great degree driven by the desire for guides to the creation of new means and methods of warfare, not just the use of existing ones. For this reason, the problems that military history and theory are supposed to address are much more formidable than had previously been the case because the difficulties and complexities of strategy and operations having been compounded manifold by those of policy and technological innovation, and their interconnections and collateral effects. These complications do not preclude productive application of history and classical theory to the consideration of current affairs, but they do require attentiveness to what is not practical, as well as what is.

The major misuses of history and classical theory with respect to current policy are three. First, deploying them as no more than marketing devices—that is, the self-conscious use of history or classical theory as polemical instruments rather than as vehicles for critical discussion. Second, exploring historical case study or classical theory with critical intent, but without equally critical consideration of the soundness of the conceptualization and research of the sources used. And third, asking history or classical theory to predict the future, which is beyond the power of even well-conceived and well-researched accounts of the past or the most brilliant intellectual systematizing.

The major uses of history and classical theory with respect to present-day affairs are also three. First, sound understanding of good history and classical theory can serve as a detector of and an antidote to the misuses of history just listed. Second, consideration of good history and classical theory can provoke questions about present-day affairs that might otherwise have gone unasked, and thus stimulate the formulation of productive lines of inquiry. And third, diligent study of history and its classical theoretical associate can encourage the development of intelligent executive capability; in other words, history can fulfill its traditional function as one of the most important agents of education. This may be less than those who must face the challenging present and uncertain future have expected from the record of the past. But as Clausewitz warned in the last sentence of *On War*, "The man who sacrifices the possible in search of the impossible is a fool."

## Notes

1. Paul Dickson, *The Official Rules* (New York: Delacorte Press, 1978), p. 58.

2. Julian S. Corbett, "Staff Histories," in Julian S. Corbett and H. J. Edwards, eds., *Naval and Military Essays Being Papers Read in the Naval and Military Section at the International Congress of Historical Studies 1913* (Cambridge: Cambridge University Press, 1914), p. 24.

3. Corbett, "Staff Histories," p. 32.

Part 2

# OPERATIONAL ARTS: CONVENTIONAL WARFARE

# EIGHT

# Future Warfare and the Principles of War

### Sir Ian Forbes

O ver the last fifteen years, following the end of the Cold War, the nature of crisis and intervention has been in a state of constant flux, seemingly defying our abilities as nations in partnership, be it in coalition or alliance, to predict and prepare in good order for the security challenges that result. The Balkans, Haiti, Somalia, and Afghanistan in particular placed unexpected demands upon our respective governments' will to react, and posed new questions for our militaries in terms of their employment, equipment, and doctrine.

Today, Iraq is taking our common thinking once again into unknown territory, as we seek to marry accelerating technology where information and precision dominate in the war-fighting sphere, with well-established doctrinal practices long associated with peacekeeping and counterinsurgency operations. Against such a background, an examination of future warfare and its relationship to the principles of war is pertinent, given that an in-depth knowledge of the latter remains at the heart of the operational art that pervades military doctrine and training.

## Context

In focusing on future warfare, it is important to recognize certain truths about the context in which it is considered, and in so doing to agree upon time frame and perspective. We live in an era of potentially deepening chaos where the regulators of international order—empires, the balance of power, and alliances—no longer guarantee security. So life is not going to get easier, and our need for a common understanding to address an increasingly complex set of security

143

issues becomes ever more crucial. Within this, American power is the dominant feature in today's international system. This will certainly remain the case for at least three decades, and most probably beyond. The remainder of the international community has yet to arrive at a common view that complements this power. This said, there is a growing realization that threats in the twenty-first century are so complex and potentially catastrophic that America should not, indeed cannot, shoulder the global security burden alone.

Notwithstanding this, America is set on a course of military transformation that will not be deviated by Iraq but informed by it. Its focus is the power of joint effect and the decisive action it promises to deliver. While it has enjoyed success in war-fighting terms thus far, U.S. force transformation will remain incomplete until it can demonstrate similar success in post-conflict scenarios. Further afield, America's allies are just beginning to embrace transformational ideas, but at what pace and to what effect, given the affordability issues involved, are indistinct at present. Thus, integration and interoperability look set to remain critical issues in coalition and alliance operations, noting the disparity in investment in military capability that will inevitably occur in the transatlantic context in the coming years. Given these many uncertainties and the unpredictability inherent within them, U.S. and Coalition operations in Afghanistan and Iraq offer the best insights for sourcing our future thinking.

## Trends

In both these theaters, involving in this new century intervention models without consent and without cease-fires in place, an analysis of conflict and post-conflict is only just beginning to emerge. Both interventions have made clear that the war-fighting phase of an operation cannot be regarded as the sum of the military process. The aftermath, with all its complexity and attendant risks, while not exclusively military, has to form part of a campaign continuum involving all aspects of national power in bringing the intervention to a successful conclusion. In both Afghanistan and Iraq, the success in the hostilities phase resulted from the application of three themes:

- The achievement of coherent effects through forces operating for the first time in a truly joint integrated way
- Decision superiority—which is a seductive slogan, but one that seeks to define a process of more transparent and rapid information exchange facilitating faster decision making
- The application of overmatching power

To focus on these in turn, the application in both theaters of a more deliberate effects-based planning process enabled by networked C4ISTAR capabilities led to faster and more effective battle-winning success with corresponding reductions, compared to previous conflicts, in the mass of friendly forces, collateral damage, and casualties. It resulted, in the words of one eminent commentator, "in a military operation, astonishingly successful, probably the most successful war ever between a democracy and a dictatorship." In it, for the first time at the front line, there was closer integration of joint forces and a relationship between land and air that represented a fundamental shift in the way ground forces will fight in the future. In the past, land commanders essentially planned on utilizing assets they owned because they were guaranteed to deliver what was wanted at the right time and in the right place. As Gen. Douglas MacArthur said, "In many situations that have seemed most desperate, my artillery has been a most vital factor." Land commanders would rarely base a plan on protection being provided predominantly by air support, Tomahawk land attack missiles (TLAM), or naval gunfire support (NGS) assets because there was no guarantee such firepower would be available at the critical moment.

In Afghanistan and Iraq, however, there was a culture of trust resulting from joint training and war-gaming that led to examples of arguably the first truly integrated employment of the full range of joint capabilities to achieve the desired effects. Execution of the plan moved beyond the tried-and-tested methods of de-confliction or coordination to something more integrated. In so doing, it stimulated a greater depth of understanding of joint capabilities and a shared war-fighting culture. In short, as a result of enhanced C4I and training, forces became more integrated and also more trusting. The outcome was battle-winning effect. The next inevitable step in this process is to move from a culture of trust to one of reliance.

Decision superiority is a description that can have varying interpretations subject to its context. For commanders in the Iraq War, it meant that a highly adaptive and collaborative command and control set of networks and systems matched with a flexible human approach allowed for greater knowledge and information exchange than planning staffs had experienced before. The result was targeting that was markedly more timely and responsive than in previous conflicts. The ability to make decisions and prepare, collaboratively and adaptively, and then to adjust the plan rapidly and responsively across the battle space led to unprecedented levels of mission success.

This was the beginning of a truly network-centric capability. It is the way that America will organize and fight in the Information Age captured by former Chief of Naval Operations, Adm. Jay Johnson, in his statement that "it represents a fundamental shift from platform-centric warfare." The concept aims to link sensors and shooters within an information loop that accelerates the decision-making cycle by flowing critical information fast to apply force in an ever more effective manner. Network-centric capability is not exclusively about networks in a technical sense. Rather, it is about how wars are fought and about how power is developed. As information accelerates and precision weapons become more lethal and reliable, smaller joint forces will wield greater power over longer distances in new venues. Put simply, network-centric warfare will exploit military capabilities by allowing nations to use traditional capabilities in new ways—the "do more with less" argument. The appeal of such a prospect to coalition and alliance partners cannot be underestimated, as it offers the best opportunity for improved interoperability and integration between partners on the battlefield, reducing the excessive expenditure associated with new platforms. Early lessons from Iraq underscore the potential of network-centric capability, but point up the fact that current technology is not sufficient to deliver the capability to its full effect. It will merit the highest priority in future warfare terms in the coming years.

And so, this brings us to overmatching power. Here, the emphasis shifts away from numbers—and a doctrine of overwhelming force—to take account of harnessing joint capability to decision superiority and thus deliver decisive effects. Improved and continuously updated knowledge of your opponent, combined with an ability to generate lethal and nonlethal effects against multiple targets simultaneously or in close succession, provides for defeat mechanisms in certain contexts that do not demand total attrition or excessive destruction of infrastructure to achieve the military aim. The argument being that knowledge and power allow for a significant reduction in the numbers of forces required.

A full understanding of the sources of enemy military coherence, for example, may allow relatively small strikes to have disproportionate effects on the ability or will of enemy forces to conduct operations. At the strategic level, such effects-based operations require close integration of military means with diplomatic, economic, and broader informational elements of national power. Whereas legacy operations focused on attrition in combating the enemy, effects-based operations focus on setting those conditions that defeat an enemy's ability to make war. In particular, they concentrate on affecting decisions and behaviors of people and organizations. Examples include the prevention, disruption, and destruction of terrorist activity; the stabilization of post-conflict environments; influencing or demoralizing the leadership of an adversary; deploying the ability to find, fix, and strike in all environments; and the ability to deter and coerce. The theory is that effects can be diverse and provide solutions to unique situations. Effects-based operations are currently in their infancy, and it will be a number of years before the concept is brought to full maturity with the changes in mind-set, new theories and methods, and empirical base being accepted and adopted in full measure. In the interim, their validity—and with it, the case for overmatching power—will remain the source of ongoing questioning and debate.

## Relationship to Principles of War

Given these trends, future war fighting (and the term *war fighting* needs to be stressed) would appear to demand fully integrated jointness, decision superiority, and, where judged credible, the application of overmatching power. These attributes were proven in Afghanistan and Iraq, and they now form the basis of conceptual work for future capabilities in NATO and coalition contexts and beyond. Do these trends change the nature of conflict? Probably they do not, but they clearly have an effect upon its conduct and execution. The principles of war that are well known to all—objective, offensive, mass, economy of force, maneuver, unity of command, security, surprise, and simplicity—would seem to endure in this future war-fighting context. Indeed, there is a strong argument for suggesting that they are even enhanced as a result of the trends described. It is for consideration that one principle—mass or concentration of force—merits review, in order to reflect overmatching power and massing of effects rather than forces outlined earlier. The remaining principles would appear to be as relevant to the future warfare landscape as those of the past and the present.

But all of this assumes that one addresses war fighting in isolation from the complexities to be faced in stabilization and reconstruction operations in which war fighting, peacekeeping, humanitarian policing and training operations, among others, become intertwined and in need of correlation, usually within an urban setting. It also omits to include the challenges associated with rebuilding societies after conflict, the nation building task, where instituting political reform to grow democratization and put in place machinery for economic development becomes the driving force in bringing about an end and an exit to the military mission. This is today and tomorrow's intervention, illustrated in stark fashion by the crisis in Iraq. Few would claim that it will be the last time that a coalition led by the United States will be engaged in such a campaign.

## The Issue of Continuum

So, future warfare, except in rare circumstances, will involve activity across a wider spectrum of activity than has previously been the case. Thus, it will demand different capabilities and practices to the prosecution of high intensity war fighting alone. The future adversary, much like those in Iraq, will be disinclined to engage an intervention force in open warfare, but rather look to its urban mass as a refuge and combat zone. And in these circumstances, commentators suggest that "Such conflicts are fed by passions and rumors that do not yield to rational measures of persuasion and control; they are apolitical to a degree for which Clausewitz made little allowance." This statement implies that the application of joint effect, decision superiority, and overmatching power are but some of the capacities and practices required.

The occupation of urban terrain facilitates an opponent's ability to control key lodgment areas, perhaps his center of gravity, while reducing the advantages in speed, precision, and mobility that a U.S.-led coalition will bring to the fight. With the added risk of casualties and prolonged campaign time lines, urban operations have the capacity to weaken coalition cohesion in a prolonged crisis.

In this, the issue is not about intensity but about complexity. Commanders during Operation Iraqi Freedom explained that on many occasions in a 50-kilometer square, the Coalition was simultaneously engaged in high intensity, high tempo armored warfare; peace support operations requiring a robust approach to framework security; and standard peacekeeping involving the distribution of humanitarian aid. Since the termination of the war-fighting phase, the crisis in Iraq has become increasingly "fighting amongst the people." This, on top of regular and irregular forces pitting asymmetric methods against the Coalition and its supporting capability. To a far lesser extent, the same is true of Afghanistan but on nothing like the same scale as the challenge facing the newly elected government of a sovereign Iraq. In short, this is the oft-quoted Three Block War in which operations take place simultaneously across the full spectrum of military activity. The aim in such circumstances is to distill the three-block scenario into a nation building operation as soon as possible, where a stable environment is the key—and therein lies the operational art.

The implications of managing this situation imply differences to conventional high intensity conflict. Principally, three-block scenarios initially require mass on the ground and effects that go beyond the application of exclusively military capability. It can be argued that an intervention where the three-block scenario is likely, which will be an inevitability where consent for intervention is not forthcoming, requires technological advantage even greater than high intensity warfare. Intelligence assessments become critical, and greater pressures to reduce vulnerabilities and casualties require optimum situational awareness. However, such operations also require the careful combination of technological advantage with the application of the human factor. Command is exercised and decisions made often on nuances, intangibles, and instincts that cannot be modeled on a computer, particularly when the tempo of operations is unrelenting. In such contexts, intelligence assessments post conflict have been criticized for overreliance on technical dominance and a lack of human intelligence and understanding of ground truth; summarized in the statement, "an institutional failure to account for the most critical dimension of the battlefield—the human one."

Additionally, these operations place greater emphasis on nonmilitary actors. The operating space is filled by a wide variety of interests, government and nongovernment in nature. Interagency coordination becomes critical, on the ground and at the operational and strategic levels, as a mixture of pan-government effects interplays with organizations ranging from commercial interest to the media. Here, indirect effects become potentially more important than direct ones. There is a shift in balance between kinetic and non-kinetic action; psychological operations and computer network interference are two examples that spring to mind. Critical in this environment is the need to ensure that the indigenous population does not become overdependent on the intervention force and frustrate mission success and the achievement of a timely exit strategy.

This is a difficult judgment, given the highly complex nature of nation building and the time-consuming process associated with developing the key components of a state that are necessary for long-term stability and growth. Korea, Cyprus, Bosnia, and Kosovo have all required significantly more time and more resources than was originally thought. Efforts in these areas, and the demand for them, will undoubtedly expand in future interventions.

## An Emerging Model

There are good grounds to suggest that a new model for intervention is emerging, following Afghanistan and Iraq. Fifteen years of Balkan, African, and now Middle Eastern experience point to the fact that conflict, and the campaign associated with it, is not only about winning or ending a war, but also about the use of military force in its aftermath to underpin rapid and fundamental societal reform. During the Cold War, intervention was focused on reinforcing the status quo. Forces deployed to Korea, Cyprus, and Palestine were tasked with maintaining a division between warring parties, not to compel resolution of the underlying dispute. Lebanon, Grenada, and Panama were undertaken to undermine undemocratic regimes and reinstall more progressive ones, rather than bringing about fundamental societal reform. But since the end of the Cold War, the intervention model that has evolved is one in which the international community, under coalition or alliance auspices, acts not simply to police cease-fires or restore the status quo but also to intervene, execute regime change, and effect a political, economic, and democratic reconstruction. Events have moved the international community toward a nation-building model in which the objective is to draw together political infrastructure as well as reconstruction and in which the measures of success are the establishment of democracy and the creation of a workable economy.

As a recent RAND study noted, each successive post–Cold War intervention has been wider in scope and more ambitious than its predecessor. Somalia, originally a humanitarian operation, migrated to democratization. In Haiti, the objective was to install a president and conduct elections. In Bosnia, it was to create a multiethnic state. Kosovo and Afghanistan moved the model further, and Iraq requires no explanation. It seems, therefore, that the task of nation building, and our collective moral responsibility for it, is an inescapable reality in this new century.

Given this, the role of the military in nation building becomes important to any debate about the principles of war, simply because that role will include the rapid establishment of a secure environment. Moreover, it will be to remove the military options available to opposing actors, primarily to create an environment in which political reform and developments can take place. It will also demand a clear distinction between security and policing functions, ensuring mutual support should circumstances require a more robust approach. Finally, the military will have a key part to play in encouraging regional stability—taking account of neighboring states and related issues. For all these reasons, it is inevitable that responsibility for security, both internally and externally, will fall initially to the military.

Such tasking raises time and resource issues of a significant nature. No post-conflict operation over the last two decades has been short lived, and the majority have soaked up large scale forces in the process. This clearly inhibits any nation's collective will to commit. Indeed, the international community is having to tackle Afghanistan in a staged approach, establishing but a modest stabilization presence with localized concentration and the intent to expand gradually as and when possible. The troop-to-population density in Afghanistan is currently 100 times less than either Bosnia or Kosovo at the same, post-conflict phase.

But the model inherent in Afghanistan and Iraq presents cause for optimism. Progress in Afghanistan is positive, with a national government being established, an Afghan Army being trained, and widening stability beyond Kabul. Although warlord and tribal influences remain endemic, there is an increasing flow of international assistance, and the beginnings of a workable judiciary and economy. At this writing, it is too early to state categorically whether Iraq is following a similar path, but recent developments with elections and military training represent significant milestones in moving the country toward improved governance and economic independence.

The issue in all this is how to define the boundaries of the role of the military in high complexity operations and post-conflict nation building. As a start, it is readily apparent that military activity is but one element in a holistic approach that has to encompass political, social, economic, cultural, legal, ethical, and moral dimensions in reconstituting a nation state. In nation building in particular, there is a need to clarify and define an exit strategy for the military dependent upon completion of its role. This is an issue that illustrates the distinction between the use of force to establish a secure environment and the establishment of the rule of law via legitimate policing. A number of emerging proposals seek to address the issue of military involvement in such contexts, notably the establishment of stabilization and reconstruction forces in national forces or alternatively the notion that European forces should focus on post-conflict and the United States on high tempo war fighting. Notwithstanding the practical advantages offered by such suggestions, significant political, economic, cultural, and moral questions will need to be answered before such proposals can be taken forward.

What is clear from events in Afghanistan and more prominently Iraq is that the military plan cannot be viewed in isolation and the military cannot be the key actor in an intervention strategy. For the emerging model to take account of a continuum of operations, including pre- and post-conflict, the strategy for the intervention has to embrace all elements of national power and involve cross government accountability. The lead authority or agency at the strategic level must be both resourced and empowered to develop strategies that are as effective at completing an intervention as they are at initiating it. This is highly complex at the single nation level, where one set of policies and practices exists; it becomes even more challenging in coalitions and alliances, given the diverse nature of different national policy paradigms. Such complexity and challenge argue strongly for the accelerated development of a coalition or alliance war-gaming process as a matter of urgency.

## Allies and Interoperability

It would seem that the oft-quoted Churchill comment "There is only one thing worse than working with Allies and that is working without them" continues to have resonance. However, given that the impetus for and investment in future warfare lie predominantly with the United States, there are profound implications for allies when operating in coalition or alliance with their American partner that impact directly on operational cohesion and effectiveness. The technological and operational gap that has been a feature of the post–Cold War era looks set to widen further unless common denominators of interoperability can be identified and shared.

Foremost among these is unquestionably C4ISR and network-centric capability. It must be assumed that the United States will be the only democracy able to make the appropriate investment in transforming its military in totality, consistent with technological advance. Consequently, some form of "plug in" or sharing offers the best prospect for interoperability and integration that will be critical if coalitions or alliances are to function successfully in future warfare contexts. Avoiding duplication will be key, notably where investment is not bringing additional capability to the coalition or alliance. Significant political, military, and commercial issues are inherent in such a process, not least in the fields of defense cooperation, technology transfer, industrial benefits, and intelligence sharing. Hard choices will have to be made: in the United States, in terms of sharing technology; in allied nations, in terms of funding and investment.

## Post Conflict and the Principles of War

The principles of war provide a formula for deploying military force to best effect. They propose a clear objective and rely on a set of military characteristics in the execution of a campaign plan to bring this about. As such, they predicate utility in a military context without political interference or constraint. In post-conflict operations—be it peace making, humanitarian provision, or nation building—no situation is militarily clear-cut, and the line between war fighting and stabilizing is inevitably ill defined. In such contexts, as we saw most vividly in Kosovo, different principles will apply to differing circumstances. For instance, a principle of concentration of force will apply to an operation such as the assault on Falluja, but it will have less applicability in other post-conflict settings. Likewise, offensive action will be relevant at times when resolve and military capability may be necessary in an escalating situation where security is in danger of compromise. Events in Bosnia and in Kosovo over the last decade are vivid illustrations of this. Indeed, it is questionable whether such a principle is more rather than less stabilizing in the post-conflict setting. In 1775, the British Philosopher Edmund Burke observed that, "the Use of Force is but temporary. It may subdue for a moment; but it does not remove the necessity of subduing again; and a nation is not governed which is perpetually to be conquered." These words are as relevant today as they were then.

Better then, that these principles—offensive action, concentration of force, and maybe even surprise—be replaced in post conflict by a single principle of synchronization of effort. It applies equally well to all levels of conflict and suits the three-block scenario. At the operational and tactical levels the importance of synchronizing military effort is well understood. At the strategic level, the principle reinforces the importance of the cohesion achieved by the cooperation and synchronization of political, military, and diplomatic actions. Hence the suggestion would appear to have wider applicability across all levels of command and the spectrum of conflict.

## Conclusion

In summary, there is a danger of drawing false conclusions for the use of military force for future crisis intervention scenarios if high intensity war fighting and its characteristics are considered in isolation. Military activity has to be seen as but part of an overall campaign, a continuum of activity, in which high intensity war fighting and high complexity post-conflict operations including nation building are synonymous and complementary. The war-fighting phase will require jointness and effects-based planning, decision superiority, and overmatching power—applications that will have significant implications for coalition and alliance interoperability and integration and consequently their military effectiveness. Development of a common net-centric capability is the key to this. However, unless and until there is a more productive partnership for addressing and sharing C4ISR technology and inter-operability issues between the United States and its allies, operational cohesion and effectiveness in coalitions and alliances will lack military robustness.

That the mission defines the coalition remains an unalterable truth in the future warfare context. In the high intensity war-fighting condition, given that Iraq did validate a concept of overmatching power, mass is the only prin-ciple of war that may be less relevant than previously. High complexity post-conflict operations including stabilization and nation building missions suggest that offensive action and surprise are principles that could be better deployed as synchronization of effort to describe the nonlethal as well as lethal aspects of these operations in addition to nonmilitary, interagency measures. It must also be stressed that any post-conflict phase of an operation will be more political in nature than the war-fighting phase, and in such cir-cumstances violation of any of the requirements offered by the principles of war cannot be ruled out.

## NINE

# Transformation
# and Operational Art

### DAVID A. FASTABEND

In reassessing something as fundamental as the *principles of war,* the problems of scope and rigor are immediate. War itself increasingly escapes both declaration and definition, and even at its most aggregated level, war is viewed through multiple lenses: tactical, strategic, and operational. Operational art—the linkage of tactical means to strategic ends—holds particular promise for our reassessment of the principles of war. If transformation will extend to the principles of war, then operational art will certainly transform as well. Therefore, this paper will examine whether and how *transformation* has changed *operational art.*

We might first blanch at any question that incorporates not just one but two of our more problematic terms. We certainly subject both "transformation" and "operational art" to a wide range of interpretation. For example, the Office of the Secretary of Defense's *Transformation Planning Guidance* defines transformation as

> a process that shapes the changing nature of military competition and cooperation through new combinations of concepts, capabilities, people, and organizations that exploit our nation's advantages and protect against our asymmetric vulnerabilities to sustain our strategic position, which helps underpin peace and stability in the world.

This is actually not a bad definition. Too bad no one really uses it! Transformation is most typically associated with the "revolution in military affairs," that is, the technological changes, organizational changes, and operational concepts we typically associate with information technology, extended range, and speed and precision. Whether or not this revolution has provoked the *appropriate* transformation in the various services is a topic for which many PowerPoint slides have been thrown into the breach, but we don't need to explore that. I accept, and I beg you to accept, the notion that transformation, as it is most commonly understood, has to do with "things like this": information technology, more effective communications, extended range, and faster speed and greater precision.

Paradoxically, operational art is more rigorously defined. Joint Publication 1-02 tells us that operational art is

> the employment of military forces to achieve strategic goals and/or operational objectives through the design, organization, and conduct of strategies, campaigns, major operations, and battles . . . operational art determines, when, where, and for what purpose major forces are employed.

It is no accident that we call this art. Lt. Gen. Leonard D. Holder, U.S. Army (Ret.) wrote that

> Theater operations fall more clearly into the domain of art than that of science. Below the level of broad principles, each situation varies so strongly in personal, geographic, demographic, historical, and economic details that the teaching of the operational art will resemble political science more than small-unit tactics.

To use the vernacular, the *ways* of operational art link tactical *means* to strategic *ends*.

## Means, Ways, and Ends

### *Means*

The techniques have changed . . . who could dispute this? We are faster, more precise, more informed. Even comparing Operation Iraqi Freedom to Operation Desert Storm, we see that we used one-third of the forces, one-half of the time, and one-tenth of the bombs. These facts get the lion's share of the attention of the transformation community, but they are really the least interesting dimensions of the problem at hand. We have to ask: Does a significant evolution in tactical means translate to a transformation in operational art? Is art transformed if Michelangelo uses a faster drying paint? Furthermore, the canvas itself is fundamentally unchanged: terrain is still terrain, and overcoming the friction of operations on and through terrain—both open and complex—continues to pose daunting physical challenges for forces that still await a "transformation in logistics."

## Ways

This is the *Schwerpunkt,* or point of main effort and focus, of operational art: the creative ways in which we combine tactical means to achieve strategic ends. This is the realm of styles of warfare, of how we visualize operations. It is the realm of the operational concept, and there has been a veritable frenzy of work on operational concepts over the last few years. But where are the emerging schools of operational art? The Impressionists? The Modernists? Where are the fundamentally new ways to visualize warfare or operations? Sadly, there are all too few. To date, we generally apply faster drying paint to draw the same type of picture more rapidly and precisely. There is all too little transformation in our operational concepts, particularly at the joint level. They are too functional, too generic. They lack "the picture," and what is operational art without the picture?

This should not surprise us. Students of art history tell me that the emergence of a new school of art generally takes quite a long time, and it is usually apparent only in retrospect rather than as it happens. New schools of art are never predicted, of course. So, although our concepts may propose network-centric warfare as an emerging school of operational art, it may be that information tools are only tactical "brushstrokes." Only over time will it be clear that there is a distinct "school of transformational operational art."

## Ends

The most significant revolutions of art are associated with a fundamental shift in the purpose of art, a transition in its role in society. Therefore, we must not overlook the ends of operational art, the strategic context in which it occurs. Although the rapidly evolving tactical tools of war draw the most attention from the "transformation community," today we find that it is the very context of operational art—our strategic ends—that are changing the most. This change, tragically, is almost totally unrecognized by transformation enthusiasts, and could potentially render much of what we understand now about "transformation" to be irrelevant.

## The New Strategic Context

Simply put, our strategic context has changed, and changed significantly. Since 9/11, the most frequently quoted passage from Clausewitz is this one, and for good reason:

> The first, the supreme, the most far-reaching act of judgment that the statesman and commander have to make is to establish . . . the kind of war on which they are embarking; neither mistaking it for, nor trying to turn it into, something that it is alien to its nature. This is the first of all strategic questions and the most comprehensive.

In any strategic context, as Clausewitz pointed out, it is essential to understand "the kind of war on which [we] are embarking." Although the fundamental nature of war is constant, its methods and techniques constantly change to reflect the strategic context and operational capabilities at hand. The United States is driving a rapid evolution in the methods and techniques of war. Our overwhelming success in this endeavor, however, has driven many adversaries to seek their own adaptive advantages through asymmetric means and methods.

Some enemies, indeed, are almost perfectly asymmetric. Non-state actors, in particular, project no mirror image of the nation-state model that has dominated global relationships for the last few centuries. They are asymmetric in means. They are asymmetric in motivation: they don't value what we value; they don't fear what we fear. Whereas our government is necessarily hierarchical, these enemies are a network. Whereas we develop rules of engagement to limit tactical collateral damage, they feel morally unconstrained in their efforts to deliver strategic effects. Highly adaptive, they are self-organizing on the basis of ideas alone, exposing very little of targetable value in terms of infrastructure or institutions.

A cursory examination of the ideas in competition may forecast the depth and duration of this conflict. The United States, its economy dependent on overseas markets and trade, has contributed to a wave of globalization both in markets and in ideas. Throughout much of the world, political pluralism, economic competition, unfettered trade, and tolerance of diversity have produced the greatest individual freedom and material abundance in human history. Other parts of the world remain mired in economic deprivation, political failure, and social resentment. Many remain irreconcilably opposed to religious freedom, secular pluralism, and modernization. Although not all have taken up arms in this war of ideas, such irreconcilables comprise millions of potential combatants.

Meanwhile, not all former strategic threats have vanished. In the Far East, North Korea's "nuclearization" risks intensifying more than fifty years of unremitting hostility, and many others pursue developing weapons of mass destruction. We confront the growing danger that such weapons will find their way into the hands of non-state groups or individuals. Armed with such weapons and with no infrastructure of their own at risk, such "super-empowered individuals" could be eager to apply them to our homeland. At the same time in our globalizing world, military-capable technology is increasingly fungible, and thus potential adversaries may have the means to achieve parity or even superiority in niche technologies tailored to their military ambitions. For us and for them, those technologies facilitate increasingly rapid, simultaneous, and noncontiguous military operations. Such operations increasingly characterize today's conflicts and portend daunting future operational challenges.

Our adversaries seek to present us with multiple strategic dilemmas that prevent us from applying the national elements of power in a coherent way to achieve strategic political ends. The dilemmas we face today are currently categorized into four areas: Irregular, Traditional, Catastrophic, and Disruptive. These four areas are not mutually exclusive, and future wars are likely to encompass two, three, or even all four of them. In the future, operational art will need to connect capabilities in ways that achieve victory against any combination of the four challenges.

Our test, then, is to prepare for the future even as we relentlessly pursue those who seek the destruction of our way of life, waging a protracted war of ideas to alter the conditions that motivate our enemies. Some might equate these challenges to the Cold War, but there are critical distinctions.

- Our non-state adversaries are not satisfied with a "cold" standoff, but instead seek at every turn to make it "hot," seeking our destruction by any means possible.
- Our own forces cannot focus solely on future overseas contingencies, but also must defend bases and facilities both at home and abroad.
- We face an evolving asymmetric threat that will relentlessly seek shelter in those environments and methods for which we are least prepared.

- Our national strategy is not "defensive" but "preventive" because some of our adversaries are not easily deterred.
- Above all, because at least some current adversaries consider "peaceful coexistence" with the United States unacceptable, we must either alter the conditions and convictions prompting their hostility, or destroy them outright.

All of this portends a significant transition in our strategic context, one where we can no longer view peace as our default condition, and war the exception. Our new reality is very different: a protracted conflict of irreconcilable ideas, where combat operations will be routine, and peace will be the exception. We face a foreseeable future of extended conflict in which we can expect to fight every day, and in which real peace will be the anomaly. This is not the strategic context for which we built our armed forces, and this new strategic context will alter every dimension of our doctrine, organization, training, and soldier/leader development.

## More Questions

Transformation has changed the *means* of operational art, but its impact on operational art per se is not yet evident. The strategic context or *ends* of operational art are completely transformed, but to date, almost everything we commonly associate with "transformation" is irrelevant to this fact. Transformation has not *yet* had an extensive impact on the *ways* of operational art. To determine if transformation will ultimately extend to operational art and therefore the principles of war themselves, many additional questions are before us.

### *What Will Not Transform Within Operational Art?*

There is no reason to suppose that operational art will not continue to

- translate strategic purpose into tactical action;
- integrate the joint, interagency, and multinational context;
- compose campaigns and major operations;
- sequence battles, engagements, and military activities;
- focus power at the decisive times and places.

Although these fundamental characteristics of operational art will persist, the specific elements of operational design may evolve. Commanders will still need campaign plans, but there will be new challenges in developing campaigns of appropriate scope, campaigns that balance simultaneity and sequence while integrating a far broader range of participants. It is less clear if those campaigns will feature "centers of gravity." The original metaphor of physics was a center of gravity, but in this complex and distributed world everyone acknowledges the necessity for multiple centers of gravity. Is the metaphor still useful? Roger Spiller has suggested that a center of gravity is not something one designates; it is, rather, something one discovers. Decisive points and lines of operation may be useful, but in the volatile, uncertain, complex, and ambiguous strategic environments before us, they may be increasingly less *spatial* and more *logical*. There is considerable discussion of effects-based operations (EBOs) and their potential to transform operational art. But again, there are many questions. What are the elements of operational design comprising EBOs? How do they compare and contrast to the classical elements of operational design? What is their relative value added?

## What Are Our Expectations for Future Operational Art?

The basic traditional framework of operational art remains intact—executing joint campaigns to achieve operational and strategic objectives—but the U.S. Army is exploring several new ideas in its capstone operational concept. These are the army in joint operations. There are many new ideas in the army's future force concept that, when enabled by the appropriate capabilities, will constitute significant changes to operational art.

- *Shaping and Entry Operations.* We could raise this time-honored idea to a new level by acknowledging its criticality for rapid intervention and for ensuring tactical successes that build more rapidly into operational victories. Commanders and staffs who think in these terms from the outset, and who more explicitly recognize the need to simultaneously set conditions for success, will significantly alter the nature of operational art. The notion of setting conditions is appropriate to setting regional security conditions prior to a conflict and equally appropriate to protracted warfare whether we are facing traditional, irregular, or disruptive challenges.
- *Operational Maneuver over Strategic Distances.* An important new element of operational art is the idea that force projection blends seamlessly, without pause, to operational employment of forces to support campaign execution. In the past, deployers delivered forces to predictable ports and

airfields at intermediate or forward staging bases where any number of subsequent actions were necessary to get them and their sustainment into the fight in forward areas. The ability to cross the strategic to operational seam would make the activity of *deployment* almost indistinguishable from *employment*.

- *Intra-theater Operational Maneuver.* One of our more ambitious expectations for future operational art is intra-theater operational maneuver, that is, the ability to vault over or around enemy dispositions rather than through them and doing so with mounted, rather than dismounted, forces. Historically, with a few exceptions (Inchon), the best examples of operational maneuver first involved major tactical breakthroughs that permitted and led to exploitation and operational envelopments. If we can achieve the routine application of three-dimensional operational maneuver, we will routinely alter the geometry of the battle space with concomitant impacts on operational art.

- *Decisive Maneuver.* For the decisive maneuver of the future, we will seek simultaneous, distributed operations over a noncontiguous battle space. This central tenet within future force concepts extends well beyond traditional operational art, which emphasized sequential operations, synchronized in time and purpose, within a linear, contiguous framework. Direct attack of elements of centers of gravity has long been possible with respect to long-range fires, but combining fires with maneuver in deep operations within a noncontiguous framework to achieve rapid decision is substantively new. To the extent that resources and the factors of the situation will allow, commanders will seek continuous operations and overwhelming operational tempo.

- *Concurrent and Subsequent Stability Operations.* Operational design no longer permits the military commander the luxury of planning first for the destruction of his adversary and then planning for the peace. The nature of victory has changed; raw kinetic power is no longer the sole ingredient for a coherent campaign recipe. Future operations will be characterized by requirements for concurrent and subsequent stability operations, the former to secure and perpetuate the results of decisive maneuver *during* the campaign, and the latter to "win the peace" once enemy military forces are defeated, to ensure long-term resolution of the sources of conflict. Although our experiences in Iraq over the last two years have drawn immense attention to this challenge, we must avoid a narrow, tunnel-vision focus. It is important to understand the relationship between these operations and major combat operations, but they should be considered in tandem.

- *Distributed Support and Sustainment.* Distributed Support and Sustainment in future operations will maintain freedom of action and provide continuous sustainment of committed forces in all phases of operations, throughout a noncontiguous battle space and with the smallest feasible deployed logistical footprint. At the operational level, distribution-based sustainment will be continuous, but sensitive to the more numerous lines of communications, shifting operational priorities and surge requirements associated with a distributed, noncontiguous battle space. To cope with them, future force sustainment units must rely on the same level of situational understanding as the operational formations they support, allowing logisticians to anticipate operational commanders' priorities rather than merely reacting to them.

- *Network-Enabled Battle Command.* Throughout the future campaign, network-enabled battle command will facilitate the situational understanding needed for the self-synchronization and effective application of joint and U.S. Army combat capabilities in any form of operation. It will permit integration of joint, interagency, and multinational efforts in heretofore impossible-to-achieve ways. As they do today, commanders will need to be prepared at any time to adapt operational plans and tactical methods to imperfect situational understanding. However, network-enabled battle command enables faster recognition of the need for those changes, and the ability to disseminate and execute those changes across the force. Thus, campaigns will lose the stodginess associated with the word *campaign,* and they will become more fluid.

Network-enabled battle command leverages the *network effect,* that is, the exponential increase in the value of a network as the number of those using it increases.[1] It extends the interconnectedness of headquarters—already significant—to the extremities of the force: individual soldiers, weapons, sensors, platforms, etc. This extended connectedness in a networked collaborative command and control environment can extend the benefits of decentralization—initiative, adaptability, and increased tempo—without sacrificing the coordination or unity of effort characteristic of centralization. Nonetheless, network-enabled battle command is commander centric, vice network centric.

## What Elements of Our Current Transformation Are Not Relevant to the Emerging Strategic Context?

In light of our persistent challenges in operations in both Afghanistan and Iraq, there is considerable debate about the relevance of many "transformational" changes to the types of warfare we increasingly anticipate. Many believe that an excessive reliance on information technology ignores the intractable uncertainty of war and attempts to reduce combat to a targeting drill. Many doubt that net-centric warfare works in a counterinsurgency environment, and that as the environment becomes increasingly networked, less obvious nodes are available for attack through effects-based operations.

Insights such as these are important, not least because our newfound capability for rapid *institutional* adaptation permits us to more quickly make the wrong decisions for a sizeable part of the force, if we are intellectually hasty. Lack of rigor will always be discovered—better it not be found out on tomorrow's battlefield. We must recognize how much of our effort is focused on today's fight, on tonight's news report, on today's adversary. We should take care not to casually discard elements of our current transformation simply because they don't solve today's conundrum.

## Can One Size Fit All?

There is considerable debate on the merits of specialized versus multipurpose forces. In a forthcoming article Bob Killebrew writes that

> Armies should have a clear idea of what they're doing when they set out to win wars. The Army's particular challenge, in the midst of rebuilding its force structures and redeploying forces to Iraq annually, is to reinvent the rules for decisive land warfare in the 21st century—to refocus an Army trained and equipped to win decisive conventional victories to one that can do whatever's necessary for strategic victory in a rapidly-expanding spectrum of potential conflict ranging from nuclear war to terrorism. Whether an army designed to win a war in thirty days can also play a decisive role in a war like the one in Iraq is a critical question. Perhaps new "operational art" will lead to forces that mix the best characteristics of precisely tailored, special operations-type units for counterinsurgency operations and firepower-intensive maneuver forces for the big wars. Certainly the joint doctrinal community will play a key role. But to the Army falls the unique responsibility for deciding how land forces "win" decisively in the 21st century.

The unanswered question is whether land forces can develop units that are adequately multifunctional or must revert to a more costly structure of special purpose forces. Single scope organizational solutions are likely to create vulnerabilities, not mitigate them.

### What Transformation in Military Affairs Is Most Likely to Render a Transformation in Operational Art?

Although it is widely accepted that today's transformation is largely driven by information technology, it is worth noting that most of the great revolutions in operational art over history are revolutions measured in time and distance:

- Gunpowder . . . extended the application of violence from arm's reach to line of sight.
- First industrial revolution (railways) . . . distributed forces throughout the battle space by land movement on *fixed* routes.
- Second industrial revolution (automotive) . . . distributed forces in depth throughout the battle space on *flexible* land routes

Current (information) revolution is generating capabilities that are faster, reach farther, are more precise, and are more efficient, but it remains to be seen if they portend a real revolution in a style of warfare. Such a revolution will more probably come with the next technical revolution in the conquering of mass over distance: mass materiel movement by air or sea, movement unconstrained by access to seaports of debarkation or aerial ports of debarkation. Such movement will require capabilities that do not currently exist: austere access, high speed shipping, super-short takeoff and landing, joint heavy lift vertical takeoff and landing—or other capabilities we can not now imagine.

### What Elements of Strategic Theory Will Be More Useful to Describe the Context in Which Operational Art Must Be Conducted?

In a recent seminar on future warfare hosted by the U.S. Army's Training and Doctrine Command, we asked each participant to sum up the observations of several days of discussion. What was the dominant theme? The need for an effective theory of conflict, one that addresses the need of our current strategic context. To understand the phenomena of "war" or "operations other than war," we must view them in their context: conflict. The dominant academic views such as realism, rationalism, or idealism present very partial perspectives on conflict—perspectives based on categories that frustrate holistic understanding in an environment of volatility, uncertainty, complexity, and ambiguity. A general theory of conflict would expedite communication and coordination between the diverse efforts of the many agencies of conflict.

## The Future of Art

I have posed far more questions than I have answered. Transformation has transformed the "means" of operational art, but it is not clear that these changes will transform operational art itself. The impact on the "ways" of operational art is not yet evident, but we have ambitions that operational art will significantly change in the future. There is little evidence of any transformational impact on the "ends" of operational art; those ends have evolved for reasons far removed from the "revolution in military affairs." Given the ambiguous impact of transformation on operational art, the impact on the principles of war is even more imponderable.

What is not in doubt is that our nation will continue to have strategic interests that it will pursue in an increasingly volatile, uncertain, complex, and ambiguous world. Although the means will change—transform perhaps—the linkage of those means to our evolving strategic ends will continue to entail imagination and innovation. It will continue to be, in a word, art.

## Note

1. This description of network effect originates from the 2005 Navy–Marine Corp FORCEnet functional concept.

# TEN

# Operational Art and Doctrine

MILAN VEGO

Even though operational art has been applied in almost all major conflicts since the Napoleonic Wars, there is no consensus on what it really is. Very often operational art is confused with strategy. Yet strategy, properly understood, has not dealt with the actual employment of forces since at least the late-nineteenth century. It was then that the lower part of strategy became what Field Marshal Helmuth von Moltke Sr. called operations, or what is called operational art today. At the same time, the upper part of strategy became more tied to policy. Many officers also erroneously believe that service doctrine is identical to operational art, but operational art is much broader and deeper than any doctrine. It is operational art that provides the broader framework in which to develop a service doctrine, and from which selected parts are integrated in succinct form into service doctrine.

Depending on its main purpose, tactical, service, and joint/combined doctrine is differentiated. Service doctrine and joint/combined doctrine require considerable input from operational art theory; otherwise, they will not serve their stated purpose. A *service doctrine* is aimed at providing broad guidelines for the employment of one's service forces to accomplish operational (or strategic) objectives through the planning, preparation, and conduct of major operations. A properly written service doctrine should explain in some detail the employment of the numbered or theater forces at the *operational level of war* as part of a joint or combined force. Because of its larger scope, service doctrine is inherently more flexible than tactical doctrine.

A service doctrine commits the entire service to the same rules, principles, and standards for the conduct of war. It explains what military problems must be solved and how they should be solved. It should explain how a particular service should plan, prepare, and execute major operations, independently and in cooperation with other services or multinational forces. It should establish a framework for tactical doctrine of individual combat arms/branches, platforms, and sensors. It should also provide direction for the future capabilities of the specific service. A service doctrine also provides a baseline for individual and unit training, and it greatly influences the professional education of officers. A service doctrine should explain the employment of service forces across the spectrum of warfare and in any physical environment. To be successful, it must be adapted to the specific operational or strategic environment.[1] Figure 1 illustrates the levels of war concept, from the tactical to the operational.

In contrast to a service doctrine, *joint doctrine* provides guidelines for employing the forces of two or more services in the planning, preparation, and conduct of campaigns or major joint/combined operations aimed to accomplish a strategic objective. The chief of general staff normally promulgates joint doctrine. It focuses on the employment of multiservice forces at the operational and strategic levels of war. Joint doctrine serves as a link between strategy and policy on one hand and operational art on the other. It should provide a framework for writing the doctrine of each service of the

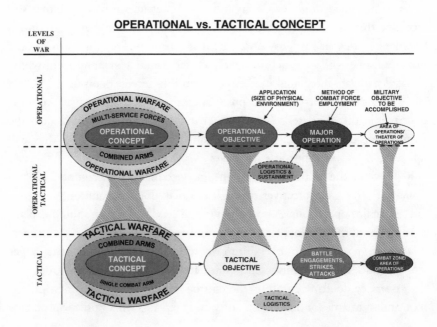

armed forces. A joint doctrine offers a common perspective from which to plan and operate, and fundamentally shapes the way we think about and train for war. The aim is to distill insights and wisdom gained from our collective experience with warfare into basic principles to guide the employment of joint forces.[2] Joint doctrine provides an authoritative and common perspective for all services. It fundamentally shapes the way all the services think, plan, and train. It is a critical factor for success in the joint employment of forces.[3] *Combined* doctrine consists of fundamental principles on the employment of multinational forces approved by the coalition/alliance partners and aimed to accomplish strategic objectives in a theater.

## Framework

A well-written service or joint doctrine should concisely and clearly explain the strategic framework or environment in which service or joint forces are to operate. Hence, the doctrinal document should briefly outline the key features of national policy, national security, and military strategy. The relationship between policy and strategy on the one hand and operational art on the other should be explained as well. The main aspects of the strategic environment—specifically political, diplomatic, military, economic, social, informational, legal, and others—provide a framework within which one's service or joint forces operate. The strategic role and missions of the respective service or the armed forces as a whole should be explained as well.

Sound joint doctrine needs to explain the spectrum of conflict, from peacetime competition to low- and high-intensity conflict. It should present the nature of war in general and the service's or joint community's consensus on what future war might look like. In addition, service doctrine should explain the views of the respective service on the nature of war in the physical medium in which its forces operate (land, sea, air).

## Operational Art Inputs

Operational art provides a major part of the input to service or joint doctrine, because ultimate success is achieved only through sound sequencing and synchronization of both military and nonmilitary sources of power. It is the skillful application of operational art that ensures proper linking between tactical actions and the aims of strategy and policy through the conduct of major operations and campaigns.

In one of its many definitions, operational art is described as the employment of one's military forces to accomplish strategic objectives in a theater of war or theater of operations through the design, organization, and conduct of campaigns and major operations. It involves fundamental decisions about when and where to fight and whether to accept or decline combat. Its essence is the identification of the enemy's center of gravity (COG) and the concentration of superior combat potential to achieve decisive success.[4] In generic terms, operational art can be understood as that intermediate component of military art, between strategy and tactics, concerned with both the theory and practice of planning, preparing, conducting, and sustaining major operations and campaigns aimed at accomplishing operational or strategic objectives in a given theater or part of the theater.

### Fundamental Warfare Areas

Properly written doctrine should explain in some detail the service consensus on the nature of warfare in the fundamental warfare areas, such as defense or offense and their variations (in land warfare), strike warfare or amphibious warfare (in naval warfare), and close air support and counter air (for air forces). It is important to note that service views on the current state of each fundamental warfare area and some trends in the future should be expressed from the operational rather than the tactical perspective.

### Objectives

One of operational art's most important inputs to doctrine deals with the strategic and operational objectives to be accomplished in combat. Joint doctrine is normally concerned with strategic objectives, while the forces of each service normally aim at accomplishing operational objectives as part of a campaign. Service doctrine for ground forces should explain the main objectives of land warfare (destruction of the enemy's army vs. capturing the enemy's territory). Likewise, service doctrine for navy and air forces should describe in broad terms the main objectives of naval and air warfare, respectively. Traditionally, the principal objectives of naval warfare are to obtain and to maintain sea control (general or local; permanent or temporary; absolute or disputed), sea supremacy, sea denial, choke point control/denial, and basing/deployment areas control. Likewise, the principal objectives in air warfare are normally to obtain and to maintain air superiority or supremacy and their opposites (counter-air superiority/supremacy). These objectives should be discussed in generic terms and not be tied to a specific opponent or situation.

### Methods of Combat Force Employment

The scale of the military objective to be accomplished determines the force size and mix required. In addition, the objective determines whether a tactical action, major operation, or campaign should be selected as the method of combat force employment. Service doctrine should focus on planning, preparing, executing, and sustaining major operations conducted predominantly either by a single service or by the forces of several services. In generic terms, a *major operation* can be described as a series of related battles, engagements, strikes, and other tactical actions, conducted by diverse combat arms of a single or several services, in terms of time and place, to accomplish an operational objective in a given theater of operations. Major operations are normally planned and are an integral part of a maritime or land campaign.

Properly written joint doctrine should focus on the planning, preparation, and execution of land and maritime campaigns. In today's terms, a *campaign* in a high-intensity conventional war consists of a series of major air/land (or land), naval, and air operations, sequenced and synchronized in time and place and aimed to accomplish a military strategic or theater-strategic objective. In modern times, campaigns have been inherently joint or combined in nature. Hence, a single service cannot plan and conduct a campaign. This, of course, does not preclude the U.S. Air Force or Navy playing in some situations a principal or even decisive role in a land or maritime campaign, respectively.

Service and joint doctrine also needs to describe concisely the type and main characteristics of operations conducted by the respective service or joint forces in operations other than high-intensity conflict—specifically, counterterrorism, insurgency, counterinsurgency, counter-drug, peace operations (peacekeeping/peace enforcement), and noncombatant evacuation operations (NEO).

### The Theater and Its Structure

The size and mix of forces determine the size of the theater in all three dimensions in which these forces will deploy, fight, be sustained, and redeploy. Hence, both service and joint doctrine should include brief descriptions of the type of theaters (land, maritime, littoral) and their subdivisions (theater of operations, area of operations, and combat zones or sectors). The focus should be on describing a theater of operations and its structure. That is where the operational level of war is conducted.

## Levels of Command and Levels of War

Another important input to service or joint doctrine is the theoretical foundation of the concept of levels of war. Hence, a brief description of what constitutes the strategic, operational, and tactical levels of war and the respective levels of command should be concisely and clearly stated. Distinctions among the levels of war are greatest in a high-intensity conflict.[5] It is there that the scope and complexity of conflict require the full utilization of all military and nonmilitary sources of power. However, this is not the case in a low-intensity conflict, where political, diplomatic, social, economic, psychological, and other factors predominate. Hence, the operational level of war in such a conflict is smaller in scope and complexity than the strategic and tactical levels.

The levels of war are determined primarily on the basis of the military objective and the methods of combat force employment selected to accomplish the objective. The levels of war should be, but are not necessarily, linked to the respective levels of command. The latter are established to accomplish the respective military objectives through the conduct of tactical actions, major operations, and campaigns. An operational commander has sufficient authority to plan and conduct major operations and campaigns, but in fact he does not always do so. It is often believed that operational art is identical to the operational level of war. However, this is not the case. While operational art is a component of military art and hence encompasses both theory and practice, the operational level of war is only concerned with the practical application of operational art. Also, operational art is not tied to any specific physical environment but is applied across several levels of war, from operational-tactical to national-strategic.

## Operational Leadership

The very heart of warfare at any level is the human element. Hence, no service or joint doctrine should fail to address in some detail this critically important aspect of war. The role of technology should be highlighted too, but it should never be given a dominant role. Sound service or joint doctrine should present the consensus on the key requirements for successful operational leadership, such as the commander's personality traits and professional education. Prominent place should be given to one of the most important requirements for successful operational leadership, "operational thinking," or the ability to think operationally versus tactically. The commander's responsibility, in peacetime and in combat, and the type and importance of decisions should be concisely explained. The tenets of successful leadership—such as the dominant role of policy and strategy, the utmost importance of firm adherence to the objective, matching the ends and means, obtaining and maintaining freedom of action, and exercising initiative—should be concisely and clearly presented. Finally, sound doctrine should highlight the critical role the soldiers, airmen, and sailors have in the successful performance of any military organization.

### Operational C2

Both service and joint doctrine need to provide an overview of the national command structure. The focus of joint doctrine should be on the theater level (theater-strategic or theater of operations), while service doctrine should center on the theater and numbered force commanders. The type and nature of command relationships (operational control, tactical control, support, etc.) and command subordination (assigned, allocated, attached forces, etc.) should be briefly explained. Good service or joint doctrine should present the tenets of sound organization. They should not prescribe what the command organization options should be.

All too often, command and control is understood as essentially dealing with command organization. While these two are related and closely intertwined, they are not the same thing. C2 is primarily a process that uses command organization in accomplishing assigned military objectives. The need for an effective leadership and the critical role of subordinate commanders and the rank and file should be highlighted. Special attention should be given to the importance of personal relationships between commanders and their subordinates. The advantages and disadvantages of centralized and decentralized C2 should also be explained. While the increasing importance in C2 of new information technologies should be emphasized, this should be balanced by the timeless importance of the human element. In sound C2, technology is always an aid, not the principal element. Technology, no matter how advanced, can never replace the human element in warfare.

### Operational Decision Making and Planning

One of the key inputs of operational art to service and joint doctrine is operational decision making and planning. Well-written doctrine should briefly describe the strategic context for planning and then present the responsibilities and authority of the operational commanders in planning major operations and campaigns. Joint operational planning obviously provides the framework for the subordinate functional or service component commanders' planning. The role and importance of the commander's long-range estimate of a situation in making sound operational decisions should be explained. Distinctions between planning campaigns and major operations planned by the respective functional and service component commanders should also be described in some detail.

Political, diplomatic, economic, informational, legal, and other considerations in planning should be briefly described as well. Special emphasis should be given to the explanation of the combat and post-combat phases. Both phases comprise a seamless whole; there is no clear boundary between them. Strategic objectives are accomplished in the combat phase. However, they must be consolidated and exploited in the post-conflict phase; otherwise, the combat phase would be fought in vain. Service and joint doctrine should describe the planning process for both the combat and post-combat phases. The critical role of interagency coordination in operational planning should be emphasized. Good doctrine should also briefly explain the products of the planning process—that is, operations plans (OPLANs), operation orders (OPORDs), and fragmentary orders (FRAGOs). Planning is incomplete if it does not include a series of supporting plans, specifically plans for deployment, logistics, deception, operational fires, civil-military affairs, and others.

The current emphasis on effects-based operations (EBOs) confuses the issue by effectively diminishing or totally eliminating the objective as the basis for operational planning. Yet the entire theory and practice of operational art in fact revolves around the operational or strategic objectives to be accomplished. The EBO proponents contend that in designing a campaign, a combatant commander or component commanders provide objectives that describe desired effects. Once these effects are defined, the planners devise a framework consisting of the elements that each effect comprises. After quantifiable measures have been applied to each effect to be achieved, tasks are assigned in various ways.[6] In other words, instead of using well-proven and effective methods determining objectives first, and then tasks and targets, the EBO proponents start with objectives, then effects, then effects' elements (in fact, targets), and then tasks. If effects-based planning became a part of operational planning, it would be extremely difficult for the commander to conduct an estimate of the situation. The content of the mission statement and the commander's intent would need to be drastically changed as well. Moreover, the linkage between the objective and the corresponding COG would essentially be broken. Yet all sound plans for a campaign or major operation are directed at the destruction, annihilation, or neutralization of the enemy's COG and the protection of one's own.

### Operational Design

A well-written doctrine should have a section on operational design. The focus of joint doctrine should be on campaign design, while service doctrine should center on the design for the respective major operations. Operational design is not a plan but a collection of related elements of a campaign or major operation, which the commanders and their staffs should consider during the planning process. Not all elements are equally important, nor need they all be considered during the planning process. Sound doctrine should present in a concise form the key elements of operational design—specifically, the desired (strategic) end state (issued by the strategic leadership); strategic or operational objectives; the balancing of the factors of space, time, and force against the strategic/operational objective; the operational/strategic axis; and the identification of critical factors and COGs. The heart of operational design is the operational idea or scheme. Good service or joint doctrine should briefly describe the main elements of the operational scheme for a major operation or campaign. Specifically, methods of defeating the enemy's strategic or operational COG, operational maneuver and fires, deception, sequencing and synchronization, branches and sequels, phasing, and reserve should be explained in some detail.

### Operational Functions

Another important input of operational art to service and joint doctrine deals with certain "functions," called theater-wide or operational functions. The principal operational functions are operational command structure, operational intelligence, operational command and control warfare (C2W), operational fires, operational logistics, and operational protection. Only the theater commander has the necessary authority and responsibility to establish, maintain, and employ these functions in support of a campaign or major operations. They also need to be properly sequenced and synchronized.

## Operational Concepts

A sound service or joint doctrine should revolve around specific operational concepts. These, in turn, are subordinate to what in the U.S. military is called an *operating concept*—an overarching structure referring in the broadest terms to how one's military forces are to operate. An operating concept is a generic term because it does not relate to any particular level of war. In fact, some operating concepts can be applied at several levels of war.[7] Depending on its purpose and scale, an operating concept can be tactical, strategic, or operational. It can be either offensive or defensive in its basic purpose. A *tactical concept* is intended for fighting battles, engagements, attacks, and other tactical actions in a given combat zone or sector. A *strategic concept* deals with the use of both military and nonmilitary sources of power to accomplish national-strategic objectives.

In the proper understanding of the term, an operational *concept* is intended for the theater-wide employment of one's combat forces. Properly understood, an operational concept should be narrowed down to deal only with the application of military art at the operational level of war. An operational concept is not related to any specific task and situation. It is intended to provide a blueprint of how one's forces should be most effectively applied to accomplish the assigned operational or, in some cases, strategic objectives. It should effectively link diverse tactical actions by individual combat arms/branches to accomplish operational or strategic objectives. The extent to which an operational concept matches its actual execution is closely correlated to strategic success.[8]

## Requirements

An operational concept should be part of a larger and broader strategic concept. It should be based on a sound evaluation of future adversaries. Not only military but also nonmilitary aspects of the situation should be taken fully into account. The operational concept should be based on certain strategic assumptions about future threats to national security and the current and projected state of technological developments. Yet good theory is the key to developing viable operational concepts.[9]

## Elements

A sound operational concept should consist of a number of functional concepts that collectively ensure that the planned operational or strategic objectives are accomplished quickly and with the fewest losses for one's forces. Traditionally, the principal functional concepts include command and control, maneuver, fires, sequencing and synchronization, sustainment, protection, and regeneration of combat power. The key for success is sequencing and synchronization of the combat arms of one or more services. Examples of functional concepts include sea-based logistical support and sustainment, maritime prepositioning, close air support, and naval surface gunfire support. Each functional concept, in turn, includes a number of enabling concepts, describing the tactics, techniques, and procedures (TTPs) of how a particular functional concept has to be carried out.

## *Tenets*

Doctrine should explicitly or implicitly describe the tenets of how a particular operational concept should be executed. This is where the human factor plays a critical role. In generic terms, the main tenets of operational concepts are initiative, depth, agility, and synchronization. Initiative is the ability to set or change the terms of battle through action. It requires commanders to maintain an offensive spirit. Depth is measured in time, distance, and resources. Agility is the ability to think and act faster than the enemy. It involves mental, C2, and organizational abilities to adjust rapidly and to use the situation, terrain, and weather to defeat the enemy. The plan must be simple and flexible so the commander can react when an opportunity presents itself.

## *Type*

Operational concepts can be historical, current, or future. As the term implies, a historical concept is one that was applied in the past. It might have been articulated as such, or it might need to be deduced. For obvious reasons, service or joint doctrine should be based on current and/or future operational concepts. A current operational concept is applied by using today's organization and TTPs. It is usually codified in existing doctrinal documents. It may encompass some emergent elements based on current practices and lessons learned. A current operational concept, when part of service or joint doctrine, should provide the basis for organization, planning and execution, education, and training.[10]

A future operational concept envisages the employment of one's forces at some time in the future. It should not be part of the service or joint doctrine unless it is validated through rigorous tests and vigorous debate and experimentation. Any future concept should start as an untested hypothesis; it should be tested to the point at which it has been validated with reasonable confidence, at which point it becomes part of service or joint doctrine. Many future concepts cannot be fully tested in a peacetime environment. By their very nature, they cannot be deduced from past practice or observed from current practice.[11]

## Features

An operational concept should be simple and flexible. Simplicity is desirable because otherwise the concept would be difficult to apply in practice and would greatly complicate force training. A lack of flexibility would mean losing the ability to take advantage of the enemy's unanticipated weaknesses or a sudden change in the operational situation. An operational concept should significantly enhance the speed of one's decisions and actions and thereby greatly limit the enemy's options. A sound operational concept should not only cause the enemy's physical dislocation but also have profound negative psychological effects on the enemy's commander and forces. An operational concept must be articulated clearly and succinctly to be properly understood, and should then be debated and tested to influence the development process.[12]

The highly successful and novel German air-land battle concept (popularly called the *Blitzkrieg,* or "lightning war") of the late 1930s was relatively simple and highly flexible. It envisioned the power of offensive action to follow the principle of running water: The enemy's resistance would be avoided whenever possible. The key was in locating the gaps in the enemy's defenses; afterward, the weight of effort would be in that area. The attacking force would then immediately exploit breakthroughs to seize and destroy important objectives. Speed, mobility, surprise, and utilization of windows of opportunity were central elements in the concept. The aim was local superiority through forward concentration. The second key element was the selection of the weight of effort (*Schwerpunkt*) and concentration of one's forces in that area (*Schwerpunktbildung*). Air and ground reconnaissance would be conducted to locate the weaknesses in the enemy's defenses. Then the enemy's flanks and rear would be attacked.

The Soviet operational maneuver group (OMG) concept developed in the early 1980s was similar in some ways to the German blitzkrieg. It aimed, by early introduction of OMGs into NATO's operational depth, to prevent establishment of an effective defense and to preempt NATO's forces from using tactical nuclear weapons. In contrast to Soviet mobile groups used in World War II, the primary aim of OMGs was to cause the physical or psychological dislocation of the enemy and thereby create the preconditions for his ultimate destruction. Encirclement was envisaged only in the case of some favorable circumstances.[13]

## Future Concepts

A properly written service doctrine should take projected capabilities fully into account as part of new and developing operational concepts. By definition, future operational concepts are untested and unproven. Hence, a given concept should be subjected to rigorous testing, experimentation, and debate, which should lead to either complete or partial validation or invalidation of the concept. Through the testing process, a future concept evolves from a disinterested hypothesis to a more assertive conclusion. Only after being tested can a future concept serve as a basis for force planning.

A sound operational concept should be based on a certain vision of the future war. At the same time, it should also reflect a historical view of war. Sound concepts are rarely derived from abstract theoretical premises. A full understanding of warfare and its future is based on the practical lessons of history. An operational concept that ignores those lessons would lose its credibility. The worst case is a concept that intentionally falsifies and misuses history in order to support preconceived theories. One should resist the temptation to present a future concept as unique because the current age is unique too. Experience demonstrates that this has rarely been the case.

Nor should the future concept be described as "revolutionary" just to make a case for its adoption. Revolutionary breakthroughs are rare—almost all technological changes have in fact been evolutionary, and they are actually the norm rather than an exception.[14] However, it is possible to have new concepts based on a proper integration of existing technologies and the human factor, as the German blitzkrieg concept illustrates. A successful future concept has invariably demanded significant changes in military organization, education, and training. In some cases, considerable changes in military culture have been required as well. Vast exaggerations of the value and importance of some future operational concept invariably damage the credibility of its proponents. They can also significantly stifle debate on the concept's merits and thereby greatly undermine its potential value.

A sound operational concept must be consistent with the nature and theory of war. The Clausewitzian teachings on the nature of war are still valid despite the passage of time. At the same time, some attributes of war are subject to change. Hence, a future operational concept should deduce these attributes without violating those that are essentially timeless.[15]

A future operational concept that claims to offer a fundamental change in the conduct of war and claims that "old" rules of warfare do not apply anymore must convincingly show that this is the case. Assertions without proofs will simply not do. The future operational concept also must strike a proper balance between the science and art parts of operational warfare; neither science nor art should be ignored. Unfortunately, proponents of new technologies tend to overemphasize the science part and neglect or simply ignore the art part. Experience convincingly shows that this wrong. War is and will always be a combination of science and art. No new technologies can change that.[16]

Despite its leading proponents' claims to the contrary, NCW contains only some elements of the future operational concept. It can also be contended that NCW does not have all the main attributes of a sound tactical concept. C2 is essentially equated to linking vast arrays of sensors and shooters. However, any effective C2 must integrate both humans and technical systems. By relying exclusively on information technology, the great danger is that the entire C2 process will simply collapse, either because of information overload or because of enemy actions aimed to disrupt, neutralize, or even destroy a selected part of the network. A sound tactical or operational concept must also provide adequate protection of one's forces and their logistical support and sustainment. This is one of the grave weaknesses of the NCW concept as practiced today. NCW can be elevated to the operational level only if effective actions are taken to remedy the existing gaps and weaknesses in its concept. The next step is to link the major tactical forces of the numbered and theater armies, fleets, and air forces for conducting major operations and campaigns. In other words, the new concept must embed key elements of operational art; otherwise, it will never reach its full potential.

A future operational concept must be firmly embedded in a proper technological context. It should not pursue new technologies for their own sake but should describe how these technologies should be properly integrated with other aspects of warfare and thereby find new methods in the conduct of war. A danger is to search for technologies that will be available beyond the time frame of the new service or joint doctrine. At the same time, a future concept should not be based on existing technologies. Hence, a proper balance must be found between what is available and what is desirable. The human element should always be put first in developing any future concept. At the same time, political, societal, cultural, and economic factors that affect war at the operational level must be taken fully into account. Any future operational concept should be based on the national way of war. The way of war is based on national culture and popular values and national experience in the conduct of war. Hence, any future operational concept must be in consonance with that larger framework; otherwise, it is unlikely to be acceptable to a service/joint community and the society at large.[17]

Not all future concepts become the heart of a service or joint doctrine; many concepts may not survive the testing and evaluation process. Yet the invalidation of a future concept should not be considered a failure of the concept development process.[18] In many cases, those elements that were proven valid through testing and experiments can become part of a revised or entirely new operational concept. Current and future concepts are not fixed but evolve over time in response to changes in the situation and one's capabilities.

In some cases, service doctrine might be promulgated even though the future operational concept has not been completely validated, or required capabilities will not be available for some time in the future. This was the case with the U.S. concept of amphibious landing operations. The U.S. Marines developed a highly successful operational concept for conducting major amphibious landings. The document, *Tentative Manual for Landing Operations,* was issued in January 1934. After a series of fleet landing force exercises, the U.S. Navy officially adopted it as *Landing Operations, United States Navy* (FTP 167). After some changes were incorporated, its title was changed to *Manual for Naval Overseas Operations* in August of that same year.

The development of the landing operations manual had started in 1931. Initial studies dealt primarily with cargo handling, signaling, and boat capacities. The navy's then-existing landing force manual provided detailed instructions for drills and formations, but not for ship-to-shore movements. Perhaps the most important source used in writing the landing manual was a pamphlet written by the U.S. Joint Board and published in 1933. This document included planning procedures, naval gunfire support, and combat landing for major amphibious operations.[19] However, it was the students of the U.S. Marine Corps schools who had done most of the preliminary work after their classes were suspended in November 1933. The students and their instructors were directed to participate in an all-out effort to write a new landing manual. The *Landing Operations Manual* was so successful that the U.S. Army adopted it in 1940. Amphibious landing operations conducted by both the Marine Corps and the Army during World War II were based on that manual.

Likewise, the U.S. Army's AirLand Battle concept did not exist in 1982 when the FM 100-5 *Operations* was published. Yet FM 100-5 proved to be a sound document that accurately predicted capabilities several years into the future. The concept envisioned aggressive action to secure and retain the initiative. The enemy force would be defeated by being thrown off balance with powerful initial blows carried out from unexpected directions. This would be followed up by actions aimed to prevent the enemy's recovery. The best results would be achieved by carrying out initial strikes against critical units and areas whose loss would degrade the enemy's coherence rather than by strikes against the enemy's first-echelon forces.[20]

In the mid- and late-1930s, despite the lack of a sufficient number of submarines and long-range and reliable communications for their effective command and control, the Germans were highly successful in developing a new and original tactical concept for employing their U-boats against merchant shipping. The idea of what the Germans called "mass employment" of U-boats (later popularly called "wolf pack" tactics in the West) conceived submarine screens vectored to enemy convoys several thousand miles away via shortwave radio by naval headquarters ashore. This idea originated in 1917 and 1918. For various reasons, the idea of operating U-boats in groups remained purely theoretical. It was not until the "strategic games" that the German Naval High Command held in 1934 that the concept received some attention.[21] After Karl Dönitz took command of the Weddigen U-boat Flotilla in 1935, the idea of employing submarine groups slowly became a reality.

The concept was tested for the first time during the large-scale *Wehrmacht* exercises held in the fall of 1937. During that exercise, Commodore Dönitz controlled his boats deployed in the Baltic via shortwave radio from the submarine tender at the naval base in Kiel. Based on the exercises conducted in 1937, Dönitz also requested that a command ship equipped with the latest communications means be built for command and control of U-boats in case of war.[22] After further large-scale exercises held in the North Sea in May 1939, U-boats conducted an exercise in group tactics off Cape Finisterre and in the Bay of Biscay. In July 1939, Dönitz (by then promoted to rear admiral) conducted a similar exercise in the Baltic. All these exercises proved to him that his concept of using U-boats in groups was well founded. Nevertheless, the German Naval High Command continued to believe that the U-boats would be employed in the next war individually, not in groups.[23] Dönitz's concept depended on a sufficient number of U-boats deployed in the operational area at any given time. Hence, he used the winter exercise of 1938–39 to insist that a successful campaign against British maritime trade would require a force of at least 300 U-boats—mainly 517- and 750-ton type boats.[24] Because of the small number of U-boats in service at the beginning of war, Dönitz was unable to apply his concept in practice until 1941.

In the 1990s the U.S. Marines developed Operational Maneuver from the Sea (OMFTS), which in many respects resembles an operational concept. However, the capabilities required to make the concept reality—specifically, the capability to insert forces and seize an operational objective far inland and to ensure logistical sustainment—are still lacking. The OMFTS envisages amphibious landing operations to accomplish operational objectives deep on the enemy's coast. Its tenets include a focus on the operational objective ashore, use of the sea as maneuvering space, high tempo and momentum, pitting one's strength against the enemy's weaknesses, integration of organic and joint forces and assets, and use of deception.[25] The assault would be launched from over the horizon, twenty-five to fifty miles at sea. Instead of establishing the first objective on the beachhead (plus a logistic base) and securing it against the enemy's counterattack, the first objective could be to attack enemy positions up to two hundred miles inland.

The critical elements of OMFTS are sea-based logistics for providing operational and tactical sustainment. The ships would be logistically supported as much as possible from ships at sea. Also, instead of the maritime prepositioning forces (MPFs) depending on secure ports and airfields in the immediate vicinity of the objective, the troops would be integrated with their equipment at sea and then moved directly from the prepositioning ships to their areas of operation ashore. Logistics would be based on intermediate-level maintenance capability, while depot-level capabilities would reside in the continental United States.[26]

Normally, future operational concepts should not become an integral part of service or joint doctrine unless they are thoroughly tested. A serious problem can occur if the theoretical foundations of the concept are far ahead of the actual capabilities. This was the case with Soviet operational concepts in the early and mid-1930s. The Soviet theoreticians were probably the best operational thinkers of the day. However, the Soviet Union was then still in the midst of an industrialization drive; it was largely backward. Hence, the country's economic capabilities were inadequate to support the conversion of operational theories into the desired performance in the field. The Soviets adopted the "deep battle" (*dubokoy boy*) concept in the Red Army's field service regulations in 1935. This concept imagined forces no larger than corps attacking the enemy simultaneously over the entire depth of his fielded forces, first to isolate him and then to encircle and ultimately destroy him. This concept was developed to solve the problem of how to restore maneuver to a battlefield that had become static.[27]

Taking the ideas of Marshal Mikhail Tukhachevskiy and Gen. V. K. Triandifilov, the Soviets further developed the concept of deep battle by adopting the concept of deep operations (*glubokaya operatsiya*). This concept, adopted in the Red Army's field service regulation in 1936, envisioned a penetration and full envelopment maneuver into the enemy's operational depth to destroy the major part of the enemy forces.[28] The concept was intended for major operations conducted by the armies and fronts (army groups). In contrast to deep battle, it also visualized the employment of ground, air, and airborne forces to launch simultaneous blows throughout the enemy's entire operational depth.[29]

### Diverse Concepts

Based on an existing or future threat, the enemy's service/joint doctrine, and the nature of the physical environment in which one's forces are to be employed, several operational concepts should be developed and embedded into service or joint doctrine. A sound service doctrine should be based on a number of operational concepts developed for each fundamental warfare area. In contrast, joint doctrine should revolve around several operational concepts, envisaging the employment of multiservice forces in both offense and defense. The employment of service forces in conducting major operations or joint forces in a campaign should be guided by the tenets of the respective operational concept.

Experience shows that it is usually a bad thing to develop and adopt a single operational concept. This invariably limits the options available to the operational commander; it also narrows one's perspective and doesn't ensure that one's forces are trained across the spectrum of conflict. The U.S. AirLand Battle concept was the keystone of FM 100-5, *Operations,* published in 1982. It was intended to fight a high-intensity conflict in Europe against numerically much superior Soviet-led Warsaw Pact forces. However, that concept is only partially applicable or completely inapplicable in other types of physical environments and against weaker enemies. In contrast, in World War II, the Allies applied different operational concepts in various theaters. The U.S. Army applied one concept in the European theater. Gen. Douglas MacArthur developed a joint operational concept in his advances during the New Guinea and Philippines campaigns. In simple terms, it envisaged the establishment of air and naval bases by land forces while friendly naval and air forces provided support and isolated the next landing objective.[30]

## Conclusion

The need for sound service and joint service doctrines should be obvious. The service and joint doctrines should be written from the perspective of operational art. Doctrine should be clearly focused on the employment of numbered and theater forces at the operational level of war. Service doctrine should focus on the planning, preparation, and execution of major operations, while joint doctrine should center on campaigns. New operational concepts, while embracing new information technologies, must at their very center be based on the human element—that is, leadership rather than management—and war fighting.

Success at the operational level of war cannot be achieved without operational concepts for the employment of one's own combat forces. A sound service or joint doctrine should contain several operational concepts covering the key areas in which one's forces would be employed. A sound doctrine for the theaterwide employment of one's forces serves as a bridge between strategy and tactics. It should be used as a baseline for thinking on how to fight. A well-written and properly focused service/joint doctrine considerably enhances the professionalism and skills of the commanders and their staffs, at all levels of command in one's service. It also considerably enhances "operational thinking" in a service. At the same time, tactical commanders need to know and understand operational art; otherwise, they will have great difficulty in understanding the higher operational commander's intent. Service/joint doctrine also enhances proper understanding of common operational terms' true meanings, thereby considerably facilitating communications within services and within a joint or combined force.

## Notes

1. George T. Donovan, *The Structure of Tactical Revolution in The U.S. Army From 1968 to 1986* (Fort Leavenworth, Kan.: School of Advanced Military Studies, U.S. Army Command and Staff College, 17 December 1998), pp. 12–14.

2. Michael C. Vitale, "Jointness by Design, Not Accident," *Joint Force Quarterly* Autumn 1995, p. 27.

3. Vinod Anand, "Evolution of a Joint Doctrine for Indian Armed Forces," *Strategic Analysis,* July 2000, No. 4, pp. 733–50.

4. Scott A. Marcy, "Operational Art: Getting Started," *Military Review* 9 (September 1990), p. 107.

5. Gary P. Petrole, *Understanding the Operational Effect* (Fort Leavenworth, Kan.: School of Advanced Military Studies, U.S. Army Command and General Staff College, 8 May 1991), p. 7.

6. David B. Lee and Douglas Kupersmith, *Effects Based Operations: Objectives to Metrics Methodology—An Example* (Vienna, Va.: Military Operations Research Society, Analyzing Effects-Based Operations Workshop, January 2002), pp. 8, 10.

7. John F. Schmitt, *A Practical Guide for Developing and Writing Military Concepts* (McLean, Va.: Hicks & Associates Inc., DART Working Paper # 02-4, December 2002), pp. 7–8.

8. David A. Fastabend, "That Elusive Operational Concept," *Army Magazine,* June 2001, p. 40.

9. This term is very often used loosely in the U.S. Navy; apparently the true meaning of "operational" is not properly understood. The basis for an operational concept is a tactical concept. However, their purposes differ, because an operational concept is applied to win *major operations* and *campaigns,* not battles or engagements.

10. Schmitt, *Developing and Writing Military Concepts,* pp. 3–4.

11. Ibid., p. 4.

12. Ibid.

13. Darrell E. Crawford, *Deep Operations In AirLand Battle Doctrine: The Employment of U.S. Ground Forces In Deep Operational Maneuver* (Fort Leavenworth, Kan.: School of Advanced Military Studies, U.S. Army Command and General Staff College, 16 May 1989), p. 7.

14. Schmitt, *Developing and Writing Military Concepts,* pp. 12–13.

15. Ibid., p. 13.

16. Ibid., p. 14.

17. Ibid.

18. Ibid., p. 4.

19. Barry P. Messina, *Development of U.S. Joint and Amphibious Doctrine, 1898–1945* (Alexandria, Va.: Center for Naval Analysis, September 1994), pp. 30–31.

20. U.S. Department of the Army, FM 100-5 *Operations* (Washington, D.C.: August 1982), p. 2–21.

21. Von der Frost, "Der Einsatz von U-Booten auf feindliche Seestreitkraefte, Verbaende oder Geleitzuege, Strategische Kriegsspiel 1934, RM 20/957, Bundesarchiv-Militarische Archiv (BA-MA) Freiburg.

22. Karl Doenitz, *Memoirs. Ten Years and Twenty Days.* Translated by R. H. Stevens (Annapolis, Md.: Naval Institute Press, 1990), p. 32.

23. Ibid., pp. 21, 32.

24. Holger H. Herwig, "Innovation Ignored: The submarine problem. Germany, Britain and the United States, 1919–1939," in Williamson Murray and Allan R. Millet, editors, *Military Innovation in the Interwar Years* (Cambridge, UK: Cambridge University Press, 1996), p. 239.

25. Headquarters, U.S. Marine Corps, Department of the Navy, MCDP 3: *Expeditionary Operations* (Washington, D.C.: U.S. Government Printing Office, 1998), p. 89.

26. Michael P. Mahaney, *Operational Durability. The Marines and Operational Maneuver from the Sea* (Fort Leavenworth, Kan.: School of Advanced Military Studies, U.S. Army Command and General Staff College, 1 May 2001), p. 29.

27. Richard W. Harrison, *The Russian Way of War: Operational Art, 1904–1940* (Lawrence, Kan.: University Press of Kansas, 2001), pp. 188–94.

28. Wayne A. Parks, *Operational-Level Deep Operations: A Key Component of Operational Art and Future Warfare* (Fort Leavenworth, Kan.: School of Advanced Military Studies, U.S. Army Command and General Staff College, 21 May 1998), pp. 24–25.

29. Harrison, *The Russian Way of War,* pp. 194–95.

30. Fastabend, "That Elusive Operational Concept," p. 39.

ELEVEN

# Rethinking Operational Art

ROBERT R. TOMES

## Introduction

It is quite possible that the motivation for the "Rethinking the Principles of War" project that led to this volume involved a degree of cognitive dissonance concerning the application of operational art's largely Cold War domain expertise to the global war on terrorism and to nation building in Afghanistan and Iraq. Those participating in its creation, for example, were part of the generation that matured the current approach to operational art, which evolved from strategic and operational challenges associated with the Cold War European theater into a comprehensive yet flexible approach to campaigning.[1]

Underlying this "rethinking" project was, arguably, a healthy introspection about the application of legacy doctrine to emerging challenges and concern that shifts in the role military force was playing in national security strategy were not paralleled by shifts in operational art and campaign planning. Discussions among defense intellectuals and theorists during the yearlong "Rethinking the Principles of War" seminar series, furthermore, surfaced questions about the state of doctrine, applicability of operational art training to nation building campaigns, and the overall state of leadership training. Regardless of whether dissonance was a factor, important questions remain about the state of training for command at the operational level of war.

As commanders surged their fleets, mobilized divisions, and launched wings to fight Operation Enduring Freedom in Afghanistan and Operation

---

All views and opinions are the authors alone and do not represent the views of any government office, department, agency, or other entity.

Iraqi Freedom, were they prepared as much as possible by the formal aspects of learning operational art? Are long-standing tactics, techniques, and procedures on which the domain knowledge of operational art has progressed applicable to post-9/11 operational challenges? Is military advice derived from existing tenets of operational art salient to civilian leaders? How does applied knowledge, meant to guide combat decision making, fare in a conflict in which ideas, public diplomacy, and nonmilitary factors are the primary resources?

This volume cannot resolve whether domain knowledge and expertise embodied by doctrine, theory, and training to lead at the operational level of war are appropriate for a global war on terrorism. It can, however, point out cracks in the canon. This chapter contributes to the rethinking effort by proposing an ambitious approach to rethinking operational art.

It assumes that, although guided by doctrine and informed by deliberate planning processes, operational art has from inception been a creative activity. In this sense, operational art is the realm of what Clausewitz called military genius. The following sections review the evolution of operational art, discuss doctrine's role in diffusing domain knowledge about leading at the operational level of war, and revisits Clausewitz's concept of military genius. An initial argument is made to rethink how the concept of military genius provides a means to bridge 1970s' origins of operational art and the need to rethink how doctrine enables the training of command intuition in the twenty-first century.

## Background on Operational Art

American military theorists adapted Soviet thinking about the operational level of war in the 1970s, refining concepts first articulated in Aleksandr Svechin's 1923 military classic, *Strategiya*. An approach to the operational art of war crystallized as part of a larger post-Vietnam renewal in U.S. military effectiveness; this occurred during a time of fundamental change in military strategy. A strategic imperative to develop a conventional alternative to a nuclear defense of Europe was emerging. A renaissance in conventional military thought resulted, and operational art became the trade of a new guild of maneuverers.

Meanwhile, Vietnam veterans channeled experiences and frustration into multifaceted reforms that, in part, aimed to formalize a process through which defense plans and theater military operations would be derived directly from national strategy. They sought a system that clarified the relationship between strategic or political ends and military means.

The generation of military officers that ascended the ranks in the shadow of Vietnam also codified a set of guidelines or beliefs meant to bound the appropriate use of military forces. Military forces would be employed only when vital national interests were at risk. Overwhelming force would be used to accomplish clear mission objectives directly related to national-level strategic objectives. An exit strategy would be defined in the commander's intent for the operation. U.S. troops would be brought home quickly. These were, of course, the core tenets of the Powell Doctrine, named for former Secretary of State and Chairman of the Joint Chiefs of Staff Colin L. Powell.

After the Cold War, an unstated corollary to these guidelines was added: American troops would not lead reconstruction or nation building efforts. Doctrine reflected this until quite recently, assigning such tasks to international organizations and giving cursory treatment to what are now called security operations.

Beliefs about the use of military force, therefore, coevolved with and in some ways shaped the evolution of operational art. The post-Vietnam generation created operational art as a domain of expert military knowledge and leadership training. Arguably, their efforts to rebuild and revitalize the armed forces depended on their ability to create a professional discipline of operational-level leaders whose culture and mind-set strongly identified with the above beliefs about the use of military force. This is, perhaps, one reason why operational art is often discussed as the art of designing and leading theater campaigns in support of defined political objectives. In the quest to align strategic means with military ends, to prevent another Vietnam, the post-Vietnam generation associated expertise at the operational level of war with a sense of accountability and responsibility regarding the deployment of military forces into combat to achieve national strategy.

The post-Vietnam revolution in military effectiveness included a commitment to career development, a newfound urgency to improve troop training, and a desire to provide realistic preparation for combat at all levels. It also represented understanding that battlefield command training required development of character, intellect, intuition, tempered audacity, and vision. Professional military education and leadership training were refashioned, with additional time devoted to decision processes, preparation for planning and staff assignments, and command of joint campaigns.

Operational art is really about training, teaching, and exercising for command at the operational level of war. What are the principal targets of such training? The most important are a bundle of cognitive processes and faculties that underwrite the outward manifestations of leadership capacity and style, the traits most often discussed among military biographers. Technical proficiency in the operational arts manifests itself as being able to creatively apply knowledge at the operational level of war. In this sense, operational art is a school of design; it is also a discipline in which expertise is built over time as a unique set of skills and knowledge in the art of war planning and execution.

As a domain of leadership development, then, learning the operational art of war has since the early 1980s meant acquiring practical knowledge about designing, organizing, integrating, and implementing theater strategies, campaigns, major operations, and battles over an increasingly complex range of scenarios.

Strategic and theater challenges drove the evolution of operational art. Within the Cold War strategic context, planning challenges became more diffuse and operational plans more dynamic. Technology reduced the warning time to react to crises abroad. Transnational threats became more strategically relevant.

Gen. Don Starry, U.S. Army (Ret.), said of the planning environment as operational art was being integrated into U.S. defense planning, "It was an era characterized by the expanded threat in Europe, a growing threat of conflict in the Third World (especially the Middle East), increasing worldwide economic interdependence, [and] greater difficulty articulating political goals for the planners who designed military activities to achieve them."[2] Challenges during the late-1970s through the mid-1980s included increased Soviet military capabilities in Central Europe; Moscow's expansionist agenda in the Persian Gulf, Africa, and the Western Hemisphere; allied doubts about the viability of the U.S. nuclear deterrent; renewed urgency to develop theater intelligence indications and warning systems; and an imperative to field nonnuclear strategic strike capabilities, including precision attack systems that would be effective at night in all weather conditions.

Initially, the domain knowledge of operational art involved rediscovery and adaptation of traditional, maneuver-focused conventional war-fighting approaches with existing, dominant strategic narratives focused on nuclear war planning. After a period in which nuclear doctrine *was* the essence of military strategy, thinking about the principles of war expanded beyond the image of theater military operations as merely a serious step on an abstract escalation ladder.

The aim was reducing reliance on nuclear weapons as the sole means of deterring Soviet armor forces from a thrust into allied territory. Starry recalls that Soviet defense planners "embraced the notion that they could fight and win at the operational level of war with or without nuclear weapons. Their preferred solution: without."[3] Planners feared that successive echelons of Soviet armor would overrun the North Atlantic Treaty Organization's front lines before the political decision occurred to employ battlefield nuclear weapons.

AirLand Battle doctrine, the first doctrinal framework associated with American operational art, underwent successive revisions in the 1980s as U.S. conventional (general purpose) military spending increased. The founding of an operational level of war involved a deliberate effort to develop and systemically diffuse practical knowledge about how to defeat a specific enemy and a specific set of adversary capabilities. It aimed to achieve strategic objectives through the flexible, tailored application of military power. Consequently, Operational art was heavily focused on learning how to decisively coordinate and apply the principal capabilities and resources from which military power was derived.

By the end of the 1980s, writings on conventional deterrence flourished and much ink was spilled on the question of fighting a conventional, theater battle that did not escalate beyond the nuclear threshold. For the first time in decades, defense planning rediscovered a theory of victory wholly based on the orchestration of nonnuclear campaigns for traditional force-on-force combined arms warfare. Within the professional military education community, operational art evolved as a dynamic framework for organizing domain knowledge addressing the how, why, and what aspects of the relationship between strategic ends and theater means. Commanders assumed responsibility for combined arms task forces and joint mission execution. The paucity of attention paid to small wars and insurgencies in the evolution of training programs for operational art and campaign planning is notable, in retrospect.

U.S. thinking about operational art also matured as part of a larger pro-integration theme, including a push toward greater interoperability. Interoperability advances included the advent of digital information technology, the first steps toward integrated theater air, land, sea, and space operations, a global command and control network, and fledgling cross-service initiatives that matured into joint war-fighting doctrine. Among the most notable outcomes of the integration mind-set was the 1986 Gold-Waters Nicolas Act.

## Rethinking Basic Premises

Operational art subsequently evolved as a process to translate strategy into action through campaign planning and maneuver while assuring that all activities are integrated, at all levels, with resource allocation and wartime decision making guided by clearly understood political objectives. It became a specific way of addressing military planning and operational challenges by designing competitive interactions with opponents that overcome space and time constraints through skillful integration of technology, operational practices, and organizational innovations.

For decades, a premise of U.S. campaign theory has been that a single regional command would be the main effort in a major conflict or activity with other unified and regional commands supporting it. Practical application of this theory does not hold in the war on terrorism. The U.S. Special Operations Command has the lead in the military aspects of the global war on terrorism. The U.S. Strategic Command leads global intelligence, surveillance, and reconnaissance (ISR) activities, has the global strategic strike mission, and recently increased its authority over global bandwidth management for command, control, and communications. The U.S. Central Command, meanwhile, is the lead for the operations in Afghanistan and Iraq—currently the most visible (but not the only) local hot spots in the war on terrorism. National, civil, and other organizations lead crucial parts of the global counterterrorism campaign in areas such as clandestine operations, public diplomacy, and border control.

Of note to current students of American military thought and defense planning, although the primary military challenge underwriting an American operational art concerned the European theater, Persian Gulf security challenges have shaped its evolution from the very beginning. Long before the 1991 Gulf War or the 2003 liberation of Iraq, increased attention to Persian Gulf security focused defense planning on understanding the requirements for moving military forces into the region and sustaining combat power.

The Rapid Deployment Force (RDF) was established in late 1979. That year, Saddam Hussein seized power, and then National Security Adviser Zbigniew Brzezinski commented that the RDF would be used preemptively "in those parts of the world where our vital interests might be engaged and were there are no permanently stationed American forces."[4] In May 1980 European allies relented to contingency plans in which the United States would re-deploy forces from NATO to the Middle East during a crisis. This was the first NATO agreement to support what post–Cold War NATO planners termed out of area operations.

The RDF evolved into the Rapid Deployment Joint Task Force (RDJTF) in March 1980, the principal organization around which the U.S. Central Command was created in January 1983. The Central Command would, over the next two decades, figure prominently in the development of U.S. concepts and capabilities to deploy and coordinate significant combat power "rapidly" to its area of operations. Meanwhile, the concept of rapid deployment was transformed into what current military thinkers term rapid dominance or rapid decisive action and became a cornerstone of what observers labeled a new American way of war.

The term *rapid dominance* and its much ballyhooed adjunct "shock and awe" entered U.S. military thought with Harlan K. Ullman and James P. Wade's widely read December 1996 monograph, *Shock and Awe: Achieving Rapid Dominance.*[5] Although shock and awe was widely discussed during the time of the U.S. invasion of Iraq, rapid dominance is actually the work's more important concept.

Rapid dominance is a relatively straightforward approach: an "ability to move quickly before an adversary can react" with "the ability to affect and dominate an adversary's will both physically and psychologically."[6] On the surface, nothing in this approach is profound in the history of warfare. What is new is the level of precision, ability to attack thousands of targets simultaneously, flexibility demonstrated by maneuver forces, and unparalleled degree of cooperation achieved among air, ground, and naval combatants. In other words, it is the migration of a previously tactical-level concept to the operational and even strategic levels; the significant shift being on seizing opportunities for victory with minimal use of direct assaults and the abrogation of attrition-minded, linear attacks.

As rapid dominance entered the lexicon in the 1990s, military theorists and defense planners continued to grapple with the implications of changes in warfare. The depth and breadth of the battlefield continued to expand. Agile maneuver forces realized greater operating range and lethality with standoff precision firepower. Decision time lines at all levels of command collapsed. An appreciation for the complexity of the modern battlefield and for the enduring challenges of battlefield decision making in the digital information age underpinned the continued evolution of operational art.

An affinity for the very idea of rapid dominance took root across the U.S. defense strategy community. Defense planners and some theorists reacted to increasing uncertainty about the nature of emerging threats with a traditional organizational preference for offensive initiative. This reinforced the idea that, if a comprehensive level of surveillance could be linked to rapid decision making and long-range precision strike capabilities—part of the process of achieving information superiority in U.S. doctrine—opponents could be subjected to a level of battlefield dominance that made their motivations and strategies irrelevant. Any action the opponent might consider could be observed; any movement of adversary forces would be exposed and therefore vulnerable to long-range, rapid, and precise strikes by U.S. forces. In reality, many targets are difficult to identify, find, and attack because they are mobile, fleeting, or hidden in urban clutter. The search for solutions tends to focus on new surveillance and targeting technologies. As many have lamented, solutions did not address human intelligence shortcomings, pursue cultural insight, or acknowledge much-needed training in cross-social understanding.

The evolution of U.S. military thought during the 1990s and early 2000s included discussion about peace keeping, nation building, and "operations other than war" (OOTW) missions. Many did predict that regional adversaries would adopt asymmetric strategies designed around or to negate U.S. military strengths. Difficulties operating in urban areas were a focus, fueling the near fetish mentality for technical surveillance and reconnaissance capabilities.

The preponderance of discussion, nevertheless, imagined conflicts involving military forces and opponents with identifiable and targetable centers of gravity. Many assumed that the military would not become entrenched in prolonged nation building efforts involving an armed insurgency. The first post–Cold War revision of the U.S. Army field manual on low-level conflicts (renamed OOTW in the mid-1990s), stated that the "norm" would be for the United States to remain neutral when deployed in a country facing an insurgency. Nation building was not perceived by many as a core mission. Others believed nation building tasks and "irregular" warfare were lesser-included contingencies easily managed with a transformed, smaller, more agile force possessing information superiority.

At a 1982 international conference on terrorism and "low-level conflict," for example, Army Chief of Staff Gen. Edward C. Meyer called attention to a similar tension between planning for low-level conflict (including counterinsurgency) and planning for the dominant planning scenarios, nuclear, and large-scale conventional conflict. Defense planners, he observed, tend to make the "greatest intellectual and physical investment in preparations for the levels of conflict we have least often faced."[7] This seems fair in light of the fact that the mission around which operational art evolved, the defense of Europe, was never implemented.

Critics will balk at this, pointing out that the evolution of operational art was a resounding success. It underwrote deterrence credibility during the

Cold War. Meyer's comments are not a condemnation of the totality of evolution of Cold War operational art and defense plans. They were, after all, meant to offset the Soviet conventional forces buildup that shaped the 1980s' threat environment. By all accounts, including archival data and statements from Soviet defense officials, American capabilities at the operational level of war did contribute to deterrence.

Among the most important programs in the 1980s was the then-classified Assault Breaker concept demonstration, which aimed to "rip the heart out of" the Soviet army's "armored and mechanized strengths."[8] It included advanced reconnaissance systems, precision weapons, doctrine, information awareness systems, and integrated air-ground operations. "To the Soviets," Dr. Norman Friedman concluded, the integration of technology and doctrine in the planned operational application of Assault Breaker concepts represented "a disaster" because it pointed to the capability to blunt Soviet armor without resorting to nuclear weapons.[9]

Operational art, in this context, became fixated on decision processes and using situational awareness technology rather than continued refinement of the underlying professional knowledge base required to build expertise on emerging operational challenges. Information dominance, a central feature of the Assault Breaker concept, would be provided through sophisticated intelligence and surveillance systems; knowledge dominance would result.

Only later did this line of thinking emerge as a potentially systemic gap in operational art, one that emerges as a recurring theme in criticism of U.S. military thought and doctrine. Now, with national strategy shifting, it may be time to rethink some of operational art's basic premises. The global war on terrorism is perhaps the first global campaign in which the nature of the conflict and the strategic repercussions of failing, in this case terrorists with weapons of mass destruction, directly challenge the traditional hierarchy of strategy, operational art, and tactics.

This points us back to the central theme of this volume: rethinking the state of doctrine and military theory.

## On Principles, Elements, and Still More Principles

The purpose of doctrine and theory is to bring order to the process of building expertise in the operational arts by structuring professional military knowledge and training. It is best done with realistic training and exercises that populate long-term memory in a fashion commensurate with the types of challenges or command tasks one expects to face during an actual campaign or battle.

Among the founding rationales for an organizing construct for thinking about a level of war between strategy and tactics was recognition that decision-making processes and the cognitive faculties underwriting them were of increasing concern to professional military education. Leaders had to be sensitized to physiological and psychological elements affecting decision making. Certainly maverick theorist John Boyd thought so: experience, knowledge of military history, understanding of the enemy, sensitivity to culture, and recognition of how time and space influenced competitive interactions were at the core of his Observation-Orientation-Decision-Action (OODA) Loop and other writings on warfare. Special Forces recruitment and training has for decades emphasized both intelligence and ability to excel under extremely stressful conditions known to disrupt mental acuity. Aviators are also trained to deal with the effects of combat and stress on their performance.

Conventional wisdom has long held that disciplined and imaginative self-study aids the preparation of military minds for the rigors of battlefield command. This has been confirmed by studies with neural imaging techniques and investigations into expert decision processes. The ability to mentally simulate situations and ponder alternatives is a key enabler in developing mental models and associated heuristics, which is how people learn to successfully and quickly gather and use information in stressful situations. The role of military theory, including the principles of war, was to frame this process and to facilitate the construction of relevant mental models. Carefully crafted historical cases and self-study remain critical to the development of analytic and decision-making skills.

Operational art training uses doctrine and theory to help leaders assimilate the body of knowledge required to succeed in leading at the operational level of war. This includes providing a body of cases, experiences, and examples that help leaders construct their own mental models and heuristics. These include analogies, procedures, rules of thumb, tradecraft, and learning the subtleties of intuition.

The operational level of war is a recognized, though abstract, intellectual space that is closely tied to a body of practice that has a specialized discourse and is carried forward by shared expectations about military education and what it means to command troops in combat. The narrative through which these expectations are carried is dominated by a tightly coupled system of joint doctrine.

Among the most important components of this narrative are the current joint operations doctrine's fundamental "elements," meant to guide thinking about combat operations. They include synergy, simultaneity and depth, anticipation, balance, leverage, timing and tempo, operational reach and approach, forces and function, and arranging operations.

In the pantheon of intellectual concepts meant to facilitate learning, these elements are intended to be subordinate to or flow from the nine principles of war. They include mass, offensive, security, simplicity, maneuver, objective, unity of command, surprise, and economy of force.

The loosely constructed (though treated as immutable) principles of war are intended to frame the study of historical cases in a way that tethers strategists, commanders, and even newcomers to military science to the broad strokes of military history. The principles of war have been largely employed to convey the "know what" aspects of deliberate learning; since the 1980s, operational art has been focused on the "know-how" and the more practical expertise required to excel in diverse operational settings. The elements of joint doctrine were intended to help facilitate this latter process.

It is useful to think of operational art as a pedagogical bridge. It links the classic approach to teaching the theory of warfare through great texts to the practical arts and related domain knowledge of theater planning, campaign plan execution, and battle staff training. Consequently, it has become the primary intellectual space within which successive war-fighting doctrines manifest themselves as frameworks to inform behavior. In terms of applied knowledge that is truly applicable, however, the real-world problem space bound by the operational level of war has grown more diverse and dynamic without corresponding changes in training, doctrine, and theory. Knowledge resources may no longer meet the knowledge needs of battle staffs. Although there is much intellectual interest in complexity and chaos as ways to think about the evolving battle space, training in the operational arts is not drawing from the large and varied literature of complexity theory as it has been applied to information management, learning processes, and decision support systems.

The principles of war are generally meant to guide (not proscribe) learning and thinking about warfare as a whole; for example, facilitating comparative study across military history. The elements of joint planning and similar lists are often used to frame debate about important aspects of the business of operational planning. But the elements themselves are not normally referenced or discussed in intellectual discussions concerning how operational art is changing or needs to change in light of emerging strategic challenges and associated campaign objectives. Paradoxically, the above-listed elements are inherently more adaptable as concepts to facilitate learning than the more frequently discussed principles of war.

A resulting problem, therefore, is that operational art as domain knowledge for campaigning is struggling to catch up with and remain relevant to the types of campaigns currently being fought. This problem is amplified when theorists and training resources attempt to patch doctrinal shortcomings by merely restating or repackaging the principles of war as a framework to approach applications of operational art to specific campaign planning issues. For example, framing practical knowledge about humanitarian assistance planning or reconstruction management in terms of mass, unity of command, or simplicity burdens rather than lightens assimilation of new decision-relevant knowledge.

Of course, if military leaders do ponder whether the "hidden hand" of theory and doctrine provides a suitable structure for applying tacit knowledge, they probably do not consciously mull the principles of war. Instead, the concern is whether the tacit knowledge or expertise that commanders tap to convey commander's intent, develop campaign plans, and lead organizations in unfamiliar tasks are the wrong *type* of domain knowledge or simply cannot be formed into an optimal mental model. This is, perhaps, the concern that spurred the cognitive dissonance mentioned in the chapter introduction.

True expertise, in battle staffs, theater command, leading maneuver elements, or other fields takes ten or so years to develop; it is only developed through experience. That is, learning by doing, practicing, or actually struggling through situations. Expert judgment and intuitive leadership across different operational contexts involves the ability to recognize and work from patterns or the mental models mentioned above. These models manifest in associations that thread across short- and long-term memory; they provide an intricate web of possible memory associations and scenarios that use tacit knowledge to identify and recommend options for decisions or action. Research suggests that this type of expertise requires exposure to around fifty thousand cases, or "experiences." For chess players, for example, this takes around ten years of progressively more difficult games; for radiologists it is diagnosing or reading out more than forty thousand films. The problem is that such expertise is not easily transferred across games or areas of medicine without significant retraining.[10]

The aggregate effects of novel planning situations, implications of failure, increased time compression on decision cycles, and distribution of operations across greater distances all strain the decision processes required to tap tacit campaign knowledge to address unforeseen military situations. Confronted with conflicting, unfamiliar, and rapidly changing information about ongoing operations, it becomes more difficult for short-term or working memory to focus the executive brain's higher decision capacities on the relevant information. Bringing to bear the data stored in long-term memory involves cues that facilitate the reconstruction of patterns and the association of assimilated knowledge with aspects of current events.

This does not mean that training future leaders in the art of command at the operational level of war must reject the training and knowledge that has proven itself in successive conflicts over the last three decades. Indeed, American operational art does tend toward adaptation and flexibility—something that prevents overly rigid and linear approaches to decision making. American culture includes traits of innovation, adaptability, and entrepreneurship. The problem, however, is that campaigning in the twenty-first century may not afford commanders the relative luxury of time to learn. And the narrative of current operational art may be overly scripted toward the orchestration of theater force-on-force battles with uneven attention to civil affairs and cultural intelligence. To be sure, the cadre of operational art experts remains to narrowly cast, with scant attention given to training nonmilitary leaders on the art of the campaign. This is proving to be a deficit in the calculus of overall effectiveness at the operational level of war in the dominant type of campaign likely to define the early twenty-first century.

All of this is muddled, arguably, by the existence of a doctrinal alternative to the very idea of a set list of principles of war at the pinnacle of joint doctrine and operational art. In addition to the principles of war, there are principles of operations *other* than war. They include objective, unity of effort, security, restraint, perseverance, and legitimacy. These principles presumably apply to countering terrorism and to nation building, which in classic definitions are not considered war proper. Such distinctions rarely matter in the real world or to those in operational commands engaged in security and stability operations. The problem now is how to adapt training and education that account for the apparent mismatch between theory and reality.

The principles of operations other than war seem more applicable than the standard list of nine to the global war on terrorism and to current counterinsurgency missions in Iraq. They also, in hindsight, seem more applicable to the first phase of Operation Iraqi Freedom, which exposes a fundamental challenge to arguments that the current list of principles of war retains currency. Unity of effort is particularly noteworthy as a principle since it logically opposes the hierarchical standard implied by the more recognized unity of command.

Why the two lists of principles? It seems that doctrinally we have yet to accept the war on terrorism and the combat associated with counterinsurgency as central to the operational level of war. Future doctrine writers will do a great service to those learning the operational art of war by removing artificial boundaries in the development of domain knowledge about campaigns at the operational level of war. Sadly, it appears this is not occurring as quickly as the emerging generation of leaders might expect in an age dominated by visions of flexibility and adaptability.

One also must wonder if there are too few principles, elements, concepts, and other ideational constructs addressing the integration of military activities with other national efforts. In the post-9/11 environment, strategic adversaries are non-state actors that target civilians. Tactical success in some strategically important areas depends on federal and civil cooperation with the Defense Department and intelligence agencies. State and local authorities are on the front lines. Theater campaigns are international relations and diplomacy intensive as national strategy hinges on the health of political-military coalitions. International planning staffs are collocated with and increasingly integrated into U.S. combatant command joint staffs.

Recent developments beg an operational art training program geared toward the "other than war" principles listed above, along with joint operational elements that can bridge between traditional approaches to campaign planning and emerging ones. This is perhaps the crux of the issue posed in the introduction: making theory a more effective guide to action. Another issue concerns the addition of an operational level of war training and education program for members of the larger interagency national security community that are collaborating, sometimes leading, important missions in the global war on terrorism.

## Revisiting Clausewitz to Rethink Operational Art

The founding of American operational art coincided with a rediscovery of Clausewitz and other founding texts in the Western military science canon. Clausewitz sought to frame his theory of war in a framework of understanding about behavior in war. Doing so required addressing all factors affecting behavior, including learning (defined by later scholars as changes in behavior, demonstrative or cognitive). Among the least cited and perhaps least understood aspect of his *On War* is his discussion of military genius.

Among military theorists, the 1990s were a decade rife with Clausewitz bashing. I believe this was unfortunate. It prevented opportunities to rethink and apply important aspects of Clausewitz's theoretical framework. Specifically, it delayed a rethinking of how military history and the domain knowledge of operational art related to the cognitive faculties underwriting true expertise in campaigning. How? Instead of reaching for a synthesis of military domain knowledge and a progressive evolution of military science to accommodate emerging training and education needs, loose criticism of the classics of military theory and history suggested that reading and understanding them at all was a wasted effort. Perhaps this is also a reflection of diminishing affinity for reading in American culture.

Regardless of the root cause, one result is diminished interest in intellectual debate about the foundations on which current theory and doctrine rest. This makes it difficult to pursue what one might regard as a traditional professional caucus on doctrine, the kind of community self-criticism and reform that transpired in the late 1970s and early 1980s. Instead, what is probably needed is a move to recast the debate about theory and doctrine within the context of operational art and using contemporary pedagogical tools to complement reading. Additional efforts should be made to teach leaders to be more conscious of psychological factors that may undermine or degrade decision-making capacities, including cognitive biases and fallacies that impede realization of much-touted visions like information dominance.

Of course, an American proclivity to treat history as a theme park rather than a learning laboratory does not help. The profession of military science needs to assimilate the grand narratives of the discipline and reapply them to current pedagogical needs, including facilitating a grand synthesis with an emerging literature on security and stability operations, active diplomacy, and the resurgence of cultural anthropology applied to military operations.

Recall that operational art is domain expertise existing in raw form as a knowledge base for campaign planning and command at the operational level of war. Tapping knowledge in different situations to overcome novel challenges requires a certain form of intuition that can be learned but is not, research suggests, immediately transferable across disciplines. Few military theorists have approached the issue of cognitive development in the systematic fashion outlined by Clausewitz.

For Peter Paret, "the role genius plays in war" is central to Clausewitz's "entire theoretical effort."[11] In *On War,* Clausewitz located the concept of military genius at the intersection of his theoretical description of the nature of war and his discussion of intellectual and emotional qualities exhibited by great battlefield commanders. Military genius is discussed in a chapter in book one titled "On the Nature of War." Its explication comes between the discussion of combat as the means of war and another chapter titled "Danger in War" that identifies the frailty of intellect and the importance of character in combat decision making. Immediately following the genius discussion comes a chapter in which combat is described as consisting of "layers of increasing intensity of danger" where "ideas are governed" by emotions and where "the light of reason is refracted in a manner quite different from that which is normal in academic speculation." In combat, he continues, "the ordinary man can never achieve a state of perfect unconcern in which his mind can work with normal flexibility. Here again we recognize that ordinary qualities are not enough; and the greater the area of responsibility, the truer this assertion becomes."[12]

Discussion of the cognitive aspects of command continues. Chapter five addressed physical effort in war, which acts to "chain the spirit and secretly wear away men's energies."[13] The next argues that extraordinary men, with quick decision-making faculties, a "buoyant disposition," and self-control, are needed to manage the uncertainty of battlefield information.[14] Chapter seven elaborates the concept of friction in war—"a force that theory can never quite define"; it has its greatest negative impact on the commander endowed with ordinary intellect and temperament.[15]

Military genius provides the pivot on which Clausewitz turns from defining war as a specific type of human behavior shaped by psychological elements to his argument about the proper elements of a theory of war (which concerns book two). Qualities of military genius are divided into two categories:

### *Qualities Related to Intellect*

- Presence of mind, strength of mind, a steady nerve
- Strength of will
- Spatial awareness
- Experience in war
- Self-awareness, reflection

## *Qualities Related to Temperament*

- Self-control, balance
- Instinct, a sense of truth
- Imagination, creativity
- Courage, boldness
- Energy in action, endurance
- Ambition (for glory or honor)

One reason why creativity and imagination are included in the qualities of military genius is to further the argument that theory cannot aim toward absolute rules for the commander. Paret seems to view creativity as a corollary to friction in war, stating that together they "help determine the character and rate of progress of military operations."[16] Whereas friction serves to constantly focus resistance on the commander, at the psychological level, imagination endows the military genius with the ability to conceptualize and implement unique solutions geared to the immediate situation.

An elegant treatment of the cognitive aspects of command for its time, the military genius framework buttresses seemingly straightforward arguments about the benefits of disciplined training and preparation of military minds with decision-making theory addressing the stresses of war on leaders. Genius buffers uncertainty, moderates friction, and tempers angst.

Clausewitz's discussion of military genius is dominated by concepts that contemporary theorists find more suitable to management theory, the leadership arts, and cognitive science. For example, he outlines what scientists today discuss as the role of the limbic system in facilitating problem solving during times of great stress and anxiety; sociologists would discuss other arguments in terms of acculturation and socialization processes. Current decision theorists and leadership gurus recognize similar concepts in emotional intelligence discussions and performance enhancement training.

The center of the limbic system is an organ known as the amygdala, which can hijack the brain and diminish cognitive faculties. Emotions and sensory input flooding into the brain through the brainstem pass first through the primary organs of the limbic system before being transmitted to the frontal lobes. A heightened state of anxiety or inability to assess unfolding events can essentially short-circuit the cognitive processes through which domain expertise is tapped to assess problems and quickly develop solutions or appropriate actions. Even experts trained in similar fields sometimes find the stress of forcing their expertise into unfamiliar decision-making situations to be debilitating. Stress does not have to be life threatening or to involve combat. It can stem from time compression, anxiety associated with uncertainty, or merely the pressures of dealing with complicated situations for which no readily available mental model exists to help deconstruct and analyze options.

This sounds very similar to the erosion of the intellect during combat or in stressful decision environs; it is the process against which we pit training, theory, doctrine, study, and practice.

Learning to cope with the stress of operational challenges requires experience that enables leaders to moderate limbic system issues (emotional learning) and to suppress cognitive biases associated with frontal lobe executive thinking processes. Such cognitive dimensions are the central elements of a larger framework in which military genius is established.

## Concluding Observations

The dissonance mentioned in the introduction may have stemmed in part from tensions between traditional (conditioned) understanding of the principles of war as a framework for applied operational art and the realities of applying the *discipline* of operational art in a strategic context requiring simultaneous, coequal, theater, and global campaigns. Perhaps it relates to tensions inherent in operational approaches and campaigns that conflate the principles of war and principles of operations other than war without clear strategic objectives to clarify campaign objectives.

"Genius, dear sirs, never acts against the rules," Clausewitz wrote around 1804 in an essay discrediting "geometric rules of war," implying that rule-based theories of war could not hope to be timeless because great commanders across historical eras inevitably displayed their genius in ways that *contradict* such rules. Genius "*rises above all rules,*" Clausewitz continued, urging theorists to "accommodate themselves to this reality by taking care not to inhibit the action of genius in their doctrines."[17] And so it is today: commanders and junior officers routinely astound doctrine experts with innovative solutions to operational and tactical challenges. Lessons learned databases and unofficial blog sites will likely contribute to much rethinking of the operational art of war and lead to revisions in doctrine.

Of course, commanders do not actually make decisions by consciously evoking the principles. They do, however, shape how combatant commanders are introduced to scenarios and even the mental simulation of options to build internal decision styles. The principles of war and elements of joint operations unconsciously influence how data are assimilated into working memory and the structure of cognitive associations commanders are able to pull from when faced with real command challenges.

Excelling at the operational level of war is assumed to be more probable if one has mastered the discipline of operational art. In reality, success is highly contingent, including variables addressing enemy capabilities, resources, timing, and chance. Still, great battle captains tend to have certain cognitive faculties and experience that provide an advantage; these are in part facilitated by the subtleties of making inference about and extrapolation from domain knowledge.

Our task is to link innovations, adaptations, and novel approaches from the campaign against terrorism with other aspects of the discipline known as operational art to preserve the rich tradition and full range of domain knowledge for future leaders. This includes challenging prevailing policies and contesting educational regimes meant to prepare leaders for command.

## Notes

1. The author first engaged in a discussion on the activity by then–Rear Adm. John Morgan when he returned from his tour as the Enterprise Battle Group Commander in Operational Enduring Freedom, 2001. Observations about dissonance, admittedly anecdotal, derived from the author's frequent interactions with other members of the Rethinking the Principles of War steering group and participation in the seminar series.

2. Donn A. Starry, "Reflections" in George F. Hofmann and Donn A. Starry (eds.), *Camp Colt to Desert Storm: The History of U.S. Armored Forces* (Lexington, Kan.: University of Kentucky Press, 1999), p. 547.

3. Starry, "Reflections" in Hofmann and Starry, p. 546.

4. Brzezinski cited in Christopher Coker, *U.S. Military Power in the 1980s* (London: The Macmillan Press, 1983), p. 31.

5. Harlan K. Ullman and James P. Wade, *Shock and Awe: Achieving Rapid Dominance* (Washington, D.C.: National Defense University Press, 1996).

6. Ullman and Wade, *Shock and Awe,* p. 2.

7. General Edward C. Meyer, "Low-Level Conflict: An Overview" in Brian M. Jenkins (ed.), *Terrorism and Beyond: An International Conference on Terrorism and Low-Level Conflict* (Santa Monica, Calif.: The RAND Corporation, 1982), p. 39. See also Steven Metz, "A Flame Kept Burning: Counterinsurgency Support After the Cold War," *Parameters* (Autumn 1995), pp. 31–41.

8. Colin S. Gray, *Strategy for Chaos: Revolutions in Military Affairs and the Evidence of History* (London: Frank Cass, 2002), p. 247.

9. Friedman cited in Gray, *Strategy for Chaos,* p. 248.

10. See Gary Klein, *Sources of Power: How People Make Decisions* (Cambridge: Massachusetts Institute of Technology Press, 1998); Gerd Gigerenzer et al., *Simple Heuristics that Make Us Smarter* (New York: Oxford University Press, 1999); William M. Goldstein and Robin M. Hogarth, eds., *Research on Judgment and Decision Making* (New York: Cambridge University Press, 1997); Scott Plous, *The Psychology of Judgment and Decision Making* (New York: McGraw-Hill, Inc., 1993); Gerd Gigerenzer and Reinhard Selten, eds., *Bounded Rationality: The Adaptive Toolbox* (Cambridge, Mass.: Massachusetts Institute of Technology Press, 1999).

11. Peter Paret, "The Genesis of *On War*" in Carl von Clausewitz, *On War,* Michael Howard and Peter Paret, trans. and eds. (Princeton: Princeton New Jersey Press, 1984), p. 3.

12. Ibid., p. 113–14.

13. Ibid., p. 114.

14. Ibid., p. 117.

15. Ibid., p. 120.

16. Peter Paret, *Clausewitz and the State: The Man, His Theories, and His Times* (Princeton: Princeton University Press, 1985), p. 375.

17. Ibid., p. 161.

# TWELVE

# From Operational Art to Grand Strategy

ROBERT R. LEONHARD

On 16 August 1870, at the Battle of Vionville-Mars-la-Tour during the Franco-Prussian War, Major General von Bredow, a Prussian cavalry commander, stormed out of the battlefield smoke and straight into a French artillery battery, capturing the guns in a glorious charge. The incident became known to history as "Von Bredow's Death Ride," a testament to the 380 dead and wounded German cavalrymen—almost half of the general's command—who paid in blood for their leader's impetuous charge. The anecdote has become a caricature of how to misinterpret history, because in the years leading up to World War I, cavalry enthusiasts pointed to von Bredow's triumph as proof of the viability and battlefield utility of well-led horse-soldiers. In fact, the incident pointed in the opposite direction, and the cavalry charge became one of many bloody anachronisms exposed on the battlefields of 1914. Von Bredow's Death Ride was shown to be what it really was—the product of several coincidental factors: a valley that created a dead spot in the French field of fire, the lack of entrenchments, and an almost accidental envelopment by the Prussians.

In a similar way modern American war fighters are looking to a glorious recent past to prop up a popular conceptualization of war known as "operational art." As a result, they are clinging to an outdated mode of thinking about warfare and thus falling short as an institution in coming to grips with the present and future. The purpose of this chapter is to explain why operational art has withered away as a bona fide discipline, and why there is an urgent need to direct our efforts toward the development of grand strategy. This is not an academic exercise. The failure to graduate to a grand strategic approach to future operations will result in an officer corps distracted by anachronism and deficient in needed skills.

In 1982 the U.S. Army published a landmark document, Field Manual 100-5, *Operations,* in which the doctrine of AirLand Battle was established. Four years later, the manual was updated, and together, the 1982 and 1986 manuals provided the foundation for a radical change in how the army (and by extension, the joint community) fought. The primary theme of the manuals was the importance of the operational level of war and the associated concept of operational art. Rather than fixating on the tactical battle, as previous versions of the *Operations* manuals had done, AirLand Battle transcended tactical battles and engagements and focused on the campaign—the building block of operational art.

It is hard to overstate the depth of the doctrinal revolution of the mid-1980s. The discovery (or rediscovery) of operational art had profound effects throughout the army and the fellow services. Officer education improved and deepened; the classics of war were resurrected and reread; training became more realistic. Officers began to study past wars not solely to elicit the glorious details of battles and bloodshed, but also to grasp the logic and dynamics of theater campaigning. Although this period produced an enormous and varied reexamination of military history, most of the army community's attention was focused on World War II and the wars that followed it. Operational art was the touchstone of professional dialogue, and the officer education system assimilated the new emphasis with the founding of the School of Advanced Military Studies (SAMS) at Fort Leavenworth and updated curricula at the War College.

It was curious, however, that although operational art occupied a central position in the army's vocabulary, the term itself remained somewhat mysterious. What exactly was the operational level of war? What was operational art? During my time as a student at SAMS and later as a war planner, I grew to believe that although the army's officer corps shared a general understanding of these terms, the taxonomy lacked precision. The operational level of war was seen as the planning level that existed between tactics on the lower end and strategy on the higher. Wedged between these two—one the business of lieutenants, captains, field grade officers, and division and corps commanders; the other the purview of the four-star generals and their civilian masters—was the operational level, whose job it was to link tactics to strategy and vice versa. In other words, we understood *where* operational art was, if not exactly *what* it was.

The 1993 version of *Operations* defines operational art as "the employment of military forces to attain strategic goals through the design, organization, integration, and execution of battles and engagements into campaigns and major operations. In war, operational art determines when, where, and for what purpose major forces will fight over time." Joint Publication 3-0, *Doctrine for Joint Operations,* provides similar wording to describe it. This description implies, as it should, a two-way flow of logic within operational art. On the one hand, it takes tactical events (battles, engagements) and arranges them into campaigns and major operations. In other words, there is a part of operational art that "looks upward" from tactics to strategy. On the other hand, operational art also is the controlling agent that ordains the when, where, and purpose of tactical events. This part of it looks downward from strategy to tactics. Thus, tactical events don't just occur and await organization; rather, they are (at least ideally) intentional events—an expression of the will of the commander.

The doctrine claims that operational art is solely about "military forces" and the design of campaigns and major operations. Conspicuously absent from this definition of operational art is a discussion of the other elements of national power: diplomacy, economics, and information. Instead, operational art is a discipline clearly and neatly contained within the military sphere. It belongs to military flag officers, and they are the keepers of the arcane art. The problem with this notion is that few officers today would accept that definition. In practice, if not in theory or doctrine, today's operational art is most decidedly involved in employing the nonmilitary elements of power. Even Joint Pub 3-0 states that the practice of operational art requires interagency cooperation and attention to political and other nonmilitary factors. We will return to this point shortly.

The other odd thing about the current definition of operational art is that it is suspiciously similar to what used to be called "strategy." In his landmark work, *Strategy,* first published in 1954, B. H. Liddell Hart, in turn quoting Clausewitz, defined strategy as "the art of the employment of battles as a means to gain the object of war . . . strategy forms the plan of the war, maps out the proposed course of the different campaigns which compose the war, and regulates the battles to be fought in each." This definition is so close to what Americans forty years later termed "operational art" as to be almost indistinguishable. The focus on campaign design perhaps delineates operational art from Hart's notion of strategy, but the difference is subtle to say the least. Is operational art merely a new term for strategy?

Strategy itself is a troublesome term. It derives from the Greek *strategos*—the warrior-politician of ancient Athens. Although the term has evolved to mean a variety of things today, it originally meant simply "what the *strategos* does." So what did the *strategoi* do? In short, they employed all elements of national power in war and peace. Pericles was perhaps the ultimate example of a *strategos*. Having conceived a unique concept for surviving Sparta's clear advantages in land warfare, Pericles had to transcend merely military ideas and instead master politics, economics, *and* military art. His idea was to dislocate Sparta's armies by bringing the Athenians entirely within the Long Wall that surrounded Athens and reached to the city's ports of Piraeus and Phaleron. Lacking the ability to breach the walls, Sparta would be unable to bring her military might to bear. Meanwhile, Pericles argued, the Athenians would retain active command of the sea, which in turn would provide the vital access to their colonies abroad. With such impunity, Athens could outlast the Spartans, whose society, built on the foundation of slavery, would eventually collapse.

In order to convince the Athenian citizens—many of whom owned lands outside of the Long Wall—to acquiesce in the temporary loss of property and the discomfort of remaining within the walls of the city, Pericles had to be a master of both politics and economics. If his argument were weak or fallacious, the military concept of operation would fail, and, as actually occurred after his death, the Athenians would instead opt for a more aggressive war against the enemy. Thus, the art of the *strategos* was to derive a plan of action that integrated politics, economics, and the military dimensions of the war. In its most basic sense, then, strategy went beyond merely military considerations.

But strategy did not stop evolving in the Peloponnesian War. Instead, the term developed over the centuries to take on a decidedly military connotation. As warfare in the Western world occupied the attention of strategic thinkers, the concept of strategy migrated from meaning what the *strategos* did to what the general in the field did: prosecute military operations. In our own day, the art of strategy has gained audiences from virtually every field—from baseball to big business—but even those who employ strategy in civilian fields tend to think of it as deriving from military roots. As the Cold War progressed, policy makers came to realize that strategy at the national level must include nonmilitary components, and the National Security Act of 1947 thus called for the administration to produce a National Security Strategy (NSS). The NSS is designed to lay out a comprehensive approach to foreign policy in general and security in particular. It includes both military and nonmilitary concepts, just as Pericles would have wished.

One step down from the NSS is the National Defense Strategy (NDS), written by the secretary of defense. The NDS articulates how the nation will achieve the goals of the NSS through military preparedness and operations, and through international alliances and other forms of cooperation. It is a key document in the development of weapons programs and force structure decisions. Closely linked to the NSS and NDS is the National Military Strategy (NMS), prepared by the chairman of the Joint Chiefs of Staff. The NMS, along with the Joint Strategic Capabilities Plan (JSCP), provides direction to combatant commanders—four-star flag officers—who command the joint forces in each major area of operations. The NMS also provides high-level guidance on desired capabilities.

Armed with this array of strategic documents, the combatant commander develops theater strategies to deal with various missions or contingencies. In this effort he is guided by joint doctrine, which defines strategy as "the art and science of developing and employing armed forces and other instruments of national power in a synchronized and integrated fashion to secure national or multinational objectives."

We have, then, an organization for modern conflict in which four-star flag officers and their staffs develop strategic plans for their theaters or regions. These plans seek to integrate all elements of national power, with an emphasis (from the combatant commander's point of view) on the military dimensions. Subordinate to this planning level is the operational level, in which commanders and their staffs practice operational art, defined as the employment of military forces, but with the doctrinal caveat that nonmilitary elements must also be included. The question remains: Is this a proper construct? The answer is no.

The reality of modern operations does not allow for an operational art that encompasses military dimensions only, with a weak admission that nonmilitary factors must be considered. Operations in Afghanistan, Iraq, and the Balkans all point clearly to the requirement for leaders at all levels to operate in the political as well as economic realms. Required skill sets for planners and operators include cultural education, language skills, and an understanding of civil policy making. The simple solution would be to redefine operational art so as to include the integration of all elements of national power, but the problem then becomes obvious: operational art would become synonymous with strategy as, in fact, it is in practice.

During the course of modern history, operational art occupied a prominent role because there was a need for an intellectual discipline that focused on the interaction of military formations on a grand scale. Over the course of the last century, armies, navies, and air forces competed with each other over vast distances, consuming huge quantities of supplies. Whenever these large formations of war took to the field, they dominated all other considerations. When war broke out, politics, economics, and cultural sensitivities became secondary matters as the science of breaking the enemy's military formations became the life or death of the nation. Theater commanders in World War II certainly had a part to play in political matters, but they were primarily focused on maneuvering divisions, air fleets, and navies. Populations hid or fled. Once a nation's borders had been violated, the political organization of the land became virtually meaningless. Farms, towns, and cities became merely the terrain over which military formations contended with each other. Operational art—a misnomer, because it actually involved both science and art—was the intellectual discipline that developed to handle this type of warfare.

In order to fully grasp the American concept of operational art, we must see beyond the official definition of it and look at how it was practically defined by those practicing it. While there has always been agreement as to the positioning of the operational level of war between the tactical and strategic, a spectrum of opinion developed as to how it related to tactics on the one hand and strategy on the other. Some describe the operational level as comprising everything that happens beyond the range of direct fire weapons. Thus, the march of formations into battle, reinforcement from one battle to another, and all actions outside the battlefield—logistical, administrative, etc.—were at the operational level. Since this concept specified the range of direct fire weapons, the implication was that indirect fire weapons—if their range were long enough to reach beyond the tactical battlefield—could be thought of as providing "operational fires." Those fires, in turn, could be "maneuvered" like formations throughout a theater as part of operational art. This "grand tactical" view of operational art tied it very closely to the tactical level, and the focus was on facilitating victory in battle.

At the other end of the spectrum were those who viewed operational art as, in the words of Richard Simpkins, "one step removed from strategy." In his book *Race to the Swift,* Simpkins described the operational level and operational art as necessarily being immediately juxtaposed to the attainment of the strategic goals—one step removed. The implication was that two or three steps removed was either bad operational art or something altogether different from it. On this end of the spectrum, the art and science of operations almost had to include nonmilitary factors. At the tactical end it did not, and thus it fitted better into the classical definition.

Between the two ends of the spectrum of operational art, most serving officers were content to know *where* the operational level was, if not exactly *what* it was. As the Cold War receded into the history books, the day-to-day reality of operations was making it clear that political, economic, and cultural factors were intruding further and further into the operational level and down into the ranks of even tactical formations. In Operation Iraqi Freedom, lieutenants and captains were routinely involved in administering local government and economies. Clearly this was a different dynamic than that of the Cold War days.

As both a company grade and field grade infantry officer in the 1980s and 1990s, I had the privilege of serving during the heyday of operational art. During most of that time, my profession focused on understanding, detecting, breaking, and defeating Soviet tactical formations. A competent combat arms officer had to be intimately familiar with how a motorized rifle regiment deployed for battle. During long, cold days at the National Training Center in Fort Irwin, California, my comrades and I would search for the enemy's regimental reconnaissance vehicles, and then, in turn, await the combat recon patrol, the forward security element, and the advance guard battalion. We all knew how many and what type of vehicles were in each of these echelons, as well as how each was used to facilitate victory once the regimental main body arrived. And while we were hammering away at simulated Soviet tactical formations, our counterparts in the division, corps, and army staffs were doing essentially the same thing with Soviet corps, armies, and fronts. This form of "operational art" was all about defeating Soviet military echelons, and it was almost entirely a military affair.

Whatever happened to the officer corps of the late-twentieth century, whose competence was in fighting the Soviet army? How relevant today are the templates we so treasured in those days? The *doctrinal* template was a graphic representation of how Soviets operated. The *situation* template applied Soviet doctrine to the actual ground conditions and provided an estimate at how the enemy would actually deploy. While our reconnaissance assets went to work to confirm or deny our situation template, the planners would be busy working up the *event* template, full of estimates on where and when the attacking enemy would commit his tactical echelons. Finally, we would put together the infamous *decision support* template and associated matrix that provided the commander with a scientifically arranged set of decision points to help him combat the enemy. As the plan came together with these tools, the fire support community would develop parallel plans that brought (hopefully) overwhelming air, artillery, and—sometimes—naval fires designed to delay, disrupt, or break enemy formations before they over-whelmed the outnumbered Americans.

The training community within the U.S. Army and the other armed serv-ices excelled at providing extremely challenging training exercises along these lines, first at the tactical level, and later at the operational level, using an increasingly sophisticated suite of computer simulations. The officer educa-tion system likewise concentrated on producing officers who were competent at breaking Soviet formations. The tacit belief was that even if we never actu-ally faced the Soviets themselves, the skills we were learning could apply to any other enemies with equal effect. The Soviets were, after all, the "worst case."

The zenith of American operational art was co-synchronous with the height of the Cold War and its most fundamental, yet hypothetical scenario, a Soviet invasion of Western Europe. AirLand Battle was developed to solve exactly that problem, although it was never put to the test in that way. Instead, American joint forces would test this and the other service doctrines everywhere but Western Europe, with largely favorable results. The same year that the revised version of AirLand Battle doctrine was published (1986), Congress passed the landmark legislation, the Defense Reorganization Act, commonly referred to as the Goldwater-Nichols Act. In short, this act man-dated the development of joint integration. It promoted the chairman of the Joint Chiefs from his former "first among equals" status to the highest-ranking military officer in the nation. It reformed officer education and assignment practices in favor of inculcating jointness in field grade and gen-eral officers. Goldwater-Nichols had profound effects, mostly beneficial, and birthed America's first truly joint military force.

The Goldwater-Nichols Act passed on the strength of a growing convic-tion among the nation's civilian leaders that the military had lost its capacity to fight competently at the operational level of war. The disaster of Desert One—the failed attempt to rescue American hostages in Iran that left eight servicemen dead—pointed to serious flaws in the organization and training of the services. The 1983 invasion of Grenada, although a nominal success, likewise exposed woefully inadequate interservice cooperation and a general ineptitude in war-fighting skills. On the heels of these setbacks, Goldwater-Nichols sought to reform the joint community, and, along with the birth (or rebirth) of opera-tional art, it resulted in clear advances in the military's ability to fight.

In 1989, the dog finally caught the car it was chasing and didn't know what to do with it. The Soviet Union began to collapse; the Berlin Wall came down; former enemies became friends; and the bipolar world of the Cold War mutated into the more complex, multipolar world of the information age. In the military realm, American joint war-fighting power reached a new height. Operation Just Cause in Panama visited an almost instantaneous catastrophe on Manuel Noriega's defense apparatus and forced a regime change. In 1991, the U.S. Army's heavy forces worked with its joint partners to turn the third largest army in the world into the second largest army in Iraq, as the joke went. Helped along by a comically incompetent Saddam Hussein, the American-led Coalition totally dominated their more numerous industrial age foes. Despite later setbacks in Somalia and an ongoing frustration in the Balkans—both of which were marginalized in doctrine by referring to them as "operations other than war"—it had become clear that the Americans were unequaled in conventional military power.

The world watched. The world took note. And then the world changed.

American operational art had achieved its zenith. It became almost inconceivable that any nation-state could generate conventional combat formations of ships, aircraft, or troops that stood any chance at all in action against the American military machine. The post–Cold War defense community took note of emerging information age technology and arrogantly believed that it could steer the course of a military revolution through white papers and the manipulation of the research and development budget. But the nation's enemies were evolving at a far faster rate, propelled not by budgets, the worship of technology, or theory, but by a deep-seated need to oppose the new Roman Empire. As the grip of the bipolar world gave way to a new world disorder, ethnic hatred, international crime, anachronistic regimes stymied by corruption, and wild-eyed ghazi warriors emerged to stand in the way of America's vision of a *Pax* Americana.

But how to oppose this Goliath? Certainly not with javelin, sword, and shield, the very weapons the giant uses to dominate his foes. Instead, the enemies reached for a sling and five smooth stones.

Enthusiasts of so-called fourth generation warfare like to suppose that the irregular warfare that emerged following the collapse of the Soviet Union was a new phenomenon that grew from a sequential development begun in 1648. According to the theory, the first three generations of warfare progressed from the initial appearance of firearms on the battlefield to the temporary superiority of firepower (American Civil War, World War I) to the development of armored and mechanized maneuver warfare. After the end of the Cold War, the theory states, irregular warriors arose to contest the conventional military power of the mighty West. This new type of warfare manifests itself in insurgencies, terror, and guerrilla tactics, as well as by political, economic, and cultural activity. In this assessment, the theory of fourth generation warfare is accurate. The flaw in the theory, however, is that this is something new. In fact, irregular warfare is as old as mankind itself.

The history of warfare, far from being a sequential development begun in 1648, is in fact a continuing cycle of conventional and unconventional, confrontation and dislocation. Dislocation, which I have defined in my previous work as "the art of rendering enemy strength irrelevant," is a constant factor in history. When a nation or group builds a strong military capability, its enemies will most often seek to build similar capability and *confront* strength with strength. The result is an arms race or a conventional war. But when the enemy believes that it is impossible to confront strength, he will instead seek to dislocate it. Dislocation tries to change the circumstances of the conflict in such a way that the opponent's strength simply doesn't matter.

The terror attacks of 9/11 were a classic, if horrifying, example of dislocation. The Islamic jihadists that masterminded the attack knew that they could not oppose the United States military. Instead, they simply drove around it, rendering it totally irrelevant to the outcome of the attack. They did not confront or defeat American operational art. They simply ensured that it wasn't in the game that day.

That is the problem currently facing the U.S. military machine. Since 9/11 the joint teams in Afghanistan and Iraq, as well as in many other lesser contingencies, have scored great successes in the war on terror. But they are doing so in spite of a doctrine that still insists on proclaiming the viability of operational art—the "employment of military forces to attain strategic goals." In fact, today's warriors find themselves practicing a different art and science of conflict, one so far removed from breaking easily templated Soviet formations that it is hard to imagine a doctrinal connection between the two.

Twenty-first century warfare demands a new discipline for conceptualizing conflict. Operational art is an industrial age idea, suited primarily for breaking enemy formations of ships, aircraft, and troops in a theater of war wherein political and cultural issues are secondary. Since it claims to be a purely military concept, its practitioners are already leaving it behind in practice, while still paying lip service to it in theory. Eventually the doctrine will have to catch up, and industrial age operational art will give place to information age grand strategy.

To see how and why this transformation must occur, we need to outline the characteristics of the current operating environment.

One of the most obvious changes in warfare has to do with frequency. I use that term to describe the rate at which events happen—from bullets spewing out of a rifle barrel to decisions issuing out of headquarters. In high-frequency warfare, events happen fast and continuously. The oft-contemplated Cold War campaign in Western Europe would have been a high-frequency affair: units moving quickly with overwhelming operational momentum; commanders making rapid decisions; joint fires creating a cauldron of destruction, and so on. America's "100 Hour" ground campaign in the First Gulf War, as well as the conventional phase of Operation Iraqi Freedom, featured such high-frequency operations—so high, indeed, that the industrial age enemy could not keep up and instead collapsed in short order.

By way of contrast the current operating environment is normally low frequency. The enemy does not attack quickly and persistently. (If he did he would be easier to anticipate and oppose.) Instead, attacks are staggered and sudden, followed by periods of inactivity. Al Qaeda is known to plan for terror strikes for years before actually executing. The enemy's irregular campaigns, whether well led or incompetent, are invariably low-frequency operations.

It is hard for a high-frequency military to fight a low-frequency enemy. High-frequency conflict puts a premium on momentum, the rapid generation of combat power, quick decision making, and high-volume logistics. But a low-frequency enemy makes all of that irrelevant. Low-frequency threats call for surveillance, patience, effective intelligence, and discriminatory application of combat power. They also call for a decision-making model that is not necessarily optimized for speed, as our current model is, but rather for accuracy.

Even more apparent than a shift in frequency is the fact that the current operating environment features "formationless" warfare. This phenomenon is one of the most revolutionary of recent changes. From ancient times, armies, navies, and, later, air forces deployed and fought in formation. Indeed, much of the military literature of the past discussed the advantages and disadvantages of various formations. Every epoch of human warfare has its representative formation: the phalanxes and legions of ancient times—the Byzantine turma, the Mongol touman, the Spanish tercio—formations used well into the modern age; Napoleonic divisions and corps—again, a system that has continued to the present day; German panzer groups, Soviet armies and fronts, and so on. Naval warfare likewise employed formations, as did air forces. Formations permitted the most efficient use of kinetic weapons and provided the most economical protection possible. Formation-based warfare required discipline and training to maintain the formations, as well as break the enemy's. Once a formation lost its cohesion, morale was shattered, and a retreat or rout would ensue. Hence, fighting doctrine emphasized the synchronization of destructive combat power aimed at achieving just that end.

Formations are hard to come by these days. Those that exist are all but irrelevant to modern conflict. Any that are foolish enough to show up within range of an American joint task force face catastrophic destruction. As long as existing regimes cling to such anachronisms, we can keep hoping they actually try to use them. But for the foreseeable future, most of our potential opponents have come to understand that investing in conventional, formation-based fighting against America is a waste of resources. Instead, they are employing formationless warfare that defies templating and befuddles operational art.

Closely linked to the departure of battlefield formations is the switch of emphasis from firepower to detection. As formation fighting dominated warfare for the past several millennia, warriors had to focus their efforts on destroying formations. Their targets were not that difficult to detect—they weren't designed to be—being organized instead for the generation of combat power. Of course, as technology advanced, formations on land, sea, and air became more difficult to detect and track at times, but in general, fighting formations depended on firepower and protection, rather than invisibility, for their survival. But once formations went away, the business of detection became much more difficult, while the act of destroying whatever was found became easier. Today's terrorists and guerrilla fighters aren't hard to kill, but they are extremely hard to find. As they have done throughout history, irregular warriors rely on stealth and the ability to blend into the background—whether human or terrain.

Again, this trend finds our doctrine of operational art ill prepared. Operational art's strength was rapid decision making, the orchestration of joint and combined arms, and the consequent generation of overwhelming combat power. But today's enemies are a different sort, and to fight them requires patience, scrutiny, and a perspicacity that can peer through the fog of complexity. A form of conflict that concentrates on detecting an illusive enemy, as well as ameliorating the political, economic, social, and cultural conditions that produce and support enemies, will find little that pertains in classical operational art.

The growing complexity of the battle space is an indirect result of American joint fires. With American air dominance virtually assured in any conceivable contingency, and with the growing destructive power of conventional munitions, adaptive enemies abandon the predictability of formation-based high frequency operations in favor of operating in the infamous "dark spaces" of the world: cities, jungles, and mountains. Maintaining secrecy and dispersion, enemies defy the targeting technology and practices of American joint forces, making air dominance much less decisive. The simultaneous presence of noncombatants and global media make it imperative that joint forces discriminate with painstaking accuracy among bad guys and good guys, bad guys and neutrals, and bad guys and guys we only suspect are bad. All this tends inevitably to slow the pace of operations, and often commanders find themselves forced into a reactive mode for lack of a visible enemy.

Finally, the future operating environment is increasingly global. Throughout the nineteenth and twentieth centuries, it was easy to conceive of separate theaters of war. For purposes of resource distribution alone, it was necessary to assign priorities to various theaters of operation, and each of these theaters could be dealt with as relatively independent entities. During World War II, for example, events in Europe did not immediately and directly affect conditions in the Pacific theater. Commanders in these areas of responsibility, although they vied with each other for resources, did not have to routinely cooperate or coordinate. Theaters of war tended to be separate and independent.

This is no longer the case. The global war on terror, for example, is truly global. The various fronts of the war—Iraq, Afghanistan, the continental United States, etc.—are intimately connected with each other. Enemy leaders in the different regions communicate with each other routinely. Events in one theater can have direct and immediate consequences on the opposite side of the globe, as when a raid on insurgents in Pakistan resulted in fresh intelligence about terrorist planning in the United States. Hence, it is not a useful construct to think about theaters of war as separate entities any longer. The global nature of warfare is a direct blow to classical operational art because the operational level of war focuses on campaigning within a single theater. The "theater campaign plans" of the past will become increasingly pointless, except as annexes to a vitally integrated global strategy.

All of this calls into question the continued relevance on a doctrinal construct that was born and bred in industrial age, formation-based, high frequency warfare. Can our current doctrine of operational art be used successfully in the current and future operating environment? Yes, it undoubtedly can. A good surgeon could probably get away with using a steak knife instead of a scalpel, but that's not the optimal solution. Instead of trying to retrofit operational art onto the current operating environment, we should break with our past and refocus our attentions on grand strategy.

Grand strategy—I use the adjective grand to distinguish between the more common uses of strategy and the construct I am discussing in this chapter—is the art and science of employing all elements of power—diplomacy, economics, military, information—in order to accomplish national or coalition objectives. In a sense, even this definition is too restrictive because I have used the traditional four "elements of national power," when in fact there are more than four. Intelligence and counterintelligence operations are major elements of national power, especially in the current operating environment, when detection is so critical to success. Culture, religion, and even philosophy can be considered among the weapons of the future as well.

To build grand strategic capability in the future will ultimately require legislative action along the lines of the Goldwater-Nichols Act. Just as Congress led the way in forcing the development of joint capabilities in the 1980s, so it will again have to act in order to facilitate the development of interagency operations. The joint task force was the action agent for operational art, but the interagency task force will be the agent of grand strategy. Rather than schooling military leaders in operational art and then foisting upon them duties and responsibilities that transcend "the employment of military forces," we should instead train and educate a variety of government leaders from the State Department, the Defense Department, the Central Intelligence Agency, and other agencies of the government to wield all elements of national power.

The British viceroy of the eighteenth century makes a good model of a grand strategist. Robert Clive, the man most responsible for bringing India into the British Empire, was a true strategist in this sense. He wielded economic power through his relationship with the British East India Company. He commanded joint forces—both army and navy—in the theater of operations. And he was a skilled and beneficent political ruler who gained legitimacy in the eyes of the Indian population through his evenhanded policies. By the time he won the Battle of Plassey in 1757, Clive had planned and led an integrated campaign that ranged from gradual economic development to desperate counterattacks, from wise promotion of virtuous leaders to outright bribery to take apart enemy alliances. Clive was not a military man who dabbled in nonmilitary affairs. He was a grand strategist who viewed the military as only one of his tools—albeit an important one.

The threats and challenges that America faces today and will likely face for some time demand a strategic approach, and the art and science of strategy will apply not only at the national or theater level, but also in the day-to-day operations at each level of command. The bifurcation of military and nonmilitary factors that our joint doctrine is based on is a destructive anachronism that stands in the way of real and sustained progress. Classical operational art, at one time the zenith of American conceptualization of war, is daily losing relevance. It will perhaps reemerge if and when a peer competitor—China or the European Union, for example—modernizes to the point that they can compete with American joint arms. Against that possibility, we must continue to develop what has been characterized as "conventional" capability. Indeed, it is our continued demonstrated capability to destroy formations that will keep those formations at bay. To take us on in that way in the foreseeable future will require a nation-state with real economic power, political integration, and a solid technology base. Until that threat comes, the operating environment cries out for strategists like Clive, not grand tacticians like George Patton or even masters of operational art like H. Norman Schwarzkopf and Tommy Franks. Like chain mail and cavalry charges, operational art must give place to modernity, and the grand strategist of tomorrow must supplant the operational artist of yesterday.

THIRTEEN

# Preponderance in Power

## Sustaining Military Capabilities in the Twenty-first Century

JAMES J. CARAFANO

Forging, deploying, and maintaining fighting forces have been and remain the lifeblood of war. Yet, remarkably, the principles of war do not reflect the imperative of creating and maintaining national military power. The intellectual divorce between the conduct of conflict and the state's power to sustain combat forces occurred over the course of centuries, the result of evolving conceptions of the public and private spheres and the military's role in thinking and planning for campaigns. As a result, the principles of war rarely encouraged thinking about conflict beyond the imperatives for winning on the battlefield.

The lack of attention that the principles confer on retaining preponderance in power makes their application to twenty-first-century wars problematic. Many of the most potentially disruptive and dramatic changes in future conflicts may be driven not by how battles are fought, but in how the instruments of power are marshaled by state and non-state actors. Further proof, perhaps, that the principles should be considered a historical relic, best left in the past. If there were ever an age when critical thinking about war could be reduced to a simple set of maxims, it is long over. Critical thinking and decisive winning in the twenty-first century require the intellectual capacity to think broadly about the nature of conflict, more widely than can be inspired by a list of aphorisms.

## Spheres of Conflict

The meanings of the principles are not precise, nor is there consensus on their definitions. Yet, the intent is clear. Their primary purpose is to encourage commanders to channel and focus combat power on the decisive points of battle.[1] A 1986 U.S. Army field manual, for example, described "mass" as concentrating "combat power at the decisive time and place."[2] The army's 2001 version of the principle is slightly broader. Mass is described as concentrating the "effects" of combat power, which presumably would include the impact of air strikes, artillery fires, and psychological operations, as well as the placement of troops on the battlefield.[3] Still, the definition of mass remains oriented on war fighting.

The shift in the army's description of mass illustrates how elastic the principles are in practice. Thus, conceptually, the objective of mass, and other principles as well, might be extended beyond combat to include logistics and other theater support operations, and perhaps enlarged to include activities far from the battle zone, such as mobilizing and harnessing science and technology or commercial activities. In practice, they rarely are.[4]

The reason that the principles of war are not applied beyond the battlefield, and why they may have little utility to future wars, has much to do with their origin. The principles emerged during a period of military history that coincided with the rise of the European state.[5] In medieval Europe, before the rise of the strong states, conflict was largely an activity of the private sphere. Sovereigns might declare wars, but they were largely fought by contract armies and private financing.[6] That changed in the early modern era— the fifteenth to seventeenth centuries.

A hallmark of the ascendancy of the nation-state was its capacity to harness the means to wage war.[7] Wresting control of the instruments of warfare from the private sector and making the management of violence a state-run activity was an imperative of the nation building process. The principal theme of Machiavelli's *Art of War,* written at the beginning of the nation-state building period, offers a case in point. Machiavelli described the challenges of controlling the tools of war for weak or emerging states struggling to make the transition from pre-modern societies to sovereign nations that could compete with regional powerhouses.[8]

The recognition of war as a single sphere, where the state played a principal role in both generating combat power and taking it on campaign, reached its apex during the Napoleonic Wars at the dawn of the eighteenth century. Napoleon harnessed the political economy, mobilized the population, and commanded armies in the field. Napoleon's competitors adopted similar practices. In turn, the experiences and interpretations of the Napoleonic wars dominated military history and theory, particularly in Britain and the United

States. The seminal influence of the writings of the Prussian military theorist Carl von Clausewitz, who authored his own version of the principles of war, is one example.[9]

Admittedly, states were far from omnipotent. In practice, throughout the era of the rise of modern Europe there was a strong private sphere, a realm of trade, industry, and independent civil society that sovereigns would often have to negotiate with in order to muster the might for battle. Nor did all wars resemble conventional Napoleonic campaigns, with armies or fleets aligned under national banners for climactic engagements on fields or at sea. As historian Jeremy Black points out, there is good reason to doubt the notion of "a linear continuum of progress towards warfare in the modern world."[10] But military theorists and their counterparts in political philosophy wrote of wars and states as idealized types and assumed that the future would be an extension of the past. Hegel wrote, for example, about the inevitable triumph of the nation-state. Likewise, military theorists assumed that the state's ability to harness the people in support of war would only grow over time. Such thinking culminated in the discussion of "total war," describing the capacity of the state to direct all the energies of society toward supporting military power, even though in practice states never achieved such a degree of control.[11]

The post-Napoleonic age saw substantial effort to codify the principles and the lessons of the successes and failures of the age's great captains. In an era when state control of warfare was considered axiomatic, it was little wonder that the commentators on the principles of war gave the instruments that generated and sustained combat power such scant attention. War planning, by and large, assumed that the state would harness the economic, demographic, and diplomatic power of the state from the onset of conflict. Theorists centered their attention on where they perceived wars were won and lost on battlegrounds and in engagements at sea.

While the Middle Ages saw war as an instrument of the private sphere, and the early modern period claimed war as rightfully the dominion of the public sphere, the twentieth century presented military planners with something different entirely. War transcended both spheres—the public and the private. States developed an unprecedented ability to draw on the productive capacity of the nation. Virtually all the growth in the modern nation-state occurred in the private sector. Tapping the manpower, wealth, innovation, and expertise of the private sector was a prerequisite for war. During World War II, for example, the United States dedicated an amazing 40 percent of its gross domestic product (GDP) to the war effort and put more than 10 million men in uniform, virtually the entire male population of the country eligible for the draft. Yet, this mobilization was done with an increasing recognition that harnessing national power required negotiation between the public and private sectors.[12]

Despite the fact that modern war had become a public-private partnership, the principles of war did not evolve to reflect this reality. Rather than emphasizing the importance of harnessing the power of the nation for conflict, the principles were interpreted ever more narrowly to cover strictly military operations, firmly within the realm of the public sphere. Over the course of the twentieth century, the military's role in combat diminished. For example, in reflecting on the difficulties of mobilization during World War I, the army established an Industrial College to train officers in mobilization and procurement policies. While a student at the college, Dwight Eisenhower wrote his course paper the War Department's role in industrial mobilization.[13] The Army War College also emphasized industrial mobilization, and the War Department war plans also addressed the issue.[14] After World War II, the Industrial College became the Industrial College of the Armed Forces. But the notion that the Pentagon should create its own military planners who understood industry, and in times of trouble would direct industrial mobilization, never gained ascendancy during the Cold War. In practice, the school merely trained officers who were familiar with industry. The idea that the military would play a significant role in directing the economy was dismissed. Generals were never expected to run a wartime economy or play a prominent role in the private sphere's part in war. Their focus was not to stray too far from the combat zone.

The increasingly narrow interpretation of the principles of war can be explained by the evolution of civil-military relations theory over the course of the modern era as described in Samuel Huntington's seminal work, *The Soldier and the State*. Civil dominance over the military was maintained by "carving off for it a sphere of action independent of politics."[15] Under the "normal" theory of civil control, a sharp division is maintained between political decisions and military operations. In other words, the military should stick strictly to military matters. Political leaders would handle larger affairs, including the negotiation with the private sphere for manpower, industrial production, and other instruments of power that might be needed to develop and sustain the means to fight.

Huntington did not invent the notion of separate spheres for policy and military action. *The Soldier and the State* codified the military's perception of its role as a professional institution that had evolved over the course of the nation's history. Huntington's writings, however, did much to sharpen the armed forces' self-image. The experience of the Vietnam War served to further solidify the military's belief that it should stick to its own sphere. Officers saw the failure in Vietnam in classic "Huntingtonian" terms. "The nation went to war," wrote H. R. McMaster, a young army officer who in 1997 penned a scathing critique of the Joint Chiefs of Staff, "without the benefit of effective military advice from the statutory responsibility to be the nation's 'principal military advisors.'"[16] The prescription for addressing political intrusion into the military sphere also remained consistent. When political leaders transgressed into decisions that properly belonged to soldiers, the generals' options were to protest or resign, not crossover into the realm of politics. Conversely, military leaders should "stay within their own lane," focusing exclusively on the war-fighting duties.

Given the military's predilection for clear "red lines" between the public and private spheres, it is unremarkable that modern military theorists made little effort to extend the principles of war much beyond the battlefield. Additionally, given the significant influence of the U.S. military thinking on civil relations with its friends and allies, it is equally understandable that the American view is close to being considered a global gold standard.[17]

## Power and the Principles in the Twenty-first Century

Notwithstanding the major intellectual obstacles to revamping the principles of war, expanding them to encompass all the elements of war, both public and private, will be a daunting task. Arguably, the factors that might have the most dramatic impact on the future conduct of war might have more to do with the ways and means by which national combat power is generated by both state and non-state actors. These factors include emerging technologies, the increasing capacity of the private sector to perform traditional military missions, and the decreasing ability of developed states to allocate productive resources to war-fighting tasks. Together, they could significantly diminish the capacity of states to shape how the instruments of war are employed and severely limit the ability of commanders to determine their own destiny in battle.

### War and Disruptive Technologies

Technology has always been a factor in shaping the future of war, but its impact is far from deterministic. Much contemporary discussion on military history and the impact of technology on military transformation misses the mark. Technology does not define future ways of war. As Williamson Murray and MacGregor Knox concluded in an anthology of the dynamics of military revolution, scientific development and new weapons systems may stimulate change, but the conduct of warfare is shaped by larger economic, political, and geostrategic factors.[18]

The impact of future technologies will likely be the same. They might unleash or accelerate social and cultural changes that reshape the nature of war, but it is unlikely they will simplify or define how combat is conducted. Technology will always be a "wild card" in war's future. Future technological change, however, will diverge in character from the experiences of the last century. Since World War II, militaries have largely pioneered the technologies that were the most critical to military competition. In the United States, for example, from jet aircraft and nuclear weapons to stealth technologies and precision-guided weapons, the Pentagon largely set the course of investments in science and technology, shaped research and development programs, and determined how disruptive new technologies would be applied to battle. The impact of the public sector defense research effort was pervasive and dramatic.[19] The twenty-first century will be different.

In the future, the private sector, not the government, will likely make the largest investments in the basic science research and product development that create the technologies with the greatest capacity to change the nature of combat. In turn, how the private sector chooses to develop these technologies, apart from the guidance or prohibitions established by governments, may determine how future conflicts are fought.

Trends in information technology development offer a clear example. During the Cold War, the government financed much of the cutting-edge research on computers and related electronics that resulted in new combat capabilities. Today, the government is virtually dependent on the private sector for advances in information technology. One of the emerging operational concepts of twenty-first-century warfare is often called "network-centric" operations. Network-centric operations generate increased operational effectiveness by networking sensors, decision makers, and forces to achieve shared awareness, increased speed of command, higher tempo of operations, greater efficiency, and a degree of self-synchronization. Network-centric capabilities, however, are being assembled with systems integration technologies, many of which are already widely commercially available, including technologies that facilitate passing high volumes of secure digital data, create ad hoc networks, integrate disparate databases, and link various communication systems over cable, fiber-optic, wireless, and satellite networks. In effect, many of the concepts for network-centric warfare and how it is being implemented are significantly influenced by how the private sector has evolved in a twenty-first-century knowledge economy.

The growing dependence of modern militaries on commercial information technologies illustrates one way in which war in the twenty-first century will be different. Emerging technologies with the greatest potential to change the nature of military competition are being spearheaded not by defense departments and ministries, but by individual entrepreneurs, multinational conglomerates, start-up companies, investors, stockholders, and Wal-Mart shoppers. Militaries are already grappling with understanding and harnessing information technologies and the prospects for cyberwarfare, but these challenges may represent merely the dawn of an age in which military competition is defined by commercial research and development and consumer choice. Several candidate technologies have already emerged that may shape the character of war beyond the capacity of the public sphere to control or even influence. Understanding how they might impact military competition could provide far more insight into fighting in the future than mastering the principles of war.

## Data Mining and Link Analysis

The next set of technologies that might dramatically affect the conduct of combat are closely related to the world's growing dependence on information technologies. We live in an age increasingly awash in commercial and government data. The commercial sector is leading the drive to provide tools to harness vast, disparate amounts of information. Data mining and link analysis technologies provide the means to exploit larger and larger amounts of raw facts and figures. Data mining is a "technology for analyzing historical and current online data to support informed decision making."[20] It involves identifying patterns and anomalies from the observation of vast data sets.

The primary goals of data mining are prediction and description. Prediction involves using some variables or fields in the database to predict unknown or future values of other variables of interest; description focuses on finding human-interpretable patterns to describe the data. Description concerns increasing knowledge about a variable or data set by finding related information.[21] This second aspect of data mining is often referred to as link analysis. Whereas data mining attempts to identify anomalies in vast amounts of information, link analysis technologies sift through databases to find commonalties, looking for relationships among individuals, organizations, and other entities. Link analysis is the process of analyzing the data surrounding the suspect relationships to determine how they are connected—what links them together.

Network-centric operations seek to exploit the capability of commercial systems integration technologies to get the right information to the right place at the right time to enable combat action. Data mining and link analysis offer the potential to create unprecedented amounts of new knowledge from seemingly unrelated bits of data. The opportunities to enhance intelligence and early warning, situational awareness, and decision making, using information on and off the battlefield, could be unprecedented, and will largely be defined by how the commercial sector elects to develop these technologies. The commercial demands and capacity for data mining are significant and growing, due largely to companies like Lexus-Nexus and ChoicePoint, which have accumulated vast amounts of private and unrestricted public sector data.

Commercial firms have already begun to offer their tools to government for national security purposes. One example is Multistate Anti-Terrorism Information Exchange Information Sharing (MATRIX), created independently after 9/11 by a private sector firm using commercial technology to provide timely information sharing and exchange of terrorist and criminal information. MATRIX is the law enforcement equivalent of an Internet search engine that accesses public and commercial databases. The MATRIX program enables data analysis with an application called Factual Analysis Criminal Threat Solution (FACTS), which combs through existing nonintelligence data sources. FACTS integrates the diverse data from different storage systems in an attempt to identify, develop, and analyze information related to terrorist or criminal activities.[22]

In contrast to the commercial sector, the military's own efforts to develop data mining and link analysis technologies have been halting and ineffectual. The Defense Advanced Research Projects Agency undertook a program to examine data mining technologies. Known as the Terrorism Information Awareness system, the project examined how the analysis of data might be used to identify or predict terrorist behavior.[23] Congress abolished the project because of civil liberty and privacy concerns.

## Biotechnology and Beyond

Beyond information and cyberwarfare, technologies that completely recast the nature of military competition might emerge. One serious candidate is biotechnology. Biotechnology is one of the fastest growing commercial sectors in the world. The number of biotechnology companies in the United States alone has tripled since 1992. These firms are also research intensive, bringing new methods and products to the marketplace every day. Many of the benefits of this effort are largely dual use, increasing the possibility that knowledge, skills, and equipment could be adapted for a biological agent program. Commensurate advances in information technologies, known as bioinformatics, are accelerating rapid advances in biotechnology.[24]

As the global biotechnology industry expands, nonproliferation efforts will have a difficult time keeping pace with the opportunities available to field a bioweapon.[25] And weapons are not the only potential contribution this sector could make to new ways of war. Biotechnology may reshape medical practices on and off the battlefield and human performance, allowing for unprecedented levels of individual achievement and endurance.

Rather than driving the biotechnology revolution, the federal government is a fairly minor customer for multibillion-dollar transnational industry. Project Bioshield, a post-9/11 homeland security initiative to develop new vaccines and other prophylaxes and therapeutics against bioterrorist attacks, offers one example. The government has allocated more than $6 billion over five years to the program. But critics complain that that sum is too modest to attract the attention of major commercial research and development efforts. Nor is the United States alone on the cutting edge in biotechnology developments. In fact, many developing nations, such as Cuba and India, have very sophisticated research and production programs.

Nanotechnology and microelectronic mechanical systems (MEMS) should also be high on the list of potentially disruptive military technologies. Nanotechnology involves developing or working with materials and complete systems at the atomic, molecular, or macromolecular levels where at least one dimension falls with the range of 1 to 100 nanometers.[26] Working at such a small scale offers such unique capabilities as being able to control how nanodevices interact with other systems at the atomic or molecular level.

MEMS are devices that integrate mechanical elements, sensors, actuators, and electronics on a common silicon chip. While far larger than the molecular-sized machines envisioned by nanotechnologies, their very small size could still revolutionize warfare. The calculus of weight, for instance, would have to be completely revisited in a world populated with MEMS technology. One projection forecasts that in excess of a thousand avionics components in an F-22 could be replaced by thirty-six MEMS-enabled components.[27] Likewise, once methods of manufacture are fully developed, the cost of these systems could be extremely low.

It is difficult to forecast how MEMS and nanotechnologies might change war, since they could potentially recast how everything is made. Current research areas include materials, sensors, biomedical nanostructures, electronics, optics, and fabrication. Materials that have been modified at the nanoscale can have specific properties incorporated into them. For instance, materials can have coatings that make them water repellant or stain resistant. "Nanoscale sensors are generally designed to form a weak chemical bond to the substance of whatever is to be sensed, and then to change their properties in response (that might be a color change or a change in conductivity, fluorescence, or weight)."[28] Biomedical nanostructures, by design, interact with people at the molecular level, allowing for targeted drug delivery, adhesive materials for skin grafts or bandages, etc. Nanoscale electronics can help to shrink computer circuits even further and to make them more efficient. Nanoscale optics allows materials that fluoresce to be tuned at the nanoscale to change specific properties under certain conditions. Fabrication at the nanoscale level offers the potential of creating devices from the atom up, as opposed to having to shrink materials down to the needed size.

According to a RAND report, there are numerous future applications for nanotechnology, although most face at least some technical hurdles. These include nanofabricated computational devices like nanoscale semiconductor chips, biomolecular devices, and molecular electronics.[29] If integrated with microsystems and MEMS, then the additional uses for nanotechnology— smart systems on a chip and micro- and nanoscale instrumentation and measurement technologies—are even greater.[30]

There are undoubtedly innumerable military applications for these technologies. Government research has pioneered developments in this area. In 2004 the federal government spent about a $1 billion on nanotechnology research. That effort is rapidly being eclipsed by private sector investments in the United States, Europe, and Asia.[31] Within the next two decades, the nanotechnology industry may mirror the business of information technology, with military uses largely shaped by how the commercial sector elects to develop the technology.

## Private Sector, Public Wars

As in the past, technology will likely not be the only factor that drives military competition. The evolving character of the private sector could be another aspect of twenty-first-century global change that dramatically influences the nature of conflict. The global free market has become a reality, and commensurate with this economic condition is the emergence of an unprecedented capacity for the private sector to expand, innovate, and adapt to market needs, including an ability to provide what once were considered military services offered solely by national powers.

The trend for militaries to increasingly outsource logistical and support functions is well established. Added to that, however, is the emerging use of private sector companies to provide traditional combat services, ranging from training soldiers to patrolling streets.[32]

The increasing importance of privatized military services was particularly apparent during post-conflict operations in Iraq. Among the many tasks that the private sector can perform, security assistance is the most essential. Establishing security is a precondition for conducting post-conflict operations. In particular, establishing effective domestic security forces must be the highest priority. Private sector firms have a demonstrated capacity to provide essential services—administrative support, training, equipping, and mentoring, among others—as well as to augment indigenous police and military units. In Iraq, these services were essential for both standing up the Iraqi security forces and augmenting the security provided by U.S. military troops. Private sector assets can assist in providing an important bridging capability during the period when American military forces withdraw and domestic forces take over.[33]

A reliance on private sector assets in war is likely irreversible. Unlike the public sector, the private sector is bred for efficiency. Undoubtedly, if left to its own devices, it will always find the means to provide services faster, cheaper, and more efficiently than governments. In addition, as governments lose their monopolies over the technologies and means to generate combat power, then their capacity to retain military prowess as essentially a public activity will also be lost.

As long as free markets proliferate, the reemergence of the private sphere of war is inevitable. Nations that seek to hold back against this trend and limit the participation of the private sector will be left behind, as they will lack the capacity to keep up with states that can harness the power of the marketplace.

On the other hand, there is good reason for liberal, developed states not to fear the reemergence of a prominent role for the private sector in war. For there is little likelihood that the private sector's place in war will attend the rise of a new middle ages with sovereigns losing their capacity to manage violence. "Capitalism," as Fareed Zakaria cogently argues, is not "something that exists in opposition to the state. . . . A legitimate, well-functioning state can create the rules and laws that make capitalism work."[34] Unlike medieval kings, modern nations can use the instruments of good governance to bound the role of the private sector in military competition.

The example of the United States illustrates the means that modern, liberal states have both to enable and to harness the commercial capabilities of warfare that might remain partially or even entirely in the private sphere. The means available to moderate interaction between the public and the private sphere include the following:

- A well-established judicial system
- An activist legislative branch with its own investigatory instruments, such as the General Accountability Office
- The "60 Minutes" factor—an independent press
- Public interest group proliferation, which provides a wealth of independent oversight and analysis
- An enabled citizenry with ready access to a vast amount of public information

These assets offer unprecedented means to balance the public and private spheres, not just to constrain government conduct, but also to limit the excesses of the commercial sector.[35] In fact, these capabilities might argue that in the long term, liberal free market democracies will prove far more effective at mastering the capacity of the private sector in the twenty-first century than authoritarian states with managed economies.

That said, the role of the private sector in war raises innumerable legal, ethical, and practical issues that have to be dealt with.[36] Marrying the private sector's capacity to innovate and respond rapidly to changing demands with the government's need to be responsible and accountable for the conduct of operations is not an easy task. It will require militaries to think differently about how best to integrate the private sector into public wars. Nor can generals do this thinking in isolation. Modern military operations are an interagency activity that requires the support of many elements of executive power.

The judicial and legislative branches of government have important roles to play as well. Indeed, many of the most important instruments for constraining the role of the private sector in war lay in their hands.

## Checkbook War

The state of public financing is a third factor that might govern the conduct of conflict far more dramatically than how any general implements the principles of war. Developed nations could find in the decades ahead that the nature of mature economies and demographics significantly constrain the amount of resources that they can dedicate to military campaigns.[37]

Government in the developed world has expanded substantially during the past century. The United States stands at the apex of this trend. One of the best measures of the burden that the federal government, as a whole, imposes on the national economy through its spending policies is the percentage of GDP taken up by outlays. During the nation's first 140 years, the federal government rarely consumed more than 5 percent of the GDP. In accordance with the U.S. Constitution, Washington focused on defense and certain public goods while leaving most other functions to the states or the people themselves.

The Great Depression brought about President Franklin Roosevelt's New Deal, a program that expanded government in an attempt to relieve poverty and revive the economy. President Roosevelt created the Social Security program in 1935 and also created dozens of new agencies and public works programs. Although the economy remained mired in depression, the federal government's share of the GDP reached 10 percent by 1940.[38]

World War II pushed the United States into the largest war mobilization effort the world has ever seen. From 1940 through 1943, the federal government more than quadrupled in size—from 10 percent of GDP to 44 percent. The enormity of this 34 percent government expansion cannot be understated: an equivalent expansion today would cost $3.9 trillion, or $37,000 per household (compared with $2.1 trillion or $20,000 per house hold in 2003). Even with a top income tax rate of 91 percent, the nation could not fund World War II on tax revenues alone. The nation ended the war with a national debt larger than the GDP (which is today three times the size of today's national debt).

Following the war, Washington's share of the economy fell back to 12 percent of GDP in 1948. Among the many ways this era changed America was by making its citizens more comfortable with expanded federal powers.

The Supreme Court, under threats from President Roosevelt, had upheld these new federal powers by broadly interpreting the Constitution's Interstate Commerce Clause. From 12 percent of GDP in 1948, federal spending began a thirty-five-year growth spurt that finally reached a peacetime record of 24 percent of GDP in 1983.

In the long decades of federal expansion from the end of World War II to former President Ronald Reagan's election, Washington expanded into several new policy areas, creating the Departments of Health, Education and Welfare (in 1953; eventually it became Health and Human Services), Housing and Urban Development (in 1965), Transportation (in 1966), Energy (in 1977), and Education (in 1979; it had been a part of Health, Education and Welfare).

Federal spending generally fluctuated at just over 20 percent of GDP throughout the 1980s and early 1990s. However, in the last few years spending has sharply increased again as the war on terrorism collided with domestic spending.

Over time, the composition of federal spending has evolved as well. Between 1962 and 2000, defense spending plummeted from 9.3 percent of GDP to 3 percent. Nearly all of funding shifted from defense spending went into mandatory spending (mostly entitlement programs), which jumped from 6.1 percent of GDP to 12.1 percent during that period.

The importance of this evolution cannot be overstated. For most of the nation's history, the federal government's chief budgetary function was funding defense. The two-thirds decline in defense spending since 1962 has substantially altered the make-up and structure of the U.S. national defense.

After twenty-eight consecutive years of budget deficits, the 1998 fiscal year ended with a $69 billion budget surplus. These budget deficits, which had reached 6 percent of GDP in 1983, were eliminated by a combination of three factors. First, real defense spending plummeted by 30 percent in the 1990s as a result of the end of the Cold War. Second, tax revenues reached their highest level since World War II as a result of the economic boom. Third, legislative gridlock between Democratic President Bill Clinton and the Republican Congress doomed most new spending initiatives and allowed spending growth to slow to a crawl.

The arrival of budget surpluses, however, saw federal spending accelerating once again. These spending increases went mostly unnoticed because tax revenues continued pouring in at a pace rarely seen in American history, culminating in a $236 billion budget surplus in 2000.

With the economic boom predicted to continue indefinitely, a January 2001 Congressional Budget Office report forecast a cumulative budget surplus of $5.6 trillion over the next decade. In response, Congress adopted President George W. Bush's tax cut proposal, which was estimated to reduce revenues by up to $1.3 trillion over the decade, and in tandem accelerated the federal spending spree. Lawmakers were not yet aware that the technology bubble had burst, the economy was falling into recession, and tax revenues would soon begin plummeting. Against this budget backdrop came the September 11 attacks. A reeling economy was suddenly faced with a massive defense and homeland security mobilization.

Between 2001 and 2004, wars with Afghanistan and Iraq were funded by a 48 percent increase in defense spending. Homeland security spending, which had not even existed as a category before September 11, leapt from $16 billion to $33 billion. The low defense spending that helped bring balanced budgets in the late 1990s was over.

Although not quite reaching the levels it did under President Lyndon Johnson, federal spending during the war on terrorism has more closely reflected the Vietnam-era spending binges than the spending restraint of World War II and the Korean War. Spending not related to defense and 9/11 increased by an average of 5 percent per year from 2001 through 2003. That two-year, 11 percent increase in non-security spending represents the fastest growth in a decade. At a time when defense and homeland security priorities require especially tight non-security budgets, lawmakers have not made necessary trade-offs, and in fact, have *accelerated* non-security spending growth.

Today, spending on defense and homeland security in the United States stands at about 4.5 percent of GDP, the highest level of investment since the end of the Cold War. Still, this represents, on average, less than half of what the nation spent before the fall of the Berlin Wall. And, unlike the Cold War period, post–Cold War defense spending is faced with unprecedented competition for federal dollars with mandatory government spending on entitlement programs.

Mandatory outlays for programs such as Social Security, Medicare, and Medicaid are, and will continue to consume, ever-larger percentages of the federal spending and total GDP. As a result, they will apply increasing pressure to crowd out the resources available to the government's traditional primary mission—providing security to the nation. And that will likely change how future wars are fought.

Nor is the United States unique in facing this dilemma. European nations, which already spend on average less than 2 percent of their GDP on defense and spend much more of their national budgets on social services and entitlements, face similar predicaments. As potential economic and military powerhouses like China and India move from the ranks of the developing to the developed, they will also confront the same kinds of challenges. In fact, China and India might encounter even more pressure to rein in defense spending since they will have much larger populations demanding higher levels of social services.

## The Demographic Dilemma

Demographic changes could well exacerbate the strain on developed economies to undertake military competition. As historian John Chambers summarized in his history of conscription in the United States, militaries are shaped as much by "trends in society as by the nature of war itself."[39] As nations develop, their population growth slows, and the average age of the population increases as, commensurately, does the cost of manpower. The result is an increasingly shrinking population available to run with the dogs of war.

In the future, the changes caused by the dynamics of demographics will accelerate, altering the character of modern military forces and their attributes as an instrument of battle. While the rate of population increases in developed countries will slow, the total size of the population will continue to grow, but less of the national polity will be suitable for military service. The cost of military manpower will also increase as armed forces find themselves competing with the private sector for talented young people. The total size of militaries in relation to the nation as a whole will likely continue to decrease in the years ahead. At the same time, as populations age militaries will likely diversify those they seek to bring into the ranks to compensate on the shrinking pool of traditional military-age males. Thus, national forces might include more women, individuals with disabilities, noncitizens, and older persons, as well as a much higher percentage of reserve component personnel. Some analysts also argue that as the military comes to reflect an ever-smaller portion of the nation, a gap will develop that could threaten the nature of civil-military relations.[40]

At the same time, the use of conscription as a form of military service could well decline. With a trend toward fielding forces armed with more technology and more sophisticated skills, short-service conscription will be seen increasingly as inadequate, not allowing sufficient time to train forces and requiring excessive costs to frequently retrain new recruits. Likewise, with militaries becoming smaller in developed nations, conscription will be seen increasingly as socially divisive since it will be difficult to equitably draw on the available eligible pool of recruits. In all likelihood, military drafts will be viewed as an inefficient and ineffective means for mobilizing manpower in developed, liberal democracies.

Finally, the impact of economic and demographic trends on the developed world could exacerbate the gap between how nations and non-state actors wage war in the twenty-first century. The span between the military capabilities of undeveloped or failing and failed states, developing nations, and the developed world may only grow in the decades ahead. As a result, the twenty-first century could well see a witch's brew of countries and non-state actors, such as transnational terrorist groups, fighting with very different means, employed in a polyglot of ways, toward a dizzying array of divergent ends. Thus, economic, cultural, and social trends could produce wars with an unprecedented level of asymmetrical engagements,[41] a further burden on the poor principles of war, maxims drafted centuries ago to govern the conduct of conventional conflicts.

## Spheres and the Future

The future "ain't" what it used to be. Most futurist projections envision tomorrow as an extension of current trends. The past, however, is not always prologue. The character of war in the twenty-first century could be significantly divergent from the inevitable march toward modern conflict that stretches from the Middle Ages to the present. War in the twenty-first century will be neither a private nor a public matter, but a civil activity that spans both worlds, with each realm having a substantial amount of autonomy and influence. Indeed, the main argument presented here is that the private sphere of warfare is on the ascendancy, destroying the nation-state's monopoly on the management of violence. As a result, consideration of military matters can't be confined to the traditional place of battle. Frameworks, such as the principles of war, will be increasingly seen as anachronistic and counterproductive. Only military thinkers that understand how factors beyond the battle shape the conduct of conflict will earn the moniker "genius for war."

## Notes

1. Paul Murdock, "Principles of War on the Network-Centric Battlefield: Mass and Economy of Force," *Parameters* (Spring 2002): 86.

2. U.S. Army, *Operations,* Field Manual 100-5, (1986), p. 174.

3. U.S. Army, *Operations,* Field Manual 3-0, (2001), pp. 4–11.

4. See, for example, a recent reconsideration of the principles in Robert R. Leonhard, *The Principles of War for the Information Age* (Novato, Calif.: Presidio Press, 2000). Consistent with past writings on the principles Leonhard is overwhelming concerned with their application on the operational aspects of conflict.

5. John Alger, *Quest for Victory: The History of the Principles of War* (Westport: Greenwood, 1982).

6. See, for example, the discussion of the challenges faced by Britain's Henry V in mustering military forces for his campaigns in France in Christopher Allmand, *Henry V* (Berkeley: University of California Press, 1992), pp. 205–32.

7. Described in Philip Bobbitt, *The Shield of Achilles: War, Peace, and the Course of History* (New York: Alfred A. Knopf, 2002), pp. 69–74. Bobbitt also describes the historiographical debate over the complex relation between military developments and the process of state-formation in Europe.

8. Niccolo Machiavelli, *Art of War,* ed. Christopher Lynch (Chicago: University of Chicago Press, 2003).

9. Christopher Bassford, *Clasuewitz in English: The Reception of Clausewitz in Britain and the United States, 1815–1945* (Oxford: Oxford University Press, 1994), pp. 181–82.

10. Jeremy Black, *Western Warfare, 1775–1882* (Bloomington: University of Indiana Press, 2001), p. xiii.

11. The debate over the utility of the concept of "total war" is a subject of some controversy. This issue was addressed in some detail in a series of workshops sponsored by the German Historical Institute in Washington, D.C., and published in a series of anthologies. See, in particular, Stig Forster and Jorg Nagler, eds., *On the Road to Total War: The American Civil War and the German Wars of Unification, 1861–1871* (Cambridge: Cambridge University Press, 2000).

12. Even in industries where the military took virtual direct control over business operations, government authority was far from absolute. Running war plants during World War II faced significant civil-military challenges. See, John H. Ohly, *Industrialists in Olive Drab: The Emergency Operation of Private Industries During World War II,* ed. Clayton D. Laurie (Washington, D.C.: Center of Military History, 1999).

13. Dwight D. Eisenhower, "Brief History of Planning and Preparing for Procurement and Industrial Mobilization Since the World War," course paper for Army Industrial College, October 2, 1931, http://www.ndu.edu/library/ic7/L32-002.pdf.

14. Henry G. Gole, *The Road to Rainbow: Army Planning for Global War, 1934–1940* (Annapolis, Md.: Naval Institute Press, 2003), p. 89.

15. Eliot A. Cohen, *Supreme Command: Soldier, Statesman, and Leadership in Wartime* (New York: The Free Press, 2002), p. 227.

16. H. R. McMaster, *Dereliction of Duty: Lyndon Johnson, Robert McNamara, the Joint Chiefs of Staff, and the Lies that Led to Vietnam* (New York: HarperCollins, 1997), pp. 324–25.

17. While the American view may be dominant there is, of course, a wide range of views on the military's proper role in civil military relations. See, Michael C. Desch, *Civilian Control of the Military: The Changing Security Environment* (Baltimore: Johns Hopkins University Press, 1999).

18. Williamson Murray and MacGregor Knox, "The Future Behind Us," in *The Dynamics of Military Revolution, 1300–2050* (Cambridge: Cambridge University Press, 2001), pp. 177–78.

19. Mark L. Montroll, "Maintaining the Technological Lead," in *Transforming America's Military,* ed. Hans Binnendiijk, (Washington, D.C.: National Defense University Press, 2002), pp. 349–50.

20. Committee on the Role of Information Technology in Responding to Terrorism, Computer Science and Telecommunication Board, *Information Technology for Counterterrorism: Immediate Actions and Future Possibilities,* eds. John L. Hennessy, David A. Patterson, and Herbert S. Lin (Washington, D.C.: The National Academies Press, 2003), p. 68.

21. Usama Fayyad et al., "From Data Mining to Knowledge Discovery in Databases," *Artificial Intelligence,* Fall 1996, p. 44. Available at http://www.kdnuggets .com/gpspubs/aimag-kdd-overview-1996-Fayyad.pdf (last accessed November 16, 2004).

22. James Jay Carafano, Paul Rosenzweig, and Alane Kochems, "An Agenda for Increasing State and Local Government Efforts to Combat Terrorism," Heritage Backgrounder #1826, February 24, 2005. Available at http://www.heritage.org/ Research/HomelandDefense/bg1826.cfm.

23. Paul Rosenzweig, "Proposals for Implementing the Terrorism Information Awareness System," Heritage Legal Memorandum, #8, August 7, 2003. Available at http://www.heritage.org/Research/HomelandDefense/lm8.cfm.

24. Bioinformatics is the use of databases and analytical tools for genome analysis and innovations in molecular biology. The potentially dramatic impact of bioinformatics is illustrated by a forecast from *The Economist* which holds that the market for products derived through computer biology is expected to be worth $40 billion in three years. One study holds that bioinformatics can reduce the cost of drug development by 18 percent and cut a year off developmental timelines. Economist.com, "The Race to Computerise Biology," (December 12, 2002), n.p., [http://economist.com/ Story_ ID=1476685]. Among its many applications to bio-warfare, bioinformatics can facilitate the identification of pathogens. See, for example, Statement of D. A. Henderson, Director, Office of Public Health Preparedness, Department of Health and Human Services (December 5, 2001), n.p., [http://www.house.gov/science/full/dec05/ henderson.htm]. Bioinformatics also holds great promise in developing therapeutic responses

to a bio-attack. For example, studies show that variations in individual responses to therapeutic drugs are affected by genetic polymorphisms (variants in enzymes caused by slightly different amino acid sequences). Pharmacogenetics employs bioinformatics to assist in decoding and mapping millions of polymorphisms across the human genome, which can provide insights into the links between disease-causing genes and drug-response genes, facilitating the development of new therapeutic strategies. Michael M. Shi, "Diagnostics meets Therapeutics: the Impact of Pharmacogenetics," *Drug Discovery Today* 7/23 (December 2002): pp. 1161–1162.

25. Jonathan B. Tucker, "Putting Teeth in the Biological Weapons Convention," *Issues in Science and Technology* (Spring 2002): pp. 71–77.

26. Daniel Ratner and Mark A. Ratner, *Nanotechnology and Homeland Security* (Upper Saddle River, N.J.: Prentice Hall Professional and Technical Reference, 2004), p. 13.

27. Shannon L. Callahan, "Nanotechnology in a New Era of Strategic Competition," *Joint Force Quarterly* 26 (Autumn 2000): p. 21.

28. Daniel Ratner and Mark A. Ratner, *Nanotechnology and Homeland Security* (Upper Saddle River, N.J.: Prentice Hall Professional and Technical Reference, 2004), p. 21.

29. Philip S. Anton, Richard Silberglitt, and James Schneider, *The Global Technology Revolution: Bio/Nano/Materials Trends and Their Synergies with Information Technology by 2015,* (Santa Monica, Calif.: RAND, 2001) MR-1307-NIC, pp. 25–28.

30. Ibid., p. 28–30.

31. Rick Weiss, "Nanotechnology in Booming Biggest in U.S., Report Says," *The Washington Post,* March 28, 2005, p. A6.

32. Described in P. W. Singer, *Corporate Warriors: The Rise of the Privatized Military Industry* (Ithaca, N.Y.: Cornell University Press, 2003).

33. James Jay Carafano, "The Pentagon and Postwar Contractor Support: Rethinking the Future," Heritage Executive Memorandum #958, February 1, 2005, http://www.heritage.org/Research/NationalSecurity/em958.cfm.

34. Fareed Zakaria, *The Future of Freedom: Illiberal Democracy at Home and Abroad* (New York: Norton, 2003), p. 76.

35. The list was adopted from James Jay Carafano and Paul Rosenzweig, *Winning The Long War: Lessons from the Cold War for Defeating Terrorism and Preserving Freedom* (Washington D.C.: Heritage, 2005), pp. 88–89.

36. See, for example, T. Christian Miller, "Where Guarding a Life Hinders Doing the Job," *The Los Angeles Times* (March 25, 2005), p. A1.

37. The section on "checkbook war" was adopted from James Jay Carafano and Paul Rosenzweig, *Winning The Long War: Lessons from the Cold War for Defeating Terrorism and Preserving Freedom* (Washington D.C.: Heritage, 2005), pp. 138–40. The

author would like to recognize the work of Baker Spring, the F. M. Kirby Research Fellow at The Heritage Foundation, who substantially contributed to this research.

38. Unless otherwise noted, all statistics in this section come from Office of Management and Budget, *Budget of the U.S. Government, Fiscal Year 2005* (Washington, D.C.: U.S. Government Printing Office, 2004).

39. John Whiteclay Chambers III, *To Raise on Army: The Draft Comes to Modern America* (New York: The Free Press, 1987), p. 276.

40. Lindsay Cohn, "The Evolution of the Civil-Military 'Gap' Debate," Paper prepared for the Project on the Gap Between the Military and American Society, Triangle Institute for Security Studies, 1999. Available at http://www.poli.duke.edu/civmil/cohn_literature_review.pdf.

41. See, for example, Stephen Blank, "Rethinking Asymmetric Threats," (Carlisle Barracks, Pa.: Strategic Studies Institute, Army War College, September 2003).

# FOURTEEN

# Teaching, Learning, and Leading

## *The New Mandate*

PAULETTE M. RISHER

*A*gile. Adaptive. Self-directed. Self-aware. Culturally facile. Moral. Techni-cally and tactically proficient. These are among the attributes of the mature military leader. This is what America expects. This is what the military needs. This is what supporting and defending the Constitution of the United States against all enemies—foreign and domestic—demands.

We recruit talented men and women into our proud, tradition-rich, and, today, heavily engaged military services. They learn to lead as all leaders do—through relationships, through the influence of good and bad role models, through trial and error, through doing real jobs, and through experiencing real hardships. And they learn to lead through formal education and training programs.[1]

In the early years of their active and reserve military careers, servicemen and -women are schooled to become technically and tactically proficient. The focus is predominantly on what educators call lower-level learning, i.e., remembering, understanding, and applying facts, concepts, and procedures.[2] The art and science of leadership are taught, but they are ancillary to skill development.

At the mid-career point, a shift to higher-order learning begins. Students are challenged to analyze, evaluate, and create while becoming metacognitive (e.g., to become aware of their own thought processes, to think about thinking).[3] Teaching the art and science of leadership becomes a deliberate, focused effort and professional military education (PME) institutions become the venue of choice.[4]

Military educational institutions have historically been successful in producing competent military leaders, so why then revisit leader education? For the same reasons we are revisiting the principles of war. The complexity, ambiguity, and unpredictability that were once thought unique to the state of war are now an everyday reality, a new steady state. Terrorism, weapons of mass destruction, conventional and unconventional threats challenge our capacity, our capability, our wills, our families, and even our most cherished values and beliefs. Technology has brought precision, speed, lethality, and an unparalleled volume of information. Many of the familiar organizational structures we have grown up with are in a state of flux. Joint, interagency, and multinational operations are assumed to be the norm and independent action the exception.

Amidst this chaos, amidst this noise there are leaders who we expect and entrust to make mature, reasoned decisions. In the hands of these men and women we place our national treasure, our way of life, and our hopes for the future. We owe them an unblinking assessment and, where necessary, committed action to transform our military educational system to prepare them for the challenges they face on our behalf. Army Chief of Staff Gen. Peter Schoomaker said it so well: "Transformation is not about equipment. It's about intellect; it's about judgment; it's about the development of leaders and soldiers. You've got to make that intellectual transformation before you can make the visible transformation."[5]

Before proceeding, some caveats are in order to help quiet the all too human mental and identity defenses that can get in the way of understanding.[6] It is acknowledged that

- The U.S. military arguably does leader development better than any other institution in the United States.
- The quality of teaching in mid- and senior-level PME institutions and service academies is comparable to or higher than that in civilian colleges and universities.
- One can be a great leader without being a student in a service academy or PME institution.
- Military educational institutions will continue to compete for resources with ongoing operations, changes in force structure, acquisition programs, and military construction. With the exception of initial investments in faculty development, every recommendation in this essay can be implemented within existing resources.
- Leadership development is not limited to those at the top or those formally designated to be leaders.
- Significant pressures are being placed on the military educational system to reduce "seat time." There is no excess capacity in today's military

educational institutions. Increasing higher-order learning for more nuanced and sophisticated leader development must not rest on adding more hours, reading more books, writing more papers. It is not about accreditation or graduate degree programs. It must come from fundamental changes in our philosophy and practice of education, not just in courses with "leadership" in the title, but across the curriculum.

This essay explores how to accelerate intellectual transformation and how to systematically realize the challenge of educating leaders *not just about what to think, but how to think*. These views are shaped by academic research in education and allied fields. They are also shaped by the author's experience as a military educator, organizational development practitioner, change agent, commander, and leader. They provide a research-based foundation for higher, deeper, and more personal leader education; propose two critical applications; and provide a map for the journey—a journey that must begin now.

## Educational Vectors

Education, like most disciplines, is in a constant state of flux. New research, new technologies, and new social and cultural imperatives have caused what some characterize as a revolution in education. Whether the changes are revolutionary or evolutionary is a matter of debate, but what is undeniable is that today's best practices differ markedly from those with which most of us, especially those at the more senior ranks, grew up.

A teacher standing in the front of the room lecturing characterized the college classroom of the past and, all too often, it remains so today. The good ones were animated and engaging. The bad ones lectured from their notes (or PowerPoint charts) in a monotone voice. We were expected to listen attentively, take good notes, speak when spoken to, and generally be the passive recipient of enlightened words. All too often we crammed for tests, regurgitated the facts and processes we memorized, and promptly flushed them from our minds because they were irrelevant. Fortunately, three powerful vectors in education today represent a radical departure from this ineffective manner of teaching and learning. It is the confluence of these interrelated vectors that holds the greatest promise for educating today's and tomorrow's leaders.

### Vector 1: Constructivism

There is a growing, interdisciplinary dialogue over the fundamental nature of knowledge and how human beings construct the knowledge by which they lead their lives. The intellectual debates are complex, steeped in various professional lexicons and belief systems, and beyond the scope of this discussion. What is relevant is that many educators have seized upon the basic tenets of constructivism and sought ways of applying them in their classrooms to increase the level and depth of learning.

Constructivist teachers strive to help students create knowledge and meaning, to make sense of their world. The uniqueness of each student, their skills, talents, and multiple intelligences (not just IQ), is recognized and valued.[7] Students are expected to play an active role in their own learning. Students are challenged not just to understand the content of the lesson but also to extrapolate it to their real world. The social context of both learning and the application of that learning are recognized and integral to the teaching process. Students are challenged to be metacognitive. The role of teacher shifts from being "sage on the stage" to "guide on the side."[8] This is the constructivist school of thought, and it is a powerful and pervasive force in education from K-12 to PhD today.

### *Vector 2: Brain Research*

While our knowledge of how our brains work is still relatively primitive, over the last few years a number of discoveries have had a significant impact on classroom practice. Effective teachers guide with deliberateness the mental processes associated with learning, from perception to attention, to working or short-term memory, and finally, to long-term memory through the linking of new learning with something that is personally meaningful. As described by Dr. Patricia Wolfe in *Brain Matters: Translating Research into Classroom Practice,* the challenge is stunning.[9] Ninety-nine percent of all sensory information is discarded almost immediately upon entering the brain.[10] Novelty, intensity, and movement get our intention, but only meaning and emotion strongly influence whether that attention endures.

Without rehearsal or constant attention, information remains in working memory for only fifteen to twenty seconds, and we can hold merely about seven (+/- two) pieces of distinct information at one time. In the traditional classroom settings where lectures dominate, 95 percent of that information will be lost within twenty-four hours, and just 10 to 20 percent of learners today are auditory learners.[11] Finally, information that passes through this gauntlet is transferred and consolidated in our long-term memory as part of our conscious (procedural) or unconscious (declarative) memory. New neural pathways are formed, old ones are transformed. Learning organizes and reorganizes the brain.[12] If we want deeper learning that will guide us beyond the classroom, then we must honor our physiology.

## Vector 3: Self-Direction

The integration of constructivism and brain research into our educational practice is necessary but not sufficient for leader development. The third element of the triad turns us more deeply inward, from how we construct our worlds and brain functions to how we understand and lead ourselves. Learning situations are constructed to give students the opportunity to reflect on their learning, to become aware of their own personality and individuality. Students are challenged to be *self-directed*. According to two leading researchers, Arthur Costa and Bena Kallich, when confronted with complex, ambiguous, and intellectually demanding tasks, self-directed people exhibit the dispositions and habits of mind required to be self-managing, self-monitoring, and self-modifying. They control their first impulse for action and delay premature conclusions. They think about their own thinking, behaviors, biases, and beliefs as well as about the effects that such processes and states of mind have on others and on the environment. They can change themselves.[13] Development of capabilities for self-directedness enables individuals not only to continue their intellectual growth beyond their formal education but also to advance the nature and quality of their life pursuits.[14]

## Moving from Theory to Practice

The integration of these three powerful educational vectors provides a research-based foundation upon which to facilitate the development of leaders, and certainly any number of military educators have recognized and embraced these educational trends. However, these new educational insights are complex and their translation into classroom practice takes time, takes new skills, takes practice, and takes will. It also takes a starting point, a place where these more sophisticated ways can be applied, tried, adapted, and adopted. Two critical leader skills, cultural competence and improving full-spectrum decision making, provide these points of departure and are discussed in some depth in the following sections.

### Cultural Competence

There is a dizzying array of definitions of culture. A well-respected international definition proposed as part of the Global Leadership and Organizational Behavior (GLOBE) research program describes culture as the "shared motives, values, beliefs, identities, and interpretations or meanings of significant events that result from common experiences of members of collectives and are transmitted across age generations."[15] Educational anthropologist Dr.

Frederick Erikson adds texture to the definition by noting that culture is in us and all around us, just as is the air we breathe. It is personal, familial, communal, institutional, societal, and global in its scope and distribution. It is also habitual and for the most part invisible. It is learned and taught outside of awareness.[16]

The experiences of the U.S. military in Afghanistan and Iraq have brought the cultural challenges facing today's military to the forefront. One has only to read Dr. Leonard Wong's poignant interviews with junior officers describing their "crucible" experience in Operation Iraqi Freedom to appreciate the cultural complexity they face every day. Many of these frontline leaders were surprised and ill prepared to meet these cultural challenges.[17]

On other fronts, the U.S. military is moving to an "expeditionary" posture. Military units are being relocated from overseas to the continental United States, and there will be a loss in the depth of collective and individual cross-cultural competence that comes from living, working, and interacting in a foreign culture over an extended period of time.

The challenges we are facing today are not unique to the military, nor are they uniquely ethnic. Management theorist Margaret Wheatly observed, "as the physical and tangible boundaries of our world shrink, the psychological boundaries of people deeply divided by race, ethnicity, ideology, politics, region, inequality, and marginality seem to be headed towards greater impermeability."[18] While these are the traditional cultural borders, military leaders are also faced with working cross-organizational borders in joint, interagency, and multinational collaborations where adapting to cultural issues can be just as daunting.

How, then, do we prepare leaders to be culturally facile? How do we provide acculturation to men and women who are deployable worldwide? We know, as Maj. Gen. Robert Scales, U.S. Army (Ret.) so aptly observed, that "the mission of acculturation is too important to be relegated to last-minute briefings prior to deployment."[19] This brings us squarely back to our military educational institutions where we must begin immediately to focus on developing cultural competency. Leadership courses are the place to begin—but not end.

Dr. Ata Karim, one of the most articulate academics in the integration of "intercultural consciousness" (cultural competency within an ethical and moral framework) into leadership education, observes that "teaching leadership theory without discussion of cultural context results in a great void in students' preparation to meet the contemporary demands and challenges facing leaders in an increasingly interdependent, diverse, and culturally pluralistic world."[20]

In a survey of research, Karim suggests there are eight skills leaders often possess, but might not use, in an intercultural-specific setting. These include

1. cognitive complexity and critical thinking
2. perspective multidimensionality
3. interpretative multiplicity
4. contextual analysis
5. affective resilience
6. tolerance for ambiguity
7. identify integration
8. cultural empathy

He also describes a developmental path, a journey from the mind-set of cultural naïveté to cultural sophistication.[21]

While reshaping leadership courses to explicitly include multicultural competency is a start, it is only when the "cultural mind-set" becomes the dominant paradigm that real transformational learning occurs. It is in this broader arena that the philosophy and practices of multicultural education may be most informative.

A leading voice in multicultural education, Dr. James Banks, proposes a five-component holistic model for multicultural education. The first three elements may serve as a blueprint for military educational institutions.[22]

The first element is "content integration," or the use of examples and content from a variety of cultures in teaching. As part of their instructional practice, military educational institutions integrate history into their curriculum as a framework for developing "military operational art and science." However, with the notable exception of the venerable favorite, Sun Tzu's *The Art of War*, there is a heavy Western bias.[23] While this is to be expected, given that the U.S. military *does* have its roots in Western Europe, it is time to consciously broaden the lens and integrate examples from other parts of the world, not just into military history and operational art courses, but also into the entire curriculum.

The second element of Banks's multicultural education model is "knowledge construction," in which teachers "help students understand, investigate, and determine how the implicit cultural assumptions, frames of references, perspectives, and biases within a discipline influence the way knowledge is constructed." This questioning of the nature of knowledge challenges the Western empirical paradigm that the knowledge produced within it is neutral and objective and that its principles are universal.[24] Explicitly and habitually addressing knowledge construction across the curriculum will not only help develop and mature culturally competent leaders, it will also develop and mature critical thinking skills.

The third element of Banks's multicultural education model is "prejudice reduction," and it fits well with Karim's "leadership developmental progression model."[25] Karim postulates nine assumptions and suppositions about human tendencies and inclinations, including the belief that "most people are culturally encapsulated and ethnocentric in their world view" and "most people perceive themselves as morally decent, interpersonally sensitive and socially just."[26] U.S. military services pride themselves on being value-based organizations, and military leaders do believe they are (and generally are) "morally decent, interpersonally sensitive and socially just." While military leaders are arguably less "culturally encapsulated and ethnocentric in their world view" than the general population by virtue of role demands and expectations, position, and opportunities, there is nevertheless still room to learn and grow.

Military educational institutions who take seriously this mandate to increase cultural competency in military leaders are not only well-advised to consider the work of such multicultural educational leaders as Dr. Banks, but also to look further into academic research for specific approaches and tools. By way of an example, Dr. Kecia Thomas describes an approach that might increase cultural competency and bring to conscious awareness how the "learned and contextualized aspects of our own culture are hidden from us because they have become taken for granted and invisible for those within a culture."[27] She advocates explicitly addressing the assumptions of "psychological privilege" and its resulting ethnocentrism that come from being a member of a cultural majority in the home culture. By "psychological privilege" Thomas refers not to "physical superiority or material wealth, but rather the ability to think of oneself as 'normal,' that is, as the way to being in one's home culture."[28] The most "psychologically privileged" groups in our society are "White and male." (Note the subtlety of wording; this is not "White male," and many of the examples in the article involve psychological privileges afforded White females.)

Thomas further postulates that psychological stress and discomfort increase when members of psychologically privileged groups enter a foreign culture with the expectation that their preferences and power will be reinforced by the world around them. For many it will be the first time they experience what Peggy McIntosh characterizes as "fear, anxiety, insult, injury, or a sense of not being welcome."[29] While minimizing psychological stress and discomfort is important for the individual, there is also evidence that this cross-cultural training and education improves performance.[30] Current research suggests that the better one performs and believes that success is possible, the greater the job satisfaction. This is not a trivial consideration for an "all volunteer" military force that serves the country by choice, not mandate.[31]

## Full-Spectrum Decision Making

Making decisions is an inherent and perhaps defining activity of leadership. These decisions might be easy or complex; pedestrian or life impacting; hurried or deliberate; individual or collective; unique or repetitive. The traditional bane of decision makers has been a dearth of information. However, today there is a plague of too much information and the label "Information Fatigue Syndrome" is finding its way into both popular and academic press.[32] Efforts to harness technology in service to better decision making is in evidence on multiple fronts and are typified by Joint Forces Command's efforts with the Collaborative Information Environment (CIE) and Operational Net Assessment (ONA).[33]

Underlying this bow wave of information, technology, and process is an assumption that good decision making is teachable and learnable. The question is, how do we in fact teach leaders to make better decisions?

The first challenge is to recognize that there is a spectrum of decision-making conditions and processes. At one end of the continuum are those decisions made at an instant, at a glance, at the unconscious level. These are the friend or foe decisions that can often have life-saving or life-taking consequences. The nature, strengths, and liabilities of this type of decision making are described by Malcolm Gladwell in his interesting and thoughtful book *Blink: The Power of Thinking Without Thinking.*[34]

At the midpoint of the continuum are those decisions that have been variously labeled "recognition-primed decision making" or "naturalistic decision making." These are the type of decisions made in situations where the stakes are high and time is short. In his seminal work, *Sources of Power: How People Makes Decisions,* Dr. Gary Kline describes how decision makers such as first responders and military leaders quickly develop mental models, extrapolate from their experience and then try the first workable solution that they can devise, and then adjust as required.[35]

At the other end of the decision-making continuum are those deliberate and often-sequential judgments made within the framework of complex decision making and planning processes. Examples include the U.S. Army's military decision-making process (MDMP), the joint campaign, deliberate and crisis action planning models, and the Joint Operations Planning and Execution System (JOPES).[36] Teaching, learning, and practicing this type of decision making is central to the curriculum of many military educational institutions, particularly at the intermediate service school level, and it should be. However, leaders must be equally skilled in making decisions intuitively in the blink of an eye, when time is short and stakes are high.

Just how does one go about educating leaders to make decisions across this spectrum? There are two answers. The first involves rendering decision making a visible and explicit topic across the curriculum. While expanding one's intellectual understanding of decision making is of value, it is not enough. To gain competence and confidence in full-spectrum decision making, the leader must have the opportunity to practice, experiment, and experience both the process and the results. The second answer can be found in what has variously been called situated or contextual learning.

Although the military has long been a proponent of exercises, case studies, and simulations, we do not always employ them for maximum effect. Jan Herrington and Ron Oliver, in an elegant research effort, summarized the current thinking about the features of situated learning environments that result in the development of usable knowledge. These features included

1. authentic contexts
2. authentic activities
3. expert performances and modeling of processes
4. multiple roles and perspectives
5. collaboration
6. reflection
7. articulation
8. teacher coaching and scaffolding
9. authentic assessment[37]

As the constructivist theory suggests, learning is not only active and contextual but it is also cultural and social.[38] Each given community of practice (including the military) has a distinctive set of shared, but sometimes unrecognized, beliefs, values, philosophy, rules of the game, embedded skills, habits of thinking, mental models, linguistic paradigms, and shared meaning.[39] Authentic situated learning environments can provide an important opportunity for the enculturation of students by giving them the opportunity to serve a "cognitive apprenticeship" as they observe and practice *in situ.*[40]

Situated or contextual learning can also force leaders into what General Schoomaker calls the "zone of discomfort."[41] Leaders gain an efficacy and agency (competence and confidence) in decision making across the spectrum that will directly affect the people, projects, and processes they will lead throughout their careers.

## An Atlas for the Road Ahead

*Agile, adaptive, self-directed, self-aware, culturally facile, moral, and technically and tactically proficient leaders; acting with competence and confidence; challenged, emboldened, and nurtured by their education.* While you might choose other words to describe the goal of leader development, the power of this vision is undeniable. However, regardless of the academic research and contextual realities of the twenty-first century security environment, without action these noble words are just another cliché.

So, then, what *is* a process for transforming this vision from cliché to reality? What *is* a strategy that can be tailored by individual military education institutions? The answer involves leadership, culture, curriculum, and change.

Military academic institutions are somewhat schizophrenic in that they exhibit the characteristics of both hierarchical military organizations *and* civilian colleges and universities. Regardless of where a given institution rests on the continuum between the organizational extremes, significant responsibility falls to the most senior leader—called, variously, president, commandant, director, or commander.

In Dr. Robert Birnbaum's interesting treatise *How Academic Leadership Works: Understanding Success and Failure in the College Presidency*, he describes two levels of leadership expected of the top leader. The first is "instrumental" leadership whereby the leader, "through their technical competence, experience, and judgment," coordinates the activities of others, makes timely and sensible decisions, represents the institution to its various publics, and copes with the everyday crises caused by environmental change and internal conflict. The second level of leadership is "interpretive" and is described in terms of the management of meaning or "altering perceptions of institutional functioning and the relationship of the institution to its environment."[42] Certainly this holds true for the senior leader within a military educational institution. Constituents and stakeholders (faculty, staffs, students, and resource providers) expect the senior leader to provide vision, direction, and energy; engaging them will be vital to any successful strategy for educational transformation.

As leadership scholars have long recognized, however, "leadership is distributed. It resides not solely in the individual at the top, but in every person at every level who, in one way or another, acts as a leader to a group of followers."[43] The reality is that no genuine, far-reaching, enduring academic change will occur without the active support of key leaders throughout the institution. Unfortunately, the more common reality is that either the senior leader has the vision but does not have the skill, patience, time, or will to take on the long view or leaders and individuals below the top have the skill and the will, but are unable to enroll those above them in the vision.

In the former instance, the senior leader can make proclamations and the workforce may pretend to believe, but it is a charade and one that engenders cynicism and debilitates the spirit and energy of the organization. In the latter, the subordinate leader or educator can lead within his or her spheres of control and influence, knowing that progress will always be suboptimal. However, as motivational writer Dan Millman so pragmatically observes, "a little bit of something is better than a lot of nothing."[44]

How, then, does one actually institute enduring change? The answer may rest in the work of Kim Cameron and Robert Quinn, who surveyed large organizations that had undertaken significant organizational change efforts. They found that only those organizations that focused on culture were successful in implementing long-lasting change.[45]

Cultural change is inexpiably linked to leadership. In his classic work, *Organizational Culture and Leadership*, Dr. Edgar Schein goes as far as to suggest that "the only thing of real importance that leaders do is to create and manage culture and that the unique talent of leaders is their ability to understand and work with culture."[46] Beyond the challenging but rather pedestrian aspects of cultural changes, there are some characteristics unique to educational settings where the "cultural assumptions behind the standard pedagogy [i.e., the art, science, or profession of teaching] are so much a part of the professional common sense that we might consider them an aspect of invisible culture."[47]

Asking teachers to step back from and look afresh at their philosophy of teaching, their teaching styles, their assumptions about the nature of knowledge, and their comfortable roles and routines almost inevitably will meet with resistance. Researchers Anthony F. Grasha and Natalia Yangarber-Hicks describe the relationship between educators and students, teachers and learners, as an intimate dance.[48] Suggesting, expecting, demanding that educators move to higher-order learning, to adopt a learner-centered constructivist pedagogy, to inculcate the lessons from brain research, to weave a multicultural mind-set and nuanced decision making into their coursework creates a new genre of music. The dance is more complex; the teacher does not always lead or even choose the music. Dance lessons are in order and a critical-path investment in faculty development is nonnegotiable for real, lasting change.

Curriculum and courseware developers are another powerful (arguably the most powerful) stakeholder group that must both lead and be led through the cultural changes. Governed by guidance and traditions of the respective services or joint governing bodies, and steeped in various academic methodologies, one can anticipate a mixed reception for far-reaching, leading-edge changes in curriculum. Many developers will welcome the changes and be excited at the opportunity to move to higher-order learning; a few will be openly hostile; and all will express justifiable concerns about workload. A short, intense refresher on curriculum development to foster higher, deeper, and more personal learning is in order. An aggressive but achievable schedule is a must. Where contract services are used, the statements of work must be modified to ensure that those hired have the requisite skills and that the unwilling or unable are removed from the contract.

The final group of internal stakeholders that must be lead through this cultural change process is the students. Being a student immersed in a constructivist curriculum—where higher-order learning skills of analysis, evaluation, and creation are the norm; where you are challenged to think reflectively, to think about how you think; where you practice unfamiliar decision making in unfamiliar ways; and where you bear increased responsibility for your own learning—is a significant emotional and intellectual event. Unfortunately, there has been a misperception in some circles that increasing the number of books to be read and papers to write inherently equates with higher-order learning. They do not. If we are serious about moving to this more sophisticated, higher, deeper, and more personal learning, the focus must be on quality, not quantity.

Based on the author's education and years as a "change agent," here is a top-level process for moving teaching and learning to higher, deeper, and more personal levels:

- Focus on organizational culture.
- Articulate a compelling and consistent vision—stay on message.
- Educate the educator. Ensure that faculty and curriculum/courseware developers have the necessary skills.
- Listen, understand, and respect stakeholder concerns—the spoken and unspoken, the rational and the irrational.
- Lead from where you are in the organization—even if you lead only yourself.
- Move boldly. Beware the "raccoons" (as one former commander of U.S. Special Operations Command termed those who view themselves as the gatekeepers of the status quo).
- Institutionalize the processes. Make it hard for someone to undo the progress.
- Formally evaluate progress during and after each key step.
- Accept the inevitable pain inherent in making significant and enduring change. The burden of leading change always rests on the shoulders of the leader until the change becomes "how we do business."

## Closing Thoughts

As we conclude, it is fair to ask, "What does leader education have to do with the principles of war?" And, as is typical of critical thinking, one question often leads to other, more fundamental questions.

The first is why do we have codified principles of war? These principles, according to Rear Adm. John Morgan and Dr. Anthony Mc Ivor, were rooted in the Napoleonic Wars, molded by the industrial revolution, shaped by such theorists as Carl von Clausewitz, and codified as enduring principles by the end of the nineteenth century.[49] As we are once again visiting the content of the principles of war, it is also reasonable to think about how we use them and if they are serving us well.

There are at least five practical functions of the principles of war. We use them as a tool for planning, decision making, sense making (understanding the fog of war), assessing, and enculturation of the next generation of planners and leaders. These are powerful reasons for keeping the principles of war and, more especially, ensuring that they remain useful and relevant to today's complex security environment. However, no matter how well written and well conceived the principles of war are, they are only of value when they are well used.

The principles of war are qualitatively and quantitatively different from other codified principles because of the way we use them. It is not an over-statement to say that the principles of war are matters of life, death, and mission success. Given the stakes, it is not the words, the content of the principles of war, that are most critical; it is their application by leaders who are agile, adaptive, self-directed, self-aware, culturally facile, moral, *and* technically and tactically proficient. Our responsibility is to challenge, embolden, and nurture those leaders educationally so that they can act with competence and confidence . . . on our behalf. And we still have much work to do.

## Notes

1. James M. Kouzes and Barry Z. Posner, *The Leadership Challenge: How to Keep Getting Extraordinary Things Done in Organizations.* (San Francisco: Jossey-Bass, 1995), p. 327.

2. Lorin W. Anderson and David R. Krathwohl, eds., *A Taxonomy for Learning, Teaching, and Assessing.* (New York: Longman, 2001).

3. Anderson and Krathwohl, *A Taxonomy for Learning, Teaching, and Assessing.*

4. Center for Army Lessons Learned, *Changing the Army: An Historical Review.* Special Report, Executive Summary. (Fort Leavenworth, Kan.: Center for Army Lessons Learned, 2004).

5. Peter Schoomaker, interview by James Kitfield, for GovExec.com, 29 October 2004.

6. Knud Illeris, "Transformative Learning in the Perspective of a Comprehensive Learning Theory." *Journal of Transformative Education* 2(2) (April 2004): 79–89.

7. Howard Gardner, "Reflections on Multiple Intelligences Myths and Messages." *Phi Delta Kappan* 77 (November 1995): 200–209.

8. Joan Davis, "Conceptual Change." *Emerging Perspectives on Learning, Teaching, and Technology,* 2001. Online at: http://www.coe.uga.edu/epltt/conceptualchange.htm.

9. Patricia Wolfe, *Brain Matters: Translating Research into Classroom Practice.* Alexandria, Va.: Association for Supervision and Curriculum Development, 2001.

10. Michael Gazzaniga, *The Mind's Past.*, quoted in Wolfe, *Brain Matters,* op. cit., 79.

11. Donna Wilson and Marcus Conyers, *BrainSMART: Strategies for Boosting Test Scores.* Orlando, Fla: Authors, 2000.

12. John C. Bransford, Ann L. Brown, and Rodney R. Cocking, *How People Learn: Brain, Mind, Experience, and School.* (Washington, D.C.: The National Academies Press, 1999). Online at: http://www.nap.edu/openbook/0309065577/html/R1.html#pagetop.

13. Arthur L. Costa and Bena Kallick, "Launching Self-Directed Learners." *Educational Leadership* (September 2004): 51–55.

14. Albert Bendura, *Self-Efficacy: The Exercise of Control.* (New York: W. H. Freeman, 2003).

15. Robert House, Mansour Javidan, Paul Hanges, and Peter Dorfman, "Understanding Culture and Implicit Leadership Theories Across the Globe: An Introduction to Project GLOBE," *Journal of World Business* 37 (2003): 3–10.

16. Frederick Erickson, "Culture in Society and in Educational Practice." In *Multicultural Education: Issues and Perspectives,* James A. Banks and Cherry A. McGee Banks, eds. (New York: John Wiley, 2003).

17. Leonard Wong, *Developing Adaptive Leaders: The Crucible Experience of Operations Iraqi Freedom.* Strategic Studies Institute Report. (Carlisle, Pa.: U.S. Army War College, 2004).

18. Margaret Wheatley, "Restoring Hope to the Future Through Critical Education of Leaders." Building Bridges 2002 Collaboration. College Park, Md.: Academy of Leadership. Cited in Ata U. Karim, "A Developmental Progression Model for Intercultural Consciousness: A Leadership Imperative." *Journal of Education for Business* (September/October 2003): 34.

19. Robert H. Scales, "Culture-Centric Warfare." Naval Institute *Proceedings* (October 2004): 36.

20. Ata U. Karim, "A Developmental Progression Model for Intercultural Consciousness: A Leadership Imperative." *Journal of Education for Business* (September/October 2003): 34–39.

21. Karim, "A Developmental Progression Model," 37–38.

22. James A. Banks, "Multicultural Education: Characteristics and Goals." In *Multicultural Education: Issues and Perspectives,* James A. Banks and Cherry A. Banks, eds. (New York: John Wiley, 2003), pp. 3–30.

23. Tzu Sun, *The Art of War.* Translated by Samuel B. Griffith (Oxford: Oxford University Press, 1972).

24. Banks, "Multicultural Education," 20.

25. Karim, "A Developmental Progression Model," 34–39.

26. Ibid., 35.

27. Kecia M. Thomas, "Psychological Privilege and Ethnocentrism as Barriers to Cross-Cultural Adjustment and Effective Intercultural Interactions." *Leadership Quarterly* 7(2) (1996): 216.

28. Ibid.

29. Peggy McIntosh, "White Privilege and Male Privilege" in *Gender Basics* (Belmont, Calif.: Wadsworth, 1993). Cited in Thomas, "Psychological Privilege and Ethnocentrism," 216.

30. Christopher P. Earley, "Intercultural Training For Managers: A Comparison Of Documentary And Interpersonal Methods." *Academy of Management Journal* 30(4) (1987): 685–98.

31. Wayne Hochwarter, Christian Kiewitz, Michael J. Gundlach, and Jason Stoner, "The Impact of Vocational and Social Efficacy On Job Performance and Career Satisfaction." *Journal of Leadership and Organizational Studies* 10(4) (Winter, 2004): 27–40.

32. John Buchanan and Ned Kock, "Information Overload: A Decision Making Perspective." Paper presented at MCDM2000 meeting of the International Society on Multiple Criteria Decision Making, Ankara, Turkey (July, 2000).

33. Joint Forces Command, Operational Net Assessment (ONA) online at: http://www.jfcom.mil/about/fact_ona.htm; and Collaborative Information Environment (CIE) online at: http://www.jfcom.mil/newslink/storyarchieve/2003/ pa121103.htm.

34. Malcolm Gladwell, *Blink: The Power of Thinking Without Thinking.* (New York: Little Brown, 2005).

35. Gary Klein, *Sources of Power: How People Make Decisions.* (Cambridge, Mass.: The MIT Press, 2000).

36. Doctrine for Planning Joint Operations (Joint Pub 5.0), 13 April 1995. Online at: http://www.dtic.mil/doctrine/jel/new_pubs/jp5_0.pdf.

37. Jan Harrington and Ron Oliver, "An Instructional Design Framework for Authentic Learning Environments." *Educational Technology Research and Development* 48(3) (2000): 23–48.

38. Marcy P. Driscoll, "How People Learn (and What Technology Might Have to Do With It)" ERIC Report No. EO-IR-2002-05. (October 2000).

39. Edgar H. Schein, *Organizational Culture and Leadership* 2nd ed. (San Francisco: Jossey-Bass, 1997).

40. John S. Brown, Allan Collins, and Paul Duguid, "Situated Cognition and the Culture of Learning." *Educational Researcher* 18(1) (January–February 1989): 32–42.

41. Schoomaker, interview.

42. Robert Birnbaum, *How Academic Leadership Works: Understanding Success and Failure in the College Presidency.* (San Francisco: Jossey-Bass, 1992).

43. Daniel Goleman, Richard Boyatzis, and Annie McKee, *Primal Leadership: Realizing the Power of Emotional Intelligence.* (Boston: Harvard Business School Press, 2002).

44. Dan Millman, *No Ordinary Moments: A Peaceful Warrior's Guide to Daily Life.* (Tiburon, Calif.: H. J. Kramer, 1992).

45. Kim Cameron and Robert Quinn, *Diagnosing and Changing Organizational Culture: Based on the Competing Values Framework.* (Upper Saddle River N.J.: Prentice Hall, 1999).

46. Schein, *Organizational Culture and Leadership,* 5.

47. Erickson, "Culture in Society and in Educational Practice," 50.

48. Anthony F. Grasha and Natalia Yangarber-Hicks, "Integrating Teaching Styles and Learning Styles With Instructional Technology." *College Teaching* 48(1) (Winter 2000): 2–10.

49. John G. Morgan, Anthony Mc Ivor, et al., "Rethinking the Principles of War." Naval Institute *Proceedings.* (October 2003): 34.

Part 3

# OPERATIONAL ARTS: IRREGULAR WARFARE

# FIFTEEN

# Rethinking the Principles of War

## The Future of Warfare

Thomas X. Hammes

With the title proposed above, the editor of this volume has challenged each author to consider how warfare in the future will be different and whether the principles of war may evolve to reflect that change. Unfortunately, the first half of the subject—principles of war—is a deeply flawed concept. We will be ill served if we attempt to apply the current principles to the future of war. Therefore, this paper will first deal with why the current U.S. principles of war concept is fatally flawed. Then it will consider the future of war—its characteristics and what impact they should have on U.S. defense thinking.

The mere phrase principles of war conjures up visions of grand but concise strategic concepts that will guide the practitioner through the incredibly complex environment of war. Further, it sounds timeless, as if these principles have been passed down through centuries, tested and proven. The very words indicate they represent distilled wisdom to use in solving today's problems. Too often they are taught as such. John Alger, in his *Quest for Victory,* noted that since World War II, U.S. military manuals have considered the principles to be "fundamental truth that is professed as a guide to action."[1] They provide a short, easy-to-follow list of steps to be taken to ensure victory.

This view of the principles as specific guidelines is certainly supported by the dictionary definition. According to Random House, principles are "1. an accepted or professed rule of action or conduct . . . 2. a fundamental, primary law or truth from which others are derived . . . 3. a fundamental doctrine or tenant . . ."[2] Clearly the use of the term *principles* indicates a search for fundamental rules that can simply be applied to war from the strategic to the tactical levels.

Unfortunately, this view of the principles as a specific guide is not supported by one of the first and certainly the most famous author to write about the principles of war. No less an authority than Clausewitz, in his introduction to his *Principles of War,* stated, "These principles, though the result of long and continuous study of the history of war . . . will not so much give complete instruction to Your Royal Highness, as they will stimulate and serve as a guide for your own reflections."[3] He saw them as a way of exploring each unique military situation "to gain a preponderance of physical forces and material advantage at the decisive point."[4]

This tension between the quest for short, specific principles suitable for U.S. doctrinal publications and Clausewitz's concept of general guidelines is present even on the "Rethinking the Principles of War" project page in the Office of Force Transformation Web site. That text describes this volume of essays as intended to "serve as a definitive text for use in the Service academies, war and staff colleges, as well as graduate programs in the private sector."[5] By referring to potential changes, the language seems to imply a quest for a new set of fixed principles.

Yet earlier on the same Web page, the project's objective is stated as a desire to find guidelines for thinking about war rather than rules for execution. "The objective should be clear—it is not to replace one set of principles, hostage to time and place, with another set equally constrained. There will be no perfect or easy answers, but the beginning can pose the right questions." This paragraph indicates the seminar is seeking an answer in line with Clausewitz's intent when he wrote the principles of war for the young prince.

This same tension between rules and guides is present throughout the historical discussion on the principles of war. While a bit confusing, this tension alone does not render the principles dysfunctional for the purposes of this essay. The Web site clearly expected such a tension. What invalidates the list widely used today is its title "Principles of War." They are really not principles of war. As expressed in FM 3-0, they are principles of battle. They deal only with the military aspects of defeating an enemy's armed forces without any serious consideration of the political, economic, and social aspects of the conflict. In short, they address how to win battles, not how to win wars. The same thing has been said about our recent war plans for both Afghanistan and Iraq. The Department of Defense (DoD) planners looked only at winning the battles, not the wars. This alone indicates we need to rethink how we educate our planners—both civilian and military.

Given the source of these principles, their focus on battle is natural. While writers have studied and tried to codify war since our earliest recorded history, the quest to find specific principles that, if followed, would lead to victory began in earnest only after the Napoleonic Wars. European authors strove to understand how Napoleon won so consistently on the battlefield. It is interesting to note that these authors did not explore why Napoleon could win all those battles and not win his wars. Thus, from the very beginning, our principles of war actually focused on battles.

In *The Quest for Victory,* John Alger traces how the concept of principles of war developed and what specific principles were proposed over time. Starting with ancient writers, he examines the various attempts, widely dispersed in time and distance, to quantify how a war is won. Then he notes the exponential jump in such writings after the Napoleonic Wars. He reveals a number of authors who developed a variety of lists in attempts to quantify Napoleon's genius. Finally, he shows the lineage of the list we now use and how it comes directly from authors writing after World War I.[6] The trauma of World War I led these authors to seek a narrow set of principles that could solve all military problems. They were pursuing a formula to prevent the insanity of trench warfare.

Besides narrowing the principles, the post–World War I authors gave them a great deal more weight. Lt. Col. Eugene Cholet, a French officer, expressed the consensus among military thinkers of the era when he wrote

1st That fundamental principles of war certainly exist.
2nd That they appear to be few in number.
3rd That they are immutable and have guided the actions of all the great leaders of war.[7]

Note the certainty. If he could just find, study, and use these fundamental principles, they would assure success in war. The "fact" that the list was short, had been used by all great leaders of war, and yet somehow had never been written down, did not discourage him a bit.

At nearly the same time, J. F. C. Fuller, the noted British historian, came to essentially the same conclusion and provided a list of "objective, offensive, mass, economy of force, movement, surprise, security and cooperation."[8] He too wrote they could guide practitioners in the successful prosecution of war. While initially enthusiastic about his principles as a practical guide for military officers, he later came to place much less faith in them. As his thinking on war matured, Fuller became concerned that people were applying them mechanistically as "rules" while he had grown to see them as only as guides for a thinking approach to the art of war. Yet, despite Fuller's later reservations, it is essentially his list—with cooperation changed to Unity of Command and Speed added—that found its way into American field manuals after World War II and has remained there ever since.

Given this history of the development of the principles of war, it is not surprising that they focus on battles rather than wars. They grew out of the Napoleonic Wars and the West's fixation on the decisive battle. They were finalized based on the "lessons" of World War I. These authors strove to find a formula that would change battles from World War I's indecisive bloodlettings back to the decisive actions of earlier eras. Unfortunately, they did not address war as a whole but only how to win individual battles. The assumption seemed to be that if an army won enough battles, it must win the war.

Consider the principles as listed in FM 3-0. As you read them note the explicit emphasis on combat. Other than the brief reference to operations other than war under *Objective,* the current principles are clearly focused on battle rather than war as a whole. Even *Objective* states that the ultimate purpose of war is the destruction of the enemy's armed forces and will to fight.

FM 3-0 states:

### THE PRINCIPLES OF WAR
The nine principles of war provide general guidance for the conduct of war at the strategic, operational, and tactical levels. They are the enduring bedrock of Army doctrine.

#### Objective
Direct every military operation toward a clearly defined, decisive, and attainable objective.
The ultimate military purpose of war is the destruction of the enemy's armed forces and will to fight. . . .

#### Offensive
Seize, retain, and exploit the initiative.
Offensive action is the most effective and decisive way to attain a clearly defined common objective. . . .

#### Mass
Mass the effects of overwhelming combat power at the decisive place and time. Synchronizing all the elements of combat power where they will have decisive effect on an enemy force in a short period of time is to achieve mass. . . .

### Economy of Force

Employ all combat power available in the most effective way possible; allocate minimum essential combat power to secondary efforts. . . .

### Maneuver

Place the enemy in a position of disadvantage through the flexible application of combat power. . . .

### Unity of Command

For every objective, seek unity of command and unity of effort. . . .

### Security

Never permit the enemy to acquire unexpected advantage. . . .

### Surprise

Strike the enemy at a time or place or in a manner for which he is unprepared. . . .

### Simplicity

Prepare clear, uncomplicated plans and concise orders to ensure thorough understanding."[9]

The only hint that wars might not be decisively ended by destroying the enemy armed forces comes in a single sentence under *Objective*. There the manual makes the brief statement, "The ultimate objectives of operations other than war might be more difficult to define; nonetheless, they too must be clear from the beginning." This is the only hint in the principles that war may take a form other than conventional, and it is not explored further.

In short, the principles of war focus only on how to win conventional battles. They seem to assume that if battlefield victories are tied to coherent campaigns, they will inevitably lead to the military defeat of the enemy and his political surrender. In the era the writers were discussing, the defeat of the army and surrender of the capital usually led directly to the surrender of the government. Unfortunately, as we have seen in Afghanistan, Chechnya, and Iraq, this is no longer the case.

In fact, at the time the principles of war were codified in U.S. doctrine, two new forms of war were evolving in very different parts of the world. In the West, *Blitzkrieg* (lightning war) was about to burst on the scene with the German invasion of Poland in 1939. While it had a dramatic impact on the way war was fought, blitzkrieg resulted in no changes to the principles. Subsequent writings about the principles of war often use examples from World War II to illustrate the importance of the principles. Unfortunately, this is the last war where destruction of an opponent's armed forces and occupying his capital led directly to a Western victory.

While post–World War II writers have consistently used examples from that war to validate the principles, the most successful practitioners of maneuver warfare in World War II categorically rejected the idea that a few simple principles could serve as guides for winning wars. In preparing for World War II, the British and American armies showed growing faith in and teaching of the principles. In contrast, the Germans clung to a case study method of teaching about war. They firmly believed that only by an in-depth and continuing study of war could one develop the judgment to make good decisions in specific situations. While general concepts could assist an educated officer in analyzing a specific situation, the fact remained that every situation was unique. No simple set of rules or principles could substitute for a true understanding of the complexity of war. Rather than memorizing and applying a list of principles, the Germans insisted that their officers must develop an in-depth knowledge of military history. They could then apply the knowledge and thought processes developed in that study to the specific, inevitably unique situation they faced.

Unlike the Germans, we Americans not only taught principles of war throughout World War II but continue to teach them up until this day. FM 3-0 still states categorically they are the "bedrock of Army doctrine." That is what makes the "Rethinking the Principles of War" project critical. Only through such an honest reevaluation of the entire concept of principles of war will we learn their distinct limitations.

In short, principles of war aren't. They are actually principles of battle and should be dropped from U.S. doctrine in favor of the deeper, long-term study that will genuinely prepare our personnel for the future. That brings us to the heart of the essay. What will future war look like and how do we adapt?

## Future War

Two major schools of thought have emerged concerning the future of warfare. They are best described by J. Arquilla and D. Ronfeldt in their article "Cyberwar Is Coming."[10] One, cyberwar envisions a high technology, short-duration war where technology is vital. Cyberwar found a home in the Department of Defense during the early 1990s. First expressed as a revolution in military affairs (RMA), it was more clearly defined in Joint Vision 2010. Since then, the same basic concept has been presented as Joint Vision 2020, network-centric warfare, and now transformation. From its first expression in 1990 up till the last quarter of 2004, this concept envisioned short wars dominated by high technology forces linked by a network that provided information dominance. This concept was consistently illustrated by four pillars (dominant maneuver, precision engagement, full dimension protection, and focused logistics) built on a base of information superiority.[11]

In this vision of future war, machines fight machines. Only in the last six months has the DoD begun to shift even slightly from this vision. An early indicator of whether the DoD is serious or not will be the final form and recommendations of the current *Quarterly Defense Review.* While the recently rewritten terms of reference show a much broader appreciation of war than transformation, there are also strong indicators that program managers and the services are modifying them to support specific, existing programs. In short, we have yet to see if the DoD can shrug off its fourteen-year high technology fantasy.

The DoD's fascination with high technology is a natural outgrowth of America's fascination with and prowess in the development, production, and fielding of extraordinary high tech weapon systems. Even our European allies admit they cannot match U.S. capabilities. Pursuing this course has led to U.S. domination of the conventional battlefield. It has even led some to state that America is so dominant on the battlefield, none would dare challenge us.

Unfortunately, this view of war has developed independently of what has actually been happening in the world. While the Germans were demonstrating blitzkrieg to the stunned Allied forces, Mao Tse-tung, in China, was quietly showing weaker powers how superior political will, applied over time, could defeat much greater economic and military power. He was developing insurgency as a winning method of war.

Since World War II, wars have been a mixture of conventional and unconventional. Conventional wars—Korea, the Israeli-Arab Wars of 1956, 1967, and 1973, the Falklands, the Iran-Iraq War, and the 1991 Persian Gulf War—have ended with a return to the strategic status quo. While territory changed hands and, in some cases, regimes changed, each state came out of the war with largely the same political, economic, and social structure with which it entered.

In sharp contrast, unconventional wars—the Communist revolution in China, the First and Second Indochina Wars, the Algerian War of Independence, the Sandinista struggle in Nicaragua, the Iranian revolution, the Afghan-Soviet War of the 1980s, the Soviet-Chechen War of the 1990s, the first intifada and the Hezbollah campaign in South Lebanon—each ended with major changes in the political, economic and social structure of the territories involved. While the changes may not have been for the better, they were distinct. Even those conflicts where the counterinsurgent "won," such as Malaya, Oman, Peru, and El Salvador, resulted in significant changes. The consistent ability of weaker insurgents to force political, social, and economic changes on their governments has made it the preferred form of war for those seeking such change.

This is the form of war Arquilla and Ronfeldt described as netwar. Also known as fourth generation war (4GW), it reflects the complex, long-term conflicts that have grown out of Mao's Peoples War. Its evolution and rise to dominance on the battlefield can be traced over the last fifty years. Practitioners of 4GW focus on the political aspects of the struggle. They neutralize superior military power and technology by disappearing into the society as a whole. They know superior political will, applied over time, can defeat greater economic and military power. Fourth generation wars are protracted. The Chinese fought for twenty-seven years, the Vietnamese for thirty years, the Sandinistas for eighteen years, the Afghans for ten years versus the Soviets, and the Palestinians have been fighting since at least 1967. Even those conflicts in which the counterinsurgent wins are long. It took the British twelve years before they could declare the Malayan Emergency over.

Our opponents also know 4GW is the only kind of war America has ever lost, and done so three times: in Vietnam, Lebanon, and Somalia. Fourth generation warfare has also defeated the French in Vietnam and Algeria; the Soviet Union in Afghanistan; and Israel in Lebanon. It continues to bleed Russia in Chechnya; Israel in both the Occupied Territories and Israel proper; and the United States in Afghanistan, Iraq, and the global war on terror.

As a result, our enemies themselves state they understand and use 4GW. The February 2002 edition of the online magazine *Al-Ansar,* which purports to be an official Al Qaeda outlet, published an article on 4GW. In it, Ubeid Al-Qurashi, one of Osama bin Laden's close aides, wrote:

> In 1989, some American military experts predicted a fundamental change in the future form of warfare.[12] . . . They predicted that the wars of the 21st century would be dominated by a kind of warfare they called "the fourth generation of wars." Others called it asymmetric warfare. . . .
>
> This new type of war presents significant difficulties for the Western war machine and it can be expected that [Western] armies will change fundamentally. This forecast did not arise in a vacuum—if only the cowards [among the Muslim clerics] knew that fourth-generation wars have already occurred and that the superiority of the theoretically

weaker party has already been proven; in many instance, nation-states have been defeated by stateless nations . . .

In Afghanistan, the Mujahideen triumphed over the world's second most qualitative power at that time. . . . Similarly, a single Somali tribe humiliated America and compelled it to remove its forces from Somalia. A short time later, the Chechen Mujahideen humiliated and defeated the Russian bear. After that, the Lebanese resistance [Hezbollah] expelled the Zionists army from southern Lebanon . . .

Technology did not help these great armies, even though [this technology] is sufficient to destroy the planet hundreds of times over using the arsenal of nuclear, chemical, and biological weapons. The Mujahideen proved their superiority in fourth-generation warfare using only light weaponry. They are part of the people and hide among the multitudes. . . .

Thus, it appears that there are precedents for world powers and large countries being defeated by [small] units of Mujahideen over the past two decades, despite the great differences between the two sides. . . ."[13]

Al Qaeda understands 4GW and is using it. Their actions in Afghanistan, Iraq, and Spain demonstrate a keen understanding of how a militarily weaker power can attack a vastly more powerful enemy. Furthermore, the anti-Coalition forces in both Afghanistan and Iraq have also adopted 4GW and have stated they will defeat the Coalition using it.

To date, fourth generation opponents have focused on insurgency. However, 4GW techniques can also be used by states. In their textbook, *Unrestricted Warfare*, two Chinese army officers envisioned the use of a wide variety of 4GW techniques to neutralize U.S. military power in the region. While China has not fought a war using the approach outlined in this work, the network of political, economic, and social alliances they are building in Southeast Asia, the Middle East, and Africa indicate they take the concepts of diplomatic, economic, financial, cyber, media/information, and network warfare seriously. During the recent increase in tension between Japan and China, a number of Japanese business and government networks came under sustained attacks.

The political, strategic, operational, and tactical approaches that have worked well for insurgents can, in new forms, be adopted by and used effectively by states.

Thus, the future of war for the United States is much more likely to be netwar than cyberwar. While the DoD would clearly prefer the network-centric vision of a short, high technology war, it has not materialized. Despite the fact that our forces quickly won the conventional battles, seized the capital cities, and routed the hostile governments in Afghanistan and Iraq, the wars have not ended. Even more awkward for the network-centric concept is the fact Al Qaeda and the anti-Coalition forces in Afghanistan and Iraq refuse to accept battle. They choose to fight us in other ways. This is not a surprising development. Since the end of World War II, insurgents have chosen to make their wars political, protracted, and networked.

Insurgents do not focus on winning battles. In some insurgencies, the actual outcome of the battles has been irrelevant. In Nicaragua, Afghanistan, Chechnya, and Palestine, the insurgents never defeated their enemy's military forces. Unlike the Napoleonic battles that led to decisive victory outlined in the principles of war, they applied political power over time to directly attack their enemy's political will. In three of these four cases, they won. The jury is still out on the fourth.

## The Underlying Nature of War

In place of principles, we should consider the characteristics of modern warfare and how we can prepare ourselves to deal with each. Even before we consider these characteristics, we have to accept that no amount of technology will eliminate the friction, fog, and uncertainty that are enduring elements of all human conflict. Understanding this fact will lead us to organize, train, and educate in a very different manner than if we adopt the network-centric-warfare concept of information superiority. If we assume information superiority, we can make do with a much smaller, less well-trained, and much less broadly educated force than if we assume the battlefield will remain a very confusing place where the other side may have information dominance. It is obviously much cheaper and easier to develop a force that "knows" it has information superiority than a force that can deal effectively with uncertainty. But it is incredibly foolish. A force designed for an uncertain battlefield will thrive if it is lucky enough to have information superiority, while a force designed for near certainty will easily be destroyed if it cannot attain that superiority.

## The Characteristics of Future War

Once we accept that the underlying nature of war will not change, we need to look at the characteristics of future war. Warfare evolves. The wars we are currently fighting and the wars of the near future have their roots in the success of specific insurgencies of the last sixty years. Since 1945, wars that have led to decisive change have been political, protracted, and, more recently, networked.

In essence, war has changed. Rather than repeating the error of seeking some short definitive list, we must instead consider the primary characteristics of future war and how we will deal with them. What follows does not pretend to be definitive. I hope I have made it clear that no such list does or can exist when dealing with an activity as complex as war. Rather, I will provide a very brief discussion of what I see as three major characteristics of future war and one or two ways they change how we must look at war. My intent is merely to stimulate the reader's thoughts about future war—and further emphasize that war cannot be reduced to a simple list but requires a lifetime of study.

## *Political*

The primary characteristic of fourth generation warfare is that it is a political and not a military struggle. Understanding that fact leads one to evaluate potential actions and plans in a very different way. For instance, speed is a critical element in conventional fights. Virtually all studies of conventional war stress that superior speed on one side provides a distinct advantage in a fight. This had led to the incredible focus on shortening the sensor-to-shooter cycle within the American technical community. Yet one has to question if such a trait is even important in the decades-long struggles that are fourth generation warfare. In such a fight, stressing speed may in fact be detrimental. If one acts very quickly and kills a "terrorist" based on a tip from an informer, and then later finds the terrorist was in fact a law-abiding citizen who the informant wanted dead, the government's credibility suffers.

The very first step in dealing with the political aspects of the war is to develop a genuine understanding of the political environment where the war is being fought. This requires a deep understanding of the culture, history, and current political structure of the area. Because modern conflicts are rarely limited to a single country, this understanding must extend to the region as a whole. For instance, to understand what is happening in the insurgency in Iraq, one must understand the political, economic, social, and religious situation in Iran, Syria, Turkey, Saudi Arabia, and the other Persian Gulf states. Obviously, this requires a long-term, interagency effort. No single agency within the U.S. government can provide the complete picture required. By default, the political emphasis of fourth generation war requires both an interagency and an international approach. This does not mean small coalitions pressured or paid to participate. It means a genuine coalition of the willing that brings the knowledge, understanding, and skills of the international community to the fight.

Building such coalitions will be extremely difficult but, if we accept that wars are political and protracted, we will understand that expending this effort up front will result in much quicker resolution in the long term.

While we will have to build alliances to deal with each conflict as it arises, we cannot wait till one takes shape to develop an exponentially greater pool of cultural, language, and religious experts within the U.S. government.

One of the most often quoted (and most often ignored) tenants of Sun Tzu is "Know your enemy and know yourself; in a hundred battles you will never be in peril."[14] To achieve this we must provide a large, flexible, redundant pool of personnel who are genuine area experts. This does not mean a few hundred Arabic experts working for the government. It means tens of thousands of experts that cover the spectrum of potential areas of conflict. Only when such personnel actively participate in every aspect of our planning and execution in a fourth generation war can we hope to "know our enemy." Further, because of the political nature of 4GW, this effort cannot be limited to the DoD. Every element of the U.S. government that is deploying personnel overseas will require the same expertise to ensure we can deal with the financial, economic, social, agricultural, police, and intelligence aspects of modern war.

Once we accept that wars require a minimum of an interagency effort and preferably a truly international one, obvious difficulties arise concerning unity of command. While a gut reaction insists that unity of command is absolutely essential in war, a careful study of history indicates this "truth" is simply not true. In a political, protracted war, unity of command is less important than maintaining the political will of the coalition. The requirement to fight as a coalition network may actually preclude a unity of command. This is okay. In World War I, World War II, Desert Storm, and Operation Iraqi Freedom, we operated as coalitions with each nation retaining final decision authority over the use of its troops.

Rather than attempting to achieve unity of command, there has to be a unity of intent and a common end state. Then each node of the network is free to apply its skill to ensuring that outcome. End state provides a more complete vision of what is desired than does a simple objective. End state provides the vision of the political, economic, social, and security status of the area of conflict at the end of the effort. It cannot be expressed as simply as a geographic objective or as the destruction of a force or regime. In keeping with the fundamentally political nature of 4GW, it must include the post-conflict political vision.

## Networked

One of the major changes in insurgencies over the last twenty years has been the evolution of networked as opposed to hierarchical insurgencies. Both Mao Tse-tung and Ho Chi Minh ran very disciplined organizations. While each had to struggle to consolidate power, once they had done so, they maintained it with ruthless control from the top. In contrast, the Sandinista, Chechen, and anti-Soviet Afghan insurgencies have been coalitions of the willing. While the groups that made up these networks all agreed that the outside power must be driven from their country, they did not have a unifying political vision for the future, once they won. In both Afghanistan and Chechnya, when it became clear the Soviets were pulling out, the various insurgent groups largely quit fighting the Soviets and marshaled strength for the coming civil war amongst themselves.

Today's Palestinian, Afghan, and Iraqi insurgents are also networks. None of them has a unifying political vision should they win. In Palestine, the initial announcement Israel would withdraw from the Gaza Strip set off a struggle amongst Palestinian groups vying to establish themselves as the government of Palestine.

In the same way, the wide spectrum of insurgents in Iraq includes Shia fundamentalists, Sunni Salafis, Sunni secularists, Islamic jihadists, and criminal elements. These disparate elements cannot share a common vision of the future of Iraq. They know if they drive out the Coalition, they will still have to fight it out between themselves to see who controls what in Iraq.

It is characteristic of insurgents that they have trouble transitioning from opposition to the government. Even Mao and Ho, who had effective, unified parties, had trouble making the transition. As insurgency has evolved into coalitions of the willing, such as the Afghans and the Sandinistas, insurgents have continued the struggle for power among themselves well after the external power was defeated.

The inherent conflicts in such networked insurgencies offer opportunities that can be exploited, but only by people with deep and current cultural and political understanding of the specific insurgencies.

In a networked conflict, the single source of cooperation between groups has been the presence of a common enemy. In contrast, cohesion within each group has been based on common beliefs and a common political goal. Yet for both inter- and intra-group conflicts, the insurgents have focused on establishing human networks based on relationships between the key leaders. They have not sought high technology, computer-based networks. While they have proven to be remarkably adept at using email, cell phones, and Web sites, they have focused on human interaction as the central element of their network.

In contrast, Western forces have tried to substitute technology for human connections. This is a fundamental difference that must be recognized in the West. Once recognized, it should result in major efforts to build similar human networks among allies and neutrals when we are fighting a networked insurgent.

### *Protracted*

Mao's first law of war was "to preserve oneself and destroy the enemy."[15] He titled one of his most famous essays "On Protracted War." The exceptional duration of fourth generation wars requires a fundamentally different approach from the short, intense wars we have been preparing for. The duration, combined with the fact that insurgents consciously target the political will of those they fight, means the United States must pay special attention to sustaining our political will over time. Political will is our critical vulnerability in a protracted conflict. We must understand this fact and make it an essential part of all planning and execution in fourth generation wars. Areas normally considered outside the purview of war planners—long-term budgetary impact on entitlement programs, strategic communication with allies, neutrals, and the American people, and many other totally new aspects, for example—must become part of the interagency planning effort.

Further, the sheer duration of fourth generation wars means we must rethink our personnel policies. In the short term, we have to abandon the belief that three-month to one-year tours are best for the personnel, civilian and military, that we commit to such conflicts. In light of the duration of such conflicts, those tours that require a real understanding of the conflict and strong personal relationships between U.S. agencies and our allies require correspondingly long tours. Those tours that impose such constant risk and pressure that six months to a year is the maximum one can expect from a person may remain short, but our default position must be to longer tours.

In addition, we must focus on early identification of those who will serve in-country and provide extensive language and cultural training prior to deployment. Obviously, we must also restructure our career patterns—not just for military but also for those civilian agencies essential to a fourth generation campaign.

## Conclusion

This project to reexamine our teaching of the principles of war comes at a propitious time. It is clear, even from this brief historical survey, that the principles of war, as expressed in current U.S. military doctrine, were developed for a very different time and very different conflicts than those we face today. As was appropriate for the time when they were developed, they are not focused on war, but on battle. That time has passed and so has the usefulness of such a list.

Virtually everyone in the defense community agrees that warfare is changing dramatically. We no longer face the massive conventional conflict we prepared for throughout the Cold War. While we have been slow and at times reluctant to adjust to that fact, insurgents in various parts of the world are providing pointed lessons that war has changed. Some writers are discussing how states can apply these lessons to neutralize the massive military power of the United States.[16]

It is time to recognize that warfare has become political, protracted, and networked. It has done so because our opponents have seen that only this kind of war works against superpowers. Potential opponents from insurgents to major powers have stated they will move warfare from a purely military-technical arena to a fight that takes place across the entire range of human activity—political, economic, social, and military fields. Understanding such a conflict requires a lifelong, focused effort by professionals across the public and private sectors. There are no shortcuts.

This article barely scratches the surface of what we must to do adjust to future war. As always, people, not technology, will make the difference. We cannot simply seek a combination of simple "principles" and technology to deal with the exceptionally complex security environment we face. We have to adjust our personnel systems, particularly as they apply to professional education and development. We have to shift our focus from pure technology to technology that supports the human elements required for the long-term, interagency efforts required to win future conflicts. In short, we have to invest in the human capital that is the real key to future warfare.

# Notes

1. John I. Alger, "The Quest for Victory" (Westwood, Conn.: Greenwood Press, 1982), p. xxii.

2. Random House Dictionary, Second Edition, Unabridged, New York: Random House, 1987.

3. Carl von Clausewitz, "The Principles of War," trans. and ed. by Hans W. Gatzke (Harrisburg, Pa.: The Military Service Publishing Company, 1942), p. 11.

4. Ibid., p. 12.

5. John Hopkins University Applied Physics Laboratory Principles of War web site, http://www.jhuapl.edu/POW/about.htm

6. The Principles of War as listed in FM 100-5 are Mass, Objective, Offensive, Surprise, Economy of Force, Maneuver, Unity of Command, Security and Speed.

7. Alger, "Quest for Victory," p. 127.

8. Ibid., p. 117.

9. U.S. Department of the Army, "FM 3.0 Operations," Washington D.C., June 2001, pp. 4-11–4-12.

10. John Arquilla and David Ronfeldt, "Cyberwar is Coming," *Comparative Strategy* November 1, 1993, Vol. 12, p. 34.

11. It is interesting to note that in the last six months, the information superiority illustration has disappeared from our publications. This may be an awareness that the claim is difficult to maintain in light of the fact we are currently fighting three wars—Iraq, Afghanistan, and the Global War on Terror—and are fighting at an information deficit in all three. Unfortunately, when the Information Dominance base is removed the four pillars are left resting in the air.

12. The western military experts quoted include Maj. Don Vandergrift (USA), Col. Douglas MacGregor (USA), Maj. Chris Yunker (USMC [Ret.]), Col. G. I. Wilson (USMC), Col. Michael Wyly (USMC [Ret.]) and the late Col. John R. Boyd (USAF [Ret.]) in their article "The Changing Face of War: Into the Fourth Generation," *Marine Corps Gazette,* Oct. 1989, and this author in his article "The Evolution of War: The Fourth Generation," *Marine Corps Gazette,* Sept. 1994.

13. Special Dispatch—Jihad and Terrorism Studies, The Middle East Media Research Institute (MEMRI), February 10, 2000, No. 344.

14. Sun Tzu, "The Art of War," translated by S. B. Griffith (Oxford: Oxford University Press, 1963), p. 84.

15. Mao Tse Tung, "On Guerrilla Warfare," translated by S. B. Griffith (Baltimore: The Nautical and Aviation Publishing Company of America, 1992), p. 53.

16. L. Qiao and X. Wang, "Unrestricted Warfare: China's Master Plan to Destroy America," (West Palm Beach, Fla.: NewsMax.com; 2002); T. Hammes, *The Sling and the Stone: On War in the 21st Century* (St. Paul, Minn.: Zenith Press, 2004).

# SIXTEEN

# Small Wars

## *From Low Intensity Conflict to Irregular Challenges*

STEVEN METZ

## Introduction

As Colin Gray reminds us, "war is war" whether "large" or "small."[1] There is no duality in the nature of strategy. Still, structure and dynamics of small and large wars do differ in significant ways. Small wars, as normally understood, pit a state (or something that looks like a state) against non-state actors, ranging from bands of terrorists, militias, and heavily armed criminal gangs to more formal organizations such as warlord armies or mature, quasi-state insurgent movements. In conventional, state-on-state war, combatants operate with common, or at least similar, "rule sets," whether formal ones such as the Law of Armed Conflict and Geneva Conventions or informal ones. Conventional, state-on-state conflicts are "intra-cultural" (with culture understood in both its organizational and anthropological meanings). Small wars are almost always cross-cultural.[2] As a result, non-state combatants are less constrained, whether out of desperation or simply because they do not accept the legitimacy of the rule sets.

Phrased differently, small war is quintessentially asymmetric. In simplest terms, asymmetry is acting, organizing, and thinking *differently* from opponents in order to maximize one's own advantages, exploit an opponent's weaknesses, attain the initiative, or gain greater freedom of action. Asymmetry can be political-strategic, military-strategic, operational, or a combination. It can entail different methods, technologies, values, organizations, time perspectives, or some combination. It can be short term or long term. Asymmetry can be deliberate or inadvertent. It can be discrete or pursued in conjunction with symmetric approaches. It can have both psychological and physical dimensions.[3]

All war is characterized by some degree of asymmetry between the antagonists, but in small wars it is extensive, pervasive, and, to an extent, defining. Four types of asymmetry are particularly relevant: asymmetry of *method* (e.g., non-state antagonists rely on unconventional methods such as guerrilla activity or terrorism); asymmetry of *time perspective* (non-state antagonists often realize that quick success is impossible and seek to extend the conflict hoping to ameliorate their disadvantages); asymmetries of *organization* (with non-state antagonists relying on cells, bands, and networks rather than formal, hierarchical organizations); and *ethical* asymmetries (with non-state antagonists usually willing to use methods considered abhorrent, illegal, or horrific and thus eschewed by the state antagonist or used only in desperation.) Ultimately, these types of asymmetry and the mixing of two (or more) types of antagonists distinguished small wars.

Like major war, the conduct of small war has evolved in the last fifty years. While the nature of warfare has always shifted over time, the pace of change has escalated immensely given the interconnectedness of the modern world and its saturation with information technology and communication. For most of history, insurgents, terrorists, and militias acted in isolation with at best a vague notion of similar organizations in other places. Today, they find strategic, operational, and tactical lessons available at any Internet cafe. And even more formal methods of exchange and learning take place under the umbrella of transnational networks like Al Qaeda.

In contrast to major conventional war, though, the evolution of small war has been more strategic than operational and technological. The change, in other words, has been in *who* fights and *why* they fight rather than only in *how* they fight.[4] Tactics and technology have shifted, but less than in major, conventional war. A company commander from World War II or Korea would have been overwhelmed by the speed, pervasive technology, and joint nature of the conventional phase of Operation Iraqi Freedom. The adjustment for a company commander from Vietnam or the Malayan Emergency to the counterinsurgency phase of Iraq would have been easier—the need for security through presence, the intelligence challenges, the difficulty of strike operations, the problems of helping create an effective local security force, and the political and psychological complexity would ring familiar. But at the strategic level, World War II and Iraqi Freedom had much in common— formally organized militaries subscribing to a common rule set using (or attempting to use) conventional combined operations. The strategic context of the Iraq insurgency, though, is very different from the strategic context of Vietnam or the Malayan Emergency.

This compressed, dialectical, and strategic evolution affects the principles that contribute to success in small wars. As in major conventional war, some characteristics of small wars and the principles leading to success have endured. Others have changed. The key is distinguishing the two.

## Evolution of the Concept

Small wars are the American military tradition. During the first century of the nation's history, only brief conventional conflicts with Great Britain, Mexico, and the Confederacy interrupted the army's struggles with various Indian tribes and confederations. Outside of coastal defense (and a short period of occupation duty during Reconstruction), pacifying the frontier was the army's most significant mission. With the end of the Indian Wars and America's emergence as an international power toward the end of the nineteenth century, the U.S. Army and Marine Corps were increasingly used in small wars and pacification campaigns abroad.[5] Modern U.S. military capabilities for small wars were born during the Philippine insurrection and other campaigns in the Caribbean and Central America prior to World War II.[6] In fact, one of the first rigorous attempts by the American military to analyze the nature of small wars and to identify the keys to success arose from Marine involvement in the Caribbean and Central America. This thinking was captured in the classic Small Wars Manual of 1940.

According to that manual, small wars are "operations undertaken under executive authority, wherein military force is combined with diplomatic pressure in the internal or external affairs of another state whose government is unstable, inadequate, or unsatisfactory for the preservation of life and of such interests as are determined by the foreign policy of our Nation."[7] Its authors sought to identify the special characteristics of small wars while stressing that the general principles and characteristics of war also apply. They wrote

> Although small wars present a special problem requiring particular tactical and technical measures, the immutable principles of war remain the basis of these operations and require the greatest ingenuity in their application. As a regular war never takes exactly the form of any of its predecessors, so, even to a greater degree is each small war somewhat different from anything which has preceded it. One must ever be on guard

to prevent his views becoming fixed as to procedure or methods. Small wars demand the highest type of leadership directed by intelligence, resourcefulness, and ingenuity. Small wars are conceived in uncertainty and are conducted often with precarious responsibility and doubtful authority, under indeterminate orders lacking specific instructions.[8]

These ideas resurfaced often as the nature of small wars, and American thinking about them, developed over the years.[9]

World War II, Korea, and the emergence of the Cold War sparked a change in the U.S. military's self-image and how Americans as a whole viewed the role of military power in national strategy. Conventional war fighting shifted from a sporadic endeavor of a military configured for small wars to the focus and *raison d'etre* of American armed forces. Small wars were seen as a secondary function, a distraction drawing resources and degrading conventional war-fighting capabilities. At its extreme, this led to an atrophy of not only the capability for small wars but also of land power in general as, during President Dwight Eisenhower's first administration, the idea that strategic airpower obviated the need for land power gained traction among strategic thinkers.[10]

Some reversal of this trend took place in the early 1960s. American strategists and policy makers increasingly believed that the nuclear stalemate between the United States and the Soviet Union, along with the conventional strength of NATO, deterred a Warsaw Pact invasion of Western Europe. This led the Soviets to adopt an indirect strategy designed to weaken the West over the long term while limiting the risk of nuclear Armageddon. Opportunities arose as the decolonization of Asia and Africa created poor, new, fragile states whose elites were often bitter toward the West and Western ideas. After the 1960s, fuel was added as radicalism from the left attained greater acceptance, legitimacy, and even panache within some segments of the West's intellectual community. Violent radicals from Che Guevara to Ho Chi Minh were considered hip rather than frightening, while anticommunist regimes were perceived (sometimes accurately so) as repressive and brutal. The Soviets and their Chinese and Eastern European allies capitalized on this opening by fomenting and supporting insurgency and terrorism against regimes affiliated with the West. In nations like France and, increasingly, the United States, talk was of strategic "death by a thousand small cuts."[11] Small wars in what became known as the Third World, in other words, took on increased global strategic significance as the possibility of all-out conventional combat between NATO and the Warsaw Pact declined.

While Soviet-backed insurgents utilized a range of strategies, the most successful by far was the Maoist People's War. That war began when a highly motivated cadre mobilized a support base among the rural peasantry using nationalism and local grievances (often including corruption, repression, excessive taxation, and issues associated with land ownership). This was particularly powerful when associated with national liberation. The Chinese insurgents, for instance, gained strength when they painted their movement as an anti-Japanese one, even though they did little actual fighting with the Japanese. The same was true of the Viet Minh.

The people's war called for a period of underground political organization followed by guerrilla warfare.[12] The ultimate objective was seizure of power and creation of a Communist state. While the insurgents were prepared for a long struggle involving occasional military setbacks, they sought to launch increasingly larger military operations. In the "pure" form of the Maoist People's War, the final phase was conventional maneuver warfare after the regime was weakened by prolonged guerrilla operations. Many of the great successes of the Maoist approach (such as China itself and Vietnam) came through conventional military victory.

Throughout the course of the people's war, psychological operations and political mobilization paralleled military actions. In fact, violence was viewed as "armed propaganda" designed for maximum psychological effect, such as demonstrating the weakness or incompetence of the regime or provoking it into excessive reactions, which eroded its support. Military actions that had maximum direct effect on the insurgents often alienated the public (as well as the international community). Violence also deterred government supporters and inspired potential insurgent supporters. The Algerian National Liberation Front, and the Viet Minh and Viet Cong, for instance, focused assassinations and terrorism on unpopular local officials and landowners. Often the regimes were blamed when they hurt civilians, whereas insurgents were not—one of the core asymmetries of insurgency is variation in expectations concerning acceptable behavior. Thus, one of the key decisions for counterinsurgents to make was whether the political cost of armed strikes against the insurgents was worth paying.

The people's war strategy also directed insurgents to develop "liberated areas" that they could administer more justly than government-controlled regions. This too was a means of psychological warfare and propaganda designed to win over "undecideds." In fact, that was the essence of the people's war and the core of its triangularity: the conflict was an armed and political-psychological competition between insurgents and counterinsurgents for the "undecideds."

President John Kennedy, recognizing the strategic threat from insurgency, instigated a wide-ranging program to improve U.S. capabilities.[13] He first formed a cabinet-level special group—the Interdepartmental Committee on Overseas Internal Defense Policy—to lay the groundwork for a unified counterinsurgency policy and coordinate the disparate elements of the government.[14] The Pentagon established the Office on Counter-Insurgency and Special Activities, headed by Maj. Gen. Victor H. Krulak of the U.S. Marine Corps, and gave him direct access to the Joint Chiefs of Staff and the secretary of defense.[15] The services integrated counterinsurgency into their professional educational systems, and established training centers for it. U.S. Army Special Forces were expanded and reoriented toward counterinsurgency assistance. Even the State Department and Agency for International Development began to take counterinsurgency seriously (albeit with less enthusiasm than the military).

Despite these efforts and America's long history with insurgency and other forms of irregular war, the United States was organizationally, doctrinally, conceptually, and psychologically unprepared for Vietnam. The army, at least at the senior level, placed little stress on such mundane but vital aspects of counterinsurgency as training the South Vietnamese security forces, village pacification, local self-defense, and rooting out insurgent political cadres, at least at the higher level. Even though a number of experts in the United States developed an astute understanding of the Vietnamese Communist strategy and organization, Washington never forced the South Vietnamese regime to undergo fundamental reform and thus it never solidified its legitimacy.[16] Army Chief of Staff Gen. Earle G. Wheeler reflected the thinking of President Lyndon Johnson and his top advisors when he said, "The essence of the problem in Vietnam is military."[17]

By the time the United States did develop an organization to synchronize the military, political, and psychological dimensions of the struggle—the Civilian Operations and Revolutionary Development Support (CORDS) program—it was too late.[18] The United States never supported CORDS to a degree comparable to the major military operations; the North Vietnamese military was thoroughly entrenched in the south; the South Vietnamese regime was widely perceived as corrupt and illegitimate; and the American public was alienated. Even though the Viet Cong were militarily crushed in the 1968 Tet Offensive and saw its political underground decimated by the

Phoenix Program (which came later), the shift of power away from the regime was irreversible and carried on by the other element of the insurgent alliance—the North Vietnamese Army.[19] In classic Maoist fashion, what had begun as a small war mutated into a conventional one.

When the United States again faced insurgency in the 1980s, it drew on the Vietnam experience to develop a "carrot-and-stick" strategy that simultaneously promoted democratization, economic development, dialogue, and defense. In addition, Washington limited where it became involved. Recognizing that counterinsurgency support was a very long-term proposition and that support by the American people and their elected leaders would have to be sustained, the United States focused on regions of intense national interest, especially Central America and the Caribbean. In addition, indirect means rather than the large-scale application of American military force were the preferred method. The 1987 *National Security Strategy*, for instance, specified that indirect military power, particularly security assistance, was the primary tool of counterinsurgency. The 1988 National Security Strategy was even more explicit, emphasizing that U.S. engagement "must be realistic, often discreet, and founded on a clear relationship between the conflict's outcome and important U.S. national security interests." U.S. activity in counterinsurgency was based on the concept of internal defense and development (IDAD) under which the host government "identifies the genuine grievances of its people and takes political, economic, and social actions to redress them." [20]

During the 1970s and 1980s, a new dimension of the small war phenomenon also emerged: transnational terrorism. Terrorism has often been associated with small wars, but three emerging trends altered its strategic role during this time:

1. The transnationalization of terrorism as groups developed affiliations and networks with each other, some engineered by the Soviets and their allies, others not.
2. The growing legitimacy of radical terrorism—at least that perceived as related to "liberation"—among elites and intellectuals in both the Third World and the West.
3. The reemergence of "stand alone" terrorism-based strategies (something which had existed in Russia and other parts of Europe in the late-nineteenth and early-twentieth centuries among various radical, anarchist, and nihilist groups).

At this stage, though, transnational terrorism had limited strategic impact. Terrorists did not have the means to instigate mass casualty attacks and thus were treated more as a law enforcement than a strategic problem. Outside of South Africa, insurgents who relied heavily or exclusively on terrorism (e.g., Northern Ireland and Palestine) met with little success.

In the decade following the end of the Cold War, ethnic or sectarian conflicts using a range of irregular forces and militias dominated small wars. For the United States and its allies, this mode of conflict was both more difficult and simpler than its Cold War analog. It was more difficult because of the time and resources required to resolve ethnic and sectarian conflict, with its intense hatred and associated humanitarian disasters. The flood of weaponry spewed forth by the end of the Cold War added fuel to the fire. But it was simpler because it was less ethically ambiguous and less likely to be seen as chic or acceptable among Western elites. The insurgencies of the Cold War normally sought to overthrow repressive local elites or outside colonizers. The insurgents were thus able to tap at least some sympathy outside their country and, among these sympathizers, to craft a degree of legitimacy. The ethnic and sectarian warlords of the 1990s could not do this and were normally reviled by all but their co-ethnics or coreligionists. This greater ethical clarity meant that it was sometimes possible to cobble together multinational coalitions to deal with the conflicts—the obstacle was a net shortage of resources to resolve the conflicts rather than ethical or political hesitation.

The contours shifted again in the opening years of the twenty-first century. While some traditional small wars similar to their Cold War or post–Cold War forbearers persisted—Nepal, Peru, Colombia, Sri Lanka, Congo, Sierra Leone, Liberia, and so forth—most derived from the failure of the Islamic world to become politically, economically, or culturally competitive in an era of globalization. The type of transnational terrorism developed by the Palestinians in the 1970s merged with insurgency, whether the Maoist variant (which sought to control territory and eventually seize state power) or the intifada form (which sought to simply wear down the ruling power through persistent, low-level violence).

Like Cold War insurgency, this Islamic-based violence, which is often described as the world's first global insurgency, has crafted a degree of legitimacy even among sympathizers who consider its methods illegitimate, in part by targeting unpopular outsiders, especially Israel, Russia, and the United States. Moreover, the use of religion shapes twenty-first-century small wars. History suggests that armed conflicts with a religious justification are often more intractable (and more horrific) than ones based purely on political ide-ology. People of differing political ideologies can compromise or even share power when necessary. Those driven by religious ideologies often feel that compromise is inherently evil. When the stakes of a conflict are transcenden-tal, solutions are difficult, perhaps even impossible.

## Enduring Essence and the Dynamics of Change

At the strategic level, American thinking about small wars evolved as the phe-nomenon itself changed. During the Cold War, of course, small wars were part of the larger superpower struggle. In the 1960s, the focus was almost entirely on counterinsurgency and the methods associated with it—IDAD. In the 1980s, the conceptualization of the threat broadened to "low-intensity conflict." Counterinsurgency was still part of this, but so was counterterror-ism, support to insurgencies attempting to overthrow pro-Soviet governments, and even multinational peace keeping. For the U.S. military, this was codified with the 1990 release of army and air force doctrine in FM 100-20/AFM 3-20, *Military Operations in Low-Intensity Conflict*. Success in low-intensity conflict, according to this manual, followed adherence to five "imperatives": political dominance, unity of effort, adaptability, legitimacy, and persever-ance. The pivotal concept was *legitimacy*. Eschewing cultural differentiation, this assumed a unitary, mirror image type of rationality in which the people of a country decide whether the government or the architects of low-intensity conflict can offer them the "best deal" in terms of goods and serv-ices, and then they support that side. As small wars changed in the 1990s, the concept of "operations other than war" replaced low-intensity conflict. This was more semantic than significant: counterinsurgency was less dominant and multinational peace keeping and counterterrorism more so.

In the post–September 11 world, "operations other than war" seemed an inadequate phrase. American political and military leaders began to talk of "irregular challenges," that is, a combination of insurgency, transnational ter-rorism with global reach and increasingly sophisticated methods (probably to include the use of nuclear, radiological, biological, or chemical weapons at some point in the near future), and transnational organized crime. The 2005 *National Defense Strategy of the United States* contends, "While the security threats of the 20th century arose from powerful states that embarked on aggressive courses, the key dimensions of the 21st century—globalization and the potential proliferation of weapons of mass destruction—mean great dangers may arise in and emanate from relatively weak states and governed areas."[21] The strategy indicated that potential adversaries are shifting away from what it called "traditional" forms of warfare (a historically inaccurate term) toward asymmetric modes including "irregular" ones that it defined as the use of "'unconventional' methods," particularly insurgency and terrorism.

While small wars have changed in the last half century, they also have retained essential features. They remain steeped in and, to an extent, defined by ambiguity and asymmetry. Both of these take multiple dimensions. The beginning and end of a small war, whether an insurgency, a terrorist campaign, or some blending of the two (perhaps linked to transnational organized crime), is often ambiguous. Unlike traditional, state-on-state wars, there are seldom discrete and identifiable start and end points. In most cases, small wars are dialectical, with an existing system of order or regime challenged and the outcome a synthesis of the two rather than decisive victory. The distinction between combatants and noncombatants and even between war and peace is gray and ambiguous. And small wars remain asymmetric in methods, organizations, ethics, and often other ways.

Ambiguity and asymmetry mean that small wars are quintessentially political and psychological. Insurgents and terrorists avoid battle spaces where they are weakest—often the conventional military sphere—and focus on those where they can operate on more equal footing, particularly the psychological and the political. Small wars remain "armed theater" where insurgents or terrorists are locked in combat with counterinsurgents or counterterrorists at the same time that each is playing to wider audiences including the population of the state in conflict and external audiences who might provide support or opposition. Small wars also remain persistently brutal. Insurgents and terrorists often use violence against noncombatants both because it is less risky than attacks on the regime itself and because of the psychological message it sends, creating fear and demonstrating the regime's inability to protect its citizens. Sometimes, counterinsurgents and counterterrorists also target noncombatants out of sheer frustration and the desire to deter support for their enemies. Finally, ambiguity and asymmetry continue to make intelligence particularly difficult in small wars. Because it is quintessentially psychological and because insurgent and terrorist movements are often organized in a cellular and networked fashion (often using kinship, local ties, or ethnicity), the methods used to collect and analyze intelligence in traditional, state-on-state war are not effective. The closest analogies are the information gathering methods police forces use against organized crime and street gangs.

In general, though, the evolution of small war has been more a matter of degree or amplification than a sea change. Interconnectedness and information technology have facilitated the linkage of insurgents and terrorists to affiliated organizations, including transnational criminal enterprises. Coalitions and partnerships that would have been impossible during the Cold War have become the norm, thus helping overcome the decline in state sponsors of insurgency and terrorism. The best example is the transnational Islamist insurgency, which includes a dizzying array of subcomponents. These same factors have accelerated the rate of learning and adaptation. And terrorism has given the architects of small wars the capability of strategic power projection, allowing them to strike enemies both near and far.

## Success

While small wars are increasing in strategic significance as the likelihood of large-scale, conventional state-on-state war declines, there is more consistency and continuity in their logic, structure, and dynamics than change. Small wars remain armed struggles that occur when a highly motivated non-state actor or coalition of non-state actors cannot attain its/their political objectives through peaceful means, but also is too weak to seek them through conventional military activity. It thus must rely on unconventional means and a protracted struggle to shift the decisive battle space from the military to the psychological. The architects of small wars assume that skill, dedication, limited constraints, and superior will can allow them to erode the strength and morale of the regime or power system they oppose, forcing either compromise or collapse.

How can this be countered? History suggests that successful efforts to counter insurgents and terrorists always reflect two foundation principles, and share an array of characteristics that may not, in the strict sense, constitute "principles" but are akin to them. In a 1995 study, a team of analysts from the U.S. Army War College Strategic Studies Institute (which included the author) noted that

> throughout history, military practitioners, philosophers, and historians have struggled to comprehend the complexities of warfare. Most of these efforts produced long, complicated treatises that did not lend themselves to rapid or easy understanding. This, in turn, spurred efforts to condense the 'lessons' of war into a short list of aphorisms that practitioners of the military art could use to guide the conduct of warfare.

... Future strategists must avoid a "cookie cutter" mentality as they create, develop, and execute strategic plans. But that fact does not diminish
the utility of having principles to assist in the creative process. Creativity, without bounds, can be a risky enterprise.[22]

To be useful, then, principles must be broad enough that they are not context
specific, but they must be focused enough to provide a useful framework for
practical action. While the two foundation principles for small wars may lean
toward the overly broad, they are necessary for American strategists, policy
makers, and military commanders as they struggle to confront irregular challenges. They are: first, *understand the conflict for what it is;* and second, *take it
seriously.*

## Understand the Conflict for What It Is

Small wars are diametrical to the preferred way of war that evolved in the
American military during the twentieth century and blossomed in the post-
Vietnam period where involvement in small wars became seen as intrinsically
damaging to the military and to civil-military relations. Throughout history,
Americans have excelled at politically and ethically unambiguous wars,
amenable to material and technological solutions and relatively short. Small
wars are none of these things. Because their essence is psychological and
political, culture matters greatly. Strategists, policy makers, and military commanders must understand and utilize the perceptions and rationalities of
other cultures rather than assuming that all people see the world and think
like Americans. History suggests that the United States does not do this well.
President Lyndon Johnson, for instance, attempted to offer North Vietnam
economic development assistance if that nation would cease support of the
Viet Cong. As the consummate Texas politician, Johnson assumed that all
political leaders thought in terms of providing goods and services for constituents and was unable to understand that the North Vietnamese leadership
considered freedom from external influence more important than material
goods. This same tendency continues today as some observers have suggested
that more development assistance in Iraq would have made the American
occupation tolerable, thus failing to understand that in the Iraqi worldview,
honor and justice are more important than material well-being; thus, even
occupation with the best intentions is unacceptable.

Other elements of the American strategic culture also pose impediments
to success in small wars. Just as Karl Marx argued that the logic of capitalism
led to the substitution of capital and technology for labor, for at least the past

hundred years, Americans have sought to substitute technology for blood and sweat in war fighting. This has borne fruit in conventional, state-on-state war fighting, thus leading to the expectation that investment in technology and military transformation will result in low casualties and quick results in all modes of warfare. But strategists and leaders must understand that just as there are industries where the ability to substitute technology for labor is limited, so too in small wars. And just as there are industrial projects that require years or decades for completion, so too the resolution of small wars once they have reached critical mass.

Put differently, the logic of conventional war fighting tends to be linear. If the enemy's center of gravity (or centers of gravity, depending on one's perspective) can be eroded, destroyed, or demolished, his ability to resist will collapse and victory will ensue. Tangible, physical actions lead to desired psychological states in a direct way. But because small wars are essentially psychological and, for the United States, cross cultural, and because they involve multiple audiences who may or may not share perceptions and worldviews, the dominant logic is, at best, only partially linear. More often, it is what Edward Luttwak called "paradoxical," where what appears to be the best or most-effective solution to any problem often is not because a thinking, planning opponent will seek to counter it.[23] Strategists and commanders must not only ascertain what they should do, but also how the multiple audiences will perceive, understand, and react to these actions. The laws of psychology are much more complex and amorphous than the laws of physics, but they provide the framework for small wars. This makes adherence to the principle "understand the conflict for what it is" immensely challenging.

## Take It Seriously

U.S. strategy today suggests that irregular challenges, while not individually as dangerous as some other types, will be the most common kind of armed conflict in coming decades. And, if unchecked, they can threaten national interests. But we may not believe our own words. Effectiveness at preventing and resolving small wars is not simply a matter of more money, more people, or new concepts; it requires a very different national security organization than the one the United States has. From the military perspective, small wars require the sustained involvement of forces—whether small or large—with a deep understanding of the conflict, its strategic context, and cultural and psychological milieus. Even more importantly, preventing or resolving small wars requires the effective application of a wide range of nonmilitary resources. In counterinsurgency in particular, the military dimension is only a part, and usually not the dominant one. Success entails building effective local intelligence and law enforcement systems, developing legitimate and effective government, providing economic opportunities, and altering perceptions, beliefs, expectations, and other social psychological factors. Given the propensity of irregular enemies to penetrate government agencies, it also entails effective counterintelligence. Since intelligence gathering emulates the fight against organized crime, counterintelligence must mirror the methods used by urban police forces, including the creation of an independent and tightly controlled internal affairs division.

Ultimately the U.S. military can support such activities, but it is not the best organization to lead them. While the military is undertaking a range of initiatives to become more adept at confronting irregular challenges, the rest of the government is making inadequate effort at best and often none at all, largely because of resource constraints. The State Department's creation of the Office of the Coordinator for Reconstruction and Stabilization in 2005 was a tiny step. Much more needs to be done. This will probably include development of a separate government agency—a standing version of the Coalition Provisional Authority—to build and to integrate capabilities. The United States is, then, not following the "take it seriously" principle of small wars. This does not bode well for strategic success.

Beyond these two foundation principles, an array of factors has characterized success for counterinsurgents and counterterrorists in the modern era.

- Shape operations and campaigns with *psychological* precision to generate the necessary psychological effects.
- Degrade the opponent's strategy rather than attrit his force.
- Be able to sustain adequate effort for years, even decades.
- Seamlessly integrate all government agencies and elements of power.
- Design and sustain effective methods for both intelligence and counterintelligence.
- Adapt at least as rapidly as and more effectively than the enemy.

Each of these merits explanation.

## Shape Operations and Campaigns with *Psychological* Precision and Focus on the Psychological Effects

Increasing precision has been an integral part of the current revolution in military affairs and the military transformation designed to capitalize on it.

Unfortunately, though, the engineers of transformation have used an overly narrow notion of precision, defining it in a kinetic sense, as identifying targets and striking them exactly where desired. Psychological precision, which is crafting an operation or a campaign to generate the desired psychological results on relevant audiences, is more difficult but also much more important in small wars. The desired effects must be psychological rather than the traditional physical effects of controlling terrain or eroding the enemy's ability to resist. For instance, in counterinsurgency, operations and strategies must seek to *fracture* the insurgent movement through military, psychological, and political means; *delegitimize* it; *demoralize* it; *delink* it from internal and external supporters; and *deresource* it.[24]

## Degrade the Opponent's Strategy Rather Than Attrit His Force

Even in conventional war, attrition was always considered a less desirable way of destroying an enemy's will to resist simply because it is inefficient. It was seen as a last resort when other alternatives were unavailable or unappealing, such as the Western Front in World War I. As the United States discovered in Vietnam, this inefficiency and even ineffectiveness is even more pronounced in small wars, in part because the "replacement cost" for an irregular warrior is lower than for a training and equipped conventional soldier. Admittedly every insurgency and terrorist movement includes diehards who simply must be killed, but throughout history successful counterinsurgents and counterterrorists tended to be those who understood the enemy's strategy and rendered it ineffective through means other than simply killing insurgents or terrorists.

## Be Able to Sustain Adequate Effort for Years, Even Decades

Military transformation has, thus far, led to a force designed as a sprinter rather than a marathoner. At the tactical level, the saying "speed kills" makes sense, but at the strategic level, particularly in small wars, the pursuit of speed, to the extent it becomes an unwillingness to sustain an effort until the underlying causes have been ameliorated, is a recipe for failure. There are many ways to sustain an effort, but success requires having one.

## Seamlessly Integrate All Government Agencies and Elements of Power

The British counterinsurgency campaign in Malaya in the 1950s, after a somewhat rocky start, became a model effort.[25] One of its defining features was the integrated unification of political, military, intelligence, and law enforcement at all levels, from the local to the national. During the early months of the Iraq insurgency, ineffective cooperation between Coalition military forces and the Coalition Provisional Authority posed problems.[26] Again, there are multiple ways of attaining seamless integration in small wars, but not doing so can lead to failure.

## Design and Sustain Effective Methods for Both Intelligence and Counterintelligence

In conventional war fighting, artillery is often called the "king of the battle-field." In small wars, with their ambiguity and asymmetry, intelligence and counterintelligence are kings. They fuel both effectiveness and efficiency. But, as Robert Steele notes, irregular enemies, which he divides into "low tech brutes" and "low tech seers," require different organizations and methods for intelligence (and probably counterintelligence) than conflicts with "high tech brutes" (rogue states).[27] While the American intelligence community is gradually shifting from a focus on support to conventional war fighting to irregular challenges, this is far from complete. History suggests that the outcome of this will, in part, determine America's success (or lack of success) at small wars.

## Adapt at Least as Rapidly as and More Effectively than the Enemy

As John Nagl points out, the ability to innovate and adapt was one of the primary reasons the British were more successful at counterinsurgency in Malaya than the American army was in Vietnam.[28] This same truth still holds for types of small war: the future belongs to the adaptable. The U.S. military's experience in Iraq suggests that it does have significant capability for innovation and adaptation, particularly at the junior levels.[29] Sustaining this, improving it, and expanding it to other elements of the government would reflect this characteristic of success in small wars.

# Conclusions

Today globalization and interconnectedness have eroded the concept of a purely national definition of security. What takes place within states is now of immense concern to those outside it. The linkages between states, their permeability, the globalization of economies, the transparency arising from information technology, and the intermixing of people around the world combine in such a way that every internal war has external repercussions.[30] Internal conflicts—which take the form of small wars—create refugee flows which destabilize neighboring states. They often spawn organized crime as rebels turn to smuggling to raise capital and acquire weaponry. As the images of internal war are broadcast or emailed around the world, awareness rises and, often with it, demands for intervention. The days are gone when tens of millions could die in civil wars with barely a mention of it to the outside. Today the world knows, even though it may sometimes choose not to act. And, internal wars themselves and the weak states or ungoverned areas they create often serve as breeding grounds for terrorism. "In an increasingly interconnected world," according to the National Security Strategy of the United States, "regional crisis can strain our alliances, rekindle rivalries among the major powers, and create horrifying affronts to human dignity."[31] What this means is that small wars now have major significance.

American strategy is still adjusting to this. Perhaps the most important element of this new strategic approach is the notion that sources of instability and proxy aggression must be ameliorated, not simply tamped down. If the actual cause of instability or proxy aggression is not addressed, the thinking goes, the problem will eventually reemerge. In discussing the Middle East, for instance, President George W. Bush stated, "The stakes could not be higher. As long as that region is a place of tyranny and despair and anger, it will produce men and movements that threaten the safety of Americans and our friends. We seek the advance of democracy for the most practical of reasons: because democracies do not support terrorists or threaten the world with weapons of mass murder."[32] A permanent solution to the aggression flowing from internal instability thus demands the actual transformation of the unstable or aggressive state into one that is both stable and willing to adhere to the norms of the international community.

With the "Reagan Doctrine" of the 1980s, American strategy integrated support to insurgents into military doctrine for low-intensity conflict. This sometimes brought stunning success as the victory of the Afghan insurgents helped speed the downfall of the Soviet Union and pressure from the Contras led to the democratization of Nicaragua. While it is unthinkable that the United States would support terrorism in the future, it is not inconceivable that it would reinitiate the policy of support to some types of insurgents when other forms of pressure on enemy states are ineffective. While unlikely, at least in the midterm, if this does occur, American thinking must evolve from purely controlling and ameliorating small wars to controlling their effects.

In any case, small wars matter greatly in the contemporary strategic environment. Effectiveness and efficiency in dealing with them are important goals for the United States. Understanding their principles does not assure success. But it is a start. As in other types of armed conflict, there are foundation and secondary principles for small wars. Violating the foundation principles is an almost certain recipe for failure. For short periods of time, American leaders and strategists may deviate from the secondary principles, but consistently doing so will lead to failure. It remains to be seen whether the United States can or will pursue the path of success or failure.

# Notes

1. Dr. Colin Gray, keynote presentation at the 16th annual Strategy Conference of the U.S. Army War College Strategic Studies Institute, 16 April 2005, Carlisle Barracks, Pennsylvania.

2. John Keegan stressed this in *A History of Warfare* (New York: Alfred A. Knopf: Hutchinson: 1993).

3. This definition is developed in Douglas V. Johnson II and Steven Metz, *Asymmetry and U.S. Military Strategy: Definition, Background, and Strategic Concepts* (Carlisle Barracks, Pa: U.S. Army War College Strategic Studies Institute, 2001), pp. 4–12.

4. This distinction was developed by Martin Van Creveld, *The Transformation of War* (New York: Free Press, 1991).

5. See Russell F. Weigley, *History of the United States Army* (New York: Macmillan, 1967), pp. 295–312.

6. On the Philippines, see Brian McAllister Linn, *The Philippine War, 1889–1902* (Lawrence: University Press of Kansas, 2000).

7. United States Marine Corps, *Small Wars Manual,* 1987 reprint of the 1940 edition, p. 1.

8. Ibid., pp. 8–9. The Manual included a brief section on strategy and what is today called interagency operations, particularly cooperation with the State Department, but most of it was tactical in focus, dealing with everything from how to cross a river to how to procure animals locally.

9. The U.S. Marine Corps is producing an updated version of the *Small Wars Manual.* As of the spring of 2005, it is in draft.

10. For detail, see Steven Metz, *Eisenhower as Strategist: The Coherent Use of Military Power in War and Peace* (Carlisle Barracks, Pa.: U.S. Army War College Strategic Studies Institute, 1993), pp. 55–63.

11. See, for instance, Roger Trinquier, *Modern Warfare. A French View of Counterinsurgency* (London: Pa Mall, 1964).

12. Underground political organization did not stop once armed conflict began but continued through the duration of the campaign in areas controlled by the government.

13. Douglas S. Blaufarb, *The Counterinsurgency Era: U.S. Doctrine and Performance 1950 to the Present* (New York: Free Press, 1977), pp. 52–53.

14. Charles Maechling Jr., "Counterinsurgency: The First Ordeal by Fire," in Michael T. Klare and Peter Kornbluh, eds., *Low Intensity Warfare: Counterinsurgency, Proinsurgency, and Antiterrorism in the Eighties* (New York: Pantheon, 1988), pp. 26–27.

15. Robert B. Asprey, *War in the Shadows: The Guerrilla in History* (New York: William Morrow, 1994), p. 736.

16. For example, Andrew Molnar, et al., *Human Factors Considerations of Undergrounds in Insurgencies* (Washington, D.C.: Special Operations Research Office of the American University, 1965, reprinted by University Press of the Pacific, 2001), provided an amazingly accurate description of the Viet Cong/North Vietnamese strategy.

17. Quoted in Asprey, *War in the Shadows,* p. 724.

18. On CORDS, see Robert W. Komer, *Bureaucracy at War: U.S. Performance in the Vietnam Conflict* (Boulder, Colo.: Westview, 1986), pp. 118–21. Ambassador Komer was director of the CORDS program.

19. The classic early work on the organization and strategy of the Viet Cong is Douglas Pike, *Viet Cong: The Organization and Technique of the National Liberation Front of South Vietnam* (Cambridge, Mass.: MIT Press, 1965). Also important is Pike's classic *PAVN: Peoples Army of Vietnam* (Novato, Calif.: Presidio, 1986). On Tet, see Don Oberdorfer, *Tet!: The Turning Point in the Vietnam War* (Baltimore: Johns Hopkins University Press, 2001). On the Phoenix Program, see Douglas Valentine, *The Phoenix Program* (New York: HarperCollins, 1990); and Mark Moyar, *Phoenix and the Birds of Prey: The CIA's Secret Campaign to Destroy the Viet Cong* (Annapolis, Md.: Naval Institute Press, 1997).

20. *National Security Strategy of the United States,* January 1988, p. 34.

21. *The National Defense Strategy of the United States,* March 2005, p. 1

22. William T. Johnsen, Douglas V. Johnson II, James O. Kievit, Douglas C. Lovelace Jr., and Steven Metz, *The Principles of War in the 21st Century: Strategic Considerations* (Carlisle Barracks, Pa.: U.S. Army War College Strategic Studies Institute, 1995), pp. 1, 23.

23. Edward N. Luttwak, *Strategy: The Logic of War and Peace* (Cambridge, Mass.: Belknap, 1987), pp. 3–65.

24. This concept is developed in greater detail in Steven Metz and Raymond Millen, *Insurgency and Counterinsurgency in the 21st Century: Reconceptualizing*

*Threat and Response* (Carlisle Barracks, Pa.: U.S. Army War College Strategic Studies Institute, 2004), pp. 25–26.

25. See Robert Thompson, *Defeating Communist Insurgency: The Lessons of Malaya and Vietnam,* New York: Frederick A. Praeger, 1966.

26. For instance, Rajiv Chandrasekaran, "Mistakes Loom Large As Handover Nears," *Washington Post,* June 20, 2004, pg. A.01.

27. Robert David Steele, *On Intelligence: Spies and Secrecy in an Open World* (Oakton, Va.: OSS International Press, 2001, pp. 86–87.

28. John A. Nagl, *Counterinsurgency Lessons from Malaya and Vietnam: Learning to Eat Soup with a Knife* (Westport, Conn.: Praeger, 2002).

29. Leonard Wong, *Developing Adaptive Leaders: The Crucible Experience of Operation Iraqi Freedom* (Carlisle Barracks, Pa.: U.S. Army War College Strategic Studies Institute, 2004).

30. The effect the interconnectedness and globalization have on the strategic environment is explored in detail in Steven Metz, *Armed Conflict in the 21st Century: The Information Revolution and Post-Modern Warfare* (Carlisle Barracks, Pa.: Strategic Studies Institute, 2000), pp. 5–25.

31. *National Security Strategy of the United States,* September 2002, p. 9.

32. President Bush Discusses Importance of Democracy in Middle East, Remarks by the President on Winston Churchill and the War on Terror, Library of Congress, Washington, D.C., February 4, 2004.

# SEVENTEEN

# Principles for the Savage Wars of Peace

FRANK G. HOFFMAN

While the Pentagon pursued a transformation agenda based on a futuristic vision of information age conflict, we have actually returned to an earlier era, that of small wars. Thus, the subject of this essay has great salience today. Small wars, or what Rudyard Kipling called the "savage wars of peace," involve campaigns in which at least one side of the conflict does not employ regular forces as its principal force, and does not fight conventionally.[1] Such wars may involve protracted and extremely lethal conflicts of the most savage and persistent violence, and they cannot be classified as small in scale or by arbitrary distinctions between high-, medium-, or low-intensity conflict.[2] Small wars can result in the defeat of major powers, destabilize governments, or result in extended and expensive campaigns with great loss of lives and treasure.

Future opponents will avoid direct and conventional conflicts with America's overwhelming military power and will purposely seek novel and asymmetric combinations of irregular warfare. Thus, it is important that we grasp the nature of this aspect of war and understand the fundamentals that guide its conduct. The purpose of this essay is to define the basic principles that should be employed to guide the design and conduct of small wars.

Wars between regular and irregular forces fill many pages in the annals of military conflict, including classic imperial actions, policing or stability expeditions, revolutionary or people's wars, guerrilla actions, insurgencies, and terrorism. Most great empires have extensive experience with this form of conflict. Roman, British, and American examples abound. Observers of fourth generation warfare and non-trinitarian conflict need to look further back into history and requalify their distinctions.[3] The tactics may be "irregular" to stubbornly conventional forces, but irregular conflicts have been with us for some time, and will undoubtedly continue into the future.[4] In fact, most analyses anticipate an increase in internal and irregular conflicts because of America's conventional dominance and a "perfect storm" of ethnic unrest, religious violence, and demographic youth bulges.[5] In this respect, in frequency but not in kind, we may indeed be facing a generational change in conflict.[6]

Governments and military institutions unprepared to take the study and conduct of small wars seriously invite defeat or at least expose themselves to a series of costly disappointments.[7] While the basics of counterinsurgency, counterterrorism, and counter-guerilla warfare are well founded historically, one strategist argues "the plentitude of actual violence contrasts sharply with a dearth of profound theory."[8] It may be more accurate to note that the study and appropriate application of existing theory, no matter how profound, has been sorely lacking. It is not the lack of a theoretical framework or historical experience with irregular foes that confounds the U.S. military today. The American military's experience is quite extensive. More accurately, it is the disposition of Western militaries, especially that of the United States, to ignore this portion of the conflict spectrum or consign it to niche units. The American military culture has pathologically resisted learning from its own experience or the experience of others.[9] The American military prefers clean and conventional conflicts, with opponents who conveniently play by the same rules.

The American conception of conflict reflects a sports analogy, with similarly equipped opponents showing up at the appointed time and place, confining the contest to agreed-upon rules and the defined field of play. Our illusion makes neat demarcations between acceptable and illegal modes of fighting, as well as between combatants and spectators. We expect a common agreement on objectives—the goalposts at the end of the field—and rules of the game, with referees to throw yellow flags for infractions. This perspective is not useful in a second small wars era.

Today's enemies are more protean or chameleon-like and infinitely less predictable than those of the Cold War. We face not a static monolithic foe, but a constantly varying admixture of participants, the very antithesis of the past.[10] In the past, we focused on an opponent who was state based, homogeneous, rigid, hierarchical, and resistant to change. But today's enemies are "dynamic, unpredictable, diverse, fluid, networked, and constantly evolving."[11] Such opponents do not lend themselves to an ingrained order of battle mentality and nice, neat templates. Our new enemies play to their strengths, not ours.[12] Abetted by alchemists of military theory and our own myths, we focus on the fights we want to conduct and have fallen in love with the wrong revolution.[13]

Small wars have unique characteristics and attributes. The U.S. Marines' seminal Small Wars Manual of 1940 concludes, "Although Small Wars present a special problem requiring particular tactical and technical measures, the immutable principles of war remain the basis of these operations, and require the greatest ingenuity in their application."[14] Yet, a close reading of the particular tactical and technical measures outlined in this historically grounded document suggests that the so-called immutable principles have to be turned on their head by the ingenious practitioner. Irregular conflicts present unique problems and require unusual solutions that diverge from conventional conflicts. In fact, most failures are attributable to the rote employment of conventional rule sets. As the British author Col. Charles Callwell noted at the turn of the last century, " . . . the conditions of small wars are so diversified, the enemy's mode of fighting is often so peculiar, and the theatres of operations present such singular features, that irregular warfare must generally be carried out on a method totally different" from conventional wars.

Accordingly, one would expect to find different fundamentals or principles applicable to the conduct of such conflicts. The remainder of this essay is devoted to the delineation of these principles. This list was built by a detailed study of Callwell and the U.S. Marine Corps' Small Wars Manual. Working deductively from the litany of successful and unsuccessful examples in Algeria, Vietnam, Somalia, Afghanistan, and Malaya, and from current American joint doctrine, a set of eight principles has been developed.[15]

This set expands and modifies current American doctrine for two reasons. First, current joint doctrine for what American military officers call Operations Other Than War (OOTW) is predicated upon cases and lessons drawn primarily from peacekeeping operations.[16] This experience is still relevant, but it is insufficient to the nature of today's networked insurgency and post-modern "warrior class."[17] Past peacekeeping operations represent, in this author's view, a narrower subset of small wars. Second, the principles that are derived largely from that experience are incomplete and do not reflect the myriad and messy challenges of the twenty-first century. A revised and broader set of principles is now required.

## Eight Principles

The set of principles presented here represent a framework for education and for higher-level planners approaching the study of a particular irregular problem.

They should not be set as a prescriptive list or a set of inviolable principles to be rigidly applied. Anyone looking for a science or strategy "replete with principles that are both immutable and deeply meaningful, only indicates by that desire a basic misunderstanding of their subject."[18] The conduct of war is best understood as both art and science. The crucial element of its artistic application is recognizing unique contexts, the contingent factors, and the opportunities to create advantage purposely by violating principles or rules when needed. As Mahan understood, art accepts the existence of principles and rules, but only as guides. Each case has its own features—which modify the application of the rule, and may even make it at times wholly inapplicable. It is for the commander to apply (or adjust) the principles and rules in each case, using what Mahan called "the greatest ingenuity in their application."[19]

The spectrum of missions that may be assigned to military forces in a small war and the continuously adaptive nature of today's adversaries preclude employing overly simplistic solutions. As in conventional conflicts, the professional judgment of a highly educated and experienced commander, aided by a similarly well-informed staff, is required. As stated in the Small Wars Manual, "As a regular war never takes exactly the same form of any of its predecessors, so, even to a greater degree is each small war somewhat different from anything which has preceded it."[20] Once again, context matters. Commanders and their planners must consciously look for both similarities and distinctions in applying historical precedents.

### Understanding

Oft quoted, but rarely understood, the wise Clausewitz once stressed, "The first, the supreme, the most far-reaching act of judgment that the statesmen and commander have to make is to establish . . . the kind of war on which they are embarking."[21] This particular judgment on the part of civilian and military leaders is difficult to establish for numerous reasons. Commanders and planners who are examining a potential contingency need to assess the nature of the conflict in very detailed terms, often with limited time to access experts or databases. Contingencies can spring up suddenly, or occur in unexpected lands, far from home. Sometimes, the statesman or the commander is confused or blind as to the true nature of the war, preferring to see the future in terms of past battles. Often, political and military leaders fail to understand each other. But more often the real problem is a lack of understanding about the nature of the opponent.

Too often, military commanders focus on symmetric elements, defining the problem and the opponent's options in purely military terms. We rarely widen our scope though to include a broader assessment of the adversary's culture. An influential strategist once observed, "good strategy presumes good anthropology and good sociology."[22]

Fundamentally, war involves an iterative competition between peoples whose behavior patterns are a result of a complex combination of factors including history and geography. Our intelligence bureaucracies have experts on the strategic and military cultures of potential adversarial states. We can name all of their major formations and quantify their principal weapons. We are order of battle oriented, focused on studying what is quantifiable and predictable about opponents in neat templates.[23]

In contrast, what is more important in small wars is a very comprehensive examination of the culture of the society or country that is the source of the conflict. Because small wars usually are interventions in an internal conflict and require efforts to reconstruct or establish political, social, and economic institutions and mechanisms, an acute understanding of the society and its culture is essential. Small wars are generally culture-intensive conflicts, and the battleground, properly understood, includes the political and psychological elements of the population and culture. We need to gain a deep and nuanced understanding of the conflict we are about to embark on, and to acquire as thorough a grasp of the nature of the adversary as possible. This includes becoming well informed about the adversarial culture and social system, not just estimates of fielded forces.

The combination of national history, myth, geography, ethnic backgrounds, and religion we know as culture is contained in the socially transmitted behavior patterns, arts, beliefs, institutions, and thought characteristic of a community or population. Culture is a complex aggregate that includes knowledge, law, morals, customs, and any other capabilities and habits acquired by a member of society. It works on many levels, sometimes overtly, other times covertly. As Michael Howard stressed years ago, wars are not tactical exercises on a larger scale. Major wars are conflicts between societies, and they can be fully understood only by understanding the nature of the social systems involved. Victory is often not defined on the battlefield itself but in political, social, or economic factors.[24]

This is not news. Again, the Small Wars Manual notes, "The campaign plan and strategy must be adapted to the character of the people encountered."[25] It is impossible for U.S. forces to succeed without an intimate appreciation of the local culture. Success requires a framework and educational background to generate an ability to think in terms of culture and to see things from the perspective of others. As one veteran of several interventions observed a decade ago:

What we need is cultural intelligence. What I need to understand is how these societies function. What makes them tick? Who makes the decisions? What is it about their society that is so remarkably different in their values, in the way they think compared to my values and the way I think?[26]

One can see the need for cultural intelligence and understanding in almost every phase of Operation Iraqi Freedom. Numerous stories and anecdotes have emerged about the pervasiveness of the effects of culture and the additional complexity it brought to the battle space. Young officers and noncommissioned officers found the Iraqi culture utterly foreign and overwhelmingly complex. Others were surprised at the importance of religion to daily life in a foreign culture.[27]

Cultural ignorance has been a challenge in the past for U.S. forces. The American–led intervention in Somalia in 1992–93 was severely undermined by a limited understanding of the clan and political framework in that impoverished country. The degree of social disintegration and the infighting that were rife in this starving East African state were beyond the grasp of almost all who were involved. This lack of understanding curtailed the design and implementation of appropriate solutions within the time and resources the international community was willing to bear.[28]

This is why there is great merit in recent calls for cultural-centric warfare in which our soldiers and sailors are prepared with an acute degree of cultural awareness and the need for "global scouts" to advance our interaction with foreign societies.[29] The military and educational reforms suggested by these recommendations are wide reaching but germane to the problems we face in the twenty-first century and critical to new joint operational concepts like effects-based operations.[30]

In short, the nature of small wars places a premium on an equally in-depth knowledge base of the host nation's society and culture. Good intelligence is always a precious commodity in small wars, largely because of the frequently remote nature of the host country, the inadequacy of infrastructure, and the lack of familiarity with the native population. But because small wars mandate an in-depth appreciation of the local culture, that is where intelligence and understanding should begin. Thus, understanding is the first fundamental of small wars.

**Understanding:** *Craft military strategy and operations based upon a detailed understanding of the enduring nature of military conflict and the specific context (cultural, social, political, military, and geographic) in which force is to be* *applied. The application of military force in all dimensions must reflect the particular character of the social system being engaged.*

## End State

The most universal principle of war across time and international variations is the principle of the objective or aim.[31] Current U.S. joint doctrine and the annals of military history in both conventional and unconventional conflict are similar in this regard.[32] The government should have a clear political objective, and this overarching political objective or aim must remain paramount and always in focus. Further, the objective should be clearly understood and credibly attainable by the ways and means allocated to the task. Agreement and understanding of the objective helps create conditions that foster unity of effort among coalition and interagency partners. Clarity also facilitates development of subordinate objectives, missions, and tasks for military planners and other participants as well. When this overarching objective is not first in the minds of all participants, there can be a tendency to adopt short-term measures in reaction to insurgent or terrorist activity.

History suggests that political objectives are not always well defined, and that the translation of political to military objectives is frequently mishandled.[33] One former American commander has said, "It's not nice and neat—for openers, you don't get a clean hard mission that tells you exactly what you're supposed to do."[34] Continued and interactive discourse between senior policy makers and military officials is warranted to clarify the intentions of policy in order for military planners to translate the aim into concrete military objectives and missions. Our understanding of the nature of war underscores defeating the will of the opposing commander and his means to resist. Military planners are used to defining military objectives, and frequently do so in terms of either defeating the enemy's main combat force or by seizing defined physical objectives. Rarely can victory be defined in purely military terms, and less frequently in small wars. Yet the translation of political to military objectives is a poorly understood and largely overlooked aspect of operational art. Civilian policy makers understand it even less, and so they defer to the purported technical expertise and professionalism of their military advisers.[35]

In small wars, neutralizing irregular forces and securing and holding specific geographical areas or cities may be necessary. But they rarely are sufficient. It may be better to think in terms of an end state rather than an objective. End state has a very definitive connotation. In conventional warfare, defeat of the opponent's military force is a clear-cut end state, but the requirement in small wars is (usually) to establish a certain set of conditions conducive to stability, local governance, and economic growth. This is why irregular warfare is often described as 80 percent political and 20 percent military. The military may be only a supporting instrument, responsible for creating and maintaining a security environment, and for providing logistical support to other government agencies.

While an objective might be misconstrued in the physical sense, an end state in small wars is something that has to be created or reconstructed over time. It is a long-term condition to be established and sustained. The end state includes functioning institutions including political and security elements, legitimate processes for transparent and accountable governance, and public participation. An end state should not be confused with a preordained exit strategy, tied to a fixed and announced schedule. Such a device is a certain hindrance to success.

An end state to a small war also includes an alteration in attitudes and perceptions. Both the general population and the disaffected element that resorted to violence must accept the new end state as an acceptable political outcome. There may be a distinct military objective in the conflict, but success will be determined largely in the political and psychological sphere, and it is best to define our objectives in those terms to ensure compatibility with overall policy. From this, planners must craft a campaign plan that links military objectives and supporting actions to concretely and effectively contribute to the attainment of the desired end state. Thus, the study and successful prosecution of small wars must establish end state as a key fundamental.

**End State:** *The assigned political aim defined in terms of the particular set of attitudes, conditions and capabilities that must be created or attained to provide governance, public services and security for the local population and the host nation.*

## Unity of Effort

Unity of command is another important principle in U.S. military doctrine. In small wars it may be entirely impracticable and beyond attainment. Not surprisingly, existing joint doctrine emphasizes unity of effort instead.[36] Unity of effort will take on added importance because of the complexity inherent in balancing the military with the political dimension. It is also further complicated by the extensive participation of various nations, other government agencies and international participants. The small wars battle space may include numerous parties including other governmental agencies, nongovernmental organizations, international organizations, and private volunteer organizations. It may also include numerous private commercial entities supporting either side of the conflict as well.[37] Not all of these parties will share U.S. interests or priorities, although they may support the desired end state and they may provide crucial resources, skills, and information to the overall effort. Many organizations will desire to overtly distance themselves from U.S. policy and want to make their political independence clear. Few will accept a clear-cut chain of command and a set of missions or assignments for which they are accountable.

Harnessing the efforts and capabilities of myriad entities toward a common goal is one of the biggest challenges and opportunities for the commander that is unique to small wars. An inability to establish unity of effort elongates the mission, exposes the intervening forces to additional risks and burdens, and may undermine the entire mission.

There are a variety of techniques for achieving unity of effort. The international community or the U.S. government may designate and empower a senior official to coordinate an international response as was done in Somalia and in Bosnia. A U.S. ambassador may employ a variation of the country team approach to manage and integrate the various national and international participants in a small war, as the United States did in El Salvador. It is possible that in some security situations a regional combatant commander or joint task force (JTF) commander may be designated as overall executive for a coalition. This approach was employed in previous U.S. nation building endeavors in Japan and Germany, as well as in Iraq. Eventually, this approach transitions to some civilian official.

No matter what methodology is selected, extensive coordination is necessary. Interagency coordination and cooperation are essential to achieving effective unity of effort and synchronizing the coherent application of all elements of national power. Political, economic, diplomatic, military, and informational efforts must be effectively balanced and coordinated. The success of the British involvement in the Malayan Emergency is instructive. Subordination of the military to the civilian and the resultant unity of effort was the key to British success. There is perhaps no better example of how a clear and logical organizational chart can have decisive results on unity of effort. Sir Harold Briggs was appointed director of operations, and recognizing the need for unified command, established a War Council that included civil, police, and military representatives and acted as a coordinating committee.[38] Coordinating committees were also established at state and district levels.

These committees provided for unity of effort, reduced duplicative operations, and facilitated the rapid exchange of intelligence, thereby significantly improving operational results.[39] Unity of effort was essential to British success there, although it took some time to create. British lessons learned from its imperial past seemed to have been misplaced after World War II.[40]

The failure to establish unity of effort between military and pacification programs early in Vietnam was a significant contributor to America's lack of success there.[41] Likewise, the same can be claimed for U.S. efforts in Iraq in 2003.[42] In this regard, the lessons of Bosnia may have much to offer to future practitioners.[43]

**Unity of Effort:** *Achieve coherency of actions by all parties involved to attain the desired end state by common efforts and purpose.*

## Credibility

Current American and British doctrine for peace support operations and counterinsurgency stresses the importance of legitimacy. Without any doubt, legitimacy is a vital principle at the strategic level. The perceived legitimacy of the indigenous government the military has been directed to support must be reestablished and maintained. But legitimacy is too often equated to approval by an international organization like the United Nations or the international community writ large.

Within the conduct of small wars, legitimacy is not a precondition but a product or result. It is a description of the end state to be achieved. The acceptance of a political solution and the establishment of political institutions as representative to maintain it must be perceived as legitimate in the eyes of the local population. A lasting peace and the reintegration/reharmonization of former insurgents and opposition forces will be impossible to achieve if the political solution is not perceived to be legitimate. Thus, legitimacy remains a strategic objective but it does not provide significant guidance to the employment of the military instrument.

At the operational level in small wars, however, credibility becomes fundamental. All actions must serve to create and sustain credibility in the eyes of the supported populace or government. The military must pose a credible force to the opposing insurgents, or when we are supporting one element of an internal conflict, we must exert our influence to establish that side as a credible military force. The force size and dispositions of our intervention force must be robust enough to be credible. When challenged, the commander must employ sufficient force to reduce the threat to the local population or his own forces, consistent with the nature of that threat and without undue collateral damage or risk to noncombatants. Both the insurgent and the local population need to perceive that the intervening military forces are, in the classic words of the 1st Marine Division commander's guidance before Operation Iraqi Freedom, "no better friend and no worse enemy."[44]

Credibility extends to the entire governmental effort, not just military forces. Commanders should ensure that all military operations, especially civil-military actions, deliver as promised. Relationships and trust are built upon credibility. Trust and mutual respect between our forces and the host nation and its representatives reinforce this credibility. At the end of the day, providing the required level of security and providing for the welfare of the population will go far toward establishing relative credibility over an insurgent group, and setting conditions for success.

Credibility is also reinforced when the intervening force acts in consonance with internationally accepted values and legal obligations. Operating within the law and our own guidelines reinforces our credibility with local leaders and the population. Regardless of the outrages committed by the insurgent or terrorist, our response must always be within lawful bounds. Governments that do not act consistently and in accordance with their own legal system automatically lose the right to demand that their people comply with the law.[45] The same is true for military forces. A failure to follow signed treaties and international law with regard to the employment of force, or the handling of prisoners, gives ammunition to the opposition in the fight for the hearts and minds of the indigenous population. Ultimately, success in small wars is tied to the generation of an overwhelming impression of credibility. This perception pulls the local population to their own government, and helps convince the irregular combatants that their cause is doomed. Once that tipping point is achieved, success is assured.

**Credibility:** *The consistent and legal application of force and the regular provision of services and assistance towards the assigned end state, and consistent support towards the host country in accordance with agreements.*

### Discriminate Force

One of the key fundamentals of this form of conflict is the concept of minimizing the use of blunt military force. It is possible to conduct a brilliant series of tactical actions with overwhelming force and firepower and lose the larger strategic goal. "In small wars caution must be exercised and instead of striving to generate the maximum power with forces available," advises the

Small Wars Manual, "the goal is to gain decisive results with the least application of force and the consequent minimum loss of life."[46] Time and time again, history shows that a lack of judgment or an excess of violence can lead to prolonged conflict.[47] This has been codified into U.S. doctrine as the principle of restraint.

The excessive application of military firepower or an imprudent ill-advised act can significantly alter the strategic situation. Firepower-intensive operations may antagonize both external and internal parties that are neutral to the insurgent, swinging support and resources to the opponent. Excessive collateral damage or accidental injuries to noncombatants will undermine the credibility of U.S. efforts to assist the host nation and make our intervention longer and more costly. The French experience in Algeria is one example of this risk, as were aspects of the U.S. involvement in Vietnam. In Algeria, the French employed raids, reprisals, and interrogations that produced a series of tactical successes. However, they failed to gain the support of the populace in the long run, and lost popular support at home at the same time.[48] The Soviets also employed more firepower than was necessary and did not adapt their tactics in Afghanistan. They then repeated their own mistakes in Chechnya.[49] In Vietnam, U.S. forces inappropriately applied technological superiority and firepower, frequently in a manner at odds with American policy objectives.[50]

The principle of restraint does not adequately capture the degree of discipline and force application needed to succeed in small wars of the twenty-first century. The concept of restraint may be very appropriate for some kinds of small wars, especially peacekeeping operations or when U.S. forces are conducting post-conflict stability operations. Care must always be taken to preserve life, minimize casualties among noncombatants, and reduce property damage. This is a moral and legal imperative, albeit difficult in practice.

However, modern small wars pit U.S. forces against acutely agile opponents with no qualms about killing innocents by the thousands. Such opponents recognize no bounds and are not easily deterred, nor can they be deflected by clever appeals to their conscience. Today's warrior class cannot be argued into submission.[51]

An element of attrition exists in most forms of combat, and has always been present in small wars. The requirement to present a credible threat of force, or even to apply deadly force if and where needed, is a regrettable necessity. Some elements in today's world cannot be persuaded or deterred from violence. The fundamental guidepost that should steer us in preparing military personnel for the dynamics of modern small wars is discriminate force. The issue is not restraint or holding back as much as finding the right balance between force and restraint based on the facts on the ground.[52]

We need to prepare our soldiers and marines, by their training and by the use of appropriate rules of engagement, to recognize those situations in which the context and the commander's intent require the application of military force. When they recognize that the situation requires the application of deadly force, they should ensure that each firefight or engagement is carried through to resolution. When the situation calls for standing down or withdrawing, the small-unit leader should feel empowered to take that action. The development of nonlethal technology is very relevant to this balance, affording the commander more options for quelling potentially disruptive situations short of employing the full weight of his traditional combat power.

Lethal force must always be applied in consideration of the wider mission and the local context in small wars. We need to establish in mental skill sets the ability to properly discriminate between situations, within the context of the chaos and uncertainty of deadly combat. Military training and educational programs must create and sustain the necessary degree of confidence and professional judgment required to apply force appropriately in small wars. This will require changes in current military training programs that focus only on how to employ firepower more efficiently.

**Discriminate Force:** *The disciplined application of the appropriate amount and type of military force to the specific nature of the contingency and the tactical situation, within the overall intent of the commander and the rules of engagement; a product of extensive training, education and discipline.*

## Freedom of Action

Security has been a well-recognized principle of war for some time. It is usually thought of in terms of securing one's base of operations or lines of communication. The purpose of underscoring security is to ensure that one's own force is not surprised by enemy initiatives. Because of the nature of guerrilla operations and the propensity for the weaker side to resort to raids and ambushes against outposts, detachments, and convoys, this principle is highly relevant to small wars. However, the purpose of gaining security has been often misunderstood. It is not just to prevent surprise or to achieve such a position of force protection that the accomplishment of the mission is sacrificed on the altar of "security at any cost."

The objective is to obtain and sustain a position of advantage in order to prevent the enemy from interfering with our main effort. The whole purpose of security is to preserve or enhance our freedom of action. We do this by reducing the vulnerability of our force to undue influence or interference, in both the physical and informational domains. Thus, the aim of creating firm bases, convoy operations, security patrols, check points, etc., is not security per se but to sustain our freedom of action vis-à-vis the government and the population we are supporting.

Of particular note, this principle should not be used as an argument for developing an isolated bastion that separates our forces from its coalition partners, its interagency teammates, or the indigenous populace. A cantonment out of touch with the local population may offer secure basing arrangements and an opportunity for our forces to rest between missions. But if it allows the adversary to control key elements of the population or critical resources, or to operate with impunity, it does not contribute to mission success. Close contact and saturation patrols may afford more force protection than intensive fortifications. The improved situational awareness and intelligence gathered through close interaction and cooperation with the populace is one way to establish security and stability for both our forces and the general populace. The coalition or JTF will always have to balance its force protection with its mission. The concept of prudent risk will guide commanders as they seek to achieve this balance.[53]

Freedom of action has a psychological benefit that is at least as important as the material gain because it gives tangible evidence of success in the minds of the populace. Most people want to be on the winning team, and if we are unable to secure a home base and freely operate throughout the area of responsibility, it is unlikely we will be successful in convincing a wavering population that we can extend the necessary security to them. Likewise, our ability to operate at will sends a signal to the populace that it should support the side of the conflict we have joined.

**Freedom of Action:** *Those measures designed to economically provide for freedom of action for friendly forces to maneuver freely throughout the operational area and to retard the movement or actions of non-friendly forces. This allows the Joint force to preclude surprise or interference, retain the initiative and exploit opportunities to maneuver temporally or spatially anywhere at a time and place of its choosing.*

### Endurance

One of the principles in American joint doctrine is perseverance. It is defined as "the measured, protracted application of military capability in support of strategic aims."[54] This principle acknowledges that small wars are rarely short affairs. The asymmetric nature of small wars often forces the weaker power into strategies that rely upon protracting the conflict in hopes of capitalizing on an asymmetry of wills. If we demonstrate through word, deed, or policy that we do not have the stomach to stay for the long haul, our adversaries employ a strategy of exhaustion to sap our will. Osama bin Laden claimed that the rapid withdrawal of American forces from Somalia in early 1994 represented a weakness in the U.S. capacity to sustain an effort.

But persistence and perseverance are attitudes. We may want to persevere but lack the national will or the institutional capacity to operate within a foreign country for the protracted nature of the conflict. We may have the will but lack the physical endurance to sustain our forces and operate in austere operating environments, or lack sufficient expeditionary forces to cover the assigned area of responsibility. This is more than just persistence, for a force can persistently apply the wrong tactics or persistently insist on employing firepower-intensive operations instead of discriminate force. We must apply both will and capacity to succeed in small wars, and must do so over an extended period of time.

For these reasons, endurance is a more appropriate fundamental than persistence. Rapid results are rare in small wars. We need to apply the approach of competitors like Lance Armstrong in events like the Tour de France. This grueling competition contains many different forms of racing; including time trials, sprints, long distance flat rides, and punishing mountain climbs. It also includes individual as well as team events, analogous to our efforts with other agencies to achieve U.S. national interests. The right tactics, good equipment, and arduous training must complement the mental perseverance of the rider. Preparation—mental, material, and physical—are at the heart of endurance. Likewise, in small wars, coalitions and JTFs must employ the right combination of tactics in different types of competition; they must employ their equipment and leverage the capabilities of the entire team, including the interagency and humanitarian relief agencies, if they are to succeed.

**Endurance:** *The combined institutional and individual capabilities that facilitate the resolute commitment of national and military resources over a protracted period.*

### Agility

Small wars place a premium on agility at three levels: mental, organizational, and operational. Yet, agility is not defined as a principle for either conventional or irregular warfare. Small wars have usually required a special mindset—akin to what the U.S. Marines call their expeditionary mind-set and cultural ethos: constantly prepared for immediate deployment overseas into austere operating environments, bringing the minimum necessary to accomplish the mission, ready to adapt to new situations, and mentally agile enough to create and implement innovative solutions to unanticipated circumstances, in cultural contexts that may be completely foreign. This is a tall order.

Mental agility is formed by the study of history, frequent exercises and simulations that test one's professional judgment against thinking opponents in tough situations. Professional military institutions need to develop and reinforce individual and collective learning for such situations. Agility is based on mental alertness and conditioning, and improves the ability to move swiftly and change direction or mode of operation on short notice, based upon pattern recognition and training.

Agility lies at the heart of the Marines Corps' "Three Block War" construct, describing the tactical complexity of having to conduct offensive, peacekeeping, and humanitarian tasks consecutively or even simultaneously.[55] Within each of these blocks, military personnel must recognize the need to adapt their tactics, techniques, and procedures on the fly. Each block requires different skill sets and different methods, and the military leader has to seamlessly alter his approach and tactics as the context changes.

In small wars, the enemy is extremely elusive, employs irregular tactics, and disperses to avoid destruction by our technological superiority. Success in these contests requires great creativity, better situational awareness, autonomy, and increased freedom of action at lower tactical levels. This enables subordinate commanders to compress decision cycles, seize the initiative, exploit actionable intelligence, and take advantage of fleeting opportunities. Small-unit actions are guided by mission tactics and decentralized means of command and control. Decentralization pushes decision-making authority and responsibility down to the lowest level necessary.

Operational agility is abetted by decentralized and distributed operations to deal with ambiguous threats and help commanders fill in the blanks that technology alone cannot resolve with the persistence, granularity, and discrimination we need for unconventional opponents.

Today's conflicts are the ultimate small-unit or squad leader's war; they demand greater levels of agility and preparation at each level. Adaptive threats will be met and overcome by an agile and distributed network of small-unit leaders who have been trained, educated, and empowered to lead.[56] This exploits one of our strongest, but least utilized, strengths, our human capital, and accelerates operational speed and tempo of operations.

At the operational level, small wars present the same challenges and the need to shift from offensive operations to stability and support operations. At the same time, we need to be alert to changes in the enemy's methods and to recognize that the nature of conflict is dialectical—the opponent has a vote too. He may not respect the American penchant for thinking of conflict in phases, our phased approach, and we need to be prepared to shift back and forth between operating modes. The United States has had problems adapting to changes in context in situations in both Beirut and Mogadishu. In these cases a lack of agility proved fatal.[57] The situation must be constantly evaluated for changes in context that may change the mission, required resources, or operations.

Organizational agility is inherent to emerging joint doctrine in the United States. The ability to rapidly reconfigure combined arms teams for deployment and employment augurs well for future small wars. The adaptive, task-organized nature of JTFs, and the ability to reaggregate or disperse based upon the situation is a classic example of organizational agility or flexibility. Given the dynamic, adaptive nature of the threat, it is likely that an effective countering strategy will require an equally dynamic and multidisciplinary organizational structure which will vary from mission to mission. In such situations, armored columns may not be at the fore. There will be times where either a civil affairs or an engineer unit must serve as the main effort, with more traditional maneuver units in support. Other times, the JTF or subordinate units may lead or be subordinated to an interagency task force, with members of the Justice Department, State Department, or intelligence community represented. The provincial reconstruction team approach used in Afghanistan and Iraq is an example of such organizational agility.

**Agility:** *The purposeful adaptation of organizational structure, mode of operations and mentality to the unique complexity, ambiguity and context of the particular contingency at hand.*

## Conclusion

In this second small wars era, military professionals will have to dust off some of the more obscure military classics, as well as throw off anachronistic conceptions about war. Dog-eared copies of Clausewitz will still be relevant, but Charles Callwell and T. E. Lawrence also have much to offer today's student of war.[58] American warriors and theorists alike would do well to understand the founding of their own country and their own participation in the "savage wars of peace."[59] Americans have quite a bit of experience to draw upon, but they are prone to quickly shrug off these irregular experiences, as interest from civilian leaders waxes and wanes, and turn again to the prevailing big war paradigm.[60]

The purpose of these principles should be clear. The conundrums and complexities inherent in the conduct of war can never be resolved by a list of fundamentals. Models and frameworks are only the starting point for an analysis of a contingency. There are no formulas for successfully outlining a strategy or artfully conducting a war, whether conventional or irregular in nature.[61] This is especially clear in regard to the latter in this new century. The nature of today's more primitive small wars and more adaptive enemies suggest that formulaic approaches could be fatal.

## Notes

1. This does not exclude examples of compound conflicts where irregular elements are used behind enemy lines (World War II) or in adjacent theaters (Peninsular campaign, Nathan Greene in the American Revolution, Lawrence with the Arab Revolt, and Mosby's Rangers in the U.S. Civil War). Thomas M. Huber, ed., *Compound Warfare: That Fatal Knot* (Fort Leavenworth, Kan.: U.S. Army Command and General Staff College Press, 2002).

2. One distinguished contributor to this volume calls these "absurd distinctions." Colin S. Gray, *Modern Strategy* (Oxford, UK: Oxford University Press, 1999), p. 274.

3. William S. Lind, Keith Nightengale, John F. Schmitt, Joseph W. Sutton, and Gary I. Wilson, "The Changing Face of War: Into the Fourth Generation," *Marine Corps Gazette,* Dec. 1989, pp. 22–26; Martin Van Creveld, *The Transformation of War* (New York: Free Press, 1991). For the latest and most detailed exposition, T. X. Hammes, *The Sling and Stone* (St. Paul, Minn.: MBI Publishing Company, 2004).

4. Data indicate that intrastate conflicts are far more frequent than interstate wars. See Richard Cincotta et al., *The Security Demographic: Population and Civil Conflict after the Cold War* (Washington, D.C.: Population Action International, 2003), p. 22.

5. National Intelligence Council, *Mapping the Future,* December 2004. Available at http://www.cia.gov/nic/NIC_globaltrend2020_s4.html.

6. Antulio J. Echevarria, "The Problem With Fourth Generation Warfare" (Army War College, Carlisle, Pa.: Strategic Studies Center, 2005). Available at www.carlisle.army.mil/ssi/newsletter/opeds/2005feb.pdf.

7. Colin Gray, *Modern Strategy,* p. 279.

8. Ibid., p. 283.

9. Bruce Hoffman, *Insurgency and Counterinsurgency in Iraq* (Washington, D.C.: RAND, June 2004), p. 7.

10. It should be noted that some Cold War–era adversaries, like the Viet Cong, were more dynamic and less predictable.

11. Brian Michael Jenkins, "Redefining the Enemy," *RAND Review,* Spring 2004, p. 17.

12. Thomas X. Hammes, "4th Generation Warfare," *Armed Forces Journal,* November 2004, p. 40.

13. Ralph Peters, *Fighting for the Future; Will America Triumph?* (Harrisburg, Pa.: Stackpole Books, 1999), pp. 22, 30.

14. U.S. Marine Corps, *Small Wars Manual* (Washington, D.C.: Government Printing Office, 1940), p. 8.

15. The foundation for the study of small war principles begins with, but should not end with, Robert Thompson, *Defeating Communist Insurgency: The Lessons of Malaya and Vietnam* (New York: Praeger, 1966).

16. The term Operations Other Than War and its abbreviation are oxymorons. Joint Publication 3-07, *Joint Doctrine for Military Operations Other Than War,* 16 June 1995, pp. II-1–II-8. These principles are also included in Marine and U.S. Army doctrine.

17. Ralph Peters, "The New Warrior Class Revisited," in his *Beyond Baghdad: Postmodern War and Peace* (Harrisburg, Pa.: Stackpole, 2003), pp. 44–60.

18. Bernard Brodie, *War and Politics* (New York: MacMillan, 1973), pp. 450–51.

19. A. T. Mahan, quoted by Jon Tetsuro Sumida in *Inventing Grand Strategy and Teaching Command* (Baltimore, Md.: Johns Hopkins University Press, 1996), p. 70.

20. *Small War Manual,* p. 9.

21. Carl von Clausewitz, *On War,* Michael Howard and Peter Paret, eds. (Princeton: Princeton University Press, 1989).

22. Brodie, *War and Politics,* p. 332. Brodie goes on to add, "Some of the greatest military blunders of all time have resulted from juvenile evaluations in this department."

23. Anthony C. Zinni, in "Non-Traditional Military Missions: Their Nature, and the Need for Cultural Awareness and Flexible Thinking," in Joe Strange, *Capital "W" War: A Case for Strategic Principles* of *War* (Quantico, Va.: Marine Corps University, 1998), p. 266.

24. Michael Howard, "The Use and Abuse of Military History," *Parameters,* Spring 1980, p. 9.

25. *Small Wars Manual,* p. 13.

26. Zinni, "Non-Traditional Military Missions," in Strange, p. 267.

27. Leonard Wong, *Developing Adaptive Leaders: The Crucible Experience of Operation Iraqi Freedom* (Carlisle, Pa.: Army War College, July 2004), pp. 7–8.

28. For a concise overview see Lynn Thomas and Steve Spataro, "Peacekeeping and Policing in Somalia," in *Policing the New World Disorder: Peace Operations and Public Security,* Robert Oakely, M. Dziedzic and Eliot Goldberg, eds. (Fort McNair, Washington, D.C.: National Defense University, 1998); F. G. Hoffman, "One Decade Later: Debacle in Somalia," Naval Institute *Proceedings,* January 2004, pp. 66–71.

29. Robert H. Scales, "Culture-Centric Warfare, Naval Institute *Proceedings,* October 2004, pp. 32–36.

30. Joint Warfighting Center, *Operational Implications of Effects-based Operations (EBO)* (Norfolk, Va.: Joint Forces Command, November 2004).

31. The most comprehensive study of principles of war, including an international perspective, is John I. Alger, *The Quest for Victory: The History of the Principles of War* (Westport, Conn.: Greenwood, 1982).

32. Joint Pub. 3-07, pp. II-1–II-2.

33. Gray, *Modern Strategy,* pp. 57–64. See also B. H. Liddell Hart, *Strategy* (New York: Meridian, 1991), pp. 338–39.

34. Gen. Anthony Zinni, "Its Not Nice and Neat," Naval Institute *Proceedings,* August 1995, p. 30.

35. See Eliot A. Cohen, *Supreme Command: Soldiers, Statesmen, and Leadership in Wartime* (New York: Free Press, 2002).

36. Joint Pub. 3-07, p. II-3.

37. Peter W. Singer, *Corporate Warriors: The Rise of the Privatized Military Industry* (Ithaca, N.Y.: Cornell University Press, 2003).

38. Robert B. Asprey, *War in the Shadows: The Guerrilla in History* (New York: Doubleday, 1975), p. 568.

39. Richard Clutterback, *The Long, Long War—Counter-Insurgency in Malaya and Vietnam* (New York: Praeger, 1966), pp. 57–59.

40. John A. Nagl, *Counterinsurgency Lessons from Malaya and Vietnam: Learning to Eat Soup with a Knife* (Westport, Conn.: Praeger, 2002).

41. Andrew F. Krepinevich, *The Army in Vietnam* (Baltimore, Md.: Johns Hopkins University Press, 1986).

42. Anthony H. Cordesman, *The Iraq War: Strategy, Tactics, and Military Lessons* (Westport, Conn.: Praeger, 2003), pp. 493–557. See also James Fallows, "Blind into Baghdad," *Atlantic Monthly,* January/February 2004.

43. See Richard L. Layton, "Command and Control Structure," in *Lessons from Bosnia; The IFOR Experience,* Larry Wentz, ed. (Ft. McNair, Washington, D.C.: National Defense University, 1998).

44. On the conduct of the Marines in the offensive phases of the war see Williamson Murray and Robert H. Scales, *The War in Iraq* (Cambridge, Mass.: Harvard/Belknap, 2003); and Bing West and Ray L. Smith, *The March Up, Taking Baghdad with the 1st Marine Division* (New York: Bantam, 2003).

45. Clutterback, *The Long, Long War,* p. 52.

46. Small Wars Manual, p. 32.

47. Anthony James Joes, *Resisting Rebellion: The History and Politics of Counter-insurgency* (Lexington: University Press of Kentucky), 2004.

48. Alistair Horne, *A Savage War of Peace: Algeria 1954–1962*, rev. ed. (New York: Penguin, 1987). See also Asprey, *War in the Shadows*, pp. 903–31.

49. Lester Grau, ed., *The Bear Went Over the Mountain: Soviet Combat Tactics in Afghanistan*, (Washington, D.C.: National Defense University, 1996); Anatoly S. Kulikov, "The First Battle of Grozny," in Russell Glenn, ed., *Capital Preservation: Preparing for Urban Operations in the Twenty-first Century* (Santa Monica: Rand, 2001).

50. See Robert H. Scales Jr., *Firepower in Limited War* (Novato, Calif.: Presidio, 1997). Scales concludes his chapter on Vietnam by noting, "If a single lesson is to be learned from the example of Vietnam it is that a finite limit exists to what modern firepower can achieve in a limited war, no matter how sophisticated the ordnance or how intelligently it is applied"; Ibid., p. 153.

51. Ralph Peters, "The New Warrior Class Revisited," in his *Beyond Baghdad: Postmodern War and Peace* (Harrisburg, Pa.: Stackpole, 2003), pp. 44–60. See also Ralph Peters, "In Praise of Attrition," in his *Fighting for the Future; Will America Triumph?* (Harrisburg, Pa.: Stackpole Books, 1999).

52. Sam Mundy, "No Better Friend, No Worse Enemy," Naval Institute *Proceedings*, April 2004.

53. American forces have been accused in both Bosnia and in Afghanistan of establishing excessively large and isolated fortresses, and employing force protection measures (body armor, tanks, and pointing weapons) that antagonize rather than generate calm among the local population. In Bosnia, some U.S. officers have openly stated that force protection was more important than mission accomplishment. Some of this derives from a low appreciation of so-called peacekeeping operations, and the rest is attributed to a risk adverse culture among military officers. Don M. Snider, John A. Nagl and Tony Pfaff, *Professionalism, the Military Ethic, and Officership in the 21st Century* (Carlisle, Pa.: Army War College, December 1999).

54. Joint Pub. 3-07, pp. II-4–II-5.

55. This construct is identified with General Charles C. Krulak, USMC (Ret.).

56. The Marines are exploring this idea with a new operating concept called Distributed Operations. See Robert E. Schmidle, "Distributed Operations From the Sea," *Marine Corps Gazette*, August 2004.

57. On Beirut, see F. G. Hoffman, *Decisive Force; The New American Way of War* (Westport, Conn.: Praeger, 1996), pp. 39–60. On Mogadishu, see Jonathan Stevenson, *Losing Mogadishu: Testing U.S. Policy in Somalia* (Annapolis, Md.: Naval Institute Press, 1995).

58. In particular, T. E. Lawrence, *Seven Pillars of Wisdom: A Triumph* (New York: Anchor, 1991). See also Basil H. Liddell Hart, *Colonel Lawrence: The Man Behind the Legend* (New York: Halcyon House, 1937).

59. Starting with Max Boot, *The Savage Wars of Peace: Small Wars and the Rise of American Power* (New York: Perseus Books, 2002); Anthony James Joes, *America and Guerrilla Warfare* (Lexington: University Press of Kentucky, 2000).

60. The author is indebted to Dr. Wray Johnson, Marine Corps University for this point. See also Metz and Millen, *Insurgency and Counterinsurgency,* p. 22.

61. Williamson Murray and Mark Grimsley, "Introduction: On Strategy," in Williamson Murray, MacGregor Knox, and Alvin Bernstein, eds., *The Making of Strategy: Rulers, States, and War* (Cambridge: Cambridge University Press, 1994), p. 1.

## Part 4

# POST-CONFLICT AND STABILITY OPERATIONS

# EIGHTEEN

# Seeing the Enemy (or Not)

## ANNA SIMONS

No matter how much money we expend in our quest for "information dominance," "persistent surveillance," and the like, we often end up understanding precious little about the non-Western adversaries or competitors we face. There are at least four sets of reasons for this: technology offers less of a solution than we presume; our values and ideals blind us to certain realities about others' values and ideals; our military, simply by virtue of being a military, is organized to excel at "conventional" war; and we are too inconsistent regarding the study of our own history, never mind other cultures.

Of course, some might argue that these attributes are precisely the things that helped us beat non-Westerners in the past. Others might insist these are the same instruments of power being wielded to great effect in Afghanistan and Iraq today. But note what else was integral to our successes, historically speaking: the absence of any deeply nuanced military-wide understanding of the enemy. To succeed in the past, we didn't have to have the sort of small-unit cultural awareness that is being trumpeted today.

So, one obvious question this chapter should raise is: Do we really need such knowledge now? A second, more academic set of questions is: What past are we referring to? and Which non-Westerners?[1] Throughout the Cold War era we won very few hot wars. Indeed, World War II marks the last time we could claim a clear victory over non-Western adversaries. Presumably, then, something, somewhere along the line must have changed. But what: the nature of war, our position in the world, or our adversaries' relations to us?

To answer such questions requires tacking back and forth between "us" and "them," and "then" and "now." In the process, it also seems useful to wonder whether we are constitutionally capable of understanding non-Westerners—never mind whether we *should* understand them. Worth examining too is the extent to which we sufficiently understand ourselves. Loren Baritz, for one, began asking questions like this after Vietnam.[2] What I want to push us to consider are the strategic implications. If we are deficient, then do we retool to overcome our traditional weaknesses, or work hard to make the most of our traditional strengths? These represent radically different alternatives. And while the safe bet might be to blend them, that would require tremendous finesse on the one hand and considerable stick-to-itiveness on the other—two attributes that we Americans often have in short supply. Nevertheless, because we tend toward optimism, I can't help but offer some interim suggestions.

## Some Definitional Issues

First, who and what should "non-Western" refer to? Ask a political scientist or economist, and "non-Western" might still apply to spaces on the map, to non-capitalist economies, and to states that lack a certain style of government. Traditionally, "non-Western" has amounted to a default category used to describe everyone and everything outside the West: all those places and spaces apart from Western Europe, Canada, the United States, Australia, and New Zealand. For anyone with an anthropological bent, however, the West itself is a highly problematic concept. What of American Indians, aborigines, and Maoris (or gypsies and Saami for that matter); should they count as Westerners or non-Westerners? Is Israel Western or non-Western? What about South Africa? Or, for those not bothered by the racial undertones of white West vs. non-white non-West, what about military historians' recent accounts extolling the virtues of Western "ways of war"? What were Hitler's storm troopers—Western or non-Western? And to which category should we assign Red armies? As for Soviet Russians, are Slavs Western or Eastern, and if they are Eastern, should they be considered *as* Eastern as the Chinese or Japanese?

One could split hairs like this for pages. The point is that labels like non-Western can all too easily obfuscate more than they reveal. Certainly at this point in time, when warlords hailing from "failed" states do cell phone business with multinational corporations three continents away, the old connotations by which the non-West was regarded as "backward and primitive" hardly pertain. Even for those portions of the non-West that remain tribal in outlook, allegiance, and organization, we'd be foolish not to acknowledge how sophisticated tribes can be. While people who belong to tightly knit local associations—tribes, clans, secret societies, or religious brotherhoods, among others—are more likely to prefer doing business person-to-person rather than by the book, so too do criminals, insurgents, and terrorists. This is one nexus that should concern us.

A second reason we should worry is that non-Western communities can now be found everywhere: Amsterdam, Hamburg, Lackawanna County, Washington, D.C. Instead of geography being the defining feature of the non-West, how people operate, with whom they interact, and why better describes the dichotomy:

| *West* | *Non-West* |
| --- | --- |
| Individuals matter more than groups; the law protects individuals | Groups (family, tribe, religion) matter more than individuals; government and law are not trusted; the group protects the individual |
| Print and electronic media are dense and diverse; rumors and misinformation *can* be dispelled | Print and electronic media are thin; rumors and misinformation may be impossible to dispel |
| Bureaucracy works | Bureaucracy is personalized and inconsistent |
| Economic relations are governed by the state; commercial transactions are formal | Economic relations are often informal; commercial transactions are informal[3] |

In the non-Western frame, people rely on systems of trust and not the rule of law, person-to-person ties and not a formal structure, and methods of communicating, banking, and doing business beyond the purview of state authorities. That such modes don't leave much of a paper or electron trail is just one of the difficulties we've discovered post-9/11. But the fact that some individuals are able to shift back and forth between these two frames poses more than just forensic problems. How do we monitor their activities, anticipate their next moves, counter them, preempt them, or catch them before an event if we're in a Western frame and they operate by non-Western rules?

This ability to operate in two radically different systems—which they manage, but we (for legal and other reasons) eschew—is not as new as we might suppose. Individual Cherokee, Delaware, Seneca, Huron, and countless others moved back and forth between non-Western village life and Western cities as early as the 1700s. In fact, were we only to pay more attention to our own interactions with the non-Westerners we've been interacting with longest—namely, American Indians—we would learn some interesting lessons. That we don't is a point I will return to. For now, let me introduce them as a foil in this discussion. Two things can be said about many of the Indians who operated across the seams at the outset of American history: they fully appreciated and respected Anglo-America's overwhelming military strength. Indeed, part of the government's strategy behind inviting so many chiefs and others to visit their Great White Father was to try to impress them too. But for that particular subset of literate, white-acting Indians whom whites could accept as civilized, the more seriously they felt themselves being taken, the more this convinced them they *would* be able to reach an accommodation with the government on behalf of their tribes. Ironically, it didn't take intimate familiarity with Anglo-American ways to convince large numbers of other Indians that the more whites there were, moving onto their lands, the more of a threat Anglo-America posed—and the more whites needed to be *op*posed, not solicited. The catch is that the uncivilized Indians who understood this best understood least about their adversary's (meaning "our") potential weaknesses.

## Encounters with Indians

The question of who understood whom better in Indian-white encounters is highly suggestive. Arguably, Indians who rejected acculturation out of hand understood what they thought they needed to about whites. Whites were encroaching, invading, and greedy. White values were incompatible with theirs. What more did they need to learn?

What these Indians didn't sufficiently appreciate was the nature of the American state, the inner workings of government, or the nuances of domestic politics. They couldn't identify how they might have been able to exert potential leverage or how to exploit latent, even blatant fissures. And it's not as though there weren't exploitable fissures. At the height of the Indian Wars in the West (1846–90), for instance, the same sorts of tensions between hawks and doves existed as do now. There was media sensationalism. There were humanitarian do-gooders. There were opportunists. Even contractors (in the form of settlers) could be found on the battlefield.

What, meanwhile, did whites understand about Indians? They recognized that there were different tribes, and even factions within tribes. They also understood Indians' material weaknesses, which they sought to take advantage of by plying them with trade goods and whiskey. But did they understand Indian values? Could they accurately identify the glue that held tribes together? Or could they account for why leaders like Osceola, Crazy Horse, Captain Jack of the Modocs, or Geronimo would so willingly fight in order to die *as* warriors?

In many regards, Indians offer a premier example of non-state actors who fought to preserve their identity, but who—by doing so, and in seeking to defend their autonomy—invariably wound up too splintered and disunited. What, after all, tied all Indians together except their opposition to whites? Nothing really, considering that Indians were communally organized and locally oriented. Only after tribal integrity itself had been shattered—only after groups of Creeks, Seneca, or Sioux (to cite just three groups) found themselves subdivided into nativist and accomodationist camps—did new syncretic pan-Indian movements gain momentum. But by then it was almost always too late.

Another difficulty for Indians was that the vanguard of the invasion was similarly splintered. Too often it was settlers and not the army that moved into and through Indian lands. This made it far too easy for Indians to respond in piecemeal fashion, killing families here or there to terrorize all whites into fleeing an area. But random violence, no matter how temporarily effective, never staved the flow. Also, the fact that it was endless streams of whites pushing the frontier forward meant Indians would have had to somehow pierce through these waves before they reached the heart of the state itself. Small wonder that the groups we remember fighting hardest rarely came to understand how the United States worked. Either they had enough to deal with just defending their homelands, or the areas into which they retreated—the Everglades, the Chiricahuas, the Lava Beds, even the Northern Plains—were so far removed from the seat(s) of white power that white machinations could only have remained opaque.

This should already hint at a major contrast between then and now.

At the same time, Indians suffered a host of other disadvantages. Demographically they would have been swamped even if scourges of smallpox and cholera hadn't swept away entire communities, thereby rearranging the local balance of power just as whites showed up. In the same way that diseases stole a march on settlers, so did horses and guns. The extent to which Indians became dependent on other nonnative products remains debatable. Arguably, dependence didn't become debilitating until local subsistence bases had been so eroded Indians were left with no choice. Choice—in many ways—is the operative word. Whenever they had a choice, significant numbers of Indians preferred to remain Indian.

What did staying Indian mean? The easy answer is: different things to different peoples. Take Modocs, for instance.[4] What made a Modoc a Modoc and not a white man would have been a subset of the same things that made a Modoc a Modoc and not a Klamath. Just before the Modoc War, Modocs in northern California looked thoroughly acculturated: they wore manufactured clothes, many worked for ranchers or miners, and they responded to Anglo names. Yet, they fled the Klamath Reservation in 1870, demanding their own. Why? One answer is they wanted to be able to be Modoc together; even if they were already enmeshed in the broader American economy as individuals, they drew their identity from being able to remain Modoc with other Modoc. Modoc would have agreed among themselves about what being Modoc meant; this is part of what set them apart. But just because Modoc were fiercely attached to their own identity didn't mean they all agreed about what they should do to save themselves. Neither did all Sioux (not everyone followed Sitting Bull into Canada). Nor did all Nez Perce follow Chief Joseph.

Before exploring the significance of corporate identity for non-Western peoples, there is one more critical set of comparisons to make. Large numbers of Americans were infused with a crusading spirit. They penetrated Indian lands; the reverse never occurred. And when whites moved in they generally did so for keeps. Settlers strove, always, to improve their lot. When they fought—or egged the army into fighting on their behalf—they did so for the sake of property. Land was to be farmed, let, or sold; no matter what was done with it, it was worth money. When Indians fought with Indians, by contrast, they rarely did so over land or money. Intertribal warfare was usually waged for individual glory, portable booty, or captives—not fixed, improvable property.

In other words, what whites were used to fighting for (and over) was radically different from what Indians typically sought. Whites and Indians organized differently, planned differently, and executed operations differently. At the tactical level, Indians were often ingenious and highly successful. But the nature of tribal society made fielding, commanding, and controlling permanent armies next to impossible. As non-state actors confronting a state,

Indians didn't possess the wherewithal to overcome the organizational and logistical disparities between soldiers and themselves. Again, too, with no direct experience of a state, they couldn't possibly begin to understand how the systems arrayed against them fit (or didn't fit) together.

The same can't be said for our current adversaries.

It is hard, actually, to imagine how the United States wouldn't have steamrolled Indians. The asymmetries we can measure were stacked entirely in our favor. But therein also lays the germ of a problem that makes our experiences with Indians far less remote than we might imagine. The fact that we could beat Indians without sufficiently understanding the nature of their attachments or the inner workings of their societies means we never had to develop these skills. Again, our numbers, technical abilities, and organizational capacity precluded us from having to. So did our inheritance.

One thing that is often cited to explain American exceptionalism is that we inherited a blessed geography: a large, richly resourced region straddling the temperate zone and protected by two vast oceans. What is seldom remarked is a second critical asymmetry. Legions of Indians had already experienced the French and then the British (in the Northeast and upper Midwest), or the Spanish then the British (in the Southeast). When faced with multiple sets of players who sought tribal alliances, Indians demonstrated just how strategically astute and sophisticated they could be. They would have been utterly familiar with competition that involved attracting and keeping followers or allies and playing both sides against the middle, since such maneuvering was just as integral to their politicking as anyone else's. We shouldn't be fooled, then, into assuming Indians *couldn't* think strategically. Instead, it was sole proprietorship that threw them for a loop. That's what the United States represented. Once Indians lost the opportunity to dicker with representatives from different empires, they lost strategic depth. Sole proprietorship was an alien concept. They couldn't have known how to understand this. Nor would they have understood why they needed to.

## Ethnographic Intelligence

We can find similar encounters on every continent. Wherever unitary powers bore down on tribal peoples, tribes ended up encapsulated, co-opted, absorbed, or marginalized; tribes rarely came out on top in war. At the same time, they never entirely disappeared. If they had, "non-Western" couldn't be applied to any of the attributes or means of operating mentioned earlier.

Two additional ironies are worth noting. Dynasties, not tribes, ruled the empires that conquered large swathes of the non-West (think Hapsburgs, Ottomans, Spain, England, France, etc.), yet empires invariably granted tribes far more leeway than did states. But then, with the breakup of empires, many institutions of state (particularly in Africa) have come to be run by tribes. This too should underscore the resilience of communal attachments and hint at how superficial the West's impact has been at the grassroots.

We know from colonial history that the British, especially, preferred employing indirect rule whenever and wherever possible. To do so required keeping local political arrangements intact, which in turn meant understanding how they operated. Here is where anthropology is often blamed as the handmaiden of colonialism. Without question, social anthropology in Britain largely concentrated on understanding social relations, social structure, methods of local organization, and social control—precisely those things the British would have needed to identify, and ultimately manipulate, loci of power in the non-Western world. But to argue as though the British weren't already adept at acquiring this sort of knowledge is ludicrous. Not only were the first Englishmen into an area usually traders or missionaries (not anthropologists), the name of the game for them was to beat the competition—whether foreign, regional, or local. This meant figuring out how to win friends and influence people and thwart rivals all in one. To succeed, they had to insinuate themselves into local alliances and allegiances. As singletons or small teams, they certainly couldn't afford to apply force. Finesse (and sometimes armed finesse), yes. But also their initial aim wouldn't have been social control. Instead it was market penetration (to make money, or converts, or both).

It only stands to reason, then, that the British imperial experience led to a different set of competencies from our foundational encounters with Indians. This may also help explain why we never adopted any of Britain's best practices. We were never consistent about what we really wanted of, for, or from Indians even as we engaged them—sequestration, assimilation, or preservation. Perhaps because our aim was to push Indians aside and/or pen them up, not govern them to some extractive economic or geostrategic ends, we never institutionalized a coherent post-conflict approach either, no matter how many Indian wars we fought. Much as with development economics today—where first export promotion, then import substitution, followed by structural adjustment, and now good governance have been pushed on developing countries—policy makers were happy to try whatever scheme the "experts" came up with next.

Unlike the British, we didn't develop an Indian Civil Service. Worse, we didn't attempt to systematically develop expertise of any sort. Granted, India was Britain's crown jewel. But even for some place like the Sudan, where Britain had no real commercial interests yet clear security concerns, the British managed to post 120 or so field officers to the Sudan Political Service, most recruited from Oxford and Cambridge. Responsible for administering a population of between 9 and 10 million, these individuals excelled at understanding how different societies were organized, knowing who was whom, how they were related, what their relationships consisted of, etc.[5] They typically paid close attention to kinship, especially in acephalous (or governmentless) societies where genealogies double as charts of trustworthiness. They also tried to figure out what went on in healing cults, religious brotherhoods, and local gatherings of all types. In short, they produced and consumed ethnographic intelligence in prodigious quantities, something one would think we'd recognize we need today if we hope to understand anything about our adversaries' means of recruiting, fundraising, or planning and executing operations, all of which piggyback on indigenous (and thus latent) means of association.[6]

As necessary as ethnographic intelligence is, it alone is hardly sufficient, however. Information does not constitute or even always contribute to understanding. To get at motivation, for example, takes more than mapping who is related to whom. The British could well afford to concentrate on the political aspects of native life. Given the imperial task of maintaining socio-political *control,* what community meant, how it felt, and why it mattered were immaterial. Nor did we Americans have to delve into the murk of Indians' values, ideals, or beliefs so long as we could win our wars through attrition. But, is control possible today? Are we willing to engage in wars of conquest *or* attrition?[7]

Since the answers to these questions appear to be "no," and since non-Westerners are in our midst and not just "out there" somewhere, logic dictates that we not only try to figure out how people associate, but also why they make the choices they do. Why do some accept some of our values and reject others? What should go without saying is that non-Westerners wouldn't *be* non-Westerners if they did share our same values: they'd be us. Or they'd be fighting to become just like us. But wanting to be us has seldom turned anyone into our adversaries. Rather, most often people have fought us in order to remain themselves (even when this includes getting stuff from us *for* themselves)—all of which brings us back to issues of identity and autonomy.

## Identity Issues

It may be too much of a truism to point out that people generally fight to improve or protect their security.[8] In the United States we tend to take it for granted that our individual security, as well as our future social welfare security, are protected by the state. People who can't count on the (or a) state generally fall back on networks they know they can trust, which makes trust paramount and communal or corporate identity key. For many peoples as well, psychic or spiritual security is tied to communal attachments; obligations to kin carry moral weight; separations we make between the living and the dead and their effects on well-being are just that—separations *we* make. These are just some of the reasons why when people feel their sense of identity is threatened, they react as if their security is at stake.

The tricky thing is identity is not equally deep-rooted or fixed for everyone. Across cultures it doesn't attach to the same things, nor does it attach people to the same things. It isn't even located in the same place. This last might seem an odd assertion, but where is the soul located? Different peoples point to different places: the liver, the stomach, the heart. Identity is not dissimilar.

For instance, in Japan identity has long attached to knowing what it means to be Japanese and understanding exactly how to act in any given situation. In Germany identity has been thought to inhere in blood. In many Middle Eastern societies, identity is determined by one's patrilineage. In the United States what would we say? We'd probably point to values.

Ruth Benedict, an anthropologist who sought to understand the Japanese during World War II, doesn't directly address identity in quite this way in *The Chrysanthemum and the Sword,* but that work does inspire the following Zen-like question: Can a Japanese be Japanese with non-Japanese? The answer Benedict suggests is no, Japanese can only be Japanese with other Japanese; otherwise, they find themselves playing tennis at a croquet match. No one else knows the rules to their game—or appreciates the skill with which they play it—so what, after a point, *is* the point?[9]

To what extent does this likewise hold true for Americans? I think most Americans, particularly those who have spent time abroad, would answer that an American among foreigners is still American. He may get homesick, but this doesn't erode or imperil his sense of his own identity or his confidence that, even as a lone individual, he is still just as American, with values still worth projecting. It could be that this confidence is born of belonging to a recognized superpower. The British never had an identity problem out in the colonies; otherwise, colonial rule would have been impossible. But it may also have to do with what identity attaches to, where it is located, and what it attaches *us* to. The preeminence of values is very different from the preeminence of location (or place), or of parentage, say. That is why, if we reframe these identity questions ever so slightly—to, What makes a Japanese a Japanese? What makes an American an American? What makes an Iraqi an Iraqi? or, What makes a Kurd a Kurd?—it should be evident that it is not the same things in all cases. The message this, in turn, should convey is simpler still: everyone *isn't* just like us.

Unfortunately, the fact that anyone can become an American leads us to assume that everyone *is* just like us, they just haven't been able to become one of us yet. It also leads us to presume that our values are universal human values. Just liberate and educate people elsewhere, grant them the same kinds of opportunities we have, and they should act increasingly American.[10] Immigration proves this to us every day. It also helps render us unreconstructable solipsists. With the best of intentions, we firmly believe "we are the world."

The problem is that at least some of what we stand for disgusts people elsewhere. This is especially true for Islamists, who aren't fueled solely or even largely by envy, no matter how much we might like to think so. Rather, to many devout Muslims our values are completely antithetical to theirs, and they want no part of our exporting, let alone proselytizing, our morality. What is sadly ironic about this is that we recognize how impossible it is to reach a satisfactory compromise over moral issues like abortion, euthanasia, or stem cell research here at home, yet we push an agenda abroad that includes gender equity, despite the fact that it—like abortion for many of our citizens—is considered not just wrong by many Muslims, but a threat to the entire community.

Because, meanwhile, we adopt a solipsistic attitude toward others, whenever we try to put ourselves in others' shoes it tends to be as individuals. After disasters (especially natural disasters), we are among the most giving people on earth; we can't help but feel for people who are hurt, hungry, sick, or poor. But that's because we relate to them as fellow human beings. Only when we stereotype are we likely to regard others as members of groups first and as individuals second. Thus, it is extremely difficult for us to appreciate that some peoples preferentially regard themselves as belonging to groups first, with the group deserving commitment way beyond what we would regard as more rational commitment to self.

Yet, this explains the actions of individuals like Captain Jack, a Modoc leader who sought to preserve peace between Modocs and whites in northern California during the late 1860s and early 1870s, and only reluctantly and against his better judgment led his people to war. To be a good Modoc leader *required* him to follow the wishes of the group, no matter how much these flew in the face of common sense and contradicted his personal convictions. In tribe after tribe we find individuals who made similar choices. Something we didn't understand in the nineteenth century, and still don't in the twenty-first, is that Indian and white conceptions of what matters can differ dramatically. A trip through most reservations today should, but probably can't, drive this home. On the one hand, most reservations bear a closer resemblance to the third rather than the first world. On the other, they can exert an invisible but powerful pull on even the most urbane tribal members. It is not uncommon for members to chuck a successful life on the "outside" in order to return to what, to non-Indians, resembles a rural wasteland.[11] But that's because what looks poor to us offers a rich sense of connectedness to those who belong—something that, since we can't see let alone feel it, we tend to discount if not dismiss altogether.

## What We Can't (or Don't) See

Connectedness reminds people who and what they are attached to, who and what they can trust, and who and what poses a threat. If we can't fully appreciate the worth of this to non-Westerners, we should at least recognize its usefulness.

Remember: extended families, lineages, clans, religious brotherhoods, and other indigenous associations represent latent networks, ideally suited for covert use by criminals, terrorists, and insurgents. Worse, they are so embedded and enmeshed that the entire social fabric would have to be destroyed before they can be dismantled. But also, people are born into them. This makes them automatically familiar and comfortable. They are resilient and robust. Not only can such networks regenerate themselves should some part be damaged or destroyed, but also the template itself can be regenerated; it is inherent everywhere that communal ties persist.

This is what we are up against today and for the foreseeable future.

Nor are we well prepared to cope. Our technology, like our analysis, suffers from our penchant to mirror image. While we are unsurpassed at using technical means for gathering intelligence on other democracies, we have been demonstrably less proficient predicting collapse or upheaval in non-

Western (and non-Western style) states—witness Iran, the former Soviet Union, even Haiti. We are worse still when access is denied—consider Iraq and North Korea. Scale this down to non-state actors and the challenges proliferate. In these cases, it is not so much access that is denied as that we've never committed the resources. We focus very little attention on university campuses, in shantytowns and slums, or along borders—places where trouble often brews. But also, it is increasingly difficult to monitor diasporas that can stretch from the Hadramaut to Houston, especially when many anti-state actors have learned the hard way that the more they use Western means of communicating (or banking), the easier it is for Western intelligence organizations to "see" what they are up to. Under such circumstances, it only makes sense for them to revert to more traditional, non-Western means—something that plays to their strengths, but also takes advantage of our limitations.

Satellites can't, after all, map moral compulsion. Also, without deep local knowledge to begin with, we cannot differentiate between normal, routine ways of doing business and suspicious activities. Finally, because non-Western means are typically person-to-person and are built on relations of *proven* trust, we can neither easily insinuate ourselves nor readily penetrate these networks on the ground. Usually we do not even attempt to do this until after groups have done something to alert us to the fact that we should already have been monitoring them, when generally it is too late and exponentially more difficult to do so.

Compounding our difficulties are nuanced modes of communication. Never mind local dialects, people intimately familiar with one another understand each other's euphemisms. They can speak in linguistic shorthand or cultural code. This too renders listening in from a distance that much more ineffective, and penetration (again) exceedingly difficult.

The technical means we currently possess were not only designed for a different array of adversaries, operating in clever but nonetheless predictable ways, but also for the first time in the history of modern weaponry it seems highly unlikely we will engineer or invent some kind of new device with which to devastate our opponents.[12] If technology can't help us make sense of the least Western aspects of the non-Western world, how can we develop technology to penetrate it (not that we're likely to want to acknowledge technology's shortcomings)? We love technology. This is a consequence of who we are, how we fight, what we value, and what's worked for us in the past.

We should seriously wonder, though, about the role technology *has* played in fighting non-Westerners in the past. It would seem that logistics and supply were vastly more critical—again, not something likely to work in our favor now. But this is conjecture. Such matters merit serious study, both for the light they might shed on the possibilities, as well as for what they could reveal about our misconceptions.

## The "Lessons Learned" Lesson

Other lessons we *have* collated. In every new set of engagements post–World War II—Vietnam, Somalia, Afghanistan, Iraq—we have discovered, after the fact, things we should have known because they'd been learned previously. What this suggests is that, though the military publishes and even distributes them, no one really learns "lessons learned."

One could argue that all armies find themselves perpetually having to relearn lessons about low intensity conflict, irregular warfare, and counterinsurgency because no army is designed—structurally or functionally—to wage non-attrition warfare against foes who don't fight back *in* army form. But up until recently, this also never posed us a serious threat. In all our previous wars with non-Western non-state actors, we could safely assume that though they might achieve tactical and operational surprise, they could never best us strategically. They didn't study us strategically; they lacked access, familiarity, and resources. That is certainly no longer true, and arguably hasn't been for some time.

Other reasons why lessons learned don't stick could have to do with our "fix it" mentality. Soldiers are inherent problem solvers. They also prefer looking forward, not backward. Then too there is personnel churn, which guarantees few rewards and scant opportunities for building up anything approaching deep region-specific or historical knowledge. Finally, we can't forget solipsism. This convinces us that, as Americans, we already know what to do. If operations go awry or winning people over takes longer than it should, that's simply because we haven't deployed our assets properly. It's inconceivable that our methods and values—including adaptability—might not work. I don't mean to suggest that solipsism is bad. Without it we probably wouldn't be as egalitarian, as generous, or as optimistic as we are—traits that describe those in uniform especially. But solipsism can be costly and does blind us to certain realities. It makes "hearts and minds," "public diplomacy," and "strategic influence" campaigns extremely appealing since these fit with not only who we think we are as a people but also whom we think other people are.

"Hearts and minds" also sounds good. It promises a benign, humane way to wage war. The problem is it seldom works. As historian Douglas Porch has pointed out, the French in North Africa waged great hearts and minds campaigns, but basically for domestic consumption and public relations back in France.[13] In Vietnam and ever since, we've sought to borrow from the French without fully appreciating which among their methods really did work. For instance, consider this assessment: "In Morocco, the ultimate testing ground of the Lyautey method, pacification came everywhere through armed and bitter contests with resistant townsmen and tribesmen. Pacification was war, not peace. Politics and economics did little to pacify the people of Morocco's cities or the tribes of the Middle Atlas *until* they were subdued by the threat or the use of force (emphasis added)."[14] Force included, by the way, denying tribes access to pasture lands.

Alternatively, we might turn to Malaysia, *the* case that gets all the attention whenever anyone discusses hearts and minds. It is clear the British won the first Malaysian Emergency and the Malaysians won the second (and there *were* two, the second having ended in 1989) by going after "stomachs and minds"—not hearts and minds. To effectively defeat the Chinese Communists required controlling food and access to food production, not so different from the French approach to pasture lands in the Middle Atlas.

I borrow the phrase stomachs and minds from Bill Donovan, founder of the Office of Strategic Services, who in 1954 said about the situation in Southeast Asia, "It is not essentially a military matter. It is a political struggle, which must be won in the stomachs of the hungry and in the minds of the people. In Washington they think that American military might is the solution to the problem, but any intelligence man knows this is not true."[15]

In point of fact, the Communists in Vietnam were promising villagers all sorts of reforms, including land reform, which implies they were addressing tangible material grievances and that the Vietnamese they were seeking to win over could be said to be already "hungry."

If we reconsider what we Americans did when we waged our guerrilla wars, we *made* people hungry.[16] We took *away* their food, their way of life, and their ability to remain self-sufficient and autonomous. Sometimes we did this consciously, as was the case when we purposely wiped out the buffalo in order to subdue Plains Indians. At other times, we simply reacted opportunistically to what was already occurring. One of the arguments Andrew Jackson made to Eastern Indians to justify their "removal" was to point out that, with game having been depleted east of the Mississippi, they could only continue being Indians by moving west of the Mississippi. He and his successors took a slightly different approach with the Seminoles, who were intractable enough to fight three wars and were not dependent on any single food source. Troops in Florida purposely went after Seminole women and children. The rationale was that this would force the warriors to give up since, without their families, what kind of life would they be fighting for? The same was done to the Apaches. In both cases victory also required penetrating the Indians' redoubts.

Nor was it small groups of soldiers who won the Indian Wars.[17] It took one-fourth of the entire standing army to hunt down forty warriors after Geronimo's last breakout. Granted, the army then wasn't as large as it is now. But even given our current precision strike capabilities, which would seem to obviate lots of men in the field, targeting itself requires precision intelligence which can only be generated by: lots of men in the field over concentrated periods of time, a few individuals in the field for considerable periods of time, or insiders who are willing to betray the group. It was the latter (Indians themselves) who were responsible for the capture of our most significant Indian foes—something relatively easy to accomplish when inter- and intratribal rivalries were public and well known. Again, consider the contrast with our current adversaries and what it says when, after four years, a $25 million reward still hasn't helped us locate Osama bin Laden.

Not surprisingly, we do still prefer hiring locals to help hunt down locals. From Vietnam through Iraq we have tried to find people who can and will do our scouting for us. And though this might seem to reflect one lesson successfully learned—this, after all, is what worked to our advantage in Indian war after Indian war—it has also meant that we've never felt the need to develop deep local expertise ourselves. Arguably, we don't bother because we never intend to stay anywhere very long; our aim, after all, is to liberate not conquer. Alternatively, this could reflect American pragmatism: the fastest shortcut to figuring out a place is to hire people who already know it. But the downside is also considerable. We then don't have the knowledge we need to affect the kind of hearts and minds campaign that those who consider themselves visionary about counterinsurgency insist we should wage.

## Hearts and Minds:
## More a Problem Than a Solution?

Unlike hearts and minds, stomachs and minds would clearly play to our strengths—in men, materiel, and technological prowess. If we were willing to engage not just in conquest but also in occupation, subjugation, and "reeducation" thereafter, destroying non-Western social structures and, thus, non-Westernness wouldn't be as difficult as many might assume. As unpleasant as it is to contemplate, it is far easier to engage in wholesale destruction, to scorch the earth, and to smash local social structures, thereby forcing people to *have* to change—the hallmarks of many wars of conquest—than it is to try to encourage people to *want* to change in the wake of surgical attacks. Arguably, nondemocratic leaders have had their minds changed thanks to precision strikes. But there is scant evidence that people can be made to give up their fight on behalf of autonomy or identity this way. Yet, because such courses of action are politically and ideologically impossible today, hearts and minds—as shorthand for counterinsurgency—is bound to continue to dominate our thinking, at least in the near term.[18]

If we examine counterinsurgency, at least three lessons learned from past efforts are considered critical to any hope for success today: unity of effort, deep local knowledge, and a small footprint. Unfortunately, achieving any of these is considerably more problematic than seems to be recognized.

Unity of effort, for instance, refers to collapsing political and military authority in the same hands; utilizing all available instruments of power (soft, hard, diplomatic, military, financial, etc.); and coordinating, planning for, and engaging in joint and seamless operations. Basically, unity of effort requires fusion. In theory, it is difficult to imagine anyone actually opposing this. But, in practice, to attain it means getting rid of the conventional/unconventional divide for starters, something proponents of counterinsurgency themselves would object to just on principle. More significantly, it would require a complete overhaul of the entire division of labor. There is no other way to ensure that complementarity—among individuals, between units, across services, etc.—would prevail over the current ethos of competition, stovepipes, and rice bowls. But even were teamwork to be institutionalized somehow, it is unclear that status hierarchies wouldn't reassert themselves and then undermine the brave, new system.

Status hierarchies represent something different from a stovepipe problem; they are also more pernicious. Consider, for instance, Special Operations Forces (SOF). For decades, Civil Affairs (CA) and Psychological Operations (PsyOps) have been treated as lesser disciplines. The status (and resource) hierarchy within SOF is direct action, unconventional warfare and foreign internal defense, then civil affairs and psychological operations. This is how resources flow: armed force trumps armed finesse trumps unarmed finesse.[19] It is easy to understand why. Among other things, direct action achieves immediate visible results, yields quantifiable measures of effectiveness, and is far more "warrioresque." Unconventional warfare and foreign internal defense are, by definition, far more protracted, open ended, and consequently messier. CA and PsyOps? Because no one is quite sure *what* they can achieve, too little confidence is placed in them, the upshot of which is they most often get called in *post*-combat. Some of this has changed in the past few years, but the status of CA and PsyOps within SOF still defies what SOF's most ardent proponents of counterinsurgency argue we need: better unity of effort *with* locals. Indeed, with that as the goal, statuses within SOF should actually be reversed. The catch is that even were CA and PsyOps to be formally granted the lead, the resources, and the manpower, the warrior/direct action bias may be too innate in soldiers and the need for soldiers to be warriors too strong in Americans for such a configuration to work, while to reorient the division of labor altogether—which is what would also be required—would take strong, unrelenting leadership.

Long-lasting leadership itself should be a sine qua non for unity of effort. But this is precluded by current career trajectories. The same holds for acquiring deep local knowledge, something possible only if people stay put for long periods of time and engage in serious, career-long study of their areas of responsibility. Administrators in the Sudan Political Service spent years and sometimes decades in the field. T. E. Lawrence, everyone's favorite exemplar of someone who developed deep local knowledge, began before he went to university. Or take Glubb Pasha. He spent twenty-six years with the Arab Legion, and when he was unexpectedly asked to leave Jordan in his twenty-sixth year of service, felt he still hadn't completed his mission.

There is no short-circuiting the amount of education, language training, and experience necessary for developing a *Fingerspitzengefühl* (intuition) for other cultures. Nor can the requisite foreign immersion be done in packs or classes and groups. This is why there is a real danger in notions like "culture-centric warfare" or the idea that with just a bit more cultural awareness training, soldiers can overcome their prejudices as well as be transformed into culturally literate strategic corporals. Without question, teaching everyone in uniform how to avoid cultural gaffes is important, just as knowing some phrases in the local language *will* help with building rapport. But to assume that any of this secures trust or will achieve fundamental sociocultural changes in the target populations is to misjudge how trust is earned in the non-West, never mind what it takes to undo social structures, people's fundamental attachments, or commitment to their identity. The only Americans likely to earn lasting trust are individuals who can commit to the same life-long attachments locals do, and via the same methods, which is one reason we should want to invest heavily—but selectively—in individuals, and not be fooled into thinking our usual mass approach will suffice.

Not only is concentrating via individuals the only way to gain deep local knowledge, but it is also the only way to keep our footprint small. The catch is this is only likely to work in permissive environments, where strength in numbers and firepower don't matter as much as they do in combat zones. Where, exactly, counterinsurgency falls on the conflict continuum has never been either clear or static. Ergo, force structure issues have always been and will likely remain sticky. Still, small numbers of truly culture-centric individuals could provide the ideal stopgap. Everyone doesn't have to be deeply culturally aware or conversant in the local language so long as we have some individuals who are. The challenge would be to carefully select, assess, and assign only certain individuals to this capability, train everyone else to recognize their usefulness, and then institutionalize their use.

There are myriad reasons—financial, logistical, practical—to take a precision-oriented rather than broad-gauged, everyone-needs-to-be-culturally-fluent approach. Looming above all of them, however, are certain realities about us and our Americanness: cultural, let alone moral, relativism is hard to square with American values. It makes no sense to send young men and women off to fight for American values, then tell them to respect the values of people whose practices or beliefs can't be squared with ours. Nor are training and education particularly effective against faith-based convictions about what is or isn't immoral or abhorrent. In other words, it seems wholly unrealistic to expect most soldiers to remain anything but solipsists, especially when solipsism is what makes them not just good but exemplary Americans.

Solipsists, it should be noted, *can't* keep their footprint small. But nor are we likely to keep our footprint small when we engage in public diplomacy and strategic influence campaigns, something most self-proclaimed counterinsurgency experts promote. Pushing democracy or women's rights is hardly low profile. It also squarely pits our values and beliefs, no matter how much we dress them up as "universal human values," against others'. What this, in turn, signals is that we consider ourselves superior—hardly the right message in a hearts and minds context, but, ironically, absolutely ideal were we only willing to pull out all the stops for twenty-first-century conquest and Empire.

In a weird twist, then, our attraction to hearts and minds may represent the ultimate proof that we don't sufficiently understand our limitations, or our past. Without question, hearts and minds buys critical force protection. That may, indeed, be what it does best. Otherwise, it represents a wholly Western concept, which in and of itself should suggest a means-ends-audience problem. Arguably, only if we understand how to disarticulate societies elsewhere are we likely to be able to prevail against non-Westerners who not only reject our advances but also seek our destruction. For this we need people with local-level expertise. Cultural sensitivity is hardly adequate. At the same time, the force applied may well have to be visceral—hitting people in their stomachs, not wooing them through their hearts.

## Yet More Unsettling Questions

Counterinsurgency is clearly appealing on a number of counts: it alludes to limited war, holds out the promise that the clever application of ideas and strategic strikes can beat crude force, and implies that there is some sort of method that, if only we adhere to it, will bring us victory. Most of the proofs for this, however, come from counterfactual rather than actual history. That alone should raise concerns. Others should come from whether we are capable of the insight, foresight, patience, and practices counterinsurgency is said to demand.[20] Or to be blunter still, never mind whether counterinsurgency *might* work, can *we* make it work? It could be that we are too constrained by what does and doesn't come easily or naturally to us *as* Americans. As Americans we may not be able to put into practice what the theory demands.

One pressing task, then, should be to gauge what we organizationally and temperamentally can and cannot do. A second task is to hook this back to what we truly need to know—and who needs to know it—when confronting non-Western adversaries (insurgents or not), all of which begs much more thorough study of Western/non-Western encounters, ours as well as others'. We leap to too many easy conclusions based on too few cases. Even in the most advanced professional military education programs, officers are exposed to the same handful of examples again and again: Malaysia in the 1950s (not Malaysia in the 1980s); Vietnam (but only U.S. and occasionally French experiences in Vietnam, never Chinese experiences); El Salvador (again, from the American perspective). The pattern is too predictable: almost nothing is studied about lessons learned from the non-Western point of view, no doubt because this is harder to come by. But also, virtually no attempt is made to ask non-Westerners how they have conducted their own counterinsurgencies. Typically, Britain and the IRA, even Russia and Chechnya, attract attention. But why shouldn't India, Turkey, or Senegal as well?

Finally, given all the advantages today's non-Westerners have over their forebears, it seems imperative to begin asking a series of hard, potentially discomfiting questions, most of which should end in "then what?" If, for instance, a "war of ideas" coupled with precision killing won't change our adversaries' attachment to their own identity or their desire for autonomy, then what? If we can't figure out how to beat people whose grievance is that we have humiliated them—and continue to do so—then what? Or, if some struggles just aren't winnable by the rules we've said we'll play by, then what?

## Notes

1. At least two sets of caveats are in order. I generalize to a far greater extent throughout this chapter than any anthropologist should, but space prevents me from citing all the cases relevant to the overall argument. Second, my intent is to be provocative rather than definitive. Much of the argument is drawn from a series of courses I teach in a program that awards master's degrees in Irregular Warfare to military officers, both U.S. and international. I continue to learn a tremendous amount from them and, as this chapter should make clear, we have far more to yet learn from our non-Western allies in particular.

2. Loren Baritz, *Backfire: Vietnam—The Myths that Made Us Fight, the Illusions that Helped Us Lose, the Legacy that Haunts Us Today* (New York: Ballantine Books, 1985).

3. This chart is borrowed from work done by Anna Simons and David Tucker, and can be found in a report ("Improving Human Intelligence in the War on Terrorism: The Need for an Ethnographic Capability") submitted to the Office of Net Assessment/Office of the Secretary of Defense, December 2004.

4. See Arthur Quinn, *Hell with the Fire Out: A History of the Modoc War* (London: Faber & Faber, 1997); Keith A. Murray, *The Modocs and Their War* (Norman: University of Oklahoma Press, 1959).

5. Frances Deng and M. W. Daly, *Bonds of Silk: the Human Factor in the British Administration of the Sudan* (East Lansing: Michigan State University, 1989).

6. As important as cross-cultural awareness, sensitivity, and appreciation are—and tremendous lip service is being paid throughout the Department of Defense today both to these aspects of 'cultural intelligence' and the need to better understand our adversaries—what 'ethnographic intelligence' refers to is decidedly different. Ethnographic intelligence can only be put together by drilling into social relations and delving below patterns of association to map actual connections between people, frequency and content of interactions, etc.—all of which requires extensive time in place and training in ethnographic techniques.

7. See Anna Simons, "The Death of Conquest, " *The National Interest,* Spring 2003.

8. People also fight for reputation and honor, which help *secure* security.

9. Ruth Benedict, *The Chrysanthemum and the Sword: Patterns of Japanese Culture* (New York: Houghton Mifflin, 1989 [1946]), p. 228.

10. Or, at the very least, they should act increasingly pro-American.

11. For a particularly striking example of this, see Wilma Mankiller and Michael Wallis, *Mankiller: A Chief and her People* (New York: St. Martin's Griffin, 1993).

12. Perhaps at some point in the future there will be some biogenetic way to specifically target—and incapacitate—individuals belonging to particular groups, though given intermarriages across groups, not even this would be neat, tidy, or conclusive.

13. Douglas Porch, *The Conquest of Morocco* (Knopf, 1983); *The Conquest of the Sahara* (Northbrook, Ill.: Fromm International, 1986).

14. William Hoisington Jr., *Lyautey and the French Conquest of Morocco* (New York: St. Martin's Press, 1995), p. 205.

15. Richard Dunlop, *Donovan: America's Master Spy* (New York: Rand McNally, 1982), p. 505.

16. Guerrilla tactics adopted during the Civil War likewise involved scorching the earth—tactics practiced by Stonewall Jackson, George Crook, and William Sherman, among others.

17. Robert Kaplan, as well as proponents of 4th Generation Warfare, has recently made such claims. See Robert Kaplan, "Indian Country," *The Wall Street Journal,* September 21, 2004, p. 22.

18. While waging a war of attrition against anything but tanks won't play in the world today, this is something that could certainly change with further terrorist attacks on U.S. soil.

19. For more on SOF status hierarchies, see Anna Simons and David Tucker, "United States Special Operations Forces and the War on Terrorism," *Small Wars & Insurgencies,* Spring 2003.

20. See, for instance, James Fallows, "Getting Out Right," *The Atlantic Monthly,* April 2005.

# NINETEEN

# A "Post-Hostilities" Moment?

MICHAEL VLAHOS

As it is used today in the American defense world, the term *post-hostilities*—or "post-conflict," or "stabilization and reconstruction operations"—seems to have meaning that is intuitively obvious and self-evident. But the term's meaning, in contrast, is actually sublime, going to the very heart of war and the American ethos.

In Old Norse the term *post-hostilities* would be called a *kenning*. Viking poetry liked to substitute a metaphorical phrase for a noun, so that a battle would be signaled by "a storm of swords." But what exactly does the "post-hostilities *kenning*" mean?

Before we can understand its full sublimity, we should know how it came to be. Thus, a story is in order—even if some might find it iconoclastic or, worse, schismatic.

## How "Post-Hostilities" Came to Be

It was like an ancient fetish, to be sure. You had only to hold the plastic three-ring binder in your hands to feel its talismanic powers. First of all, it had heft: it was thick plastic with that unforgettable 1970s' smell, and it was emblazoned with the new camo scheme. Forest camouflage. European camouflage.

Vietnam was behind us. A new age had begun. Here, with the special authority that hope brings to determined new enterprise, was its sacred text. Field Manual 100-5: *"Operations."*[1]

No more COIN, no more JKF School, no more Phoenix Program or MACVSOG. The new "Operations" was peppered with reassuring pictographs of cozy German villages and pine forest battlefields. From now on *this* would be our war.

And the Soviets—faithfully aggressive codependents—obliged us: boasting of a "protracted conventional phase" and the video-bluster of a thousand-thousand armored vehicles pent up in Saxony and Prussia, eager for a quick thrust to the Rhine.

After Vietnam this was the dream of a dream of war. It was to be fought on the field of a hundred heroic battles past. The enemy had overwhelming force, casting us like Henry Plantagenet at Agincourt—or even the three hundred Spartans—in the most romantic role war can offer. Against all odds, with Soviets signaling they would hold on nukes, this was the role of a lifetime.

How we rose to the occasion! It was not only a chance to escape the past but also a chance for future transcendence, perhaps even for apotheosis in war.

And what's more, we fought that war—*and won it!*

No, the manicured landscapes of western Germany stayed pristine. The war was fought in a strangely suitable way, onstage. In a series of flamboyant war games, in which the Soviet General Staff sat as crestfallen audience, the great Byzantine-Communist Empire was defeated in open battle.

I was there, young onlooker, somewhere in the heady haze of the early 1980s. I was at the big happening thing, the big game, "The Global War Game," that late–Cold War coronet ensconced at the Naval War College in Newport. There in the converted yellow brick barracks called Sims Hall, the United States defeated the Soviet Union, toe-to-toe, in vicarious battle.

It was the game aftermath I remember so well. Presiding over the hot-wash was Bing West: reveling in the glory and the carnage. We had flanked the invincible Group of Soviet Forces. They had pushed straight ahead, taking enormous losses, while the American Army in Bavaria thrust deep around and up into Czechoslovakia, moving fast toward Berlin, cutting them off. It was the crowning stroke, a latter-day Cannae. I shivered as the ghosts of Clausewitz and Napoleon entered the room, adding Olympian approval to that long-ago late afternoon.

This is a remembrance that strikes to the heart of America's current military confusion. It is not that we have forgotten what war is; it is rather that we have forgotten how much we miss the war we want. And, moreover, how little we reflect on what this means.

War is first and foremost all about the society that wages it. War is an amazing collective ritual. Armies may fight it, but it is not theirs to own. War is for all of us: it is society's most electric celebration of shared identity.

This is the unrecognized principle of all war. Moreover it is a "first things"—*a priori*—principle from which all others flow. War is first about us, and only then about "the enemy" we fight.

So it isn't simply that the U.S. Army likes a good battle. All Americans desire a good battle. Yet our national ethos is built around not just good but *great* battles, and around a very particular kind of war. Call it *Redemptive War,* in which great battle breaks down evil's will to resist, so that a suddenly former enemy can embrace its own redemption. The idea that battle is to destroy the dark elements in the enemy soul, thus cleansing it so that it may be uplifted to the good, is the florescent imagery of America's great wars. Hence, strategic bombing came across in postwar newsreels as a full cleansing, so that a new, democratic Germany could rise from the ruins.

Battle is the American instrument to annihilate evil.[2] American battle is not about killing: it is a majestic utility of virtue. Better even if the dark side is destroyed without great loss of life. But then in Vietnam, darkly, it was us who killed and killed to no purpose. More bombs than in all of World War II could not break evil in battle. Only people died.

It wasn't just the army that hated the Vietnam experience: *we all hated it.* We are a straight-ahead nation, and we desire virtuous, straight-ahead battle. *Decisive battle.* Without such good battle we are suddenly at a loss. We feel vulnerable and weak.

And everybody knows this.

That is why vicarious triumphs like the Global War Games of the 1980s were so important. They symbolized both the return of the decisive battle and U.S. military reputation. But there was a downside to absolute success. Victory in the Cold War also threatened to strip meaning from war. In our majestic moment of triumph we feared that great battle's catharsis was gone. Gone for sure was our dark colleague in superpower, its place taken by pitiful charges. The American military suddenly had to reconcile itself to a fatiguing future of mere military management. It was a lesser future of peacekeeping, peacemaking, and other oh-so-lesser included realities: Operations Other Than War (OOTW). There was only a single precious reprieve: that lovingly reconstructed celebration of Old War we called Desert Storm. But in the end it took on the sad nostalgic trappings of a last hurrah—the last tank-on-tank battle, the last "*masse de manoeuvre,*" the last full gathering of division and corps and army in a grand theater of war.

But even though others could see our unease in operations without *grand bataille,* still none rose to take true advantage. There was something of a debacle in Mogadishu, but it came with a small, accompanying divine grace: a sacrificial firefight.[3] So, after all, Americans left with their honor. Bosnia and Kosovo and Haiti and East Timor followed, sans fight, sans anything, so that the prospect of future battle shrank to the barren rim of end-of-millennium imagination.

But we came to our own rescue with characteristic American enterprise and invention: the self-fulfilling theology of "military transformation." Transformation was the siren song of our deepest desire. Rather than a philosophy or a plan or even mere doctrine, it was a magnificent military assertion that, once said, was so. The idea—so seductive in those early Internet days when America seemed destined both to rule and to change the world forever—paid homage to Ovid. The U.S. military was entering a *metamorphosis:* we were becoming Gods of War.

Transformation was not related to old peacetime campaigns for military improvement or reform. Transformation was a cultural promise: that new-found divinity would let us keep the *Redemptive War* we desired—where there was no fight that we could not remake to our taste. We could take out evil anywhere and everywhere we found it lurking. Furthermore, transformation's technology had made destroying evil more humane. Hence, "shock and awe," designed to destroy evil's claim on a society: not by killing people but by killing its authority over them.

Thus, 9/11 arrived as realization: a new panorama for military catharsis. Gods of War could find this in two world arenas. In one, transfigured Rangers and SEALs on laser-designating horseback—"super-empowered individuals"—could overthrow primitive ruling regimes, like Kipling's *Man Who Would Be King,* with local irregular forces reverently in tow. In the other—the classical clash of armies—they would, of course, reign serene.

But Afghan adventure was only a taste of things to come. For the "super-empowered" had bigger fish to fry. There was a world to tame, and battles aplenty in prospect.

Iraq in part was to show how Gods of War could truly fulfill the American Way of War. Those who planned the invasion framed it beneath the shimmering nimbus of World War II—and so it was. What would come after battle also took on a familiar glow: liberation, reconstruction, and the fruits of American democratic life.

We got our battle, vicariously shared through "embedded" video eye while it lasted. But then celebratory story was succeeded by unwanted events. After battle—after victory—came something else. It was not exactly "post-hostilities."

# Why Is "Post-Hostilities" in Iraq Not War?

Last year the Defense Science Board (DSB) was tasked to focus its summer study on "transition to and from hostilities." It was finally clear in the new year 2004 that something had gone terribly wrong in Iraq. Since the invasion had been such a smashing success, obviously preparations before and follow-up after were to blame, rather than hostilities—the *real war*. We had failed simply to properly execute the "before" and "after" parts of the war. The DSB study was created in part to show us how to do things right the next time.

As the panels of the study met through the spring, the Sunni insurgency in Iraq grew. Then, briefly but frighteningly, it was joined by Shia insurgency. The entire American enterprise in Mesopotamia was suddenly in grave danger. In August, when the DSB met to gather its combined briefing for the Secretary of Defense, the Shia insurgency flared up yet again. The "post-hostilities phase" was on its way to claiming 90 percent of all U.S. casualties in Iraq.

But the DSB tasking from above did not allow giving up the idea that war—real war—was still the central, truly significant event of the entire American experience in Iraq. "After war" was necessarily a lesser "post-hostilities environment" that was proving troublesome because we had neglected to properly prepare for it. By learning the correct "lessons" from Iraq, we reassured ourselves, the next invasion or true "war" we undertook would necessarily have a more recognizable "post-hostilities" character, meaning there would be no heavy fighting like in Iraq.

The question seems impossible to contain: How could the piece of a conflict that has claimed more than 90 percent of U.S. casualties be described as "post-hostilities"? If the United States invades and occupies another society, leading to an ongoing conflict, how is it possible to define only the initial phase of the conflict as true "hostilities"?

The answer is to be found in the deep needs of American ethos, its still-living memories of Vietnam and the Cold War, and transformation's powerful promise.

The Cold War's successor marketing campaign—*Transformation*—promised that the nation could be defended in Olympian fashion. Wars would be brief, casualties almost unfelt, and "evil" enemies at risk of instant "shock and awe" if they dared challenge us. The majesty of such power made it seem unimaginable that U.S. forces would ever again be subject to the horrors of "guerrilla war" like in Vietnam, where undiscoverable insurgents might kill American soldiers with impunity.

Doctrine split easily to deal with two different but cleanly defined threats. High-intensity war would be impossible because no regime anywhere would hazard battle with Americans. Thus, Iraq invasion instantly satisfied all expectations. A Republican group even hawks a T-shirt, with appropriate graphics, that reads: "Iraqi roadkill—It's what's for dinner."[4] Ralph Peters, clear mirror of the *Zeitgeist,* wrote:

> The basic lesson that governments and militaries around the world just learned was this: Don't fight the United States. Period. This stunning war did more to foster peace than a hundred treaties could begin to do. Consider the fear and impotent anger would-be opponents of the United States must feel today? The Iraqi defeat was a defeat for every other military in the world—in a sense, even for our allies, whose forces cannot begin to keep pace with our own.[5]

"Low-intensity conflict" likewise was also to be different. Transformation promised dominance and control over rogue groups and terrorists as it did over rogue regimes. As Tom Barnett, a high courtier of the transformation enterprise,[6] wrote: "As a matter of effectiveness, cost, and moral preference, operations will have to shift from being reactive (i.e., retaliatory and punitive) to being largely preventative. . . . The ultimate attribute of the emerging American Way of War is the superempowerment [sic] of the war fighter— whether on the ground, in the air, or at sea." Gods even in the granular battle space of the one-on-one: "In the increasingly transparent battle space, the speed and access of our networked forces open the way to profoundly altering initial conditions of conflict, developing high rates of change that cannot be outpaced, and sharply narrowing an enemy's strategic choices."[7]

In this sense, no insurgency could ever threaten us like Vietnam, because it could never coalesce. There would be no hiding from America's all-seeing eye. New War was thus still uniquely American in character. Battles big and small would still have that special quality of overwhelming evil, and decisively finishing it.

Above all, transformation did not allow for insurgency after war.

Thus, we refused for months to acknowledge insurgency in Iraq. We assured ourselves that these were "dead-enders" who would be mopped up in short order. How could they possibly survive against America's super-empowered, and their all-seeing eye?

Iraq was intended to showcase transformation's new "American Way of War." Transformation doctrine declared that such was our mastery that not only the outcome but also the very boundary of combat action itself was ours to command. Therefore, once we had defined "hostilities," everything after must by all rights be "post-hostilities."

Yet even though the opposite happened, Defense Department Power-Point briefings still refer to U.S. operations in Iraq today as "post-hostilities."

## If It Is Not War, Then What Is "Post-Hostilities"?

To some, we have returned to the landscape of Vietnam: a nightmare place from which we sought so earnestly to escape. But Vietnam was ultimately a war with pitched battles, sieges, and armored blitzes. Thus, the kenning of "post-hostilities" should not look to Vietnam. There, insurgency and "guerrilla" operations *were adjuncts*—in the end, mere accompaniment to a largely classic war narrative.

If "post-hostilities" in contrast is understood as a kenning for "managing the mess of a society whose former way of life has been suddenly upended," then it is a pretty good metaphor for Iraq. From this vantage, then, the 2004 DSB Summer Study was acknowledging that "post-hostilities" in Iraq had been poorly planned and managed. Poor management led to organized insurgency that might have been avoided. This implies that U.S. mismanagement put the "hostilities" in "post-hostilities." It was our doing, or our *not doing*, that let a "mess" get so violent that it came to resemble real war.

But this explanation—satisfying as it is to an America-centric story—has little to do with what was actually going on with Iraqis themselves. American invasion was from the beginning less liberation than it was "creative destruction." With an old order at last wiped away, real Iraqi identities began to take political shape. New Muslim politics in post-Saddam Iraq are the politics of armed resistance.

Yet this was not simply rebellion against American occupation. Fighting and face-offs and demonstrations were directed as much at other groups as against us. It can be argued that insurgency here is actually a form of negotiation to establish boundaries for new political relationships. In this context U.S. forces have become part of the political fabric of Iraq.

A celebrated example is the long fight between Moqtada al-Sadr and U.S. forces. Many soldiers and many more fighters died in battle, but battle itself was a negotiation. Both his initial insurgency in April 2004 and his later stand in Najaf that summer had complex political goals, or as Bartle Bull wrote: "Muqtada was fighting for tactical advantage within the Shia community, seizing momentum from the older, conservative clerical establishment—and all the while earning cross-sectarian credit as Iraq's most vocal anti-occupation nationalist."[8]

Likewise, the Association of Muslim Scholars—the main Sunni leadership group—makes clear that much of the Sunni insurgency represents not simply self-serving terrorists or old Baathists, but authentic political formation. Abdul-Salam al-Kubaisi says this to *Al-Hayat:* "If the Shi'a secure a majority of seats in the Parliament, they must consider the sacrifices of Arab Sunnis against the Americans," pointing to the fact that "the resistance was behind the pressure on the Americans to create a Governing Council, then an Iraqi Government, and finally to organize the elections." Kubaisi asserted that if it were not for the Sunni resistance, U.S. position would have been different in dealing with all issues and forces.[9]

Kubaisi is saying that armed resistance created the conditions for Iraq's political deliverance.[10] Moreover, he asserts that independence (from U.S. occupation) was sought both by Shia and Sunni, and that both have resisted U.S. occupation—of course, much of its success is due to Ali a-Sistani's popular mobilization: *that forced the United States to grant a real election.*

Moreover, Kurdish behavior fits this framework of autonomous political emergence through resistance. If anything, the Kurds are well ahead of Sunni and Shia armed politics. They have no mere militias but a well-heeled army of "Kurdistan," the Peshmerga. Furthermore, they take full advantage of their leverage with the United States. They know how much we need them in our ongoing negotiations with both Sunni and Shia. The Iraqi newspaper *Al-Hayat* has reported that Abdul Karim Mahoud al-Muhammadawi, leader of the Marsh Arab Hezbollah party in the United Iraqi Alliance, complains that Kurds are using strong-arm tactics. The Kurds seek to dominate northern Iraq, with the Peshmerga legitimated as its military authority. They demand full control of Kirkuk. Thus, they must bend local Turkmen and Arabs to their political will. Here is military negotiation achieved without fighting, but simply through its threat. Nir Rosen captures the feeling of Turkmen feeling the Kurdish squeeze: "I asked [Omar Khattab, Iraqi Turkmen Front] why he had a small army in his office. Was he expecting violence during the election? 'God willing,' he said, 'there will be violence. We are expecting it. You think we will keep silent about the 108,000 Kurds? Civil war has to happen, but we won't start it. Why do you think we were cleaning our weapons?'"[11]

This is very old-style politics, the kind that we can look up in histories of the fourteenth-century Levant, or alternatively in what was once called "Yugoslavia" or in Moldova or Lebanon or, for that matter, Afghanistan today. Or in Iraq:

where Americans are simply part of the jostling political mix. Hence Al-Muhammadawi (of United Iraqi Alliance's Hezbollah party) could naturally go on to suggest, "America will use the Kurdish bloc as a card with which to pressure those parties, with the direction of which Washington does not agree."[12]

All this is to say that "post-hostilities" is not war. But neither is it mere management after the battle, as transformation doctrine had hoped. Neither will properly manicured "post-hostilities" look more like it should the next time, when "we get it right." Iraq is not "post-hostilities" (non-war) that degenerated into something more warlike because it was botched.

Rather, it is the new way of life, and the new main occupation of the U.S. military. It is not the "mess management" that comes after war. It is the central activity that replaces war, both Old War (Napoleon) and New War (transformation). Instead of the kenning "not war," why not define "post-hostilities" this way: "negotiating a relationship with a society through an emphasis on military means."

Of course, this destroys the original "post-hostilities" design. "Post-hostilities" above all was configured to preserve the sacred centrality of real war. Preserving real war means both protecting its unique character and enshrining it as the decisive agent in larger national objectives. In this paradigm, "post-hostilities" is the finishing and completing phase, *but real war is the decisive phase.*

But Iraq spoiled the necessary and ratifying contrast between "hostilities" (war) and "post-hostilities" (the after-war). Indeed, after two years of acute military engagement, the Iraq war-*after-the-war* has clearly been the central experience, and what is achieved or not achieved here will be remembered as the decisive part of this historical event.

Americans still earnestly desire decision to be achieved in war. The after-war phase therefore must represent the fruits of victory offered by war. Thus, baring the true character of "post-hostilities" in Iraq is deeply troubling. It is more than troubling because it announces not simply a failure of postwar management, but also the failure of victory in war, and thus of war itself.

The "celestial" phase of real war in Iraq was not decisive—*it was insufficient to secure victory.* What we still call the "hostilities phase" was, in actuality, little more than an entry phase. It served to establish Americans in Iraq and thus enabled us to begin negotiating relationships with a variegated Iraqi society.

The job of the DSB 2004 Summer Study was to preserve real war by also reauthorizing a reliable war model for future American action. The study briefing works hard to do this by urging that more effort be made in the pre- and post-hostilities "phases."[13] But these phases are carefully described as "transition to and from hostilities." They are identified as critical and necessary, but also transitional and supportive. In this way, war's sacred status is reaffirmed by the cleanly framed *and separate significance* of its flanking sidekicks, "pre" and "post."

Yet it was the "hostilities" that became subsidiary: critical and necessary, but also transitional and supportive. While "transformation" was recodifying and re-anointing war, war itself was being reassigned as transient, subsidiary military activity. A world transformation had moved decision to the very place where it was not supposed to be: in the period we call by the kenning "post-hostilities."

## War to Be Replaced
## by "Post-Hostilities" Environments?

America's military future is a future, since the last hurrah of Desert Storm, of "negotiating relationships through an emphasis on military means." More simply we could call it *engagement-that-is-not-war*. These sorts of relationship-engagements have included Somalia, Bosnia, Kosovo, East Timor, Haiti, Afghanistan . . . and Iraq.

Each has required an entry phase, as transformation doctrine always promised, opening the door to relationship. In a few cases, like Serbia, the entry itself ensured the relationship—perhaps because that relationship needed no more than to break the crust of an eroding regime. In other cases, like Somalia, the entry was no more than splashing ashore; but the relation-ship, never really enjoined, was botched, leading to debacle.

Afghanistan and Iraq seem to point more authoritatively to a pattern of future entry. Imagine a Syrian intervention. How about U.S. occupation of select parts of a former Saudi Arabia? Taking over from a collapsing Hashemite kingdom in Jordan? Picking up the pieces after Pharaoh in Egypt? Sorting through a decomposing Pakistan?

Each of these possibilities—*none of them wholly unlikely*—means sharp and overwhelming armed entry, quiet or noisy. But even at its shrillest, like Iraq, it will still be just an entry. It is what comes after that will count for everything.

And what comes after, even along a broad spectrum of "what," will not simply be "irregular warfare" or "nation building" or "counterinsurgency" or "operations other than war." It will not be a fourth or fifth or "X" generation of war either. It can, of course, be called any and all of these things, but it will be much more, and often much less, than war by any definition.[14]

So how did we get here?

Arguably, we saw it coming. We have talked endlessly about the world change that has brought it, but we have simply not put the pieces together.

For example, there has been much debate about "globalization" and the corollary "decline of the nation-state." Yet we have not taken the full measure of what this means to a military—and American national—mission built around the decisive cleansing of war.

Thus, we gave world change a title—"globalization"—that made it seem straightforward. We might confess that it would be messy and protracted. But its very inevitability as a ruling concept assured us that it was something that could and must be managed. Conflict? Of course! Yet another challenge for the American mission!

Correspondingly, we saw the ebbing of the nation-state and assumed that new conflicts would be "asymmetrical." It would be war as "nonclassical war"—in terms of tactics and means only. But what we missed was that nonclassical tactics and means would serve wholly different—though hardly new—human relationships. The politics, not just the fighting, were different.

National wars over about three centuries became a mature genre. They were highly tuned, tightly run enterprises tied increasingly to abstract narratives and the needs of great national institutions. But the appeal of the genre collapsed at the end of the last century. Elaborate military enterprise became distasteful to the very societies that once lovingly sacrificed themselves in its name. Big war long ago ceased to serve tribal celebrations of identity—at least for those "tribes" we call nation-states. It may linger in the resumes of some successful big states like India, China, and the United States, but only as hollow advertisement.[15] Because the emotional stake in national war moreover shifted dramatically from glorification to lamentation.[16] The spirit of the age has turned against national wars.

Yet at the same time, real tribes—bands and clans and family, people bound by blood and faith—have somehow returned full force from the backrooms of the pre-modern. Their stake in identity has seemingly boundless energy. Tamil suicide bombers and Chechen sharpshooters, penitent Shia and ascetic Arabian "07"—these are the armed forces of future engagement and future relationship, whether we seek to integrate them or to contain them.

Local reassertion means local relationships and local military engagements. This is not classic insurgency anymore either. Insurgent tactics and techniques abound, but they are not tied to organized causes necessarily coalescing into full nation-states.

A recent exchange between the "Fox All-Stars" on the Fox News channel illustrates neatly the conceptual difficulty for us. The question was raised as to whether Sunni fighters in Iraq amounted to a real insurgency. No, they all agreed, because there was no shadow government standing behind the fighters, and thus there was no legitimate political movement that would form a new national regime if their goals were won.

Political formation below the national level is somehow still illegitimate if it is not "nation anointed." Tribes and fighting groups and associations of clerics do not represent legitimate politics.

*But they do*—and when we intervene to overthrow a longstanding old-style national regime, we actually abet new local politics, including armed local politics. As Gods of War we can enter most societies and overturn their national framework of administration and authority. Yet, at once, other politics start to spring from the ground up. In Iraq and elsewhere, the United States is the midwife of local political evolution.

Thus, our military-based relationships in Iraq—*"friendly" and hostile alike*—have been almost exclusively local relationships as well. We can call our hostile relationships "war" if we wish, yet they are not so different from the context, texture, and goals of our "friendly" relationships—and often those friendly relationships involve incipient conflict barely avoided, or even outright and desperate fighting.

It is not about destroying evil anymore. It is all about the relationships. And the fighting is the grist of negotiation.

## Lessons for American Strategists and Soldiers

See the larger reality.

Tell the American people (and admit to yourselves) that Old War is dead. And so is New War. Military operations have transcended the transformation paradigm, which, after all, was just *après* Cold War narcissism, dressed up as hi-tech.

Truth is, the "American Way of War" is no longer ours to demand. We Gods of War can no longer command the sort of war we wish to fight. We are to that degree diminished; and so, softly, we are no longer gods.

We Americans must now deal with "engagement that is not battle," and which in its very totality does not add up to war. Those cherished things (battle and decision by battle) are now withheld from us, partly because those we fight understand that to give us these would also be to give us back victory.

But most of the new world of "engagement that is not battle" is no one's choice. It is a consequence of world developments. Traditional battle in formal, drawn-up ritual between national societies—*with perhaps a couple important holdouts*—is both unwanted and unsustainable. But even these holdouts reserve classical war as a billboard form of symbolic legitimacy rather than an available utility. Any old style war, if truly engaged, would finish them off.

So Old War is still remote possibility. But this is not what should interest us. We should look instead at what Old War today would do. Its final flourish would have every prospect of dramatically accelerating the very changes that its fortunate absence continues to disguise: namely, the emerging dynamics of community and new identity. Only the *absence* of Old War can perpetuate its symbolic nation-state system claims. The absence of Old War helps to uphold the Old System itself.

This suggests that even as national administrative claims on people remain, *the national claim on identity ebbs.* Yet at the same time, more intimate local identity revives, and armed human conflict shifts to local struggles and local-to-larger political discourse. The very intimacy of armed engagements built on person to person makes it seem much more like premodern "war"—which is to say, the gang-and-retinue jostling so common before the rise of modern states.[17]

Those terrible bureaucracies that could suck the human energy out of whole societies have, after all, been with us for only two or three centuries.[18] The truly terrible twentieth-century forms, from Soviet to Third Reich, had the greatest suction, of course, but were also the most evanescent. Yet even at their most efficient, modern bureaucracy deadens community. And it is community that we see rising everywhere.

Even were we to marshal technology eons ahead of all others, we could not overturn such change as is sweeping across human culture today. We can choose to fight others as often and for as long as we wish, but we cannot make the world submit to the war we want to fight.

Yet there is a practical difficulty even in this realization. The "American Way of War" through two centuries of heroic sacrifice and cherished story has come to define us. There is even a cable TV channel called the War Channel, devoted to replaying the fighting scenes embedded in American ethos.

How can we leave behind our icons of "war" and "battle" and hope to fight effectively as American soldiers? How can we hope to be effective soldiers—in the strange world emerging before our eyes—if we continue to cling to them?

## Notes

1. United States Army, Field Manual 100-5: Operations, 1976.

2. Religion as the subtext of the American Way of War is a difficult theme to frame, because it makes some uncomfortable. Nevertheless it is an historical and con-

temporary reality, as Ernst Lee Tuveson brings to light in *Redeemer Nation: The Idea of America's Millennial Role* (Chicago: University of Chicago Press, 1980). The current president carries forward the war imagery of Lincoln and Wilson (FDR and Reagan had a lighter Calvinist touch!). There are several enduring rhetorical components of *Redeemer War,* suggested in Michael Vlahos, "Religion and U.S. Grand Strategy," *The Globalist,* June 8, 2003. Available at http://www.theglobalist.com/DBWeb/StoryId.aspx?StoryId=3230. In this context Peter Singer points out in *The President of Good and Evil* (New York: Dutton, 2004) that Bush talked about evil over three hundred times between January 2001 and June 2003, using the word over one thousand times, *almost always as a noun.*

3. Fittingly, even reverently memorialized in Ridley Scott's film, *Black Hawk Down.*

4. Selling briskly at the 32nd Annual CPAC, February 17–19, 2005.

5. Ralph Peters, "A New Age of Warfare," *New York Post,* April 10, 2003, quoted in Michael Vlahos, "Defeating the Gods of War," *Culture's Mask: War and Change After Iraq,* Baltimore: Johns Hopkins University Applied Physics Laboratory, September 2004. Available at http://jhuapl.edu/POW/library/culturemask.htm

6. Barnett here recreates "courtier" in its original Renaissance sense, as described by Castiglione in his classic of 1528, *The Courtier.* Castiglione's portrait revolved around the idea of *spezzatura:* "a sophisticated naturalness, a mastery that seems unstudied." See Richard Stengel, *You're Too Kind: A Brief History of Flattery* (New York: Simon and Schuster, 2000).

7. Arthur K. Cebrowski and Thomas P. M. Barnett, "The American Way of War," *Transformation Trends* (Washington, D.C.: Department of Defense, 13 January 2003). Available at http://www.oft.osd.mil/library/library_files/trends_165_transformation_trends_13_january_2003%20Issue.pdf

8. Bartle Bull, "The Rise of Shia Iraq," *Prospect Magazine,* November 2004. Available at http://prospectmagazine.co.uk/article_details.php?id=6515&category=&issue=496&author=2206&AuthKey=02383c1590b03fa070e7d61b90035334

9. *Al-Hayat,* 9 February 2005, translated by Gilbert Ashcar, in "Informed Comment: Thoughts on Middle East, History, and Religion." Available at http://www.juancole.com/2005/02/ams-on-its-role-in-forcing-elections.html

10. Armed resistance as a touchstone for coalescing community identities was underscored by a number of reports in the spring of 2005. For example, when Sunni leaders met in Baghdad in March: "But as the conference went on, it became a virtual rally for the insurgency, with tribal sheiks and clerics alike speaking scornfully about the government and constitution. Again and again, the speakers praised the resistance, and drew loud applause from the audience gathered in the auditorium of the Babylon Hotel." Richard F. Worth, "Sunni Leader Vows Support for Insurgents," *New York Times,* March 29, 2005; or President Jalal Talabani's quip that, "Iraqi forces, *the popular forces* and government forces, are now ready to end the insurgency and

end this terrorism"—the popular forces of course representing the civil militias of the Kurdish and Shi'a communities—meaning, the legitimization of the armed resistance; Jim Muir, "Iraqi militias 'could beat rebels,'" *BBC News,* April 18, 2005. Available at http://news.bbc.co.uk/2/hi/middle_east/4454985.stm

11. Nir Rosen, "In the Balance," *The New York Times,* 20 February 2005, sec. 6, p. 30. Available at http://www.nytimes.com/2005/02/20/magazine/20ELECTION. html?ex=1109826000&en=9c73418a6c648edb&ei=5070&pagewanted=1&oref= login&oref=login

12. *Al-Hayat,* 28 February 2005, translated and excerpted in "Informed Comment: Thoughts on Middle East, History, and Religion. Available at http://www.juancole.com/2005/03/negotiations-continue-on-formation-of.html

13. Defense Science Board Summer 2004 Summer Study on Transition to and from Hostilities, Office of the Under Secretary of Defense for Acquisition, Technology, and Logistics, December 2004 (Washington, D.C.: Department of Defense, 2004). Available at http://www.acq.osd.mil/dsb/reports/2004-12-DSB_SS_Report_Final.pdf

14. Michael Vlahos, *Two Enemies: Non-State Actors and Change in the Muslim World* (Baltimore: Johns Hopkins University Applied Physics Laboratory, January 2005). Available at http://jhuapl.edu/POW/library/Vlahos_Two_Enemies.pdf

15. This is especially true for China, where highly paraded military symbolism is an essential element in continuing regime legitimacy.

16. In fact Transformation ideology offers an almost primal guarantee that future war will be free from national-level sacrifice and lamentation. Peters and Barnett are especially good at presenting the new war-compact.

17. See for example, an entire study on this way of life, Matthew Innes, *State and Society in the Early Middle Ages: The Middle Rhine Valley, 400–1000* (Cambridge: Cambridge University Press, 2000).

18. Yes, even here, as evoked so hauntingly in Randall Jarrell's "The Death of the Ball Turret Gunner" [(New York: Farrar, Straus and Giroux, 1981), p. 144]:

From my mother's sleep I fell into the State,
And I hunched in its belly till my wet fur froze.
Six miles from earth, loosed from its dream of life,
I woke to black flak and the nightmare fighters.
When I died they washed me out of the turret with a hose.

# TWENTY

# Rethinking and Rebuilding the Relationship between War and Policy

## Post-Conflict Reconstruction

BATHSHEBA CROCKER WITH JOHN EWERS
AND CRAIG COHEN*

## Introduction

When President George W. Bush stood on the USS *Abraham Lincoln* and announced the end of major combat operations in Iraq, he may not have been wrong, but the banner celebrating "mission accomplished" was, as it turned out, premature.[1] Two years hence, the United States is still embroiled in Iraq as well as Afghanistan, and only just beginning to develop civilian and military doctrines for rebuilding failed states and war-torn societies. If September 11 has come to represent the immediacy and gravity of twenty-first-century security threats, that May 1 address could well symbolize the sheer complexity of meeting those threats. Making the United States safe in the coming decades will take more than technological and strategic innovations in hunting terrorists or confronting rogue regimes. It will also require conducting successful post-conflict operations.

*Bathsheba Crocker is Fellow and Co-director of the Post-Conflict Reconstruction (PCR) Project of the Center for Strategic and International Studies (CSIS), a bipartisan think tank in Washington, D.C. John Ewers is a Marine Colonel and the Commandant of the Marine Corps Fellow at CSIS for 2004–5. Craig Cohen is a Researcher in the PCR Project. The authors would like to thank Jonathan Hadaway, Christina Caan, Aidan Kirby, Mike Eros, and Jenny Urizar for their research assistance.

The ongoing difficulties in Iraq and Afghanistan and continuing loss of lives and resources have belied the notion that the United States "won" both wars, and has caused many in Washington to broaden their common understanding of mission success to include "winning the peace." It is odd that a nation with a military establishment so enamored of Carl von Clausewitz—who observed that "war is merely the continuation of policy by other means"[2]—would be caught short in planning for post-conflict challenges. Yet as the U.S. Army's Third Infantry Division's after-action report on the Iraq invasion suggests, the U.S. military entered Iraq in the spring of 2003 with little if anything in the way of a plan for Phase IV operations.[3] What began as Phase IV warfare, or "transition,"[4] has become—in Iraq particularly—a painful reminder that postwar duties, sometimes called nation building, are not optional, but rather are integral to the military's mission to fight and win wars. A recognition of this requires both a broadening of the definition of the military's core missions and an understanding that winning the peace goes well beyond coercive engagement.

"Nation building" has returned to the center of the national security debate in U.S. foreign policy circles. Ungoverned territory—a result of either war or state weakness—has proven dangerous to the interests of the industrialized world. Afghanistan stands as the most prominent example of a state failure halfway around the world that directly threatened American lives and interests. It is likely that post-conflict operations will be a necessary and unavoidable part of virtually all military operations and interventions in which the United States participates in the foreseeable future. Technological innovations toward network-centric warfare or advanced weapons systems will have limited applicability in dealing with this challenge.

How, then, do we build for success? Enhanced capacity to fight insurgencies and build institutions and enhanced coordination between U.S. departments and agencies are likely to get us only so far. True success will be a product of some heightened level of "jointness" among the defense, diplomatic, and aid communities, as well as between the military and civilians, between the United States and international partners, and between international and local actors. In rethinking our notions of warfare, we must find a way to avoid seeing it as an isolated act. Rather, we must reintegrate war fighting into the fabric of the policy of which it is part.

This chapter examines developments in theory and practice of post-conflict reconstruction—one of the crossroads of war and policy. It outlines evolving doctrine, highlighting both the requirements for success and limitations of engagement. It then surveys some of the recent trends of activity within the U.S. government and abroad to build post-conflict capacity. Finally, it evaluates whether we are doing enough to meet the strategic need, and what further action is required. Is it possible to develop a more integrated strategic approach to both war and policy?

## Evolving Theory and Doctrine

Multiple terms have been used to describe operations that generally fall under the category of post-conflict reconstruction. These include nation building, state building, stability operations, stabilization and reconstruction, security and stability operations, transition, transitional authority, trusteeship, Phase IV warfare, peacekeeping, peace building, and peace operations. The manner in which these terms are commonly used as synonyms for one another reveals the lack of clarity of the ultimate vision underlying post-conflict operations.

Throughout American history, nation building has at certain times been a critical element of U.S. national security strategy. As far back as 1818, John Quincy Adams warned of the dangers posed to the United States by ungoverned, "derelict" provinces, including the future state of Florida under Spanish rule.[5] The rationale, strategies, and tools used by the United States to reconstruct the South after the Civil War, to intervene both in its own hemisphere and in the Philippines in the late-nineteenth century, to rebuild Germany and Japan after World War II, and, more recently, to try to bring stability and democracy to Panama, Somalia, Haiti, Bosnia, and now Afghanistan and Iraq are strikingly similar, although they also demonstrate a clear evolution of both thought and policy over time.[6]

Nation building in recent American parlance has essentially become synonymous with state building, defined as "the creation of new government institutions and the strengthening of existing ones."[7] The goal of nation building is not to "impose common identities on deeply divided peoples," but to "organize states that can administer their territories and allow people to live together despite differences."[8] The perceived danger of ungoverned territory since September 11 has focused American attention on increasing the enforcement capacity of states in order to prevent instability and to deny safe haven to terrorists.

The United States is not the only international actor that has led nation building operations. The United Nations has taken a lead role since the end of the Cold War in, inter alia, Namibia, Cambodia, El Salvador, Mozambique, Eastern Slavonia, Sierra Leone, Kosovo, and East Timor. United Nations (UN) missions tend to be undermanned and under-resourced in relation to U.S. efforts, and to operate in less demanding circumstances (but arguably with greater success). The best chance of success in nation building has come through joint efforts involving the United States and the international community, as in Bosnia.[9]

The UN does not use the term nation building; instead, it refers to such efforts as peace operations. UN doctrine on rebuilding failed states and war-torn societies has emerged over the past decade out of the "1992 Agenda for Peace" and its 1995 supplement, the "Report of the Panel on UN Peace Operations in 2000" (the Brahimi Report), the 2004 "High-level Panel on Threats, Challenges and Change," and the March 2005 report by the secretary-general, "In Larger Freedom."[10] These reports reveal a clear transition toward state-based solutions. During the 1990s, "the state remained largely on the periphery of post-conflict policy discussion,"[11] but recently the UN has increasingly come to see the necessity of building capacity across virtually all aspects of state functioning—from providing security to delivering basic services, from creating jobs to ensuring justice. The UN has sought to deliver a coherent response under the leadership of a single special representative to the secretary-general (SRSG) who can integrate humanitarian and development coordination structures into the broader political and peace-keeping mission.[12]

The UN and the United States have come to share a common view that both national and collective security depend in part on stabilizing and strengthening fragile states.[13] Or as one study puts it, "A world of capable, efficient, and legitimate states will help to achieve the goals of order, stability, predictability, and national and human security."[14] The U.S. government officially refers to such missions in the aftermath of conflict or state collapse as stabilization and reconstruction operations. These operations typically last five to eight years; since the end of the Cold War, the United States has begun one every eighteen to twenty-four months.[15]

Two questions frame emerging post-conflict doctrine. First, what should be the goal of such operations—in terms of the ultimate vision of success, the level of capacity sufficient to play a catalytic role, and the types of capacity that are most closely correlated with progress? Second, how best to achieve this goal? A consensus is beginning to emerge on the answers to these questions.

Typically, the ultimate goal of stabilization and reconstruction operations is to create a peaceful, democratic state with a market economy, though there is broad agreement that international actors must be realistic about what is actually achievable in the near and medium terms—a "minimally capable state," for instance, that can restore internal security and public safety.[16] One may ask, however, whether this is a bold enough vision. The goal of a "minimally capable state" is likely to yield a minimum commitment of resources, and unlikely to fundamentally transform the societies in question.

One might think that local actors are in favor of such a "light footprint" approach, on account of their sensitivity to sovereignty issues, but even such outspoken critics of the international community as Ashraf Ghani, former Afghanistan minister of finance and now president of Kabul University, has argued that international actors should build state capacity across ten functions: security, rule of law, redistribution of public finances, human capital development, the rights of citizens, infrastructure development, regulation of natural resources, creation of an economic market and jobs, public borrowing from international institutions, and fostering an accountable, democratic government.[17] Ghani's list is no minimalist vision intent on curbing international interference. The May 2002 joint publication of the Association of the U.S. Army (AUSA) and the Center for Strategic and International Studies (CSIS), *Task Framework on Post-Conflict Reconstruction,* groups these basic functions into four pillars: (1) security; (2) justice and reconciliation; (3) social and economic well-being; and (4) governance and participation.[18] Progress must be made in all four areas to have a chance of reaching a stable outcome.

A seeming contradiction has thus emerged in which international actors have increasingly taken control of the full gamut of state functioning (referred to by some as neotrusteeship),[19] while maintaining the goal of building local capacity.[20] One difficulty with this approach is that a state is not just a technical matter; it is a highly sensitive, political institution. There may, in fact, be some capacities that simply cannot be built from the outside. Worse still, international actors often undermine the very state-building processes they are trying to encourage by formulating parallel sets of rules and authorities apart from the state.[21] The international community is thus "limited in the amount of capacity it can build," and "complicit in the *destruction* of institutional capacity."[22] This effort to find a balance between imperialist impulses and building local capacity has been referred to by one Afghanistan-watcher as "nation-building lite."[23]

It is not only a question of how much of an international presence is proper, but also of the character of what is being built. Is the goal in

Afghanistan "building order in anarchic zones where terrorists find shelter," "creating a state strong enough to keep al Qaeda from returning," or "getting the guns out of warlords hands and opening up space for political competition free of violence?"[24] The goal should determine our response, though too often it does not.

This is so in part because once one gets beyond vague—albeit laudatory—generalities like reducing the sources of conflict in a society or planting the seeds of democracy, the goals become extremely difficult to pin down in terms of operational and tactical requirements, particularly in the absence of shrewd political analysis. What are, in fact, the true sources of conflict, and how best to deal with them? What do the seeds of democracy look like before they sprout? In Afghanistan, for instance, will allowing warlords to help reestablish the Afghan state (the manner in which most European states developed) yield a more stable outcome than confronting them militarily? In Iraq, how does one go about building local capacity while fighting a deep-rooted insurgency?

In part because of this complexity and uncertainty, the U.S. military has historically been wary of taking on nation building functions. Although the military was largely responsible for the successful reconstruction efforts in Germany and Japan after World War II, nation building in such places as Vietnam came to be viewed as "mission creep" that hindered war-fighting efforts in the short term and capabilities in the long term. There is an apparent lingering tension in the military among the recognition that some involvement in state building is unavoidable, the experience that it is extremely difficult to extract forces from stabilization operations once they've been inserted, and the sense that units may lose their comparative advantage or combat readiness as time goes by without the ability to transfer responsibility to civilian or local actors. Since Afghanistan and Iraq, however, the military is increasingly coming to view postwar duties as "part of the military's mission to fight and win wars,"[25] and moving farther away from the notion that such duties constitute mission creep to be avoided at all costs.[26]

The National Defense Strategy (NDS) of 2005 identifies stability operations as significant to three of the eight key operational capabilities that are the focus of defense transformation: denying enemies sanctuary; improving proficiency against irregular challenges; and increasing capabilities of international and domestic partners.[27] The 2005 NDS also aims at reconfiguring military forces and planning templates to conduct "extended stability operations involving substantial combat and requiring the rapid and sustained application of national and international capabilities spanning the elements of state power."[28] Ultimately, the NDS recognizes that stability operations are essential to achieve far-reaching objectives of campaigns to win decisively and "bring about fundamental, favorable change in a crisis region and create enduring results."[29]

The tenor of the 2005 NDS is echoed in a (draft) Department of Defense (DoD) Directive on Stability Operations, based on a 2003 Defense Science Board study.[30] Identifying stability operations as a "core U.S. military mission," the directive defines such operations as "military and civilian activities conducted in peacetime and across the full range of conflict to establish order in states and regions."[31] The immediate goal of stability operations is "to provide the populace with security and restore essential services," with a long-term goal of developing "indigenous capacity for securing essential services, a viable market economy, rule of law, democratic institutions, and a robust civil society." The directive establishes, as DoD policy, that such operations are to "be given priority and attention comparable to combat operations, and to be explicitly addressed and integrated across all DOD activities including doctrine, organization, education, training and exercises, material, leadership and personnel development, facilities, and planning."[32]

The policy contained in the NDS and the draft DoD directive, which is also reflected in the DoD's approach to the 2005 Quadrennial Defense Review,[33] places less emphasis on waging conventional warfare and more on dealing with insurgencies, terrorist networks, failed states, and other nontraditional threats.[34] The policy aims to create a future standard that judges unit readiness for war not only by traditional standards but also by the number of foreign language speakers in its ranks, cultural awareness, and ability to train and equip local forces.[35] Beyond that, it recognizes the importance of a multidisciplinary approach that incorporates all elements of national power by mandating, for example, that military stability operations plans be coordinated to the extent practicable with other agencies and integrated with U.S. government stabilization and reconstruction plans.[36]

Since the fall of Saddam Hussein, the Office of the Secretary of Defense, the Joint Staff, and the military services (primarily the U.S. Army and Marine Corps) have also undertaken initiatives that reflect the new focus on stability operations. For example, in January 2005, the secretary of defense approved a Joint Operating Concept (JOC), which contemplates that stability operations will be applicable during all phases of major combat operations—precrisis, during combat, and postwar. Security, initial humanitarian assistance, limited governance, restoration of essential public services, and other reconstruction assistance are identified as the focus of stability operations. At the same time, a draft revision of Joint Doctrine for Operations replaces a chapter on Military Operations Other Than War (MOOTW) with chapters on Major Combat Operations and Stability Operations (as well as Homeland Security).[37]

Both outside and inside the defense community, there is a broad consensus that the military is absolutely necessary to postwar missions, but it is also insufficient to bring about success without a coherent strategy that spans the Defense Department, the State Department, USAID, and other civilian agencies and capacities. Security is the sine qua non of post-conflict reconstruction, but establishing security does not alone provide a sustainable outcome or viable exit strategy. More important in the long run is building enough local capacity to provide security with only minimal international assistance, and the military certainly plays a role in that regard. But a functioning economy, a rule-based legal and political system, and some level of social well-being are all vital to ensuring that the gains made by heightened security endure. The tasks associated with these requirements normally fall outside the ambit of the military's expertise and capacities.

While the military is critical to preparing the groundwork for reconstruction, for example, by confronting spoilers and insurgents, enforcing curfews, demobilizing militias, de-mining, and providing security for elections,[38] the military is the first to accept that they are "not the best suited institution for reconstruction, governance and economic policy."[39] Some sort of joint military and civilian effort is thus required, with improved planning for postwar strategies, enhanced capacity of civilian agencies, and improved coordination and interagency workings. There is a real need to think more in terms of interagency jointness rather than merely improved coordination—along the lines of how the Goldwater-Nichols Act of 1986 restructured the armed forces to achieve interservice jointness.

Stabilization and reconstruction must become a core competency of the DoD, the State Department, and USAID if the U.S. government is to be effective in meeting the threats of the twenty-first century. The management discipline the military brings to combat operations should be extended to post-conflict activities,[40] and it should serve as a model for joint military-civilian strategies. Interagency task forces—both at the National Security Council and regional levels—stabilization and reconstruction joint commands deployed concurrently with combat units, rapidly deployable interagency crisis action teams, and a fund for stabilization and reconstruction or "countries in transition" would all go far toward improving operational capacity.[41]

As with any mission, though, the key question for post-conflict operations is who is in charge. True unity of command between civilians and the military has so far proved elusive in American operations. Australia, for its part, has done a better job of creating a single leadership model with their intervention in the Solomon Islands. Most agree that there ought to be a more centralized authority within the U.S. government (in headquarters and on the ground), although this could potentially interfere with the process of creating a more significant role for local actors.[42] A few have taken up the idea of creating a cabinet-level position of director of reconstruction.[43] A clearer demarcation of authority within the international community is also likely to lead to greater success—a lead state or regional organization rebuilding collapsed states has proven vital in the past.[44] In short, leadership matters, but it must strike the right balance of direction and cooperation—among agencies, military and civilian actors, governments, and people on the ground.

Straightening out lines of authority is only one part of the equation. The level of effort put forward—both in terms of time and resources committed, including the number of troops—is a key determinant of success or failure.[45] One analysis has placed the need for global peacekeepers at any one time at two hundred thousand out of a pool of six hundred thousand.[46] On the civilian side, there is a need for a post-conflict reconstruction team with the capacity to work on issues such as rule of law, governance, disarmament, demobilization, and reintegration (including job creation). There is also the additional need for a reserve of civilian police to deploy on international missions. USAID currently maintains fewer than two thousand full-time direct-hire personnel, and the State Department lacks a true operational planning capacity.

In sum, conventional wisdom holds that if the United States and its international partners could, for post-conflict operations, develop the following, then winning the peace might be easier:

1. a greater jointness of action
2. clearer lines of authority
3. better-informed leadership and strategic planning (including analysis)
4. a more rapid response capability
5. a greater and more sustained commitment in terms of resources and time

Others dispute this contention—primarily on two grounds. The first argument questions whether the political will actually exists to do post-conflict work at the level it takes to succeed. Often our attention span is too short, building civilian capacity is rarely considered a priority, and nation building is still viewed in various quarters as a project of liberal internationalists, imperial adventurers, and/or naive social engineers. The second argues that we lack the tools for state building, or that certain types of state capacity may not be imposable from the outside.[47]

Recently, there has been a focus on prevention as a more cost-effective and politically palatable way of preventing war and state failure.[48] Prevention proponents argue for promoting opportunities for broad-based growth and poverty reduction, supporting legitimate and democratic institutions, and creating effective U.S. assistance to police and military forces. The United Kingdom government's *Investing in Prevention* argues, for instance, in favor of:

1. developing a common diagnostic framework to identify causes and dynamics of instability
2. creating a strategic approach to reduce instability with a coherent package of development, political, and security engagement
3. creating a set of priority initiatives for the international community to take over the next five years
4. taking measures to strengthen international systems to prioritize resources and develop coherent strategies.[49]

In other words, focusing on crisis response (stabilizing situations after they have erupted) is only one part of a broader package of preventing instability through a combination of both "hard" and "soft" power.

But problems with prevention, both theoretical and practical, exist as well. First, if the international community does not know how to do state building, there is no guarantee we know how to do prevention, which ultimately includes many of the same objectives and tasks. In fact, the UN's record on prevention has been poor. Roughly one-half of all peace agreements break down within five years, with the parties reverting back to war.[50]

Second, donors have not shown a willingness to invest in prevention. Is it feasible to expect a county spending money on fighting a fire to pay for fireproofing at the same time? This may help explain why the United States is willing to spend hundreds of billions of dollars in Iraq, but reluctant to fund preventive initiatives at even a fraction of the price.

Third, there is the sovereignty obstacle to preventive intervention in countries on the verge of collapse or complicit in massive human rights violations. Even supporters of the emerging norm of the responsibility to protect would likely be unwilling to have the norm applied to their own internal affairs.[51] This notion of conditional sovereignty is opposed by most in the developing world when the West (and particularly the United States) is the intervener.

Where does this leave us? Most likely, we are left with the "least bad option": responding to crises when they represent a direct threat to international peace and security (or U.S. national security), and trying to forge a greater unity of approach when the time comes; recognizing that winning the war only gets us part of the way home, and that the strategies and competencies required for stabilization and reconstruction operations and fighting insurgencies will be different than those required for traditional warfare; and, in the meantime, investing in some preventive measures to try to forestall state failure.

One area that serves both post-conflict and preventive purposes is strengthening democratic institutions, including security sector reform. Forging partnerships with indigenous militaries to become more accountable to national parliaments and civil society helps to avoid the types of predatory behavior that undermines democratic rule and subverts the state-building process.

## Building Post-Conflict Capacity in the United States

The continuing challenge in Iraq has focused U.S. attention on the need to develop greater capacity in rebuilding failed states and stabilizing war-torn societies. Despite stirrings in the military, Defense Department, State Department, USAID, and Congress, however, action to develop jointness within the U.S. government has been in short supply.

The most tangible commitment made by the U.S. government to reconfigure post-conflict operations was the creation of the Office of the Coordinator for Reconstruction and Stabilization (S/CRS) within the U.S. Department of State in the summer of 2004. The mission of S/CRS is to lead, coordinate, and institutionalize U.S. government civilian capacity to prevent or prepare for post-conflict situations, and to help stabilize and reconstruct societies in transition from conflict or civil strife so they can reach a sustainable path toward peace, democracy, and a market economy. Its core objectives include working across the U.S. government and with the world community to anticipate state failure, avert it when possible, and help post-conflict states lay a foundation for lasting peace, good governance, and sustainable development.[52]

The tasks associated with S/CRS's mission and core objectives are numerous: monitoring, early warning, and planning; mobilization and deployment management; development of skills and identification of resources; establishing a corporate memory to learn from experience; and coordination with the international community.[53] More specifically, the office has or will:

- Identify vulnerable states at six-month intervals for intensive interagency planning
- Deploy "Humanitarian, Stabilization, and Reconstruction Teams" (HSRTs) to Regional Combatant Commands to participate in post-conflict planning where U.S. military forces are likely to be engaged
- Stand up Interagency Country Reconstruction and Stabilization Groups (CRSGs) in Washington and within regional bureaus to promote joint approaches
- Develop and train an Active Response Corps of State Department officers to deploy as first responders, and establish a Standing and Technical Corps for immediate deployment
- If funded, manage a Conflict Response Fund for stabilization and reconstruction[54]
- Develop planning, essential tasks, and metrics frameworks for post-conflict operations
- Lead post-conflict gaming exercises
- Serve as a liaison with key international partners such as the United Nations, European Union, International Financial Institutions, G-8, regional organizations, and country partners[55]

For its part, the DoD has been receptive to S/CRS's new role. The clearest sign of this is the recent reports that the DoD has offered to help fund S/CRS to the tune of $200 million.[56] The 2005 NDS sets forth that the DoD should cooperate with the new office "to increase the capacity of interagency and international partners to perform non-military stabilization and reconstruction tasks that might otherwise often become military responsibilities by default."[57] The DoD has also established the Defense Reconstruction Support Office to coordinate the reconstruction efforts in Afghanistan and Iraq. The office's functions include representing the DoD "in interagency foray on pertinent operational matters."[58] One might reasonably question, though, whether the DoD position, as reflected in the NDS and various other implementing documents, represents a commitment to civilian-military jointness in post-conflict operations, or a continuing reluctance on the part of the military to engage in such operations.

After the first year of S/CRS's existence, moreover, it remained unclear whether the office is properly situated within the State Department, and the U.S. government more broadly, or whether it will be sufficiently funded to perform its tasks and meet its mission of building civilian capacity and coordinating post-conflict operations. Housing the office within the State Department may ultimately serve as a barrier to truly joint action, as the office may lack the necessary clout within the interagency system to enforce jointness, and other agencies may be willing to come only so far toward facilitating the work of a State Department office. Congress has been reluctant, to date, to open its coffers to S/CRS, as they continue to be drained by costly engagements in Iraq and Afghanistan. A congressional willingness to fund the creation of an Active Response Corps, however, would be an important early indicator of whether S/CRS will have the tools it needs to succeed.[59]

Within USAID there has been a growing acceptance of the need to focus its assistance and programming on rebuilding failed states and stabilizing war-torn societies. Gone are the days when assistance was largely considered an apolitical enterprise, focused on large infrastructure projects, or primarily on children, education, and health. USAID's January 2005 "Fragile States Strategy" makes the argument that the president's 2002 National Security Strategy elevated development to a third pillar alongside defense and diplomacy, and that much of the development effort ought to be focused on meeting the unique demands of fragile states. It outlines four specific elements of its strategy: better monitoring and analysis; priorities responding to the realities on the ground; programs focused on the sources of fragility; and streamlining operational procedures to support rapid and effective response. The strategy's overall goal is to reverse the decline in fragile states and to advance their recovery to the stage that transformational development is possible. Much of this work focuses on addressing ineffective and illegitimate governance, an area that USAID has considerable experience in through its work in crisis states.[60]

For its part, Congress has in its midst a number of bills to increase post-conflict capacity, but none have received the support necessary to pass into law. The Lugar-Biden bill is the most notable effort to build operational readiness within civilian agencies for post-conflict operations.[61] Other legislation introduced has included bills to: encourage the establishment of a UN Rapid Deployment Police and Security Force; facilitate the establishment of a UN civilian police corps for international peace operations; increase the U.S. capabilities to provide reconstruction assistance to countries or regions impacted by conflict; amend the War Powers Resolution to require the president to include a post-conflict strategy in his report under the resolution; establish a return of talent program to allow aliens legally present in the United States to return temporarily to their country of citizenship to participate in post-conflict activities; and improve the U.S. government's coordination in identifying and responding to weak and failing states and the conduct of stabilization and reconstruction operations. Nevertheless, nothing of significance, other than $80 million appropriated in fiscal year 2005 for the Global Peace Operations Initiative (funding for training of foreign country forces to participate in international peacekeeping operations) has been passed into law.[62]

The DoD continues to receive the lion's share of national security funding, and within the DoD, stability operations are receiving heightened attention at the operational, tactical, and service levels. Much of this at present is taking the form of a shift in training toward post-conflict scenarios. U.S. Joint Forces Command (JFCOM), which initiated the stability operations JOC, is making efforts to ensure that ongoing and future military actions reflect integrated stability operations whose tasks include protecting the stabilization force, developing and sharing intelligence, neutralizing spoilers, protecting emerging government leaders, protecting critical infrastructure, supporting reconstruction operations, and contributing to winning the information battle.[63] The JFCOM Joint Training Directorate training approach, which has included exercises to train the U.S. forces rotating in to Multinational Forces Iraq, focuses on political, military, economic, social, infrastructure, and information.[64] And the JFCOM Multinational Interagency Experimentation, J-9, has continuing "multinational experiment exercises" scheduled that are intended to develop a mature concept of operations or template for a lead-nation-neutral multinational interagency group that can be employed by future coalition actors to coordinate planning and leverage capabilities of relevant civilian agencies and nongovernment and international organizations.[65]

Initiatives on stability operations by the U.S. Army and Marine Corps also illustrate the sense of urgency behind DoD efforts. Recognition of the significance of stability operations at the Secretary of Defense level, provoked by strategic considerations, has coincided with an increased tactical focus triggered by the immediate requirement to rotate units into Iraq, and to a lesser extent Afghanistan, to conduct stability operations. Thus, stability operations training programs were developed at the joint and service tactical command level, and the overall effort continues to be informed by an information cycle that incorporates lessons learned into training almost as soon as events giving rise to the lessons occur. Moreover, current operations and lessons learned have had a profound impact on the continuing reassessment of all professional military education and training, both joint and within the services, based in part on the growing recognition that stability operations are but a part of the military's response to twenty-first-century threats.[66]

It is worth noting, however, that one of the continuing challenges associated with the strategic recognition that war is not only combat (and that strategy is not only war)[67] is figuring out how to strike an appropriate balance among legitimate, though competing core capabilities. For example, when the DoD Strategic Planning Guidance initially directed the army and marine corps to either create standing units focused on stability operations or develop the capability to rapidly assemble modular force elements with the same effect as standing units,[68] the army opted for the flexibility of its Unit of Action Brigade Combat Teams[69] by enhancing the stability operations' relevant capabilities already organic to those units and by adding to expertise from other units.[70] Indeed, in both the U.S. Army and the Marine Corps the consensus appears to be that flexible units with capabilities to conduct the full spectrum of military operations, including post-conflict operations, rather than designated reconstruction units, are the appropriate response to the increased priority of stability operations. While this conclusion is reasonable on its face, it remains to be seen whether military or service cultures' reluctance to allow stability operations into the mainstream will prove a significant obstacle to the future development of optimal military capabilities in this area.[71]

It also remains to be seen how all of these pieces will fit together in a live operation. Should S/CRS coordinate the large on-the-ground presence of the DoD and (to a lesser extent) USAID in post-conflict interventions? Will enough civilian capacity be created (and resources devoted) to shift the post-conflict center of gravity away from the military and toward a civilian approach? At what point in time should this handover take place, and what are the metrics for making this calculation?

The United States may have been remiss in failing to comprehend the complexity and significance of post-conflict operations before May 2003, but the situations that developed in Afghanistan and Iraq prompted swift reassessment. As Coalition forces began to focus on tracking down insurgents and building the capacity of locals, the United States had to redefine strategic victory beyond annihilating the enemy. In the process, it became clear that the kinds of coordination and capacity it needed for unconventional conflicts and postwar reconstruction were lacking.[72] In response, the U.S. government has begun to move with a greater sense of urgency to identify post-conflict reconstruction requirements and build capacities to fill the gaps.

It is telling that two years after President Bush's "Mission Accomplished" speech, he gave a talk at the International Republican Institute outlining the need for better postwar planning: "You know," the president explained, "one of the lessons we learned from our experience in Iraq is that, while military personnel can be rapidly deployed anywhere in the world, the same is not true of U.S. government civilians."[73] The president lauded S/CRS's efforts to create a civilian Active Response Corps to deploy as "first responders" to crisis situations, backed up by budget requests of $24 million for S/CRS and $100 million for a Conflict Response Fund in fiscal year 2006. "If a crisis emerges, and assistance is needed, the United States of America will be ready," the president stated to applause.[74] Clearly, post-conflict reconstruction has become a key component of the U.S. government's civilian and military transformation efforts, from the president on down, but there is still a long way to go before the approach has been fully integrated into thinking, planning, capacity development, and, ultimately, operational success.

## Recent International Trends

Post-conflict reconstruction operations have a greater chance of success if they are multilateral in character, capitalize on international core competencies, and represent an efficient division of labor.[75] It is obvious that U.S. efforts to rally international support for the reconstruction of Iraq were perhaps doomed from the start by a U.S.-led invasion that created (or at least accelerated) the need for such efforts. It is also apparent that the rift created impedes cooperation between the United States and international partners on post-conflict operations outside of Iraq and may pose theoretical as well as practical obstacles to international cooperation on reconstruction. Nevertheless, international post-conflict reconstruction operations and capacity building continues.

Taking May 2005 as a snapshot, the UN was engaged in seventeen peacekeeping operations involving close to sixty-five thousand military and civilian personnel. In the wake of the Brahimi report, the UN initiated a number of measures to improve peacekeeping. These measures included creating the UN Standby Arrangements System, based on conditional commitments by member states of specified resources to be made available within agreed response times;[76] and SHIRBRIG, a multinational, Danish-led brigade that can be made available to the UN as a rapidly deployable force for up to six months to support certain peacekeeping operations.[77]

As part of its reform effort, the UN is currently in the process of creating a Peacebuilding Commission and Peacebuilding Support Office that can focus on the immediate challenge of post-conflict operations. The secretary-general has followed up on the UN High-level Panel's suggestion to create new institutions that can pull together post-conflict expertise in a more coherent manner and focus the Security Council's attention on the early transition period. Questions still to be determined are whether the commission will have access to a separate peace-building fund, as well as the mechanisms under which those funds will be allocated. Negotiations are likely to extend into 2006 before the commission is fully operational.

Turning to regional organizations, the European Union (EU) defense ministers have committed themselves to establishing, by 2007, thirteen rapid reaction battle groups (of around fifteen hundred troops each) ready for deployment within five to ten days. Although the lack of strategic transport, particularly airplanes—only the United Kingdom has any strategic lift capacity—will be a continuing challenge for the EU's Rapid Reaction Force (EURRF), the force has already deployed in support of Operation Concordia in Macedonia and to the Congo; a nonmilitary deployment of EURRF formed the Police Mission in Bosnia. The EU also has a European Gendarmerie Force, the product of a 2004 agreement between EU defense ministers from France, Italy, Spain, Portugal, and the Netherlands, which will be capable of deploying a nearly three thousand strong gendarmerie worldwide (eight to nine hundred ready within thirty days; twenty-three hundred reinforcements on standby). The force is intended to support post-conflict peacekeeping, maintaining public order, and other policing duties. The EU does not contemplate sending the force to active conflict zones like Iraq.[78]

NATO also has stability operations capabilities, as was initially demonstrated its decision to assume the leadership of the International Security Assistance Force (ISAF) mission in Afghanistan in 2003. The NATO Response Force (NRF), billed as capable of conducting the full range of military missions (presumably including stability operations) is scheduled to have a full operational capacity by October 2006 with about twenty-one thousand personnel.

The NRF is not intended to be a permanent standing force; rather it is comprised of national force contributions that will rotate through periods of training and certification as a joint force.[79]

Other international organizations with post-conflict reconstruction capabilities include the Organization for Security and Cooperation in Europe (OSCE), which has asked member states to create national ready rosters, including civilian police, which would enable the OSCE Crisis Management Center to deploy Rapid Expert Assistance and Cooperation Teams of civilian and police experts to assist with pre-crisis management and post-conflict reconstruction.[80] The African Union (AU) conducted its first independent peacekeeping operation with its twenty-six hundred strong mission in Burundi, which began in 2003, and deployed an observer mission in Darfur in 2004, with initial troop strength of about three thousand. Other African activities include the Economic Community of West African States Monitoring Group, which has deployed in Liberia, Guinea-Bissau, and Sierra Leone. The African Peace Facility was established 2004 to support African-led, -operated, and -staffed peacekeeping initiatives in Africa and is expected to contribute to strengthening the capacity of the AU to undertake peace operations.[81]

The leading post-conflict capacity builders among individual states include the United Kingdom, Canada, and the Nordic countries. The UK has established a Post-Conflict Reconstruction Unit whose tasks include developing strategy for post-conflict stabilization, including linking military and civilian planning, working with the international community, and the planning, implementation, and management of UK contributions to post-conflict stabilization. Britain has also established a section responsible for Peace Support Operations at the Joint Doctrine and Concept Centre at the Ministry of Defence. The Prime Minister's Strategy Unit has also spent 2004–5 looking at instability in failed states and war-torn societies.[82] In CANADEM, Canada now possesses a national level reserve of civilians for rapid mobilization in support of national and international operations. Canada is also home to the International Association of Peacekeeping Training Centers, a voluntary association of centers, institutions, and programs dealing with peace operations research, education, and training.[83] The Nordic countries routinely contribute to various peacekeeping training programs and have established NORDCAPS, for military cooperation, including peace support operations among Nordic countries.[84] SWEDINT develops and coordinates Swedish contributions to peace support operations and has a Partnership for Peace training center.[85] South American countries, including Argentina,

Brazil, Paraguay, Chile, and Uruguay, are building capabilities, not least by leading MINUSTAH, the UN's stabilization force in Haiti.[86]

A key question that is likely to dominate debate in the coming years is whether the UN will be the single legitimate means of intervening in post-conflict countries, or whether it will be but one delivery mechanism among many—including regional organizations, coalitions of the willing, and uni-lateral intervention. The Security Council impasse over Kosovo and then Iraq, and the subsequent interventions, provide two very different models for considering whether the world community will see military engagement that transpires outside of the UN framework as legitimate and effective. As regional organizations increase their capacity to stabilize and reconstruct failed states and war-torn societies, it remains to be seen whether powers such as the United States or within Europe will seek Security Council approval before marching ahead. The more the UN is sidelined, however, the more likely it will be that interventions will be viewed by local actors within the context of "great power" or regional interests and rivalries, potentially to the detriment of the ultimate mission goals.

## The Way Forward

Post-conflict reconstruction is here to stay. While it may be unlikely that in the near term the United States will bring the full force of its military to bear on a country as it has in Iraq and Afghanistan, it is reasonable to expect that in the current global security environment, the U.S. government and its international partners will invest billions—if not hundreds of billions—of dollars in the coming years on trying to rebuild failed states and to stabilize war-torn societies. Afghanistan has shown us all too clearly the dangerous interconnection between ungoverned territory and threats to international security, including the capacity to strike at key U.S. interests.

In post-conflict operations, it is critical that mission priorities and lines of authority are clear, and that a greater jointness of planning and action is achieved. The doctrine that governs these operations must be more far-reaching than traditional military doctrine if it is to succeed. It must see war fighting as an extension of a broader policy that requires stabilization and reconstruction to help ensure security, and it must build civilian capacity to stand on an equal footing with military readiness. The more agencies, depart-ments, and even governments and regional and multilateral organizations can function as a single team with a single goal, the greater the chances of success.

Of course, operations on the ground do not take place in a vacuum, but are carried forward by enormous bureaucracies with differing perspectives, interests, and equally large and complex sets of objectives. There may always, therefore, be a high degree of redundancy and working at cross-purposes in any operation. One of the primary impediments to success in post-conflict operations, however, has been a lack of willingness on the part of the defense, diplomatic, and aid communities to conceive of the mission in a truly integrated way. This is starting to change, and a number of important strides forward have occurred within the U.S. government—as exemplified by the creation of the new Office of the Coordinator for Reconstruction and Stabilization—but the changes to date, both within the U.S. government and among our friends and partners, remain insufficient to the challenges before us.

It is time the U.S. Congress perceived stabilization and reconstruction as a core national security challenge of the twenty-first century, and passed into law the necessary legislation to catalyze the change in mind-set taking root within the U.S. government: that our national security depends on our ability to get post-conflict operations right. It is also time for the military and civilian sides of the U.S. government to take the next step forward—achieving the level of jointness, unity of command and purpose, and pooling of resources that will determine success.

There will be other Iraqs, Haitis, and Sudans, and our success in transforming these societies (whether through coercive force, capacity building, or both) will largely determine the world we live in for generations to come. There may be real limits to what is achievable, especially in the near term, but progress can be made. Success will not merely be a function of new weapons systems or the creation of interagency coordinating bodies. The military can no longer be the default option for nation building solely because it has the resources and is the most prepared for global crises. Success will be a function of our political will, our ability to work in a truly integrated fashion, and our ability to cooperate more effectively with international partners. The doctrine is being written—both within the U.S. government and internationally. It is time we did a better job of comparing notes and began operating from a single blueprint.

## Notes

1. On the same day, the President made the announcement aboard the *Abraham Lincoln* about Iraq, the Secretary of Defense made a similar statement about

Afghanistan from Kabul—presumably no coincidence. The orchestration of announcements does not make either false. Major combat operations were (arguably) at an end in each country. The simultaneous announcement does, however, suggest a wishful Administration view of the significance of the moment and may provide insight into its apparent failure to grasp the complexity of the post-conflict phase.

2. The actual quote is "war is not a mere act of policy, but a true political instrument, a continuation of political activity by other means." Clausewitz, *On War,* 87.

3. U.S. Army, *After Action Report.*

4. The four phases of military operations are deter/engage, seize initiative, decisive operations, and transition. Joint Publication 3-0, *Doctrine for Joint Operations.*

5. Gaddis, *Surprise, Security, and the American Experience,* 17.

6. Dobbins et al., *America's Role in Nation-Building.*

7. Fukuyama, *State-Building,* ix.

8. Ottoway, "Think Again: Nation-Building."

9. Dobbins et al., *The UN's Role in Nation-Building,* xxxviii.

10. UN, "Report of the Secretary-General: An Agenda for Peace"; UN, "Supplement to an Agenda for Peace"; UN, "Report of the Panel on United Nations Peace Operations"; UN, "A More Secure World"; UN, "Report of the Secretary-General: In Larger Freedom."

11. Jones, "Aid, Peace, and Justice in a Reordered World" in Donini et al., 209.

12. Ibid., 209.

13. USAID, "Fragile States Strategy"; UN, "Report of the Secretary-General: In Larger Freedom."

14. Chesterman et al, *Making States Work,* 1.

15. DoD, *Report of the Defense Science Board.*

16. Hamre and Sullivan, *Toward Post-Conflict Reconstruction,* 85.

17. Ghani, "Making Peacebuilding Work."

18. AUSA/CSIS, *Post-Conflict Reconstruction Task Framework,* 3. The State Department's new Office of the Coordinator for Reconstruction and Stabilization has adapted the AUSA/CSIS Task Framework for its "Post-Conflict Reconstruction Essential Tasks" (April 2005). S/CRS uses five pillars: security; governance and participation; humanitarian assistance and social well-being; economic stabilization and infrastructure; and justice and reconciliation.

19. Fearon and Laitin, "Neotrusteeship and the Problem of Weak States."

20. Fukuyama, *State-Building,* 40.

21. Ghani speech.

22. Fukuyama, *State-Building,* 39.

23. Ignatieff, "Nation-Building Lite," 26.

24. Ibid.

25. Carafano, "Pentagon and Postwar Contractor Support," 1.

26. DoD, *Report of the Defense Science Board.*

27. DoD, *National Defense Strategy,* 12–16.

28. Ibid., 17.

29. Ibid., 16–17.

30. DoD, Directive No. 3000.ccE. In fact, the draft Directive appears to have been viewed initially as the platform intended to elevate stability operations to the same level as major combat operations within DOD. Due to a protracted process of trying to "clear" the Directive within the Department of Defense, however, the 2005 NDS was published before the Directive, which was still in draft as of May 2005. Arguably made somewhat pro forma as a result, the Directive will provide more detailed guidance on stability operations within DOD.

31. Ibid.

32. Ibid.

33. Ibid.; Graham, "Pentagon Prepares to Rethink Focus"; Jaffe, "Defining Victory."

34. Graham, "Pentagon Prepares to Rethink Focus."

35. Jaffe, "Defining Victory."

36. DoD, Directive No. 3000.ccE.

37. Joint Publication 3-0, *Doctrine for Joint Operations (Draft).*

38. Von Hippel, *Democracy by Force,* 177–78.

39. Adams, "Force Planning and Strategy."

40. DoD, "Report of the Defense Science Board."

41. See recommendations of the following reports: CSIS, *Beyond Goldwater-Nichols;* Eizenstat et al., *On the Brink;* Orr and Forman, "Funding Post-Conflict Reconstruction," in Orr, *Winning the Peace;* DoD, *Transition To and From Hostilities;* Binnendijk, *Transforming for Stabilization and Reconstruction.*

42. Hamre and Sullivan, *Toward Post-Conflict Reconstruction,* 85.

43. Orr, "After the War, Bring in a Civilian Force"; Fukuyama, *State-Building;* Eizenstat et al., *On the Brink.*

44. Fearon and Laitin, "Neotrusteeship and the Problem of Weak States," 29.

45. Dobbins et al., *America's Role in Nation-Building.*

46. O'Hanlon and Singer, "The Humanitarian Transformation."

47. Ottoway, "Think Again: Nation-Building"; Fukuyama, "Nation-Building 101."

48. Eizenstat et al., *On the Brink;* Prime Minister's Strategy Unit, *Investing in Prevention.*

49. Prime Minister's Strategy Unit, *Investing in Prevention.*

50. Jean-Paul Azam, Paul Collier, and Anke Hoeffler, "International Policies on Civil Conflict," 2.

51. UN, "A More Secure World."

52. See the Web site for Office of the Coordinator of Reconstruction and Stabilization, http://www.state.gov/s/crs/.

53. Ibid.

54. The Stabilization and Reconstruction Civilian Management Act of 2004 (S.2127) (the Lugar-Biden bill) proposed the creation of such a fund. Since 2004, the Bush Administration has requested $100 million for a complex emergencies fund, but the Congress has so far proved unwilling to fund the request. The Administration's fiscal year 2006 budget request continues the request, at $100 million for a Conflict Response Fund that would be managed by S/CRS.

55. See the Web site for Office of the Coordinator of Reconstruction and Stabilization, http://www.state.gov/s/crs/.

56. Robinson, "When the Fighting Ends."

57. DoD, *National Defense Strategy,* 15–16.

58. Deputy Secretary of Defense Memorandum, "Establishment of the Defense Reconstruction Support Office."

59. The Bush Administration asked for $17 million for S/CRS in the 2005 Supplemental. The version of the bill passed by the House provided $3 million for S/CRS. The Senate Appropriations Committee approved $7 million, but the entire Senate approved the full $17 million request. The final conference agreement was $7.7 million. Refugees International, "Mixed Outcome."

60. USAID, *Fragile States Strategy;* UN, *In Larger Freedom.*

61. United States Congress, *Stabilization and Reconstruction Civilian Management Act.*

62. The United States has committed to funding the Global Peace Operations Initiative (GPOI) at $660 million over five years, to increase U.S. assistance for peace operations in Africa and elsewhere. "At the end of 2004, Congress provided GPOI funding in action on the Consolidated Appropriations Act for FY2005 (H.R. 4818/P.L. 108-447) . . . which provided that '$80 million may be transferred with the concurrence of the Secretary of Defense' to the Department of State Peacekeeping Operations account." President Bush and other G-8 leaders announced their commitment to the GPOI at the June 2004 G-8 summit. Serafino, "The Global Peace Operations Initiative."

63. Conlin, "DoD Concept for Stability Operations."

64. Peot, "Executive Primer."

65. Kearley, "Stability Operations Concepts and Capabilities."

66. For example, ongoing stability operations have led to immediate curriculum changes in the Marine Corps' career, intermediate, and top level schools, and to initiatives to enhance language training and create a cultural center of excellence.

67. "National Security is *more* than war and . . . war is *more* than combat and . . . combat is *more* than shooting." Pudas, "Transforming National Security."

68. The directive is contained in SPG 06-11, a classified document referred to, inter alia, in Combined Arms Center (CAC) WARNO 1 to OPORD 05-007A (Army Focus Area Stability Operations), Fort Leavenworth, Kansas, December 23, 2004.

69. Moentmann, "U.S. Army in Stability Operations."

70. Moentmann presentation.

71. Ricks, "Army Contests."

72. Graham, "Pentagon Prepares to Rethink Focus." Also see AUSA/CSIS, "Play to Win."

73. President Bush speech, May 18, 2005.

74. Ibid.

75. Dobbins et al., *America's Role in Nation-Building.*

76. See United Nations Department of Peacekeeping Operations. Available at http://www.un.org/Depts/dpko/dpko/bnote010101.pdf.

77. See SHIRBRIG Home page. Available at http://www.shirbrig.dk/ index.htm.

78. John Chalmers, "EU Reaches Battle Groups Milestone," *Reuters,* Nov. 22, 2004; EU launches crisis police force, BBC World Edition page Friday, 17 September, 2004. Available at http://news.bbc.co.uk/2/hi/europe/3665172.stm.

79. Donna Miles, "NATO Response Force Ready for Duty, Rumsfeld Says." *American Forces Press Service,* June 27, 2004. Available at http://freerepublic.com/focus/news/1161217/posts; also see NATO website, http://www.nato.int/issues/nrf/.

80. See OSCE Fact Sheet. Available at http://www.osce.org/publications/sg/2004/11/13137_53_en.pdf.

81. See International Peace Operations Association page. Available at http://www.ipoaonline.org/news_detailhtml.asp?catID=3&docID=92#top; see also African Union. Available at http://www.africa-union.org/home/Welcome.htm.

82. See Ministry of Defence, Joint Doctrine and Concepts Centre. Available at http://www.mod.uk/jdcc/pso.htm; also see PCRU page. Available at http://www. postconflict.gov.uk/.

83. See CANADEM Web site. Available at http://www.canadem.ca/about.htm.

84. See Nordic Coordinated Arrangement for Military Peace Support. Available at http://www.nordcaps.org/start.htm.

85. See Swedish Armed Forces International Centre (SWEDINT). Available at http://www.swedint.mil.se/article.php?dontaddcount=1&id=8562.

86. See United Nations Stabilization Mission in Haiti, Friday, 27 May 2005. Available at http://www.un.org/Depts/dpko/missions/minustah/.

## Works Cited

Adams, Gordon. "Force Planning and Strategy: Keeping the Horse Before the Cart." Based on a panel presentation before the National Defense University Joint Operations Symposium, *Meeting Key U.S. Defense Planning Challenges,* November 16, 2004. Available at http://www.ndu.edu/inss/symposia/joint2004/adamspaper.htm.

Association of the United States Army (AUSA)/Center for Strategic and International Studies (CSIS). "Play to Win." Washington, D.C.: AUSA/CSIS, 2003. Available at www.csis.org/isp/pcr/playtowin.pdf.

————. "Post-Conflict Reconstruction Task Framework." Washington, D.C.: AUSA/CSIS, 2001.

Azam, Jean-Paul, Paul Collier, and Anke Hoeffler. "International Policies on Civil Conflict: An Economic Perspective," December 14, 2001. Available at http://users.ox.ac.uk/~ball0144/azam_coll_hoe.pdf.

Binnendijk, Hans, and Stuart Johnson, eds. *Transforming for Stabilization and Reconstruction Operations.* Washington, D.C.: Center for Technology and National Security Policy, National Defense University, 2004.

Bush, George H. Speech to the International Republican Institute, May 18, 2005, Washington, D.C. Available at http://www.whitehouse.gov/news/releases/2005/05/20050518-2.html.

Carafano, James Jay. "The Pentagon and Postwar Contractor Support: Rethinking the Future," The Heritage Foundation, no. 958, (February 1, 2005). Available at http://www.heritage.org/Research/NationalSecurity/em958.cfm.

Center for Strategic and International Studies. "Beyond Goldwater-Nichols: Defense Reform for a New Strategic Era," March 2004. Available at http://www.csis.org/isp/bgn/docs/bgn_ph1_mb.pdf.

Chesterman, Simon, Michael Ignatieff, and Ramesh Thakur, eds. *Making States Work: State Failure and the Crisis of Governance.* Tokyo: United Nations University Press, 2004.

Clausewitz, Carl von. *On War.* M. Howard and P. Paret, eds. and trans. Princeton, N.J.: Princeton University Press, 1976.

Conlin, Christopher C. "DoD Concept for Stability Operations," Multinational/Interagency Experimentation, Joint Futures Lab (J-9), U.S. Joint Forces Command. Presentation given at PKSOI Stability Operations Conference, Carlisle, Pa., December13, 2004

Deputy Secretary of Defense Memorandum. "Establishment of the Defense Reconstruction Support Office." Washington, D.C.: Department of Defense. May 2, 2005.

Dobbins, James, Seth G. Jones, Keith Crane, Andrew Rathmell, Brett Steele, Richard Teltschik, Anga Timilsina. *The UN's Role in Nation-Building: From the Belgian Congo to Iraq.* Arlington: RAND, 2005.

Dobbins, James, John G. McGinn, Keith Crane, Seth G. Jones, Rollie Lal, Andrew Rathmell, Rachel Swanger, and Anga Timilsina. *America's Role in Nation-Building: From Germany to Iraq.* Arlington, Va.: RAND, 2003.

Donini, Antonio, Norah Niland, and Karin Wermester, eds. *Nation-Building Unraveled? Aid, Peace and Justice in Afghanistan.* Bloomfield Conn.: Kumarian Press, 2004

Eizenstat, S., J. E. Porter, and J. Weinstein. *On the Brink: Weak States and U.S. National Security.* Washington, D.C.: Center for Global Development, 2004.

Fearon, James D., and David D. Laitin. "Neotrusteeship and the Problem of Weak States," *International Security* 28, no. 4 (Spring 2004): 5.

Feil, Scott. "Building Better Foundations: Security in Post-Conflict Reconstruction," *Washington Quarterly* 25, no. 4 (Autumn 2002): 97.

Fukuyama, Francis. "Nation-Building 101," *The Atlantic Monthly* 293, no.1 (January/ February 2004): 159–62.

———. *State-Building: Governance and World Order in the 21st Century.* Ithaca: Cornell University Press, 2004.

Gaddis, John Lewis. *Surprise, Security, and the American Experience.* Cambridge, Mass.: Harvard University Press, 2004.

Ghani, Ashraf. Remarks at the Center for Strategic and International Studies, "Making Peacebuilding Work," Washington, D.C., March 11, 2005.

Goldstone, Jack, and Jay Ulfelder. "How to Construct Stable Democracies," *Washington Quarterly* 28, no.1 (Winter 2004/2005): 9–20.

Graham, Bradley. "Pentagon Prepares to Rethink Focus on Conventional Warfare," *The Washington Post,* January 26, 2005, A02.

Hamre, John, and Gordon Sullivan. "Toward Post-Conflict Reconstruction," *Washington Quarterly* 25, no. 4 (2002): 85.

Ignatieff, Michael. "Nation-Building Lite," *New York Times Magazine* 151, no. 52193, (July 19, 2002): 26.

Jaffe, Greg. "Defining Victory: As Chaos Mounts In Iraq, U.S. Army Rethinks Its Future," *Wall Street Journal,* December 8, 2004.

Joint Publication 3-0, *Doctrine for Joint Operations,* Washington, D.C.: Department of Defense, 10 September 2001.

Jones, Bruce. "Aid, Peace, and Justice in a Reordered World," in Antonio Donini, Norah Niland, and Karin Wermester, eds. *Nation-Building Unraveled? Aid, Peace and Justice in Afghanistan.* Bloomfield Conn.: Kumarian Press, 2004

Kearley, Phil. "Stability Operations Concepts and Capabilities Emerging From JFCOM/Joint Experimentation." November 4, 2004. Available at http://www. act.nato.int/organization/transformation/cde04presentations/4novstabilityopsbo/ kearly.pdf

Locher, James R. III. "Taking stock of Goldwater-Nichols." Joint Forces Quarterly (Autumn 1996): 10–17.

Mallaby, Sebastian. "Foundation for a Nation," *The Washington Post,* October 29, 2001.

Miles, Donna. "NATO Response Force Ready for Duty, Rumsfeld Says." *American Forces Press Service,* June 27, 2004. Available at http://freerepublic.com/focus/ news/1161217/posts.

Moentmann, Jim. "The U.S. Army in Stability Operations." Presentation given at PKSOI Stability Operations Conference, Carlisle, Pa., December 3, 2004.

O'Hanlon, Michael, and P. W. Singer. "The Humanitarian Transformation: Expanding Global Intervention Capacity," *Survival* 46, no. 1 (Spring 2004): 77–100.

Orr, Robert C., ed. *Winning the Peace: An American Strategy for Post-Conflict Reconstruction.* Washington, D.C.: CSIS Press, 2004.

Orr, Robert. "After the War, Bring in a Civilian Force," *International Herald Tribune,* April 3, 2003.

Orr, Robert C., and Johanna Mendelson Forman. "Funding Post-Conflict Recon-struction," in Orr, ed., *Winning the Peace: An American Strategy for Post-Conflict Reconstruction*. Washington, D.C.: CSIS Press, 2004.

Ottaway, Marina. "Think Again: Nation Building," *Foreign Policy*, (September/October 2002), http://www.ceip.org/files/Publications/2002-10-15-ottaway-FP.asp?from= pubauthor.

Pei, Minxin, and Sara Kasper. "Lessons from the Past: The American Record on Nation Building," Carnegie Endowment for International Peace Policy Brief, May 24, 2003.

Peot, Marc. "Executive Primer: Effects-Based Operations," Joint Warfighting Center (JWFC), Deployable Training Team, February 2005. Presentation provided by JWFC.

Prime Minister's Strategy Unit. *Investing in Prevention: An International Strategy to Manage Risks of Instability and Improve Crisis Response,* February 2005. Available at http://www.strategy.gov.uk/downloads/work_areas/countries_at_risk/report/ index.htm

Pudas, Terry J. "Transforming National Security." Presentation given at PKSOI Sta-bility Operations Conference, Carlisle, Pa., December 13, 2004.

Refugees International. "Mixed Outcome for Peacekeeping in Emergency Supple-mental Bill." Available at http://www.refugeesinternational.org/content/article/ detail/5736/.

Ricks, Thomas E. "Army Contests Rumsfeld Bid on Occupation," *The Washington Post,* January 16, 2005, A16.

Robinson, Linda. "When the Fighting Ends," *U.S. News and World Report,* May 30, 2005.

Serafino, Nina M. "The Global Peace Operations Initiative: Background and Issues for Congress," Congressional Research Service, February 16, 2005.

United Nations. "A More Secure World: Our Shared Responsibility," *Report of the Secretary General's High-Level Panel on Threats, Challenges and Change,* (2 December 2004). Available at United Nations Website http://www.un.org/ secureworld/.

———. "Report of the Panel on United Nations Peace Operations," (Brahimi Report) UN Doc A/55/305-S/2000/809 (21 August 2000). Available at United Nations Website, http://www.un.org/peace/reports/peace_operations/.

———. "Report of the Secretary-General: An Agenda for Peace: Preventive Diplo-macy, Peacemaking and Peace-keeping," UN Doc A/47/277—S/24111 17, (12 June 1992). Available at United Nations website, http://www.un.org/ Docs/SG/agpeace.html.

———. "Report of the Secretary-General: In Larger Freedom: Towards Develop-ment, Security and Human Rights for All," (20 March 2005). Available at United Nations Website, http://www.un.org/largerfreedom/contents.htm.

——— "Supplement to an Agenda for Peace: Position Paper of the Secretary-General on the Occasion of the Fiftieth Anniversary of the United Nations," UN Doc A/50/60—S/1995/1 (3 January 1995). Available at United Nations website, http://www.un.org/Docs/SG/agsupp.html.

United States Agency for International Development. *Fragile States Strategy,* PD-ACA-999. January 2005. Available at http://www.usaid.gov/policy/2005_fragile_states_strategy.pdf.

United States Army. *Third Infantry Division (Mechanized) After Action Report Operation Iraqi Freedom,* July 2003.

United States Congress. *Stabilization and Reconstruction Civilian Management Act of 2004,* Senate Resolution 2127, February 2004.

United States Department of Defense. "Report of the Defense Science Board 2004 Summer Study on Transition to and from Hostilities," December 2004. Available at http://www.acq.osd.mil/dsb/reports/2004-12-DSB_SS_Report_Final.pdf.

———. Directive No. 3000.ccE (Draft), February 28, 2005.

———. *The National Defense Strategy of the United States of America,* March 2005.

Von Hippel, Karin. *Democracy by Force: U.S. Military Intervention in the Post–Cold War World.* Cambridge: Cambridge University Press, 2000.

# TWENTY-ONE

# Principles for the Use of the Military in Human Security Operations

## Mary H. Kaldor*

Human security is about the security of individuals and communities rather than the security of states. Human security is usually considered to combine both physical and material security, both "freedom from fear" and "freedom from want," and is, therefore, widely described as a "soft" security approach. However, in large parts of the world, individuals and communities live in fear of being killed, tortured, raped, or abducted by both state and non-state combatant groups; the job of establishing "freedom from fear" in such places has to be regarded as a "hard security" task involving the use of the military. Because the priority is saving lives, human security operations may be more risky for soldiers than contemporary war fighting.

A great deal has been written about *whether* interventions are justified in situations where civilians are threatened by genocide or massive violations of human rights. But much less has been written about *how* such interventions should be carried out. It is often assumed that the use of military force is justifiable if the goals are worthwhile (*jus ad bellum*). Military force is often treated as if it were neutral, a "black box" to be employed when other methods of achieving a particular political goal fail. However, the methods adopted must also be appropriate and, indeed, may affect the ability to

---

*This paper draws heavily on the work of the Study Group on European Security Capabilities, which I convened. The final report is called *A Human Security Doctrine for Europe* and was presented to Javier Solana on September 15, 2004. In particular I draw on background papers written jointly with Marlies Glasius and Brigadier General Andrew Salmon respectively.

achieve the goal specified. In other words, the *how* is as important as the *why*. An intervention that is aimed at ending human rights abuses or creating the conditions for democracy has to be humane in means as well as goals, and that has far-reaching consequences for the conduct of military operations.

The use of the military in human security operations, therefore, involves quite different principles than those that apply in classic war-fighting operations, not least because the military needs to be configured in new ways. In this chapter, I will start by describing the changed strategic environment in which the military is likely to be deployed. I will then set out the central principles that apply to a human security approach.

## The Changed Strategic Environment

Human security operations refer to situations where large-scale violations of human rights are taking place. Actually, this is the case in most contemporary wars, which are quite different from what we consider to be classic war, that is, war fought between opposing sides by regular forces.[1] Contemporary conflicts are akin to the types of political violence variously described as "small wars," "limited wars," or "low-intensity operations."[2] European armies have considerable experience of engagement in such types of violence, both as a consequence of their colonial past and of the counterinsurgency wars fought after 1945. Much of this experience is relevant in contemporary situations. Ideas contained in the U.S. Marine Corps' *Small Wars Manual* (written in the 1930s) or the British doctrines developed as a result of General Templar's experiences in Malaya are clearly relevant. But some important differences exist that have to do both with the changing character of this type of warfare and, more importantly, with the changed goals of intervention.

The novel features of contemporary "small wars" can be identified by tracing the evolution of such wars since 1945. There has always been "partisan" or "guerrilla" or "irregular" warfare, but up to 1945 it was largely seen as an adjunct to a conventional military campaign. After 1945, thinkers like Mao Tse-tung and Che Guevara developed the concept of revolutionary warfare as a way of getting around massive concentrations of conventional military force. What was new about their ideas is that this type of warfare was seen as part of a coordinated military and political campaign aimed at capturing state power. The idea of revolutionary warfare was to exploit the political strength and military weakness of revolutionary groups by avoiding head-on clashes with conventional military forces and by building their political support within the population and wearing down the enemy through harassment and attrition. In wearing down the enemy, the revolutionaries tended to focus on strategic targets—important officials, communication towers, etc.—although they also engaged in terrorist actions, that is, violence aimed at creating fear and intimidation. By and large, the revolutionaries had negotiable goals: the end of colonialism, for example, the capture of state power, or social redistribution.

Since the end of the Cold War, revolutionary warfare has mutated into "new wars," where armed groups involving a combination of regular and irregular forces are engaged in campaigns usually for nationalist or religious causes. Generally, these wars take place in regions where the state is weak; thus, the aim is to capture or control parts of the state apparatus. The "new warriors" differ from earlier revolutionaries in certain important respects. Like the revolutionaries, they try to control the population, but they do so through creating fear rather than through consent. Hence, their strategy is based on violence against civilians. They might try to control the population through displacement, for example, ethnic cleansing. Or they might create spectacular events through suicide bombing. Moreover, their goals are much less amenable to negotiation—the elimination/expulsion of people of a different ethnicity or religion or the construction of a state based on an exclusive religion or ethnicity or war itself, for example, jihad.

Consequently, the key features of the changed security environment are

- Networks of combatant groups involve both state and non-state actors that are often loosely linked through modern communications like mobile phones or the Internet, and that are often difficult to distinguish from the civilian population.
- Most violence is directed against civilians rather than against opposing combatant groups.

A human security operation aimed at reducing the threat of terrorism—that is to say, violence against civilians and creating the conditions for the rule of law and democracy—has to uphold legal and democratic norms in its conduct. Such operations are somewhere between classic peacekeeping and counterinsurgency, but different from both. Classic peacekeeping operations were international and therefore did not distinguish between the citizens of different countries; that is, they were based on an assumption of equality between states or between parties to the conflict. Generally, the job of peacekeepers was to monitor cease-fires and to separate warring parties. Because their mandates tended to be quite restrictive, they were often unable to protect civilians from human rights violations. Counterinsurgency, on the other hand, was about defeating insurgents. Even though counterinsurgency doctrine often emphasised the protection of civilians or the need to find a political settlement, the priority placed on defeating insurgents often led, and still leads, to the excessive use of force and increased political polarization.

Thus, in peacekeeping operations the priority was peace, and in counterinsurgency the priority was military victory, and both took precedence over human rights. In humanitarian operations, the priority is the maintenance of human rights and the protection of individuals from threats to their security. The aim is to defend individuals rather than political power. The starting point is the equality of human life, and this means that even unintentional civilian deaths as a result of the imprecise application of force to achieve a local objective—that is, collateral damage—is not only potentially counterproductive but also increasingly illegitimate.

Human security operations in which the security of individuals takes precedence can also be described as law enforcement operations as opposed to war-fighting operations, even though some of the skills applicable to war fighting will have to be deployed. Within democratic societies, public security is about the security of individual citizens. The same approach needs to be adopted in human security operations, even though they may take place in other countries. Wars represent an interruption in the normal functioning of the rule of law, even though other laws apply, for example, International Humanitarian Law (the "laws of war"). In wars, individuals are subordinated to the collective warring parties.

Human security is about the establishment or restoration of a functioning rule of law. There is a framework of international law, human rights law, humanitarian law, and international criminal law, as well as the laws of host and sending nations, which exists to guide military behavior as well as the behavior of local people. Those responsible for causing insecurity are treated not as collective entities but as responsible individuals. This means that terrorists (they are breaking the law and effectively should be treated as criminals just like anybody else who takes innocent human life), war criminals, and drug traffickers are subject to a legal procedure. This also means capturing them—killing should be the last resort—and prosecuting them in court after consideration of the evidence—part of the due judicial process.

The example of the British Army in Northern Ireland is instructive in this respect because it is probably closest to the kind of human security operation I am proposing. Northern Ireland was a learning process for the British government because it was effectively a "new war" on British territory. When the army was initially deployed in 1969, the Catholic community perceived it as a savior. However, the army could not maintain its neutrality and impartiality because it was operating in support of the civil authority, the local Unionist-controlled Northern Ireland administration, which was, itself, part of the sectarian conflict. As a result, the army was unable to stem sectarian violence against Catholics, which contributed to the militarization of the IRA. The use of force, including the shooting of unarmed civilians on "Bloody Sunday," the dismantling of No-Go areas, and the intelligence and interrogation techniques, inflamed the situation, were criticized, and remain controversial.[3] Between 1969 and 1974, security forces killed some 188 people; 65 percent of those killed were unarmed civilians.[4] This period was the bloodiest period of the whole Northern Ireland conflict, accounting for 90 percent of all deaths—many more people were killed by Loyalist (Protestant) and Republican (Catholic) para-militaries.

After 1974, a new policy was adopted known as "normalization," "criminalization," or "Ulsterization." The emphasis was placed on police primacy in maintaining order—protecting civilians and dealing with insurgents. Captured terrorists were to be treated as criminals rather than enemies. The aim was to create space for political and economic reforms to improve the situation gradually. The job of the armed forces was to support the police. This approach lasted until the Good Friday Agreement, in April 1997, which largely ended the violence. It was an approach that did succeed in containing violence. Over the period of 1969 to 1997, as a whole, some 4,000 people were killed, 350 by security forces. But some aspects of the policy provoked controversy; for example, the legal framework was contested and sometimes ambiguous, and undercover operations by the Special Air Service where IRA militants were killed led to accusations of a "shoot to kill" policy.[5]

Despite weaknesses, the experience of Northern Ireland and the emphasis on law enforcement, criminalization, and police primacy does have considerable relevance to human security operations. Typically, these are situations, like Northern Ireland, where "the dividing line between a low-intensity conflict and an extended emergency have become blurred."[6] The aim is to develop a set of principles for the military to use in these in-between situations.

## Principles

Human security operations are usually considered appropriate for post-conflict situations and are, as in this book, relegated to the section on post-conflict. However, in the "new wars" it is often difficult to distinguish between different phases of conflict. The conditions that cause conflict—fear and hatred, a criminalized economy that profits from violent methods of controlling assets, weak illegitimate states, the existence of warlords and paramilitary groups, for example—are often exacerbated during and after periods of violence and there are no clear beginnings or endings. The situation in Israel/Palestine, for instance, was supposed to be "post-conflict" after the Oslo Accords, but reverted to being in the midst of conflict. The emphasis on means also applies to the use of force at the height of violence—the phase usually described as "conflict." How military means are used during a conflict has a critical impact on trust and legitimacy and the ability of military forces to carry out human security operations in the aftermath.

### *Human Rights*

The primacy of human rights is what distinguishes the human security approach from traditional state-based approaches. Although the principle seems obvious, deeply held and entrenched institutional and cultural obstacles have to be overcome if this principle is to be realized in practice.

In the past, interventions have been less than careful about human life partly because they took place in far-away countries with much less attention from media or from NGOs, but perhaps more importantly because of the military logic of counterinsurgency operations. In counterinsurgency campaigns, protection of civilians was (sometimes) emphasized, not so much as an end in itself but in order to undercut the insurgents' infrastructure and because the civilian population was an important source of intelligence. In other words, protection and control of the population was a means to an end, which was defeating the insurgents. Although the British Army emphasised "hearts and minds," coercive methods of control were frequently used, including resettlement, area destruction, and control of food supplies (methods adopted by today's insurgents). Moreover, the emphasis on firepower and technology by some armies—for example, the French in Algeria or the Americans in Vietnam—often meant heavy loss of life.

In humanitarian operations, protection of civilians is an end in itself. The aim is to prevent attacks on civilians and to uphold human rights, even though it might also help to separate the population from armed groups or to gain intelligence. The reason why the British intervention in Northern Ireland is a good example is because the inhabitants of Northern Ireland were British citizens (and voters) and, therefore, their protection had to come first. In effect, this principle implies that everyone is treated as a citizen.

The primacy of human rights also implies that those who violate human rights are treated as individual criminals rather than collective enemies. The establishment of the International Criminal Court, the ad hoc tribunals for Yugoslavia and Rwanda, and the special tribunal for Sierra Leone were largely predicated on the idea that seeking accountability for human rights violations only from states has done little to improve observance of human rights norms, and assigning individual criminal responsibility might be more effective. In the case of terrorism, it is equally important to realize that the perpetrators are individuals and the means of curbing terrorism should be tailored to that insight.

The implications of the primacy of human rights for the use of military force can thus be summarized as

- *Minimize all casualties.* The lives of soldiers cannot be prioritized over the lives of civilians. Of course, soldiers have a right to self-defense, and, also, there are situations where it is permissible to kill somebody who is trying to kill a third party. But any operation has to take into account the necessity to minimize casualties.
- *Protection of civilians has to take priority over defeat of insurgents.* Of course, protection of civilians might require the defeat of those threatening their security. But it is a matter of tactics as to how the priorities are established.
- *Arrest of criminals rather than defeat of enemies.* The aim should be to capture insurgent/terrorists and bring them to justice, rather than kill them.

### Political Legitimacy

The end goal of a human security operation has to be the establishment of legitimate political authority capable of upholding human security. The rule of law and a well-functioning system of justice are essential to guarantee the safety of individuals and communities. Legitimate political authority does not necessarily need to mean a state; it could consist of local government or regional or international political arrangements like the international administrations in Bosnia Herzegovina or Kosovo. Since state failure is often the primary cause of conflict, the reasons for state failure have to be taken into account in reconstructing legitimate political authority.[7]

A legitimate political authority can only be established on the basis of a political process recognized as legitimate by the local population. This means that the local support and consent for human security missions is critical to their success. It also means that the role of the military in a human security operation is, therefore, to stabilize the situation so that space can be created for a political process. This is much more important than winning through military means alone. In some cases, military victory may simply be beyond reach—every excessive use of force further inflames the situation. In other cases, short-term military victory can be achieved, but the cost in terms both of casualties and political legitimacy is too high. Israeli forces, for instance, have succeeded in slowing down the rate of suicide bombing, but this has not led to any resolution of the conflict; indeed, it has only inflamed more passion on the Palestinian side. Military victory may mean that stability can be sustained only through massive repression and coercion.

To ensure that the goal of restoring political authority is kept at the forefront of any operation, there must be clear political authority over the command and control of human security missions. This means that there needs to be a close linkage between policy makers and those on the ground, with the former having ultimate control over operations. Someone with a sense for the politics of both the sending states and the host society, with easy access to policy makers, and with a receptivity to local political actors should lead human security missions. This will guarantee a close and iterative linkage between policy and operational strategy, and the choice of ways and means to achieve the right effect on the ground. Politicians have to understand the political effects or consequences of deployment; likewise, those that are sent on missions should understand the political objectives they are working to achieve so that they can design their strategy, with the right set of effects being realized on the ground. Of course, this is a point that has always applied in warfare and has been emphasized by many of the great military strategists, including Clausewitz. But it is easy to neglect once the logic of deployment takes over, and it is not always integrated into actual operations, especially on the military side.

To summarize, the principle of political legitimacy implies

- *"Hearts and minds."* Treating local people with respect and dignity is a crucial element of a human security operation.
- *Stabilization rather than victory.*
- *Clear political authority over all operations.*

### Legality

Legality is closely related to legitimacy. First of all, any human security operation has to have a legal foundation in international law. This normally means that it has to take place within a multilateral framework on the basis of formal approval by multilateral institutions—normally the United Nations.

Second, the human security operation has to be carried out in accordance with the law and in support of the establishment of law and order. Thus, unlike classic wars where the main legal guidelines are to be found in International Humanitarian Law ("the laws of war") that applies largely to states and to armed opposition groups, this means that armed forces also have to act within the framework of some kind of domestic law that applies to individuals. In authoritarian states, the law may not be in accordance with international norms, and in failed states, the law may have broken down. This is why human security operations need to operate within some kind of legal framework that specifies the priority given to different bodies of law and what to do when legal regimes conflict.[8] At present, no such framework exists, so there is an urgent need to develop such a framework at the international level. Soldiers need to be actively involved in support of the police and civil authorities, and for this, they need to know something about the legal framework in which they are operating; for example, they need to be trained in gathering evidence necessary to secure prosecutions and convictions in court.

## Intelligence and Information

Gathering and disseminating accurate information is key to the success of all operations. In human security operations, the best source of information is the local population. It is through good knowledge of local practices that it is possible to identify insurgents or those who assist them. There needs to be an ongoing process of consultation and dialogue with people on the ground for early warning, prevention, learning, and feedback during deployment and for the measures needed to ensure redundancy of missions. Human intelligence based on engagement with local people can be supplemented by other intelligence methods (technology and espionage) but should increasingly be considered the centerpiece of intelligence.

By the same token, the local population needs to be clearly informed about the purpose of the mission and any particular operation. In general, coordination and trust are best served by a principle of openness and sharing of information, unless there are very specific reasons for secrecy, for example, the arrest of war criminals.

Both intelligence and information imply an important role for the media. The media should not be considered a difficult part of the strategic environment. On the contrary, the media role is critical in explaining the mission, in providing a forum for local engagement, and in accurately reporting, which facilitates the accountability of military actions—something critical for legitimacy and trust.

### Civil-Military Coordination

Clearly, cooperation with police forces follows from the emphasis on the role of the military acting in support of the establishment of law and order. But cooperation with other civilian experts is also important. These may be legal experts—judges, for example, or forensic specialists. But they may also include humanitarian agencies, development experts, accountants, and administrators.

Although this paper focuses on physical security, or "freedom from fear," in contemporary conflicts it is often difficult to distinguish between physical and material security, or "freedom from want." Typically, combatant groups are financed through criminal activities—looting, pillaging, illegal checkpoints, hostage-taking and kidnapping for ransoms. Sometimes, ordinary criminals join insurgent groups as a cover for their illegal activities. Moreover, combatant groups may control the provision of humanitarian assistance through their own NGOs.

One reason why the local population provides support to combatant groups is because it is the only way to survive. In contemporary conflicts, unemployment and displacement are often very high, and young men may have no choice but to join criminal gangs or paramilitary groups.

This is why a range of such tasks as generating legitimate ways of making a living, institution building, the provision of public services, and support for civil society are all necessary elements of human security operations. Human security missions need to include people who can help undertake these tasks—police and civilian experts. There also needs to be proper coordination; both military and civilian actions need to come under a common political authority.

Of course, the composition of any human security mission will vary according to the mission. Some missions, in very volatile situations, need to be very robust and to emphasize the hard end of security. In others, for example, prevention of conflicts, a security presence might be largely civilian.

## Conclusion

What is new about the principles above, in comparison with either principles for classic war fighting (operational art) or counterinsurgency/small wars is the primacy of human rights and the emphasis on the rule of law. Although for the time being human security operations are likely to be considered a subset of other types of military operations, in my view, these principles are applicable to the main contingencies in which military forces are likely to be employed. This is because in much of the world there is no clear distinction between war and peace. The kind of violence that is experienced today in Iraq or Afghanistan or in large parts of Africa or Central Asia is much more common than the kind of large-scale violence involving regular armed forces that was typical of the twentieth century and is increasingly considered unacceptable by global public opinion.

These principles have several implications for military organization and culture.

First of all, the military have to be prepared to take on new tasks. They need to be able to prevent looting, for example, when there is an insufficient police presence. They need to be ready to help with reconstruction tasks or to accompany children to school or local women to the market when the situation is dangerous. They need to be able to help the police gather evidence or even to do this themselves. Many military planners are nervous about what they call "mission creep"—going beyond what are perceived to be military tasks. What is needed is a redefinition of the "mission" to include those elements that are essential for a restoration of law and order and for human security.

Nor is it merely a matter of undertaking new tasks. It is also about the qualities of the individual soldier, that is, his or her mental training and ethics. It is fashionable to argue that the warrior ethic has been replaced by more mundane occupational and individualistic incentives and that soldiers are more risk averse than in earlier periods.[9] This type of operation is very challenging and may indeed require many of the traditional attributes of soldiers—"offensive spirit," courage, heroism, discipline, or strength—though these may need to be combined with such traits as patience, intelligence, sensitivity, independent responsibility, and flexibility that are often considered more the preserve of civilian occupations.

Second, these operations will also have to involve civilian and police capabilities, which have been created, trained, and exercised with the military. Again, much of the contemporary literature emphasizes the "civilianization" of the military. But this usually refers to the use of civilians in logistics and support. I am talking about development experts, humanitarian aid agencies, civil society specialists, legal experts, and so on. These joint security and reconstruction teams—what are often described as "nation-building" teams—must be used to working together and be prepared for deployment.

In a number of countries, considerable thinking is taking place about how to bring about these changes. The need for military-civil cooperation and for stabilization capacities is currently on the agenda of both NATO and the European Union. Several individual governments have established interagency units to plan this type of operation, for example, the Post-Conflict Reconstruction Unit in the United Kingdom and the Office of the Coordinator for Reconstruction and Stabilization in the United States. But these are still considered add-ons to the primary combat missions of armies, and the principles of human security have not yet been fully absorbed. Moreover, as I argued above, human security applies to all phases of conflict, not just the aftermath.

Finally, politicians are going to have to change the way they perceive the role of the military. The implementation of this vision of security will demand considerable political will: a willingness to put soldiers in situations where the protection of innocent lives comes before force protection. In this kind of intervention, a soldier risks his or her life to save others, and although these interventions are limited, they may be more risky than, for example, long-distance air campaigns. In fact, the public may be more prepared than politicians realize for this type of operation. Many individuals themselves volunteer to work in human rights organizations and are prepared to risk their lives. The growth of a worldwide media has, of course, affected public attitudes and has raised awareness and concern about human rights violations in other parts of the world.

## Notes

1. See Mary Kaldor, *New and Old Wars: Organized Violence in a Global Era.* (Cambridge U.K.: Polity Press, 1999).

2. The term "small wars" comes from the classic text by C. E. Caldwell, *Small Wars: Their Principles and Practise,* first published in 1896. See also Max Boot, *The Savage Wars of Peace: Small Wars and the Rise of American Power* (New York: Basic Books, 2002). The term "limited wars" was used in the 1960s and 1970s to deal with threats that were not nuclear. See Robert Osgood, *Limited War: The Challenge to American Strategy* (Chicago: University of Chicago Press, 1957). "Low-intensity operations" was the term fashionable in the 1980s. The classic text was Frank Kitson, *Low Intensity Operations: Subversion, Insurgency, Peace-keeping* (Harrison, Pa: Stackpole Books, 1971).

3. Peter Pringle and Philip Jacobson, *Those are Real bullets, Aren't they?* (London: Fourth Estate, 2000).

4. Fionnuala Ní Aoláin, *The Politics of Force: Conflict Management and State Violence in Northern Ireland* (Belfast: Blackstaff Press, 2000).

5. For example, the procedures used to deal with captured insurgents represented considerable modification of normal law and the notorious 'Diplock' Courts did away with juries and allowed confessional based evidence, which accounted for the majority of convictions. Also, the IRA always insisted that they were political and not criminal and the hunger strikes in 1981 to achieve political status in prison mobilized considerable political support for the IRA. For accusations of a "shoot to kill" policy, see, for example, Peter Taylor, *The Brits: The War against the IRA* (London: Bloomsbury, 2001).

6. Ibid., p. 225

7. See Herbert Wulf, "The Challenges of Re-Establishing a Public Monopoly of Violence" (forthcoming 2005), in Glasius and Kaldor, ed., *A Human Security Doctrine for Europe: Project; Principles; Practicalities* (Oxford: Routledge, 2005).

8. See Christine Chinkin, "An International Law Framework for a European Security Strategy" in Marlies Glasius and Mary Kaldor, eds., *A Human Security Doctrine for Europe: Project; Principles; Practicalities* (London: Routledge, 2005).

9. See for example, Morris Janowitz, *The Professional Soldier: A Social and Political Portrait* (Chicago: University of Chicago Press, 1957); Charles Moskos, "Towards a Post-Modern Military: The United States as a Paradigm," in Charles C. Moskos, John Allen Williams, and David R. Segal, eds., *The Postmodern Military: Armed Forces after the Cold War* (Oxford: Oxford University Press, 2000); Christopher Coker, *Humane Warfare* (London: Routledge, 2001).

# TWENTY-TWO

# The Role of Nonlethal Weapons in Future Military Operations

JOHN B. ALEXANDER

War is about application of force to impose your will on the enemy. That does not necessarily mean killing the adversary. Carl von Clausewitz noted that in his landmark book, *On War,* first published in 1831.[1] Much earlier, Sun Tzu wrote that it was better to keep a nation or army intact than it was to destroy it.[2] Even so, after more than two millennia, that notion is still contrary to what many people believe. Over time, however, the efficacy of controlled application of violence has been proven repeatedly.

During the past decade nonlethal weapons (NLWs) have entered into the war fighter's vernacular, albeit on a limited scale. We will know they are truly inculcated in military operations when NLWs cease to have a separate identity. They will be known simply as weapons available for the commander's use. Only then can they assume their rightful spot on the force continuum.

## No Road to Hell

From inception of discussions pertaining to NLWs, both semantics and the diversity of technologies have plagued all who have contemplated the subject. Nonlethal is far from a perfect word to describe these weapons, as it is inevitable that a few people will die when they are used. It is the intent of the user that is germane. Also, they span a wide range of technologies and thus do not fit concisely into conventional predesignated categories. The most important contribution that nonlethal weapons may make is not the plethora of unique systems. Rather, it is a generic conceptual innovation associated with providing commanders with new options that facilitate mission accomplishment without the long-term consequences associated with collateral fatalities. In short, it is not the weapons themselves but how we think about them that is important.

When NLWs first were formally introduced into the military's lexicon, several dire predictions were made about the effect such weapons would have, both militarily and politically. None have happened. Nearly every study or article concerning nonlethal weapons devolves into emotionally laden discussions. Frequently devoid of facts, those discussions do little to advance the field. Therefore, I have taken this opportunity to address those issues and set them aside so that readers can focus on the important issues related to the weapons and their use.

First and foremost is the inevitable semantic discussion as to the efficacy of the term *nonlethal weapons.* The reader should be assured that every aspect and nuance of the terms has been thoroughly debated and there is nothing new to be added. While an imperfect word, *nonlethal* has stuck for military applications. It is recognized that some level of fatalities will occur with any system devised by man. The important difference between lethal and non-lethal systems is that NLWs are designed with the *intent* of allowing application of force while minimizing the probability of fatalities.

Among the most prominent claims was the *slippery slope* argument. That argument inferred that NLWs would cause more wars because the decision to enter would be easier. This line of reasoning suggests that elected political leaders would abdicate their responsibilities and go to war because of the availability of NLWs. However ludicrous that might seem, it has been espoused in documents from several august bodies.[3] Since their fielding, there have been two major conflicts: Afghanistan and Iraq. There have been internal or border conflicts in or between more than forty countries. At least a dozen countries have experienced significant terrorist attacks. Finally, four major nation-state threats are currently on the horizon: India-Pakistan (a perennial problem); North and South Korea (always volatile); China and Taiwan; and the nuclear aspirations of Iran. While NLWs have been employed in several conflicts, they have neither caused nor exacerbated any of them. Therefore, based on reality, the slippery slope issue is demonstrably moot.

Another complaint was that NLWs would start a new arms race.[4] This argument too pales in the light of reality. The total military NLW arms budget for the entire world is about $50 million (U.S.). In 2004, the largest share— $44 million—was allocated by the United States. When compared with other arms development and operational expenditures, the cost of NLWs is nearly insignificant. As a few examples

- Operations in Afghanistan and Iraq cost $5 billion per month
- Missile defense (which is even more controversial) is $9 billion annually
- One F-22 fighter aircraft costs about $300 million (six times all NLWs)
- Ten M-1 tanks (less than a single company) is about the same as all NLWs
- One nuclear carrier costs $4 billion to build and $22 billion for the life cycle
- The US R&D budget is $16 billion, thus NLWs are 0.3 percent of that segment
- The U.S. Defense Department budget is about $380 billion, thus NLWs are 0.012 percent of the total

All of these data clearly indicate that no NLW arms race has developed. Further, it would require massive increases in funding for NLWs before any arms race might emerge. That simply has not happened nor is it likely to in the future.

Yet another ubiquitous issue is that NLWs may be used as instruments of torture. It has been vociferously argued that Tasers, and other electrical shock systems, bear some special responsibility for their potential misapplication. However, this discussion misses the entire point about torture. In fact, any device can be used to torture a person. The most frequently used torture device is a cigarette, but we do not hear cries to remove them from the market (although we should for health reasons). North Korea is known for using a simple hammer to crush knuckles, but there is no ban on hammers.

Many other common devices are used for torturing victims. These include canes, whips, and burning, cutting, or sharp tools. Even water is used effectively as are sound and light. Hand-crank generators and telephones long preceded Tasers for use in torture by electrical means. The point is simply *human intent is the key issue.* NLWs should not be viewed differently from any other inanimate objects, all of which may be improperly used to torture victims. It is the person, not the instrument, who is to blame. Therefore, this issue too should be removed from consideration of NLWs.

There are no perfect weapons. It is true that some serious injuries and deaths have occurred commensurate with the use of certain NLWs. However, the number of deaths has been greatly exaggerated by the media and opponents of NLWs.

As an example, Amnesty International repeatedly claims that seventy-four (now more) people have died after being shocked by Tasers. What they fail to mention is that autopsies conducted by independent medical examiners only very rarely (three cases) include the Taser as contributing to the death. Further, they have NEVER reported that the Taser was the primary cause of death. Most frequently, the people have died from a large overdose of drugs. It was the overdose that often caused the bizarre behavior that led the police to use the Taser in the first place. People die after reading newspapers and watching television news, but that does not imply a causal relationship.[5]

There have been similar misleading reports about the use of oleoresin capsicum (OC), better known as pepper spray, causing deaths. There is one known case in which a suspect died as a result of police use of OC. In that case, it was the accelerant used to propel the OC, not the pepper spray itself, that actually caused the fatality. The reports attributing in-custody deaths to OC are quite simply false. In all cases, death occurred for another reason, such as positional asphyxiation, in which the individual was hog-tied after the arrest had been made. That restraint procedure has been abandoned and OC was not a contributing agent.[6]

It is well known that there were seventeen fatalities from "rubber bullets" used by the British in Northern Ireland. Those deaths all occurred between 1973 and 1994. During operations in that country more than 125,000 baton rounds were fired. That puts the probability of death at 0.00014 per round fired. While extremely few in number, the deaths did cause a great deal of bad publicity. In 1994 a new round and launcher was induced into the field. Since that time, many thousands of baton rounds have been fired without a single death. That clearly indicates that good systems can be improved.[7]

It is extremely important to note that the vast majority of deaths or serious injuries come from inappropriate use of NLWs. Most frequently, this is a training problem. The importance of adequate training prior to fielding cannot be overstated. The U.S. military, and several NLW manufacturers, has designed extensive training programs. These always should be used prior to placing the weapons into service. Then, once they are deployed, rules of engagement must be firmly established, followed by constant supervision from senior leaders.

To put NLW deaths in perspective, it is useful to review other unfortunate incidents. For example, during the same twenty-year period that the 17 rubber-bullet deaths occurred, 449 children, ages three to six, died from misuse of party balloons. In fact, 17 of those deaths happened in 1987. Despite this number of accidental deaths, there is no outcry to ban balloons. The comparison with swimming pool drowning is even starker. In the United States alone, each year an average of 1,150 children under the age of fourteen die in pools, while nearly 5,000 more are hospitalized from near drowning. While the media and many organizations urge caution, none suggest that swimming pools should not be used.

Some people suggest that NLWs do not have sufficient data to support their use. There will always be room for a better understanding of the effects of NLWs. But that is true of nearly every technology or human endeavor. While a great deal of data has been gathered about the effects of firearms over the past two hundred years, scientists and medical specialists continue that process. Certainly the potential to learn more about lethal weapons has not stopped employment of them. The same should be true for NLWs. While it is essential that the fundamental effects be understood before any NLW is deployed, there are sufficient data to support those that have been accepted to date. Contrary to assertions by some opponents of NLWs, a considerable amount of scientific measurement has been made and the effects are fairly well known. That information and evaluation-collection process will continue as long as NLWs are being used. The results of those studies will be used to improve future systems.

There are a few additional issues that the reader should keep in mind when considering NLWs. The most important is, "Compared to what?"[8] When addressing the development and use of NLWs, the yardstick for measurement should be comparing them against the alternative weapons or options available. Too frequently opponents of NLWs compare them to an altruistic absolute marker such as zero deaths. Some argue that the use of pain for behavior modification is unacceptable. Unfortunately, that often leaves officials with two equally unacceptable alternatives: do nothing or use deadly force. Therefore, it is essential that these discussions be bounded by practicality, not theoretical hyperbole.

As seen earlier, many of the early theoretical arguments against certain NLWs have been disproved. The common factor in all of these is the emotional basis driving them. Proponents discuss facts as a counterbalance. But facts have not always prevailed in the debate. There is an unmitigated need to distinguish between facts and emotion when addressing these concepts.

Political correctness abounds in international discussions concerning NLWs. The issues focus on artificially imposed legal constraints based on anachronistic views of technology. Recent world events have demonstrated conclusively that reevaluation of these constraints is both necessary and over-due. Strict adherence to outdated regulations will result in the unwarranted deaths of many innocent people.

## The Road to Peace

Peace support operations, including intervention in Somalia, Haiti, and the Balkans, brought requirements for nonlethal weapons to the fore. For years many military leaders have eschewed such missions. Even President George W. Bush, when running for office the first time, indicated that America should not get involved in nation building.[9] But the reality is that however distasteful to our leadership, peace support operations have become an essential part of our ability to project power. And, they will continue to play a significant role in the future.

It is easier to say we won't get involved in foreign internal disturbances than it is to avoid accepting action or responsibility for inaction. Such was the case in Rwanda in 1994, when the world ignored the genocide of nearly a million Tutsis at the hands of their fellow countrymen who happened to be Hutus.[10] Years later, President Bill Clinton apologized for our lack of intervention at a point when minimal force could have saved thousands of lives.[11] Public awareness of the importance of having emergency-intervention capability was raised significantly with the release of the Academy Award–nominated movie *Hotel Rwanda*. This movie depicted the violence of that period. Less clear was how a small United Nations presence was able to maintain local pockets of calm until UN staff were ordered to leave the area.[12] It is highly likely that a few companies of light infantry would have made a dramatic difference. In such operations, having nonlethal weapons, in addition to traditional lethal capability, would have gone a long way in changing the outcome. Clearly, in the future there will be requirements for U.S. intervention in peace support operations. As such, NLWs will be an important part of the force mix deployed.

## The Road to Baghdad

Operation Iraqi Freedom was characterized by fast-moving, extremely violent action. In the initial phases of the conflict there was rapid transition from all-out attack to stability operations. This rapid alteration in mission proved former U.S. Marine Corps Commandant Gen. Charles Krulak correct when he had described it as the *Three Block War*.[13]

## Umm Qsar

The liberation of Umm Qsar is one example. Located just a few miles from the Kuwaiti border, the city fell to Coalition forces within a few hours. Quickly, operations were under way to clear the area of mines and open the port facilities located there. That was very important, as Umm Qsar is the only deepwater port in Iraq. Operating from that port would mean the Coalition had the ability to bring supplies directly into Iraq without transiting another country.

Located so close to Kuwait, Umm Qsar had been left without adequate supplies for some time before the invasion. Therefore, almost immediately humanitarian efforts were needed to feed the local people and provide other basic services to the community. A problem that quickly arose was security for troops who were handing out supplies. Starving people are hard to control. Great care must be taken to protect both the recipients and the providers of essential supplies. This would have been an ideal situation for use of NLWs. Instead, the world saw near chaos as television crews filmed desperate people swarming onto ill-prepared supply vehicles.

## Najaf

On April 3, 2003, a remarkable event took place in Najaf. The 2nd Battalion, 237th Infantry of the 101st Airborne Division entered the city with the mission to contact Grand Ayatollah Ali a-Sistani. The purpose was to assure him and the local residents that the U.S. military would respect the sacred places located there. However, the local population heard a rumor that the forces were coming to take over the sites and arrest the leaders. Quickly, an angry mob of several thousand people filled streets and blocked the advance of the troops.

The battalion commander, Lt. Col. Chris Hughes, made a quick decision that made television footage around the world. Instead of forcing the advance, he ordered his troops to "take a knee" and point their weapons downward. He also encouraged them to smile even though they were directly facing an obviously hostile crowd. Then, Colonel Hughes had his troops tactfully depart the area. His actions that day won praise from President Bush, who said Colonel Hughes handled the situation with "skill and honor."[14]

Colonel Hughes had very few options available to him in Najaf. His battalion had overwhelming firepower, but direct confrontation could easily have devolved into a massacre, a very public one. Crowd-control options were limited to intimidation or use of helicopter rotor wash to move people along or confrontation. In this case, tactical disengagement was the best option because of the limited toolkit available at that time.

### Baghdad Museum

Dawn on April 5, 2003, saw the first Thunder Run into Baghdad, a city of nearly five million inhabitants. A relatively small force of thirty tanks and fourteen Bradley Fighting Vehicles dashed through the city taking the defenders by surprise. For the next few days intense fighting swept through the area as the 2nd Brigade, 3rd Infantry Division (Mechanized) fought off fanatical resisters, many of whom conducted suicide attacks against the U.S. forces. Less than a thousand U.S. combat soldiers breached the city's defenses and held on to key positions until reinforcements arrived. However, during the fighting the brigade's command post was severely damaged, units were surrounded, and fuel and ammunition ran dangerously low. The fight was far tougher than what news reports led the outside world to believe.[15]

Barely had the fighting subsided when large-scale looting began. Among the places hit was the famous Baghdad Museum, which contained thousands of rare artifacts spanning eight thousand years of the history of mankind. Precious material from the Sumerian, Babylonian, and Assyrian cultures had been housed there; the museum was world renowned for its unique collection. In short order the vaults were gutted and a furor heard around the world. Many historians and archeologists blame the United States for not adequately preparing to protect the museum once it had been captured. They stated that the U.S. forces should have been aware of the cultural value of this site.[16]

In response, military spokespersons noted that the size of the force did not allow for diversion of resources. Military planners were sensitive to the value of the museum, and care was taken not to damage it during the bombardments. The museum had been closed to the public for several years, and there was doubt regarding whether or not Saddam Hussein's people had already removed the most valuable items in the collection. While many artifacts since have been recovered, a considerable number are still missing.

Operational realities prevented deployment of any troops to the area. There is, however, an NLW alternative that could have preserved the integrity of the building. Foams are available that could have been used to seal the entrances, and even inundate exhibits if necessary. The operation would have taken less than an hour to perform. The cleanup process would have been long and painstaking, but in the end, the artifacts would have been accounted for, if they were still located in the vaults when the attack took place.

What the operations at Umm Qsar, Najaf, and the Baghdad Museum have in common is that they point toward a requirement to have NLWs more readily available. Without distracting from the exigencies of the combat mission, prepositioned packages of NLWs should be placed much farther forward in the supply chain. We now know that military planners expected Coalition troops to be greeted as liberating forces, not conquerors. While support did not emerge as anticipated, the expectation should have suggested that NLWs would be needed very quickly once areas were secured. Under any set of circumstances, it was reasonable to anticipate that maintaining order would be a difficult task, and the long-term goal of winning local support begged for minimizing collateral casualties.

### The Green Zone

After suppressing military units in Baghdad, Coalition forces established the Green Zone from which they hoped to operate with a relative degree of safety. Suicide attacks commenced during the initial fighting and have never ended. A favorite tactic is for a driver to bring an explosive-laden vehicle as close as possible to his target and then detonate it. This low-tech precision-guided munition has proven to be very effective, and there does not seem to be a problem in recruiting volunteers for these martyrdom missions.

Of course, attempts have been made to prevent suicide bombers from getting to their objective. Barricades are routinely established some distance from the entrances. Still, high-speed drivers seem willing to risk the attack. The usual response is to open fire with small arms, which results in the death of the driver, and/or premature detonation. In some cases, however, the speeding vehicles were not bomb laden, and the shooting led to the deaths of innocent passengers. These incidents attracted media attention and caused further distrust by Iraqi civilians.

These attacks are not unlike the December 12, 2000, bombing of the USS *Cole* as it was docked in the harbor in Yemen. Though ship defense is the responsibility of the host nation, there were few options available to the crew. They were not allowed to use deadly force until hostile intent was demonstrated. Unfortunately, the intent came in the form of a massive blast that killed seventeen American sailors and wounded thirty-nine others. In response, several studies were conducted, and that incident was one focal point for the Naval Review Board study of nonlethal weapons.[17] The studies led to substantial changes in force protection for ships in foreign ports and recommendations for new NLWs to be developed and deployed.

For point defense, both on land or at sea, there are several nonlethal weapons that can be effective, and more will be added. A major concern has always been the ability to warn intruders and determine their intent before they get close enough to cause damage. For fixed road locations, the Portable Vehicle Arresting Barrier system has been fielded. This retractable system can be activated remotely and acts as a snare for speeding vehicles. New acoustic capabilities have been developed by American Technology of San Diego, California, that allow projection of a very narrow beam with high clarity. Once employed, there is no doubt that the persons approaching have been warned. The volume can be quickly increased to intolerable levels. A more advanced weapon is the Active Denial System (ADS). This is a millimeter-wave weapon that can project a beam for hundreds of meters. The ADS penetrates the skin only slightly, but it causes instant and intense pain. It will prove very effective in point defense situations or in creating an invisible barrier where needed.[18]

## The Road Ahead

The creation of both the Department of Homeland Security and a new Northern Command in the Defense Department are an acknowledgment that our concepts of national security have changed radically. No longer do we view the enemy solely from a foreign perspective. Now we are dealing with the realities that some battles will take place on American soil. As roles and responsibilities are forged, a previously discerned pattern will become even clearer. That is, military operations will often look more like law enforcement while, conversely, police interventions have an increasingly military texture. The difficulty in early discrimination between terrorist events and violent criminal activity ensures increasing juxtaposition of enforcement options. Such situations will exacerbate existing tensions between the need for security and the protection of individual rights.

The adverse effects of the attacks on September 11, 2001, were not limited to the casualties at the World Trade Center in New York and the Pentagon in Washington, D.C. In fact, some terrorism experts believe that the main target of those attacks was the American economic system. Obviously, collapse of the economy did not occur. Still, the immediate impact was severe in many sectors of our business world, including airlines, hotels, and other travel- and tourist-related enterprises.

When people feel threatened, they are less likely to travel away from home, especially if it is discretionary. They are less likely to participate in group entertainments, and some fear congregating in large crowded events that might be targeted by terrorists. Following the 9/11 attacks the sale of private weapons increased significantly because some people felt a need to be able to protect their homes. While these fears may not seem rational, they were grounds for millions of people to act upon.

Gun ownership in homes with small children has been a contentious issue for many years. It has been repeatedly demonstrated that kids, too frequently, are able to access the weapons, even though parents believe they are safely hidden. Tragically, every year a number of deaths occur. NLWs offer safe options that, while not related to military operations, accommodate both home security and child safety.[19]

Private security personnel in the United States outnumber sworn law enforcement officials more than five to one. Increasingly, the public is relying on these private security agencies for protection in public places. The standards for recruitment and training of personnel vary significantly. Some are fully qualified and have extensive military or law enforcement training. The vast majority of security personnel belong to agencies that define their mission as "gates, guards, and guns." They have minimal training and are often provided more for show or to take notes should an incident arise. Access to NLWs increases the ability of these security personnel to use force in a reasonable manner.

As we derive global power based on the strength of our economy, any action that contributes to those goals is beneficial. In reality, much of our economic security is tied to our emotional sense of physical security. The future dictates that security must be omnipresent and must frequently be in places where, traditionally, armed force is not appropriate. Therefore, increased availability of NLWs will likely enhance our economic status.

### Prison Riots

Prison riots are a significant problem in several South American countries, as well as elsewhere around the world.[20] Overcrowding, inadequate food, poor training of guards, and frequent commingling of visitors and inmates produce a volatile mix that often ignites. In many of those riots, substantial numbers of people have been killed. Too frequently, guards and other innocent people get caught in the middle of the altercation.

In addition, the United States now incarcerates the largest number of prisoners based on the percentage of population of any country in the world. We too have overcrowded conditions in many penal facilities that periodically lead to violent confrontations. All are not quickly put down. Occasionally, such standoffs have lasted for several days.

The U.S. military is being challenged with the necessity to hold substantial numbers of prisoners and detainees. One of the confounding issues in the global war on terror will be the detention of people in a conflict with no end in sight. American forces in Iraq have been criticized for using deadly force during prison riots.[21]

While such NLWs as Tasers, stun shields, PepperBall, and sprays are commonly used in prisons, there is more that can be done. Whenever fatalities occur in any prison the presiding officials usually are castigated. Both military guards and law enforcement jailers need enhanced capabilities to maintain order and reduce friction with the public.

### Role of Nonlethal Weapons and Terrorism

Porter Goss, Director of the Central Intelligence Agency, advised Congress that Al Qaeda and Islamic terrorism are the most significant threat to America. When questioned at the hearings, he stated that he could not guarantee that weapons of mass destruction would not be used in the United States.[22] As a nation at war, the United States must reasonably question whether or not NLWs play a role in our response capability. In fact, it is because of terrorism that I argue they are needed now more than ever. Countering terrorism also dictates that we rethink some of our fundamental biases concerning innovative weapon systems. Here, I specifically mean we must expand our ability to incapacitate groups that contain both terrorists and innocent civilians.

The present laws and treaties concerning them were drafted based on the inhumane applications of gas during the trench warfare of World War I. In general, it is easier to ban technology than it is to chastise unacceptable behavior. Today, senior officials frequently respond to the political climate by deciding that these technologies simply cannot be safely developed, much less deployed. Such restrictions, issued by administrative fiat, close down promising avenues of research and do so often without serious examination of the broader issues. In my view, these sweeping prohibitions, so tenaciously professed by politicians and legal experts, are both anachronistic and unrealistic.

We do not set our traffic speed limits based on the automotive technology of the early 1900s. Why, then, do we insist on limiting weapons development based on the technology of that period? Great advances in chemistry have been made in the past century. Just as the twentieth century was known for physics, the twenty-first century will be known for biology. Therefore, it seems highly improbable that biological weapons will be kept off the battlefield. Most ironically, it would be through development of these now-banned chemical and biological nonlethal weapons that many lives could be saved. Our politically correct and dogged adherence to anachronistic jurisprudence will result in substantial numbers of unnecessary deaths.

Terrorist attacks of the past decade have changed the geopolitical landscape and necessity for expanded use of force options. Mass hostage events, such as occurred in the Moscow theater in 2002, and the school takeover in Breslan, Russia, in 2004, dramatically emphasize the requirement for weapons that can facilitate discrimination between friend and foe via sorting after rapid incapacitation. Discussion of the efficacy of such weapons will breed debate that is urgently needed and must focus on facts, not emotion. The time for resolution of these contentious issues is now; future attacks are being planned even as this document is being prepared.

### Strategic Implications

Early intervention of NLWs, such as pulsed power weapons employed against communications targets, degradation of transportation systems, or information warfare attacks, can have strategic consequences. They clearly connote that we not only have the means to employ force, but we also have the *intent* and *will* to do so. These weapons provide senior military and political leaders with options to send unmistakable messages to potential adversaries. In the event war does break out, the enemy's capability to prosecute war will have been degraded.[23]

Timing of strategic NLW intervention is a key issue. Criticism of such techniques suggests that the use of these NLWs would precipitate the conflict. Therefore, they should be used only in situations in which conflict is likely and when the adversary is appropriately advised of the consequences that an attack might bring. As with NLWs at lower levels of military intervention, they should never be employed without adequate, overmatching lethal force in support.

Another strategic application is the ability to apply force early, with minimum probability of endangering noncombatants. This is necessary in order to quickly contain a situation while preventing it from escalating into a more significant event.

## Summary

For more than a decade NLWs have been slowly advancing both technologically and operationally. They are not panaceas, but they have been proven to be effective in several conflicts. True integration will be achieved only when they lose special identity and are routinely included in operational planning. The dire consequences predicted by opponents of NLWs have simply not come to pass. It is time to get past semantics, arguments about slippery slopes, and whether or not NLWs are inherently evil.

Military planners and law enforcement officials now know that NLWs offer use of force options across the entire spectrum of conflict. There is a need to include them as part of force packages that are readily available to forward troops as the rapid changes in the Three Block War are realized.

Finally, we must reconsider blanket bans on weapons simply because they are generically categorized as chemical, biological, or electromagnetic. When considering your position on the entire panoply of nonlethal weapons, I recommend asking the following question: If your children were to be held hostage by terrorists, what systems would you like to see in the hands of their rescuers?

# Notes

1. Von Clausewitz, *On War*, translated by J. J. Graham (London, 1908).

2. Sun Tzu, *The Art of War*, translated by Thomas Cleary (Shambhala Publications, 1988).

3. Malcolm H. Wiener, Chairman, "Non-Lethal Technologies: Military Force Options and Implications" Council on Foreign Relations, New York, 1995.

4. John B. Alexander, "Non-Lethal Weapons: Perspective and Reality," Presented to the German Bundestag, 10 November 2004.

5. Amnesty International, "United States of America. Excessive lethal force? Amnesty International's concerns about deaths and ill-treatment involving police use of Tasers." 30 November 2004.

6. John B. Alexander, *Winning the War: Advanced Strategies and Concepts for the Post 9/11 World* (St. Martin's Press, 2003), pp. 23–28.

7. Colin Burrows, "Operationalizing Non-lethality: A Northern Ireland Perspective," *Journal of Medicine, Conflict, and Survival* 17, no 3 (July–September 2001).

8. John B. Alexander, Future War: *Non-Lethal Weapons in Twenty-First Century Warfare* (St. Martin's Press, 1999), pp. 179–88.

9. Michael Ignatieff, "Nation-Building Lite." *New York Times Magazine* 28 July 2002. Available at http://www.mtholyoke.edu/acad/intrel/bush/lite.

10. William Ferroggiaro, "The US and the Genocide in Rwanda 1994." The National Security Archive, 20 Aug 2001. Available at http://www2.gwu.edu/~nsarchiv/NSAEBB/NSAEBB53/.

11. "Clinton Meets Rwanda Genocide Survivors." CNN, 25 March 1998. Available at http://www.cnn.com/WORLD/9803/25/rwanda.clinton/.

12. *Hotel Rwanda,* United Artists, December 2004.

13. Gen. Charles C. Krulak, "The Strategic Corporal: Leadership in the Three Block War," *Marines Magazine,* January 1999.

14. Ryan Chilcote, "Commander shows restraint, prevents unnecessary violence," CNN. Available at http:/edition.cnn.com/SPECIALS/2003/iraq/heroes/chrishughes.htm

15. David Zucchino, "The Thunder Run," *Los Angeles Times Magazine,* 7 December 2003.

16. Piotr Michalowski, "The Ransacking of Baghdad Museum is a Disgrace," History News Network, 14 April 2003.

17. Naval Science Board, "An Assessment of Non-Lethal Weapons Science and Technology," National Research Council, 2003.

18. Alexander, *Winning the War,* pp. 15–17, 33–36.

19. John B. Alexander, "The Taser Alternative," *The Washington Post,* 1 October 1999.

20. Bill Cormier, "Argentina Prison Riot Leaves Eight Dead," Associated Press, 11 February 2005.

21. Reuters, "US Guards Shoot Dead 4 Inmates in Iraq Prison Riot," 31 January 2005; Associated Press, "Group Condemns Fatal Prison Riot Shootings," 8 February 2005.

22. Porter Goss, "DCI's Global Intelligence Challenges Briefing," U.S. Congress, 16 February 2005.

23. Alexander, *Winning the War,* pp. 168–78.

## Part 5

# INTELLIGENCE—WINNING THE SILENT WARS

# TWENTY-THREE

# Rethinking War and Intelligence

## WILLIAM M. NOLTE

One of the aspects of war open to rethinking is intelligence. Such a review is required, first of all, because of a range of shifts in the technological and geopolitical environments in which American (and other) instruments of war and policy operate. But this rethinking must go beyond the question of intelligence as an instrument of war. As suggested by the debate over the 9/11 Commission's recommendations and legislation to reform or reorganize the U.S. intelligence community, the United States is rethinking the idea of intelligence itself. While some of this thinking suggests that intelligence can at some fundamental level be divided into military and nonmilitary intelligence, this chapter will suggest that in reality, intelligence exists to serve policy, and that the divisions of effort that accompany attempts to manage intelligence are more organizational than fundamental. In intelligence as in other areas, Clausewitz's insight into the continuity between war and policy remains valid.

As Michael Herman and others have pointed out, "intelligence," defined as knowledge or information acquired from or about another country or group—secretly, openly, or through some combination of the two—has long been a part of both the civil and military affairs of state.[1] Famously, Sun Tzu dedicated a chapter of *The Art of War* to spying, classifying spies as "native,

---

The views expressed herein are those of the author and do not necessarily reflect those of the National Security Agency, the Department of Defense, the Office of the Director of National Intelligence, or any other component of the United States government.

inner, turned, living," and—in a particularly unfortunate categorization— "dead." Beyond that discordant note, *The Art of War* makes the use of information and related topics (deception, perception management, etc.) a recurring theme. Other classic references to what we think of as intelligence deal with the secret or even the occult, as in the biblical account of Daniel and his ability to read the writing on King Belshazzar's wall.[2] But in most times and in most places, especially in the military dimension, battlefield commanders want information—with little regard for the process that provides the information, except to the degree that source or process can add to the reliability or timeliness of the information involved. If reliable information on an adversary's location (or disposition or intent) comes from a special or secret source, that is fine. If it comes from a farmer who had watched the enemy march past a few hours before or from a newspaper, that is equally acceptable. "Intelligence" did not acquire its heavily specialized association with the secret or clandestine until the twentieth century.[3]

Within the last hundred years or so, however, the equation of intelligence with information has been significantly supplanted by a definition that equates it with "secret intelligence" and the specialized organizations that collect, process, and produce such secret intelligence. The trend in this direction was accelerated with the end of the era of Great Power condominium, during which time, with limitations, observers from neutral armies accompanied those of belligerents, openly gathering and reporting "intelligence." With technological revolutions in shipbuilding and in the new technologies of wireless communication and aviation, this age of genteel openness was certain to end. The Russian Revolution, and the replacement of the czarist regime by a closed revolutionary state that saw its purpose as the destruction of the other Great Power regimes, made the development of larger, more complex "secret intelligence" organizations an imperative, even in open, democratic states.

At the start of the twenty-first century, we may be witnessing a revolution as sweeping as that which took place a century ago. To some degree, this may involve something of a counterrevolution, one that deemphasizes (though does not supplant) "secret intelligence" in favor of greater reliance on open information and expertise—even if reasons of state require that information openly available should, once compiled and translated into plans of action, then take secret form.[4] The possibility exists that in addition to, or as an extension of, a revolution in military affairs (RAM) we are in the midst of a revolution in intelligence affairs (RIA), one in which source remains a factor and in which secret sources retain value. It may also be a time, however, in which the complications and costs associated with secret sources erode some measure of their value in relation to timely and relatively unencumbered (in terms of handling restrictions) open source information and expertise.[5]

This is not to suggest that such a revolution would (or could) sweep away all continuity in military affairs, intelligence, or statecraft. No revolution ever does. The French Revolution did not produce, at least immediately, liberty, equality, and fraternity. It produced, instead—as have other revolutions—a despotism more efficient and therefore more thorough than the one it swept away. A twenty-first-century "revolution" in intelligence as a component of war—the avoidance of war, the conduct of war, or the reconstruction after war—will no doubt produce new ways to collect, process, and communicate information. Whether it will result in better decision making at the tactical or strategic level is an open question. In some cases, it will; in others, it won't.[6]

## American Intelligence, 2005

The Intelligence Reform and Terrorism Prevention Act of 2004 is now in place. By mid-2005, American intelligence will enter the Director of National Intelligence (DNI) era, superseding that of the Director of Central Intelligence (DCI). Will the reform act bear out the hopes of its supporters that a restructuring of the intelligence services will make the nation safer and make American intelligence more effective and more efficient?[7] Or will its critics be proven correct in asserting that the act will at best not improve things or at worst prove harmful to American national security?[8] Will American intelligence actively pursue the integration of traditional "foreign" intelligence into a homeland security environment? Or will we continue to think in terms of "foreign" and "domestic" categories?

The answers will probably lie somewhere between potential extremes, in one dimension. In other ways, the questions themselves and the choices they impose will appear, over time, inadequate. It should be reasonably certain, however, that the intelligence reform act will not be the last chapter in the realignment of American intelligence to deal with a volatile operational and political environment. It is a commonplace to say that the National Security Act of 1947 created the Cold War national security establishment. In reality, however, the 1947 act merely began a process that continued, even at the legislative level, through the Goldwater-Nichols Act, passed in 1986, only a few years before the collapse of the Soviet Union. Beneath that level, reform, reorganization, and general "tweaking" with the Cold War security apparatus continued to the collapse of the Soviet Union and into that conceptual void we knew as the "post–Cold War era."

Drawing from this example, the American public should not presume that a single legislative action has aligned American intelligence to deal with the security realities of the next ten, fifteen, or twenty years.[9] If anything, the operational and technical environments for American national security in the first two decades of the twenty-first century promise to be more fluid and more volatile than the United States encountered during the Cold War, and our responses—including structural responses—will need agility commensurate with those environments. The Intelligence Reform and Terrorism Prevention Act of 2004 will prove successful if it first "does no harm," and second begins a process of conscious and continuous realignment with the challenges—current, emerging, and not yet apparent—that we will face in the decade and a half to come. How should we rethink war and intelligence in this context? What issues must we face as part of a continuing process of reform and redirection, one that promises more evolution than revolution?

Let us return to an issue alluded to at the beginning of this chapter, that is, "military intelligence versus nonmilitary intelligence," "Department of Defense (DoD) versus non-DoD intelligence," and other variations on the theme. It is important that we do so by realizing that the theme itself is significantly artificial and potentially dangerous, especially if these distinctions are seen as realities (in a Platonist sense) rather than operational and organizational constructs. Once upon a time, in the communications intelligence sphere, for example, communicating entities (ours and those of our adversaries) employed "dedicated" circuits. That is to say, if you found a certain link or circuit and discovered it was conveying the diplomatic communications of the party involved, you could reasonably surmise that the content of the information was diplomatic in nature. In such an environment, an organizational structure that divided resources into "military" offices and "diplomatic" offices made sense. To a great degree, the passing of this communication environment required adjustments in the organizational response to that environment.[10]

Moreover, time and distance—more precisely "the death of distance," to use Frances Cairncross's still wonderfully evocative term[11]—have changed the information relationship between tactical users and national users. In Horatio Nelson's day, the information he "collected" firsthand was neither available to, nor would it have been useful to, in most instances, the gentlemen of the

Admiralty. By the time information they collected reached Nelson, on the other hand, it may have remained interesting, but it was almost certainly not actionable.[12]

All of this changed dramatically in the twentieth century. By the second half of that century, President Lyndon Johnson was capable of directing both strategic and tactical activity in Southeast Asia. How he did both, whether he did both, and how that affected American performance in Vietnam have been the subject of discussion ever since, some of it useful, some less so. Almost any observer of this discussion should be prepared to stipulate, however, that the ability to merge (or confuse) strategic with tactical and the ability to merge it well did not develop at one and the same time. The first was largely a technical or technological phenomenon. The second was organizational and even conceptual, and therefore, almost by definition, more difficult. One can suggest that the United States and the rest of the world's defense and security services have been struggling with this new reality ever since. Moreover, the more global one's responsibilities are, the greater the problem.

In the end, the Cold War intelligence system both succeeded (in some areas) and fell short (in others) because national and departmental intelligence, military and nonmilitary intelligence, or strategic and tactical intelligence simply do not exist as separate realities. Nor do they exist in some mythic "balance." They exist in tension, with simultaneous stresses pulling in a range of directions. At any given moment, the outcome of those tensions may seem unfair, ineffective, or even dangerous to the national interest. But a system that acknowledges those tensions and focuses on managing them will perform far better than one that pursues either/or solutions to problems and issues that call for answers that more closely resemble "both" or even "all of the above."

To cite one example, the National Security Agency (NSA) was created in 1952 after a previous effort to integrate the cryptologic services of the U.S. Army and Navy, and the embryonic cryptologic service of the Air Force, failed. Without going into details, the Armed Forces Security Agency failed not just because it did not provide sufficient integration of the military cryptologic elements, but because it provided insufficient assurance to non-DoD users of signals intelligence and other cryptologic services that their interests were being served by something called the "Armed Forces" Security Agency.[13]

For most of its history, NSA functioned as a national agency, largely dominated by a civilian professional service. Tactical signals intelligence

(SIGINT), it could be argued, was something of a stepchild, though in the context of the national strategies of the 1950s and early 1960s, it is hard to argue that NSA was out of step with a national security establishment premised on the idea of strategic nuclear retaliation. Tactical intelligence or, more specifically, tactical cryptology, was neither more nor less disadvantaged, compared to more strategic targeting and diagnostic efforts, than tactical airpower was disadvantaged compared to the needs of the Strategic Air Command.

By the 1990s, with the passage of Goldwater-Nichols and the passing of the Cold War, and with the ongoing technological revolution in computing and communications, new circumstances had come to challenge the national character of the NSA. No longer was the world divided between "strategic" communications targets using high-level encryption across dedicated circuits. By the late 1990s, NSA was describing encryption—at relatively sophisticated levels—as "ubiquitous." And more and more of the world's communications, civil and military, tactical, and strategic, governmental and private, were carried across a global information network that truly was a network. Subnetworks by the hundreds of thousands (if not millions) linked into this net.

In this environment, where does one draw the line between national communications and tactical communications?[14] The short answer is that lines can be drawn, but no one should exaggerate the degree to which they represent reality more than administrative and operational efficiency and convenience. The information that could not reach Nelson from the Admiralty can now reach latter-day Nelsons, and vice versa. The question remains, however, how much information can—and should—be acted on at the national level and how much should be held or considered within the purview of the man or woman "on the scene"? Anyone who thinks issues such as these can be fully resolved in advance, from above, and more or less "for all time" is missing one of the fundamental realities governing the relationship of intelligence and war (and intelligence and diplomacy, and intelligence and disaster relief, etc., etc.) in the twenty-first century.

## Beyond Connecting Dots:
## A New Landscape for Intelligence

"We failed to connect the dots." This phase, with competition only from "We got it all wrong," is likely to be the catch phrase used to describe the role of intelligence in the 9/11 attacks and the First Iraq War. This is true even if the accusation in the latter case would suggest too many connections based on too few dots. As in other public events and controversies, phrases—for example, Watergate: "What did the President know and when did he know it?" and the 1992 election: "It's the economy, stupid."—have the capacity to compress large and complex events into readily memorable terms.

They simultaneously contain, however, a capacity to simplify and distort issues that need at some point to be dealt with in their full complexity. The 9/11 attacks are likely for many years to come to be one of the great cases in intelligence studies. How was the intelligence on terrorist groups collected, analyzed, and disseminated to policy makers in the period before 9/11? What measure of warning did intelligence provide that terrorism could be moving from attacks on American interests overseas to attacks on the American homeland? How do we assign responsibility in cases of this sort to (a) policy makers, (b) intelligence providers, or even (c) overseers of intelligence?

Others will (and have) attempted to answer these and related questions. The purpose here is to raise the issue of whether connecting—or failing to connect—the "dots" of available information is a satisfactory, adequate, or even useful device to help us understand the role of intelligence in the failure of American government to prevent "the next Pearl Harbor" that had, for half a century, been the *raison d'être* for the instruments of American national security.

To the point, "connecting the dots" will be described here as both inadequate and potentially misleading. A discussion of why this is so may help illuminate a more useful typology for describing the process(es) by which masses of data and information come together to form intelligence. It is, to borrow a phrase, "more than the dots, stupid."

The first inadequacy of the dots image is its emphasis on the dots themselves. A century and a half ago, Leopold von Ranke dominated a generation of the European historical profession, emphasizing the view that new methods of research would permit historians for the first time to recreate the past "as it actually was."[15] This exercise in literalism (which influenced the new American graduate programs in history, including Johns Hopkins and one of its first doctors of philosophy, Woodrow Wilson) was well intentioned and provided some needed correction to excessively literary or filiopietistic schools of history. It also reflected the desire of historians and others to link with the all-powerful aura of science and the scientific method. Not for the last time, history as a profession yielded to methodological envy.

All that said, von Ranke was wrong. It is not possible to recreate the past "as it was" because the recreated past cannot exist apart from the recreator's motives and perceptions of the past. Apart from bare facts (Pearl Harbor was attacked on December 7, 1941) our efforts to experience the past take place within the construct described by John Lukacs, a more recent historian, as "historical consciousness."[16] Intelligence—for this purpose meaning the collection, processing, and analysis of data to support public policy decision making—takes place within such constructs. "What do you want to know?" cannot be separated from "Who wants to know, and why?"

At a recent international conference, an American intelligence analyst, defending the paper he had just presented, suggested with some pride that he and his colleagues worked in "fact-based intelligence." Well. Though this is certainly a preference to a professional ethic founded on "fiction-based intelligence," it comes very close to being a distant or even inadvertent descendent of von Ranke's goal of telling history "as it was." Facts are important in analysis, but they are not all that is important. Even that popular American Rankian, detective sergeant Joe Friday—he of "Just the facts, ma'am" fame—knew that building a case at some point meant the synthesis of the facts of the case with such constructs as motive and opportunity.

One could argue that such constructs are, after all, reflected in the "connecting" part of the phrase. But assembling and structuring information is more than a mechanical or linear process, involving imagination and other factors that cannot be covered within the idea of "fact based."

The points shown here

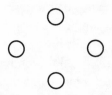

can be connected hierarchically;

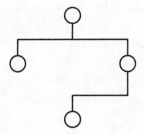

or as a network;

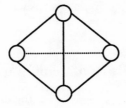

or in a number of other forms, including the traditional (and dated) intelligence production cycle.

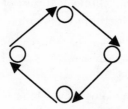

The point is that *how* the dots get connected may be at least as important as the dots themselves, and the method or template used may be less a matter of facts than a matter of perception. The elements of that perception may be technical, cultural, linguistic, or organizational. Forty years after its original publication, Roberta Wohlstetter's *Pearl Harbor, Warning and Decision* remains one, but by no means the only, of the classic studies of context and structure in information analysis.[17] Her analysis of the intelligence problem preceding Pearl Harbor as being the inability to differentiate signals from noise remains an evocative image.

At the risk of adding yet another, let me suggest that building or extracting intelligence is like placing overlays on a map. The data points (or dots) are important, as are the lines that can be used to connect them. But the map itself is not a passive or inert element in the process. It and the information it conveys are important elements in the integration of all the data.

Like it or not, the world may forever operate on the divide between "operational" successes and intelligence failures, whether the operations in question are military or diplomatic. But a richer understanding of the methodology and even the theory of intelligence should suggest that a knowledge of the "topography" on which a particular issue is being engaged is more important and more useful than merely "connecting the dots" in some fraudulent effort to reduce inherently fluid processes, frequently if not always dependent on unpredictable human choices, to dangerously mechanistic, even deterministic calculations.

## Failure of Imagination

In describing the "failures" associated with the 9/11 attacks, the 9/11 Commission described them as "failures of imagination," while further noting that "imagination is not a gift usually associated with bureaucracies."[18] The situation is more complex (and somewhat more dire) than that; one purpose of bureaucracy is to restrict if not eliminate the use of imagination. Bureaucratic solutions and bureaucracy itself are tools to limit discretion in response to what is presumed to be regular activity which a corporation or a public wishes to be handled in regular fashion, that is to say, according to the rules. Or, as the phrase goes, "by the book."

The purpose (or at least a purpose) of bureaucracy is to limit in advance the range of procedural outcomes in regularly or routinely occurring cases. Whether the goal is efficiency (to save time and effort in determining outcomes) or integrity (to ensure that every case is judged on the same criteria and receiving, at least roughly the same outcome), imagination and bureaucracy are concepts that exist at least in great tension and arguably in direct conflict with one another.

One of the realities that must be confronted is that America has taken intelligence, once a "secret service" existing at the limits of law, often operating as an extension of the crown and therefore beyond the direct or daily oversight accorded more "normal" public functions, and turned it into a bureaucracy. More accurately, it has turned intelligence into a large, complex, and expensive set of bureaucracies.

Much of the result of this is desirable. For the first time in human history, intelligence exists largely as a "normal" instrument of government. Gone are the days when Allen Dulles and Senator Richard Russell could meet in private on a single afternoon and, over a bottle of bourbon, determine the Central Intelligence Agency (CIA) budget. Chiefs of Station, once the last examples of "ministers plenipotentiary," operating far from home and with great autonomy, now know much more direct supervision from the dreaded "headquarters." This is not unlike a process that happened earlier to ambassadors, but the reality is no less disruptive and unpleasant for those caught between the tradition in which they were raised professionally (or which, at least, they heard recounted) and a very different new reality.[19]

## Institutionalizing Imagination

In addressing this dilemma, the 9/11 Commission has argued that the 9/11 attacks succeeded, not because the methods for detecting and warning of surprise attack developed after Pearl Harbor failed, but because—at least in part—those warning methodologies "were not really tried."[20] In effect, a methodology designed to confront the extraordinary event had given in to the tendency to reduce all phenomena to preexisting categories, to permit routine to eliminate curiosity, and to eliminate from consideration the non-linear event. In effect, warning became yet another part of the bureaucracy.

How should American intelligence address this issue? The 9/11 Commission, though it performed a significant service by raising this question, barely addressed its answer. It may be true that the methodologies developed in the Cold War to deal with surprise, with emerging issues, and with "warning" in general were not effectively applied in attempting to prevent the 9/11 attacks. The more difficult truth is that it is at best uncertain whether their application would have prevented those attacks. In the end, this is one of history's unknowables.

However much we would wish that the methods derived in response to the Pearl Harbor surprise would have prevented the attacks of 9/11, such a success would not have removed concern that those methods, designed to operate in the Cold War environment, are fundamentally incompatible with the emerging security environment of the twenty-first century. If one assumes that among the critical considerations of the security designs of the period after 1945 were (a) a threat from a peer adversary, in which (b) deterrence of that adversary, including nuclear deterrence was a key strategy, and (c) in which critical technologies, involving communications technologies as well as weapons technologies, were available only to the superpowers and their allies and surrogates, then one must ask how many of these considerations are likely to apply to the twenty-first-century environment, however ill defined that may be. If one adds yet another consideration, that public concerns about the growth of the national security state would ultimately produce a demand for a clear and high barrier between domestic and national security, yet another barrier to the effective integration of information came into being, not through habit or malfeasance but by design. If at some point the public wants that design altered, so be it. But a national bout of amnesia on the reasons behind the earlier design is neither honest nor useful.

## What Is to Be Done?

### Persistent Issues in Rethinking Intelligence: A Twenty-first-Century Vision of Information Management and Security

One of the first things new employees of the intelligence agencies (and other national security organizations) receive on starting their careers is a briefing that discusses the classification system. Most Americans, if they know of this system at all, know it generally as a process for determining that information is either secret or top secret. A relative few among the public may know about confidential as a classification.

Relatively few—though from one perspective, to be noted below, not few enough—know much about the labyrinth of special compartments and handling restrictions "above" top secret; or, as it sometimes seems, above, beneath, around, and through the basic classification levels. One thing is certain: the American process for managing, that is to say, simultaneously releasing and protecting national security information, is seriously out of date and out of step with twenty-first-century information technology, practices, and needs.

How could it be otherwise? The basis of the current system dates from World War II and is premised on three considerations: (1) that the amount of information to be protected is relatively small; (2) that the number of users needing access to this information is limited; and (3) that the standard medium for storing and moving this information is paper. *These conditions simply no longer apply.*

Calling for a fundamental review of information management and information security is not the same as prescribing a twenty-first-century solution to these persistent problems. Bluntly put, this author does not have such a prescription in hand.[21] Nor is it intended to diminish the efforts of hundreds, even thousands, of men and women who work diligently and professionally within the existing system to make it serve the national security needs of the United States. Without question, current efforts to improve information sharing within the national security community, including the state, local, and even corporate officials who must be drawn into this process in a homeland security environment, will produce results. The system as deployed in 2007 or 2008 will no doubt reveal progress made since 2005. The problem remains that measuring how we perform in this area in 2008 against how we did in 2005 is not the right metric, which must be an assessment of how well we simultaneously protect and disseminate information in 2008, given the information demands and systems of that time and the national needs for the appropriate management ("hold," "release," or something in between) decisions consistent with the national security interests of the United States. Something more basic than iterative reform is sorely needed.

### The Total Information Environment: Facing up to the Reality of "Open Source"

Such a fundamental review will also require a reassessment of how we address the question of whether intelligence equals secrets or whether intelligence equals information, some of it secret, some of it not. One can argue that the modern intelligence apparatuses of the United States and other nations have never disregarded open source information. The United States has the Foreign Broadcast Information Service (FBIS), and its intelligence agencies, like those of other nations, pay large sums of money for open newspapers, magazines, conference reports, and the like.

But FBIS can only be a partial solution to a larger problem. It is, to use a tired but useful image, yet another stovepipe. If the "intelligence" issue of the twenty-first century is information—how to acquire, assess, store, and forward it—more than secrets, how can we treat 98 percent (or so) of the information available to inform decisions as yet another stovepipe? How can we say to a single (and rather small) component of the U.S. intelligence community: "Here's a problem that affects all of us, and it's a really big one, by the way. But you figure it out." In the Cold War, with its emphasis on access to what were largely targets that worked very hard to deny access to information about them—not just the size and capabilities of their armies, but also the location of highways—secretly acquired information had unique privileges. And it always will. Nothing in this chapter is intended to minimize the role or value of such information in a world in which access to significant information is denied by some of the world's most dangerous countries and organizations.

But the balance of information has shifted. Years ago, Sherman Kent estimated that 85 percent or so of the information needed by decision makers was available through open source. Let's assume that the information revolution has increased that to 98 or even 99 percent. That is a measurement by volume. Intelligence agencies could (a) dispute that reality, (b) pout and drift into denial, or (c) operate within it. (The correct answer is [c].) First of all, even assuming the open source advantage to be 99 percent by volume, let's further argue that the secret intelligence represents 10 percent of the information available to decision makers, *by value*. Is that sufficient to sustain the intelligence agencies? It is if policy makers are convinced that information worth ten times (in value) its place (by volume) is worth paying for. And if the people who, by virtue of position and access, have control of that secret information also demonstrate mastery of the other 99 percent and present their unique information effectively within that larger context, which I will call the *total information environment*.

Let's look, by comparison, at the securities industry. Suppose one could present to one of the major banking or investment firms a technology or process that would allow them to master all the information available to their competitors. Beyond that, however, the process offered access to a unique stream of information not available to any other firm. The first assurance required, of course, would have to be that this unique information was not "insider information" or anything else that would get everyone involved long prison terms in a federal penitentiary for violating the Securities and Exchange Act.

That hurdle aside, how many of those firms would then take their privileged information and discard or underutilize that which is commonly available to their competitors? The guess is that none of them would be so foolish, because their information power within the total information environment would be equivalent to their unique information leveraged on the public information.[22]

Intelligence professionals need to understand the enormous shift that has taken place in the balance of secret versus open information. This is, moreover, no longer an issue of open source *information* alone. It is about open source *expertise*. Many American journalists, academicians, business professionals, and others spend more time "on the ground" or in conversation with persons in foreign places of concern to the United States than do their counterparts in the intelligence profession. American intelligence, in the services and in "civilian" agencies, has done a better job in the last decade in recognizing and reacting to this development. *But we have not done enough fast enough*. And the external metric, that is, what does the environment permit and require, is the only one that counts. No one cares that we are doing better than we did in 1991. Nor should we.

## *Warning/Emerging Issues:*
## *Thinking for the Nonlinear Event*

One value of a more energetic interaction with open source expertise will be a renewal of the functions traditionally labeled warning. In the Cold War, this largely took the form of defining indicators that suggested preparation for hostile military activity.

In the twenty-first century, the need to warn is less that of keeping watch over the Fulda Gap than of being in position to deal with a broad range of emerging issues. In the midst of an intense campaign against terrorism, one that is not likely to end soon, warning in a more traditional context (though against nontraditional threats) remains vital.

But the Cold War, in retrospect, confronted the United States with a relatively limited range of threats. We were always focused on the Soviet Union and its Warsaw Pact clients. But we also focused, at different times and to different degrees, on such areas of concern as China, Cuba, North Vietnam, and Israel's Arab opponents. We also dealt with "wars of national liberation" and political stability issues across much of the world. But in the end, all our conventional military adversaries were largely variations—in equipment, doctrine, organization—on the Soviet theme. Many, though not all, of the insurgencies and stability issues were likewise extensions of the Cold War confrontation with the superpowers.

The warning environment, 2005–20, is likely to make such an environment look stable by comparison. We could move collection capability and analysts from the Soviet target to the Egyptian or North Vietnamese target knowing that the targets provided a significant measure of commonality. What, on the other hand, if the world's great concern in 2010 has shifted from Islamic movements and terrorism to naturally occurring pandemic disease? How much crossover capability would we have? How much would we need?

Stipulating once again that the more traditional "warning" function, focused on terrorism, will remain a highest national priority for years to come, we need to address the issue of emerging issues in fully nontraditional ways. Perhaps the model intelligence agency for 2020 is not the CIA or NSA, but the Centers for Disease Control.

Whether in military or diplomatic (or otherwise nonmilitary) intelligence, core values for the twenty-first-century relationship between intelligence and crisis will be curiosity, imagination, and flexibility. Lessons learned exercises conducted on the American experience in Iraq have and will point to a number of shortcomings, including some lingering issues involving cultural and linguistic expertise. But American (and global) response to the South Asian tsunami represents, on the other hand, a remarkable exercise in institutional and professional adaptation. While we study both the successes and the hard lessons of Iraq, we need to do so in the context that recognizes capabilities as well as shortcomings.

## Training and Education

The most basic truth about rethinking the principles of war for the future of intelligence in warfare is that we simply must invest more in the training and education of the intelligence professions, uniformed and civilian. To their credit, the armed services have traditionally done so. In the civilian services, however, this is not the case. The CIA has long practiced intensive (and expensive) training of its case officers, though even this experience is heavily concentrated in the period immediately after entrance on duty. Where are the mid-level or senior-level experiences equivalent to staff or war colleges? They simply do not exist.

On the analytic side, we have been even more foolish. To a great degree, we took the skills analysts brought with them from their universities and rode those skills until the analyst retired. To its credit, the CIA began to address this with the Sherman Kent School, an initiative undertaken, not after the budget tap was reopened after 9/11, but in the lean years of the late 1990s. That was a heroic act of leadership. The Joint Military Intelligence College, for its part, has been raising the flag of professional intelligence *education* (not training) for decades. But we simply have not done enough.

## 1. Training for the Way We Operate and for a World of Nonlinear Events

First of all, training and education in American intelligence resemble much of everything else in American intelligence: anything we do vertically is better than almost anything we do horizontally. We simply do not train for the way we need to operate or claim to wish to operate. Our powers at the community level to encourage, let alone require, "jointness" are limited. Or have been, in the DCI era.

We simply need to do a better job in this area. One place to start is an open act of public service plagiarism: we need to take a hard look at the professional military education system and adapt it, shamelessly, to the intelligence community. This will not prove to be a one-for-one transfer. And it will not prove to be as easy as deleting the word military and substituting the word intelligence. But we can do much of that. And we should.

One of the last actions undertaken by the former Community Management Staff was the promulgation in December 2004 of a directive creating a National Intelligence University system, responsible for coordinating the community's education, training, and "related research" activities. This effort is designed, not to replace the existing schools and colleges in the various intelligence services, but to bring them into a network, aligned as well with other federal degree-granting institutions (for example, the service and national war colleges) and the broader academic community. In the new DNI structure, administration of the university system will be the responsibility of an office of education and training, the first time these important issues have been raised to this level in the American intelligence establishment.

We also need leadership that understands the need for career development paths and that investment in career development takes time. An average military officer can easily spend, in the course of a full career, two to four years in full-time professional education and training. His or her civilian counterpart can often put a supervisor into fibrillation by suggesting enrollment in a two-week course. The intelligence community needs, for want of a better term, to suck this one up, adapt some system of staffing levels that assumes key people will be gone for extended (but, one presumes, predictable) periods. Management will simply have to deal with this reality. That, not focusing on today's *production* of yet another report or two, is what they get paid to do. Or should get paid to do. If we are to produce anything, at least on the analytic side, it should be expertise, not more material for someone's in basket.

## 2. Indoctrinating Lessons Learned: Implementing Hard Truth

We need an effective lessons learned capability. Some may note that the U.S. Army's Center for Army Lessons Learned (CALL) is a "1980s" model, and therefore potentially too far along in its productive life. When Rajiv Gandhi was prime minister of India, he once promised to bring the villages of India into the twentieth century, leading one wag to suggest that most of the villagers would settle, at least temporarily, for the amenities and services of the nineteenth. From the point of view of a set of agencies that, individually and collectively, have never taken to heart the introspection necessary for effective lessons learned methodology, CALL looks pretty good.

The first step in adopting (and adapting) such a model is not methodological, but professional and ethical. We must accept the reality that hard introspection is what real professions accept as part of their identity and their claim to professional status. Whether this is pilots undergoing debriefings or physicians conducting a morbidity conference, the best professionals accept self-assessment—even public self-assessment—as a norm.

We may discover that CALL is not so much an obsolescent model (from this observer's view it seems to be holding its age well) but that its value to us is limited because it operates within a single institution with an officially prescribed set of doctrines. In this session, efforts of the Joint Forces Command to operate across services may prove more instructive. Whichever model or models we choose, and however we determine to adapt them to our needs at this and future times, we need to ensure that the lessons learned process does not end with publication of a study. We need to ensure a path for the reinvestment of such studies in the training and education programs of the community. Americans are often uncomfortable with the word *indoctrination,* but that is precisely what this is about. It is about continuing to communicate the approved way of handling an issue or instance, always with the thought that what may be communicated at a given time is the message that the approved doctrine has been changed to meet new conditions.[23]

## Final Thoughts

From the legislative perspective, rethinking the American national security establishment may look like a "once in a generation" opportunity. As briefly noted above, this is an illusion. If we are to be effective in rethinking war, intelligence, or the interaction of the two, we must make the process of revision and renewal a constant part of our activities. As technical and operational environments shift, as they will, the alignment of the instruments of American national security must shift with them.[24] This is a process more like sailing than navigating a 747, that is, it is not a "set it and forget it" responsibility. It requires constant observation of the environment, especially changes in that environment, and deft reactions to those changes. Only through resignation to that hard and sometimes tiresome reality can the United States be assured of an intelligence profession aligned with the country's needs and fully compatible with its values. Only then can intelligence participate fully in the ongoing rethinking of the principles of war in the twenty-first century.

## Notes

1. Michael Herman, *Intelligence Power in Peace and War* (Cambridge: CUP, 1996), ch. 1.

2. The Revised Standard Version makes clear Daniel was providing not just a *reading* but also an *interpretation,* thus striking a blow for all later generations of analysts.

3. It was not uncommon in the 19th century to name newspapers "The Intelligencer," in a meaning largely synonymous with "reporter."

4. The real issue here is not secrecy but confidentiality, of a sort common outside the national security community, in any exchange between professional advisor or counselor and client.

5. See William Nolte, "Keeping Pace with the Revolution in Military Affairs," *Studies in Intelligence,* Vol. 48, No. 1, 2004.

6. Although many of the factors driving fundamental, even disruptive, changes in the relationship between intelligence and the military and civil officials who use it in decision making will affect other states and even non-state actors, this chapter will focus on changes taking place in the American example.

7. The use of the term "intelligence services" over "intelligence community" is intentional, but it is not intended to be dogmatic. This chapter will use both, though with the thought, if not the hope, that "community" as a metaphor for the collection of American intelligence agencies will at some point be relegated to refer only to the agencies as they existed in the Cold War and its immediate aftermath.

8. See, for example, comments by former CIA officials Vincent Cannistraro and Melvin Goodman, and former NSA director William Odom, in "U.S. Intelligence Shake-Up Meets Growing Criticism," *Boston Globe,* 2 January 2005.

9. See Steven A Cambone, statement before the Senate Government Affairs Committee, 26 August 2004.

10. An even more structured division of effort was common in the Armed Forces Security Agency, NSA's predecessor, where it was assumed that the communications of armies could (or should) be processed by U.S. army personnel, naval communications by naval personnel, and so on. It was only through the influence of individuals like RADM Joseph Wenger that NSA began, shortly after its creation in 1952, to assign and locate personnel largely without regard to service affiliation.

11. Frances Cairncross, *The Death of Distance: How the Communications Revolution Will Change Our Lives* (Cambridge, Mass.: Harvard Business School Press, 1997).

12. See John Keegan's chapter on Nelson in his provocative though uneven *Intelligence in War: Knowledge of the Enemy from Napoleon to Al-Qaeda* (New York: Knopf, 2003).

13. I am using here the conventional judgment that AFSA "failed," simply because it is the simplest way to explain its supersession by NSA. The reality is more complex, and in a fuller exposition, I would argue that AFSA was not a failure, but on the contrary was an essential step between fully separate service or departmental cryptologic activities and a more integrated national effort. As Christopher Andrew has wryly noted, the United States achieved strategic cryptologic failure in 1941 with only two cryptologic agencies, those of the Army and Navy, but that in 1950, with the outbreak of the Korean War, it took four agencies—Army, Navy, Air Force, and AFSA—to achieve the same dismal result.

14. Or, for that matter, between "foreign" communications NSA is chartered to "attack," and "domestic" communications, which NSA is enjoined from attacking.

15. *"Wie so eigentlich gewesen."*

16. John Lukacs, *Historical Consciousness or the Remembered Past* (New York: Harper and Row, 1968).

17. Roberta Wohlstetter, *Pearl Harbor, Warning and Decision.* (Stanford: Stanford University Press, 1962). Other works, using different images and metaphors, include Thomas Homer-Dixon, *The Ingenuity Gap* (New York: Knopf, 2002).

18. The 9/11 Commission Report: The Final Report of the National Commission on Terrorist Attacks upon the United States, p. 344.

19. One of the issues created by the establishment of the Director of National Intelligence is the relationship of the DNI with CIA station chiefs around the world. In the DCI era, station chiefs represented the DCI as the heads of CIA's field offices and also served routinely as the senior intelligence community officers in the countries in which they were stationed, arguably an extension of DCI authorities that have now passed to the DNI. Can or should these responsibilities now be split? Do the new circumstances permit or require dual "reporting chain"? Providing answers to these and related questions will be among the most important issues of the first year or two of the DNI era.

20. The 9/11 Commission Report: The Final Report of the National Commission on Terrorist Attacks upon the United States, pp. 347–48

21. The absence of a prescription should not delay our confronting the diagnosis. When President Kennedy committed the United States to a lunar landing, he did not announce the plans for the lunar lander.

22. Because, one hopes, secret information has *value* that exceeds its *volume,* the value of the total intelligence provided by intelligence services should reflect the multiplier of that value.

23. The South Asian tsunami represents a potential case study of enormous value in "information" support to national leaders, involving as it has the full range

of national security tools (civil, diplomatic, military) and the range of information sources reporting on the event.

24. In this and other respects, one of the most important provisions of the Reform Act of 2004 may be the Joint Intelligence Community Council (Sec. 1031). A cabinet-level body, chaired by the Director of National Intelligence, empowered with "evaluating the performance of the intelligence community," among other functions, should give the DNI, in *cooperation* with other stakeholders, a powerful tool for redirecting American intelligence, bypassing or overruling as necessary the sometimes excessive autonomy claimed in the previous era by individual intelligence agencies.

# TWENTY-FOUR

# Beyond Intelligence Reform

## The Case for a Revolution in Intelligence Affairs

### DEBORAH G. BARGER

Military thinkers and strategists are grappling with the unintended con-sequence of fielding the world's most powerful fighting force—adversaries who refuse to play by American rules of warfare. According to Col. Thomas Hammes, U.S. Army, in his recent book, *The Sling and the Stone,* "The fact that only unconventional or fourth-generation warfare has succeeded against superpowers should be a key element in discussing the evo-lution of war."[1]

It has become obvious to many that the U.S. military no longer has the luxury of focusing solely on defeating an enemy's conventional military forces. Intelligence, long an integral component of war fighting, will become ever more important as fourth and even fifth generation warfare changes the nature and focus of what constitutes "security." The task list for national, as well as military, intelligence in support of stability and security is already so long that the real issue is how far such capabilities can be stretched before they break.

Just as the military is rethinking the basic principles of war, so, too, must the intelligence community rethink the business of intelligence.[2] A growing number of intelligence professionals believe that the accelerating pace of change in the security environment, the opportunities offered by new tech-nology and operating procedures, and the rising chorus of new intelligence consumers demand nothing short of revolutionary change across the vast U.S. intelligence enterprise.[3] All of the changes of the past decade, including those that will be implemented as the result of the new intelligence reform legisla-tion, the "revolutionaries" would argue, will not amount to an intelligence community optimized to address the threats and opportunities of tomorrow.

Without a "revolution in intelligence affairs," or RIA, grounded in sound strategic thinking and operational reassessment, and designed to change the entire U.S. intelligence *system,* three unsatisfactory outcomes could result:

- Change will come about incrementally through traditional planning, programming, and budgeting processes, virtually ensuring mismatches between rapidly changing external realities and slowly adapting intelligence capabilities; or
- Commissions, legislative reform groups, and others will force change, but necessarily limit their focus to glaring problems and quick solutions; or
- Fundamental changes will occur, chaotically and incoherently, as competing intelligence users and agencies push forward proposals that address their own parochial concerns, with little regard for synergies or common interests.

The advocates of an RIA believe that the conditions are ripe for the U.S. intelligence community to step back, question how it is doing things across the board, and consider fundamental change in key areas. Certainly, if the intelligence community were to recreate itself *tabula rasa,* it would no doubt create different structures and working arrangements, different operating procedures, and different investments in skills and technologies. The new legislation gives the new Director of National Intelligence (DNI) a mandate to radically improve the performance of the intelligence community, and the authority to actually implement an RIA, if one can be conceived.

The RIA proponents seek to create intelligence capabilities that can both shape, as well as adapt, to the rapidly changing security environment. Many in the security arena are beginning to believe that the only alternative to a continuing series of military interventions is to proactively address the global conditions that lead to them. So, too, must the U.S. intelligence community play a far more aggressive role in anticipating crises overseas and influencing events to preempt either terrorist attacks on the homeland or conflict overseas.

## What Is an RIA?

This essay will posit that the RIA is not to be the intelligence community equivalent of the RMA—the revolution in military affairs—that unfolded over the past three decades. Critics of the RMA will undoubtedly note its

shortcomings as a conceptual approach that was driven largely by technological opportunities that would provide the United States with a dominant military advantage on the battlefield. Fixation on technology is to be avoided in
the RIA.

Rather, the RIA is to be a framework for bringing about profound change
in the intelligence enterprise. It is a means of assessing intelligence as a complex system, objectively evaluating proposals for change (rather than through
a political or parochial lens), and putting in place a stream of activities that
will help the intelligence community succeed in implementing needed
changes, not once, but continuously. The RIA would drive fundamental
change in both culture and process. It would ensure that the intelligence
community both shapes the global security environment and morphs as
needed to stay one step ahead of looming threats.

The last RIA took place between 1945 and 1956, when the United States
first accepted the need for a permanent peacetime intelligence apparatus. The
years immediately following World War II were a period of revolutionary
change in U.S. intelligence, driven by concern over the fragmented nature of
intelligence activities during the war years and the failure to warn of the
impending Japanese attack on Pearl Harbor in 1941. The response of the
government was to centralize the coordination of intelligence activities, create the Central Intelligence Agency (CIA) as the cornerstone of that centralization, and base the entire structure on a new statutory foundation—the
National Security Act of 1947. There were many fits and starts during those
early years, but serious strategic thinking and a spirit of innovation inspired
the early founders and yielded an intelligence community extraordinarily well
suited to navigate the shoals of the Cold War. The threat of weapons of mass
destruction in the hands of terrorists and the global digital revolution are
changes of a magnitude similar to those that drove the last RIA. Now another
RIA is needed.

The current Secretary of Defense, Donald Rumsfeld, is convinced that
the transformation of intelligence is essential to the successful prosecution of
future conflicts. In 2003, he created a new undersecretary of defense for intelligence (USD[I]) under whom "all intelligence and intelligence-related oversight and policy guidance functions in the Office of the Secretary shall be
organized." The USD(I) has paid particular interest to the transformation of
national intelligence activities that will support a transformed U.S. military.
With the creation of a director of national intelligence (DNI) whose explicit
charter is to oversee the national intelligence effort that includes support to the
military, how much of intelligence transformation will be driven by the DNI,

rather than the USD(I), remains to be seen. An argument can be made that the Department of Defense (DoD) alone cannot, and should not, drive the RIA.

## Why the Department of Defense Should Not Drive the RIA

The primary reason why the DoD should not drive the RIA is that an imbalance in the use of intelligence primarily as an instrument of war rather than as an instrument of peace is likely if it does. Of greatest concern is that the intelligence capabilities most useful for anticipating crises and supporting diplomatic efforts to ensure the peace may atrophy in the face of increasing demands during the lead up to conflict, during conflict, and during stabilization after conflict. It is well to remember that the *raison d'être* of the peacetime intelligence apparatus that came to be known as the intelligence community was to provide "strategic warning" of possible threats to the United States and its interests.

Even as the United States faces challenges imposed by new forms of warfare, the intelligence community can never abandon its traditional warning mission. This means the intelligence community must continue to "cover the globe" in addition to addressing crises or conflict. Furthermore, the new "tactical" warning missions associated with homeland security are likely to be every bit as demanding as the traditional strategic warning mission related to potential conflict. Intelligence capabilities that will identify terrorists, track weapons of mass destruction, warn first responders, and protect critical infrastructure may well be different from those that support the troops in the field. There will be an increasing demand for both—thus the transformation of national intelligence cannot be driven solely by the needs of military intelligence.

Some would argue that the RIA for military intelligence was begun some time ago. The DoD has already begun a serious reexamination of its roles and missions in the global war on terrorism, even as it struggles with new concepts to deal with insurgencies in Afghanistan and Iraq. The military services recognize that shorter decision cycles and swifter reaction times require closer integration of intelligence and operations and are anxious to "remodel" defense intelligence. Thus, establishing who will be in charge of developing and implementing the RIA is essential. Otherwise, the intelligence community will continue its pattern of pulling in fifteen different directions at the same time. The new intelligence reform legislation makes clear that the DNI would lead an RIA, certainly one that affected national intelligence community elements.

The intelligence reform legislation does leave enough ambiguity over who controls what in national intelligence to ensure that the DNI and the secretary of defense must continue to work together. The DNI must continue to view the military as his largest, if not most important, customer. The DNI will need to transform national intelligence if his military customers—who are transforming themselves to fight insurgents and other fourth generation combatants—are to have their needs met. To do so, he needs the support of the secretary of defense. The DNI must also assist the secretary in leading the DoD's transformation of joint and tactical military intelligence in order to meet the specialized needs of the war fighters. The RIA will be most successful if the *entire* intelligence enterprise is addressed holistically.

## The Revolution in Military Affairs as Transformation Model

Despite the fact that the DoD should not drive the RIA, the DNI could do well to emulate some aspects of the RMA as a transformation model. As high concept and organizing principle, the RMA did much to stimulate the strategic thinking of the military.[4] In its implementation, however, the RMA was often more reactive and opportunistic than strategic. It is important to learn from both its successes and failures.

The genesis of the RMA is usually traced back to Marshal Nikolai Ogarkov, a brilliant strategic thinker and head of the Soviet General Staff, who first wrote about a "military technical revolution" in the 1970s. Ogarkov believed that "history's linear evolution is occasionally interrupted by rapid discontinuities" and that a major change in warfare had already begun with technological breakthroughs—such as long-range strike—in U.S. conventional force capabilities. Intrigued by Ogarkov's theory, Andrew Marshall, director of the DoD's Office of Net Assessment, began to look at whether certain technological innovations over the course of history gave a lopsided advantage to one combatant over another. Over time, he came to conclude that it was not technology alone that provided a decisive advantage; rather, it was the ability of a military force to transform its operations and organizations to leverage what technology offered that provided the advantage.

In addition to the contributions of Marshall's Office of Net Assessment, several other transformational streams fed the RMA debates over the next two decades. Through these multiple efforts, one can trace the development of new theory, strategy, doctrine, and innovation that provided the solid intellectual foundation for the RMA. A retired U.S. Air Force colonel named John Boyd, along with his band of "acolytes," led the first stream of revolutionary activities in the late '70s and early '80s. Boyd promoted the application of scientific and engineering knowledge to human behavior in war. His loyal followers, later known as the "reformers," were well-placed Pentagon insiders who built alliances with the media and with Congress to press for change. Robert Coram argues that Boyd was the originator of a new theory of maneuver warfare that consequently led to the development of new military strategy and doctrine. [5] Many of the concepts Boyd developed more than a quarter of a century ago were brought to bear during the Gulf War in 1991 and Operation Iraqi Freedom in 2003. His theories also led to innovations in operations and tactics that would enable the U.S. military to undertake a new approach to fighting wars.

Not long after John Boyd completed his thesis on "Patterns of Conflict," a second stream of transformational activity, led by Gen. Donn Starry and other U.S. Army officers, undertook the difficult task of trying to change the culture and behavior of the army. Doctrine, Starry believed, was the key to changing institutional culture.[6] Starry's task was to write new army doctrine that would allow the service to fight and win a conventional war in Europe after its unhappy experiences in Vietnam. He talked to the troops on the ground in Europe, gave them a draft of his ideas, and constantly asked for feedback. He was successful, he believes, because the troops convinced themselves that the new doctrine was a good idea and it was *their* idea. Starry believes another major reason for the success of AirLand Battle doctrine was that he resisted having the army's Training and Doctrine Command write the doctrine. For doctrine to take hold, he argues, the people who *teach* the doctrine should *write* the doctrine. Starry also helped reform the army's military school system so that it could both write new doctrine and teach the army how to fight with it.

Both Boyd and Starry recognized that a military transformation required going back to first principles and challenging deeply held convictions as well as changing the military's doctrine and culture. Both led coalitions for change composed of outsiders (primarily from Congress and the media) and Pentagon insiders. Both welcomed debate and alternative views. Both ignored the traditional procedural route to change, knowing that they would hit an institutional brick wall. Both sought and found support from the men and women in the field who would prove to be the most important advocates for change.

The third stream of revolutionary change came about with the enactment of the Goldwater-Nichols legislation in the mid-1980s. Goldwater-Nichols helped improve the military's ability to operate effectively as a joint force. It strengthened the operational war-fighting commands and improved officer education. The result was better military performance, strategy making, and contingency planning both in operations and peacetime activities. The Goldwater-Nichols Act also strengthened the authority of the secretary of defense and chairman of the Joint Chiefs of Staff, and made clear the responsibilities of the military service secretaries in support of the defense secretary (objectives not unlike those that drove the intelligence reform legislation).

Despite the difficulties in passing this legislation,[7] the enactment of Goldwater-Nichols supports the argument that transformational change can be accomplished even when there is strong institutional resistance. This reform "from the outside" approach allows for the tackling of prerogatives that insiders hold most dear—thus leading to more substantial reform than insiders would ever agree to take on. The four years of rancorous debate that preceded its passage, and the years of resistance to some recommendations afterward, demonstrate that without the cooperation of the institution itself, the fight is likely to continue long after passage, and implementation will be arduous at best.

In sum, the lessons that were learned by the military during its thirty-year experimentation with an RMA can inform the intelligence community's consideration of revolutionary change. The first lesson is that a period of serious strategic thinking is essential before a large institution attempts to change itself. This is neither a coherent nor a collegial process. The RMA debate was highly contentious, with seven or eight competing schools of thought. But the result was a richness of new ideas, with the best ones emerging from the crucible. Institutions serious about such change must allow for creative tension between new ideas and the status quo.

A second lesson is that accomplishing the revolution is neither quick nor painless. Revolutions in large institutions do not happen overnight, as the structure, systems, practices, and culture of bureaucracies tend to impede change rather than facilitate it. It takes time for large institutions to adopt emerging technology and formulate new organizational and operational concepts. It takes time to build consensus both inside and outside an institution. It takes time to train the leadership in new ways of behavior, and for them in turn to teach the next generations of leaders. Development of a doctrine that is shared across organizational boundaries is essential to changing behavior.

The challenge is to generate a sense of urgency to get started and to accept that it may take a while to get it right.

Third, the military learned that it is essential to create a culture of continuous reassessment and self-improvement. The challenge for U.S. intelligence will be to speed up the cycle that today would take many years to accomplish. The process of transformation needs to become part of the marrow of the organization—an ongoing, continuous process that will keep it moving faster than any adversary can keep up.

Fourth, and perhaps most important, as the transformation process unfolds, someone always needs to keep his or her head above the fray and focused on the ultimate objective. This person can be an employee in a small office like Andy Marshall, be a flag officer like Donn Starry, or be an important congressional overseer like Jim Locher—the official title is unimportant. If it is not the DNI, it must be clear that the "architect" of the RIA has the full endorsement of the DNI. Whoever it is must be around long enough to maintain focus and continuity throughout the most difficult moments of the change effort.

The RMA did not proceed from a highly regimented and structured process within the Pentagon. Instead, it evolved from a tradition of individual action, intellectual rigor, and military scholarship embedded in the military system. To some extent, these forces of change naturally converged, but Pentagon leaders and other advocates amplified that convergence to bring about the improved performance displayed in integrated military operations in Iraq in early 2003. This ability to harness decades of transformational change strategy and planning, and then to achieve an effective military advantage through implementation of that planning, is perhaps the most important lesson learned from the RMA model.

## Key Steps to Bringing on the Revolution in Intelligence Affairs

Common threads can be discerned from both the successes and the failures of past revolutions in bureaucratic institutions. Reorganization almost never tops the list of necessary precursors to success, nor does a large infusion of money (although neither necessarily leads to failure either). Organizational complacency, leadership dissension, poor communication, failure to anticipate and remove barriers to change, lack of leadership vigilance, and lack of continuity are among the elements that have typically led to failure in other institutions. Some, if not all, of these elements have been present in past efforts to reform intelligence.

The construct for the RIA comprises two distinct but related elements: a cultural shift that sees the intelligence community embrace the need to change, and a procedural shift that enables the community to objectively evaluate alternative responses to change and to incorporate them in a continuous manner. Both are necessary in order for the community to become an adaptive enterprise.

The *cultural* element involves reshaping the intelligence community's reaction to the pressure to adjust to changing circumstances. Historically, the primary cultural driver was the bureaucratic tendency to "circle the wagons" whenever criticized and to defend existing organizational boundaries and purviews whenever change was recommended. To bring about an RIA, the workforce must be willing to question the status quo and encourage their leaders to change. The leadership must encourage constructive criticism and be willing to investigate alternative solutions. All must appreciate the value these efforts can bring to the collective endeavor.

The *procedural* element involves the creation of mechanisms to bring about systematic changes to the form and the function of the intelligence community. Without the means to affect physical change, the cultural dimension of the RIA will be a meaningless exercise. Under the RIA, the necessary components for evaluating and incorporating change are as follows:

- A coalition of advocates to sponsor change-related efforts
- A finite list of key players with well-defined roles in the tactical execution of change-related activities
- The development of new theory, strategy and a formal "doctrine" to serve as a strategic foundation for institutionalizing a response to change
- Innovations in technology, operations, and organizations driven by new strategies
- Objective evaluation of revolutionary change proposals and the use of experimental designs and metrics to objectively assess alternative futures
- Development of new incentives and rewards to institutionalize a dynamic-change management approach

### Cultural Change

The intelligence community must learn to continuously transform itself—sometimes quite radically—if it is to stay ahead of a rapidly changing security environment. This means that at all times there are two competing streams of change activity—one evolutionary and the other revolutionary. The institution should move from one stream to the other based on the amount of flux in the larger national security environment. Today in the intelligence community the change process is not unlike most government bureaucracies—largely incremental. It focuses on what is "doable," given budgetary and procedural restraints, rather than on what needs to be done.

To change this, the intelligence community needs to ensure it has a sufficient number of people in the workforce who have an inherent sense of when trouble is looming over the horizon and can identify a major discontinuity in its early stages. These are people who have a healthy preoccupation with anomaly and surprise because there are opportunities to learn from them. These "worriers" trigger the need to switch the change process from the evolutionary, incremental approach to the revolutionary "we need to seriously correct course" approach.

The intelligence community must identify the activists in its midst, encourage and protect them, and empower them to go beyond just challenging the status quo so that they may translate their concerns and solutions into action. Ideally, a small group of these change agents would report directly to, and be protected by, the DNI and his leadership team. Such a group or office would be small and unobtrusive, and it would *not* be part of the normal planning, programming, acquisition, and budgeting processes. Nonetheless, it would regularly challenge the decisions made through those processes. If this activist group is doing things right, there will always be creative tension between what it is promoting as alternative ways of doing business and what the status quo is trying to protect. Its impact on planning would be based on the merits of its arguments, not its place in the hierarchy.

One of the primary tasks for this activist group would be to encourage the incubation of new ideas. Today, many groups serve as these "incubators" in the intelligence community or challenge the status quo. However, few of these organizations are plugged into the main decision-making processes within the intelligence community and therefore operate largely on the periphery. The activist group must regularly interact with *all* intelligence consumers. It would also need to tap into scholars, academics, think tanks, government labs, industry, and others who could supply innovative ideas that might not be generated from within the institution.

The activist office would also focus on tracking change within intelligence community organizations. This does not necessarily mean that large amounts of quantitative data or metrics must be tracked. It might simply mean setting milestones and seeing that they are met. If there is an intelligence failure, even one that doesn't make the press, the group should immediately get busy and conduct a critical self-assessment and find out why it happened. To assist the DNI in implementation, this office would help build continuous change into all intelligence processes—from the gathering and analysis of data, to the sharing of intelligence, to the management and governance of intelligence, to the protection of intelligence sources and methods. The goal is to help the entire institution become comfortable with ambiguity, uncertainty, and rapid change.

While legislation or presidential directives can get the ball rolling, the intelligence community itself must embrace continuous change. Those on the outside will never be able to determine if needed changes are taking hold until it is perhaps too late and another intelligence failure has occurred. An approach that establishes a healthy "coalition" of insiders and outsiders is likely to have the best chance of success in a bureaucratic enterprise like the intelligence community, and has the added benefit of involving the intelligence workforce in shaping its own destiny.

### Procedural Change

One of the valuable lessons learned from the RMA process is that new theory, strategy, doctrine, and innovations in technology, operations, and organizations must be the foundation of any intelligence transformation. If sufficient progress is not made in these six areas, then the RIA will be unrealized. Changes can certainly be made, but not changes that are likely to lead to dramatic improvements in capability and performance.

**1. Theory** attempts to discern, as CIA historian Michael Warner argues, the "intrinsic something" that defines intelligence.[8] There has been very little theoretical work done in the area of intelligence, however, largely because the data needed to prove or disprove a theory are often unavailable to the scholars and academics who study them. Going back to first principles—what intelligence is, and what its roles and missions are—is the necessary first step in the RIA. It is imperative to clarify whether the overarching mission of intelligence is substantially changing. What, specifically, is the role of intelligence in homeland security? Where does its mission begin and end? Is it a higher priority than support to U.S. troops overseas? Is it a higher priority than support to diplomatic efforts in the Middle East?

A revolutionary new theory of intelligence would, at a minimum, challenge the existing theory behind the creation of centralized intelligence in

1947. Some of the assumptions made then, which have been challenged in the recent intelligence reform debate, are that information pertaining to U.S. citizens should never be the purview of the intelligence community, and that a "coordinator" of intelligence can bring about systemic changes across the national intelligence enterprise. Revolutionary theory might go further and challenge the notion that greater centralization of intelligence will lead to better performance. New theory might posit that U.S. intelligence must always be organized and behave like the adversary. Given today's adversaries, this would mean decentralized and ephemeral, and difficult both to detect and to deceive.

Theory must remain closely tied to the historical record. The historical record of U.S. intelligence is becoming clearer as more information becomes declassified and more scholars and historians take up the challenge of drawing theoretical inferences from the data. Without serious consideration of the past and its lessons regarding intelligence successes and shortcomings, there will be no solid analytical basis for serious debate about potential paths for the future.

**2. Strategy** is simply the means to achieving an objective. Dramatic changes in the external national security environment should automatically lead to concomitant changes in intelligence strategy. After a new National Security Strategy is published, a new *intelligence* strategy for meeting new mission priorities should follow. The new intelligence strategy should shape the doctrine and the innovations necessary to meet the strategy.[9]

The DNI and his leadership team should develop new intelligence strategies, which are then shaped by the DNI's designated "architect" who understands both the parameters of the current system and the desired end state. The strategies—both mission-oriented and "business" strategies—must begin with a vision of the world ten to twenty years in the future, and must clearly describe desired outcomes. They should not be constrained by the budget in its early stages. They should include a view of the transformation of existing institutions, the creation of new capabilities, and, in some cases, the replacement of existing organizations that will inevitably become less relevant as the RIA progresses.

This will allow the intelligence community to end its current practice of devising individual, competing strategies among the intelligence agencies. An overarching intelligence mission and business strategy is the only way that intelligence can work as a "system of systems" without one set of activities confounding the others. While a group in the intelligence community already exists to do collective strategic planning, real strategy remains a prerogative of the individual agencies that continue to compete with each other for resources, status, and primacy. The DNI and his leadership team must embrace the RIA strategy and communicate it to every member of the intelligence community workforce. It must be at the center of every mission and business decision made thereafter.

**3. Doctrine,** as defined by the U.S. military, is a body of principles by which national security elements guide their actions in support of national objectives. "Doctrine" is not a word typically associated with the national intelligence community, but as the examination of the RMA has shown, if an RIA is to be attempted, a new set of guiding principles must be developed and then widely communicated and explained.

Something akin to community-wide doctrine has been a missing component of reform efforts to date. When changes have been made by legislation or by new leadership, individual intelligence agencies and components have typically been left to interpret those changes on their own. When doctrine is developed, shared, and understood, subordinate units have the flexibility to improvise, as long as the activity is within the leadership's strategic intent, which allows them to operate more quickly and efficiently. This flexibility is critical to "getting inside" the adversary's decision cycle by anticipating what he will do before he does it. Intelligence analysts and operators must also have the ability to improvise if they are to adjust rapidly to unfolding events.

Some schools and programs within the intelligence community teach the art and science of intelligence. But these schools do not teach or develop common intelligence community-wide doctrine or tradecraft, even in the wake of a new national security strategy or major changes in intelligence legislation. The new "intelligence university" framework referred to in the intelligence reform legislation may ultimately play an important role in the development of national intelligence doctrine.

**4. Technological innovation,** or the innovative application of existing technology, is often the catalyst of revolutionary change. The recognition that a mission-related challenge cannot be met often drives this type of innovation. The intelligence community has traditionally been good at developing innovative technology (but not always as good in changing the business practices that would make it most useful). The pace of technology development is now so rapid that it presents government institutions with the difficult challenge of rapidly changing their operational and organizational structures to keep up with what technology has to offer. Off-the-shelf technology does not always meet the intelligence community's specialized needs, and valuable time is spent customizing commercial products. New technology applications are often developed in organizational "stovepipes" and not shared with others who could make use of them, or worse, are duplicated unnecessarily and often at considerable expense. The new director of science and technology in the Office of the DNI could play a very important role in the RIA. The position could be charged with scanning the horizon for new technological opportunities and providing a community forum for technologists to interact with analysts, linguists, case officers, line managers, and others.

The intelligence community must emulate, as much as possible, the culture that rewarded experimentation and risk during the first RIA following Word War II. Congress must be a willing partner in this process and not punish those who experiment if the experiment does not pan out. There must also be a way for the intelligence community to move on quickly if the technologies picked for experimentation prove to be the *wrong* ones. In the past, declining budgets led the intelligence community to hang on to legacy systems and incremental improvements longer than it should have. More experimentation, more often, and on a smaller scale, is essential to keeping pace with technological innovation and developing breakthrough applications.

**5. Operational innovation** within the intelligence community can be defined as significant change in intelligence processes and activities. The challenge is more difficult when attempting to change functions and behaviors in more than one agency simultaneously, and has proven to be virtually impossible when attempting such change across the entire community. Information sharing is a case in point. Up until ten years ago, sharing intelligence information rapidly across many agencies was a fairly difficult technological exercise. Today, technology, for the most part, provides the means to make this happen. When information sharing and analytical collaboration do not happen today, it is largely because operating procedures have not changed enough to establish new behaviors and change the way people do their work.

An example of a new operational innovation, therefore, might start with the determination at senior leadership levels that information technology *will* change the way the intelligence community operates and, from this point forward, information sharing and analytical collaboration *will* be a central part of the intelligence mission. Making this happen requires more than a decree from above. It means shifting the doctrine of "need to know" to "need to share," and then developing new procedures that will reflect this doctrine—and ensuring they are followed. It requires changing the mind-sets of many managers, data collectors, and analysts (who sometimes feel that collaboration with others impedes or detracts from their work, rather than enhances it). To change such attitudes, managers must ensure that they reward only the new and desired behaviors.

For the intelligence community, operational innovation must focus on changing and perhaps rethinking core functions, such as foreign intelligence collection, analysis, clandestine activities, data processing, dissemination, and research and development. It might also profitably reexamine "business" functions such as strategic planning, program and cost analysis, systems requirements and acquisition, performance evaluation, personnel management, and security. Each of these activities, practices, and procedures should be examined in light of new or even extant technologies that offer new ways of doing business. Focused management attention and rewards for new behaviors are essential to implementing and completing the transformation process.

**6. Organizational adaptation** is usually the default solution for those attempting to "shake up" the system. Over the past decade, much of the debate about how to "reform" intelligence has centered on reorganizing the intelligence community. New reorganization schemes that group activities by intelligence missions or by transnational topics that cross multiple agencies are part of the new reform legislation. True organizational innovation, however, requires much more than "rewiring" the organizational charts. If a true revolution in intelligence affairs is to take place, and new missions, roles, and functions are to be established, some organizations should grow, some should stay the same, and some should go away, depending on their relevance to fulfilling new tasks.

It would be far better to focus the energy of an RIA on devising new strategy and doctrine and other innovations than to begin with reorganization. Let new forms follow new functions. Until these first steps are completed, it is difficult to know if some organizational elements need to remain intact and if there might be good reasons why they cannot or should not be changed. Because reorganizations cause so much anxiety and disruption in the workforce, they should be undertaken only if absolutely necessary.

The U.S. intelligence community must undergo a revolution in intelligence affairs to address the difficult choices that confront the community as it prepares to meet the future. It should be noted that just as the RMA took a generation to unfold, so will an RIA—thus, all the more reason to get started. Without a more strategic examination of the intelligence apparatus and how it should evolve culturally and procedurally as a coherent system of systems, recent changes in intelligence policy and legislation may well prove insufficient, and not advance the community's ability to continuously adapt to emerging needs.

## Notes

1. Thomas X. Hammes, *The Sling and the Stone* (St. Paul, Minn.: Zenith Press, MBI Publishing, 2004), p.4.

2. By law, the Intelligence Community comprises fifteen intelligence organizations. They are the Central Intelligence Agency; the Defense Intelligence Agency; the National Security Agency; the National Geospatial-Intelligence Agency; the National Reconnaissance Office; the national intelligence elements within the Departments of State, Homeland Security, Treasury, and Energy; and the national intelligence elements within the Army, Navy, Air Force, Marine Corps, Coast Guard, and FBI.

3. In addition to the President (the "first" customer), other consumers of intelligence outside of the military include other heads of departments and agencies of the Executive Branch (State, Homeland Security, Justice, etc); members of Congress, and "other such persons as the DNI determines to be appropriate."

4. For a sampling of the different arguments regarding the benefits and shortcomings of RMAs see Colin S. Grey, *Strategy for Chaos: Revolutions in Military Affairs and the Evidence of History* (London: Frank Cass Publishers, 2000); and MacGregor Knox and Murray Williamson, *The Dynamics of Military Revolution* (Cambridge: Cambridge University Press, 2001).

5. Robert Coram, *Boyd: The Fighter Pilot Who Changed the Art of War* (Boston: Little Brown and Company, 2002).

6. Donn Starry, interview with author, January 22, 2003.

7. It took nearly five years to pass the legislation. Many senior DoD civilians and senior military officers opposed it. For an inside look at the difficulties in passing this legislation see James R. Locher, *Victory on the Potomac; The Goldwater-Nichols Act Unifies the Pentagon* (College Station: Texas A&M University Press, 2002).

8. Michael Warner, "Wanted: A Definition of Intelligence," *Studies in Intelligence,* Central Intelligence Agency, Vol. 46, No. 3, 2002.

9. Strategy—and strategic management and behavior—should not be equated with strategic planning. Strategy is less detailed than strategic planning but more ambitious in its scope and direction.

# TWENTY-FIVE

# The Weakest Link

### *Intelligence for Preemptive and Preventive Military Action*

RICHARD L. RUSSELL

Those few intelligence officers who have troubled themselves to study Carl von Clausewitz's *On War* should find it a humbling experience. Clausewitz held little regard for the role of intelligence in warfare; he devoted a mere chapter of two pages to the issue. And what words he did spare on intelligence are few and not very flattering to an intelligence officer's ego. "Many intelligence reports in war are contradictory; even more are false, and most are uncertain."[1]

Intelligence officers today might challenge this dismal assessment by arguing that technological strides in satellite imagery, communication and electronic intercepts, and computer power have vastly improved the quality of intelligence and its role on the battlefield since Clausewitz's time. Bold and ambitious intelligence officers might even go as far as to argue that today's technology has substantially lifted what Clausewitz called the "fog of war" from the battlefield to give combatant commanders and policy makers a near "picture perfect" view of adversaries.

Policy makers have indeed placed—or misplaced—a great deal of faith and even reliance on intelligence for national security policy in the aftermath of the September 11, 2001, Al Qaeda attacks. The critical importance of intelligence to American statecraft has come to the fore as the United States grapples with its new, uncertain, and risky security environment. Specifically,

---

The views expressed are those of the author and do not reflect the policy or position of the National Defense University, the Department of Defense, or the U.S. Government.

an American consensus has emerged over the concern that nation-states or transnational terrorist groups could use weapons of mass destruction (WMD)—chemical, biological, radiological, and nuclear weapons—to strike American territory, citizens, and interests. More controversially, President George W. Bush has ardently argued that given the potential death and destruction that WMD attacks could wreak on the United States, the commander in chief can no longer afford the risk of waiting and absorbing the first blow from a WMD-armed adversary. Accordingly, President Bush articulated in his National Security Strategy a policy of preemptive and preventive strikes to protect American national interests.[2]

The articulation of a policy of preemptive and preventive strikes generates considerable partisan debate in Washington. Democratic critics charge that Bush's policy threatens to destabilize the international system by giving international legitimacy as well as incentive to all nation-states for waging preemptive and preventive war. What these critics forget or conveniently overlook, though, is that Democrat President Bill Clinton launched cruise missile strikes against Al Qaeda–related targets in Afghanistan and Sudan in 1998 in part to preempt Al Qaeda's suspected planning for chemical weapons attacks against the United States.[3] The Clinton administration, moreover, contemplated preventive strikes against North Korea's nuclear weapons program, but lacked specific intelligence needed for military targeting of Pyongyang's nuclear weapons infrastructure.[4] Partisan politics aside, the stubborn reality is that future American presidents, whether Republican or Democrat, are likely to need viable policy options for militarily striking adversaries preemptively or preventively, even if they are armed with WMD.

Intelligence will be the load-bearing pillar of the American campaigns against WMD-armed adversaries. Without high-quality and timely intelligence reporting and analysis, the policy of preemptive or preventive military action will simply not be feasible. Much ink has been spilt on the pros and cons of preemptive and preventive military action in the counterproliferation campaign, but without timely and accurate intelligence, the debate becomes academic. If the United States does not have sufficient intelligence to know the "what, when, where, and how" to attack an adversary's WMD capabilities—including WMD weapon stocks, production facilities, and delivery systems—American precision munitions will stand idle because neither preemptive nor preventive military options will be viable for the commander in chief. And while much debate, research, and thought has gone into the transformation of the American military to meet the challenges posed by new security threats, no comparable effort has been made to examine how intelligence collection and analysis needs to keep pace with threats to enable military options for the president in the information-technology era.

This chapter aims to fill in some of this intellectual gap. It examines the track record of American intelligence, principally produced by the Central Intelligence Agency (CIA), in monitoring WMD programs. The CIA has traditionally performed a unique role in informing presidential decision making, although this relationship has been cast in doubt with the recent creation of the director of national intelligence (DNI), who is to serve as the president's chief intelligence adviser. The survey finds that the CIA has habitually suffered major failings in gauging the WMD programs, most notably against Iraq in the run-up to the 2003 war. These intelligence failings were due in large measure to poor human intelligence (HUMINT) collection and shoddy analysis, areas that cannot be remedied by the creation of the DNI by itself. This chapter recommends numerous measures for increasing the quality of HUMINT and analysis produced by the CIA. These are essential measures if American intelligence is to move in lockstep with the transformation of the American military to give the commander in chief options for acting preemptively and preventively against WMD-armed adversaries.

## Assessing Intelligence Performances Against WMD Targets

Is American intelligence up to the load-bearing task for the post–September 11 national security? A cursory examination of its performance over the past decade against the WMD challenges shows that it is not.

American intelligence performance in assessing Iraq's WMD programs prior to the 2003 war was a catastrophic intelligence failure, and arguably one of the greatest intelligence failures since the CIA's inception in 1947. In an October 2002 National Intelligence Estimate (NIE), the CIA judged for President Bush that Saddam's regime was aggressively reconstituting its nuclear weapons program and had active biological and chemical production lines, as well as significant biological and chemical weapons stores.[5] In the now infamous exchange, Director of Central Intelligence (DCI) George Tenet personally told President Bush that the case against Saddam and his WMD activities is "a slam dunk." What is not publicly touted is the president's critical appraisal of the CIA's WMD case on Iraq that triggered Tenet's "slam dunk" remark. Bush said to Tenet, "I've been told all this intelligence about having WMD and this is the best we've got?"[6]

Despite Bush's reservations about the quality of CIA intelligence on Iraq's suspected WMD capabilities, and although the president used a variety of public justifications for waging war against Saddam's regime (such as Baghdad's links to international terrorism), Saddam's active and robust WMD capabilities stood head and shoulders above other justifications. CIA intelligence reporting and analysis contained in the October 2002 NIE was also funneled into Secretary of State Colin Powell's February 2003 presentation to the United Nations (UN) Security Council in an effort to sway international official and public opinion toward the U.S. strategic objective of ousting Saddam's regime. Secretary Powell masterfully delivered his presentation that Iraq was actively reconstituting its nuclear weapons program and producing, weaponizing, and stockpiling chemical and biological weapons in violation of UN Security Council Resolutions upon which the 1991 Gulf War's ceasefire was based.[7]

Post-2003 war investigation on the ground in Iraq revealed a substantially different picture of the status of Saddam's WMD and delivery programs from that painted by the prewar NIE and Secretary Powell's UN presentation. Although the Iraq Survey Group (ISG) working for DCI Tenet determined that Saddam's activities in the fields of ballistic missiles and unmanned aerial vehicles were generally consistent with the NIE, it discovered via investigations and debriefings of Iraqi military officers and scientists that Saddam's chemical, biological, and nuclear weapons programs had been abandoned since the mid-1990s in part because of an Iraqi fear of detection by the international community.[8] Iraq was neither producing nor weaponizing stores of biological and chemical weapons. Most significantly, the ISG determined that Saddam's nuclear weapons program was in ruins as a result of the 1991 war and subsequent UN weapons inspection team endeavors to dismantle the program.[9]

The CIA had been profoundly wrong in its assessments of Iraq's WMD programs largely because of incompetent HUMINT collection operations that over-relied on a few, poorly qualified Iraqi defectors, coupled with intelligence analysis that leapt to conclusions that went well beyond what intelligence "evidence" supported.[10] As the Senate Select Committee on Intelligence determined, "Most of the major key judgments in the Intelligence Community's October 2002 National Intelligence Estimate (NIE), *Iraq's Continuing Programs for Weapons of Mass Destruction,* either overstated, or were not supported by, the underlying intelligence reporting. A series of failures, particularly in analytic tradecraft, led to the mischaracterization of the intelligence."[11] The ISG findings exposed to the daylight the dark underbelly of failings in

CIA intelligence that was hidden prior to the war. The NIE's little-noticed caveat—"we lack specific information on many of key aspects of Iraq's WMD programs"—turned out to be a major understatement. And the CIA's gross overestimation of Iraq's nuclear weapons–related activity in 2003 probably reflects, in some measure, an analytic overcompensation for the gross under-estimation of the scope and progress of Iraq's nuclear weapons program in the run-up to the 1991 Gulf War.[12]

The failures on Iraq are by no means isolated events in the CIA's performance against WMD. The CIA failed to warn American policy makers in 1998 of India's nuclear weapons tests that led to reciprocal tests by Pakistan and set South Asia into an overt nuclear weapons race. Adm. David Jeremiah, U.S. Navy (Ret.), a former vice chairman of the Joint Chiefs of Staff, was tasked by DCI Tenet to review that intelligence debacle. Jeremiah concluded that the intelligence community's shoddy performance resulted from analysts who were stretched too thin, satellite collection that was vulnerable to simple deception, and HUMINT that was seriously limited. The conventional mind-set, namely, that India would not test nuclear weapons and risk nega-tive international reaction, prevailed. Moreover, the National Intelligence Officer (NIO) for Warning who sat on the National Intelligence Council that produces NIEs proved incapable of fulfilling his central task to be an effective devil's advocate to counter-prevailing, and profoundly wrong, conventional wisdom at the CIA.[13]

The CIA also failed to gauge WMD activities surrounding Pakistan's nuclear weapons program. CIA defenders hail the disruption of A. Q. Khan's international nuclear weapons production supply ring as a major intelligence coup that led Libya to publicly admit and surrender its nuclear and chemical weapons programs to the international community. That claim does not hold up under close scrutiny, however. Khan's network had been operating for at least ten years before it was ostensibly shut down after the United States put diplomatic pressure on Pakistan's President Musharraf. The network had more than enough time to provide Libya with an enormous infrastructure for building nuclear weapons, including centrifuges for enriching uranium. Khan's network managed to nimbly evade CIA collection efforts against Libya. "The program was much more advanced than we assessed," according to Robert Joseph, former National Security Council director for counter-proliferation.[14] The $100 million deal included Chinese blueprints once given to Pakistan for a nuclear warhead that could be mounted on a ballistic missile, and the CIA also failed to detect that Khan began selling nuclear technology to Iran in the late 1980s.[15]

The Pakistani government, and Musharraf in particular, probably knew that Khan was operating his supply ring despite their public denials. It does not take a great deal of imagination to suspect that Khan was enriching Pakistani military and intelligence service coffers with his nuclear deals in exchange for either turning a blind eye to his activities or even actively facilitating them. These suspicions appear more concrete in light of Musharraf's full pardon of Khan and his refusal to grant the United States or the international community access to Khan to question him. That refusal should raise alarms that Khan's network is still in operation and even more deeply hidden and dispersed to avoid CIA detection.[16]

The CIA also appears to be falling down in the critical task of identifying what other countries the Khan network might still be supplying with nuclear weapons–related equipment and expertise. Leading suspect countries include Syria, Saudi Arabia, and Egypt. Libya's surrender of its nuclear weapons program revealed a clandestine uranium enrichment construction program in South Africa that apparently was undetected by the CIA. A private company, which included some individuals who had been involved in South Africa's past clandestine nuclear weapons program, was manufacturing a plant designed to operate one thousand centrifuges for enriching uranium for shipment to Libya. Once assembled in Libya, the plant could have produced enough weapons-grade uranium for several nuclear bombs per year.[17] These uranium-enrichment kits would be ideal for countries—Syria, Saudi Arabia, and Egypt, for example—that are looking for a shortcut from large nuclear energy–related infrastructure to procure fissile material for nuclear weapons.

Intelligence performance has been dismal in Iran where Tehran's suspected nuclear weapons program went undetected by the CIA for about twenty years.[18] Khan's network appears to have been instrumental in providing critical components to Iran's program, especially centrifuge technology. The massive scope and sophistication of Iran's centrifuge program was revealed in August 2002 by Iranian dissidents and subsequently verified by International Atomic Energy Agency (IAEA) inspections.[19] The IAEA determined that Iran had been violating its obligations under the Non-Proliferation Treaty (NPT) to notify it of uranium-enrichment capabilities for about twenty years. For all that length of time, the CIA appears to have utterly failed at keeping tabs on Iranian nuclear capabilities progress judging—much like Sherlock Holmes's "dog that didn't bark"—from the absence of any public disclosures of such concerns over the past two decades. American intelligence officials have said that they had no evidence during the 1990s that Iran was receiving aid from Pakistan, and one senor intelligence official acknowledged "a fairly major failure, despite the fact that we were watching Iran and Pakistan quite closely."[20]

CIA assessments of North Korea's nuclear weapons program have been of questionable quality. In 2002 the United States was surprised to discover that North Korea had turned to the Khan network for uranium-enrichment capabilities while their plutonium program was ostensibly suspended after being detected in the early 1990s.[21] A scholarly survey of publicly available intelligence assessments of North Korea's nuclear weapons capabilities, moreover, shows that CIA estimates have been erratic and inconsistent. As Asian security expert Jonathan Pollack explains, the CIA in January 2003 told Congress, "North Korea probably has produced enough plutonium for at least one, and possibly two, nuclear weapons," a notably less confident position than assessments in 2001 and 2002 that Pyongyang already possessed one or two weapons.[22] The shifting sands of the CIA's assessment of North Korea's nuclear weapons arsenal—whether potential or actual—no doubt led to frustration among intelligence consumers. Sen. John McCain, for example, who serves on a committee appointed by President Bush to examine the intelligence community's performance against WMD targets, publicly commented in late 2004: "We know very little more about North Korea and Iran than we did 10 years ago. This agency [CIA] needs to be reformed."[23]

The hazy intelligence assessment of North Korea's nuclear weapons stockpile is reminiscent of the CIA's incomplete assessment of South Africa's past nuclear weapons program. Although the CIA had long suspected that South Africa harbored a clandestine nuclear weapons program, it was only after South Africa publicly declared in March 1993 that it had secretly built six nuclear weapons during the 1970s and 1980s that a definitive idea of the South African stockpile emerged.[24] The CIA failed to acquire specific intelligence that would have been needed to militarily move against South Africa's nuclear weapons inventory, much as it appears to have failed today to identify—much less locate—North Korea's nuclear weapons inventory.

The CIA's substantial intelligence failures and weaknesses in gauging the nuclear weapons programs in Iraq, Iran, India, Pakistan, Libya, North Korea, and South Africa over the past two decades portend an even greater string of intelligence failures regarding the chemical and biological weapons programs of adversaries, although these failures have yet to come into the public light. As a "rule of thumb," chemical weapons programs are easier to conceal than nuclear weapons programs because they can more readily be embedded and hidden in civilian economic infrastructures, such as pesticide, fertilizer, and pharmaceutical production facilities. And biological weapons programs require little infrastructure in comparison to nuclear and chemical weapons programs, making them the most difficult to detect via satellite imagery upon which the CIA overly and excessively relies in gauging WMD threats, as the Jeremiah investigation revealed in the aftermath of India's nuclear weapons testing. Detection of chemical and biological warfare programs, as well as nuclear weapons programs, must in no small measure rely on high-quality HUMINT sources with access to the clandestine programs. This is precisely the type of intelligence the CIA has systematically failed to deliver in the past two decades against the WMD targets.

## Diagnosing the Sources of Failure

How is it possible that an intelligence community that is equipped with the world's best technology, manned by about two hundred thousand employees, and funded annually with $40 billion could perform so badly against a critical American national security threat? [25] No matter what the source of intelligence community failures in tracking WMD, the remedy certainly does not lie in conducting business as usual in the high-risk post–September 11 environment.

American government officials and the public in the aftermath of 9/11 have concentrated on bureaucratic or top-down approaches to fix intelligence in general and the CIA in particular with the creation of the DNI. But the DNI post by itself will do nothing to fix the root problems that lie deep within the CIA and are manifest in poor HUMINT collection operations and shoddy intelligence analysis. These core failings will need to be addressed, not by rewiring bureaucratic diagrams, but by changing the institutional culture and business practices at the grass roots level at the CIA.

### Spying to Steal WMD Secrets

How have Iraq, Iran, India, Pakistan, Libya, North Korea, and South Africa managed to outwit the American intelligence community? A hefty share of accountably for these intelligence failings should be placed at the door of the CIA's Directorate of Operations (DO), which is responsible for the placement of America's spies in the world to acquire secrets that other states want to hide from American policy makers. The DO has consistently failed to deliver what American policy makers need most. Secretary Powell, for example, spent endless hours at CIA headquarters preparing his intelligence brief to the UN Security Council in February 2003. [26] In his UN presentation he cited HUMINT reports that Iraq was manufacturing biological weapons using mobile labs, but these reports did not pan out. CIA human reports also were apparently used to inform the poor assessments that Iraq held large weapons stock and was reconstituting a nuclear weapons program.

That CIA had no spies working inside Iraq on WMD since Saddam Hussein unceremoniously threw UN weapons inspectors out of Iraq in 1998 is one of the most a damning findings of the report on American intelligence failures in the run-up to the 2003 war just released by the Senate Select Committee on Intelligence (SSCI). And SSCI, judging from my own seventeen-year career at the CIA, has precisely diagnosed the root cause of the CIA's failure: "a broken corporate culture and poor management," which "will not be solved by additional funding and personnel."[27]

So what were the CIA's case officers in the DO doing to redress the lack of HUMINT in Saddam's Iraq? The answer is not much. The Senate discovered that "when UN inspectors departed Iraq, the placement of HUMINT agents and the development of unilateral sources inside Iraq were not top priorities for the Intelligence Community." Apparently, the DO was "whistling past the graveyard" and hoping against the odds that a crisis in Iraq would not yet again emerge there to expose the gaping hole in American HUMINT, a hole that had never been patched since the First Gulf War.

The DO's failures in Iraq for more than a decade were only the latest manifestation of America's inability to tap human agents to report on the innermost workings of hostile regimes, as well as states plotting courses at odds with American policy interests. The Joint House-Senate Commission investigating the failure of the CIA to warn of the September 11 attacks found that the CIA suffered from an overreliance on the HUMINT sources from other intelligence services, and the CIA had few to no unilateral human agents around the Al Qaeda organization.[28] Admiral Jeremiah likewise found that the CIA suffered from poor human sources in South Asia despite the precarious nuclear balance in the region. The DO, by the account of former DCI Robert Gates, also failed to penetrate the inner workings of the Soviet Union's political apparatus during the entire Cold War. Likewise, the DO failed to penetrate the ruling regimes of Moscow's clients in North Vietnam and North Korea while the United States fought wars with each in the 1950s, the '60s, and the early '70s, much as the DO failed to penetrate Iraq more recently, and was outwitted by East German and Cuban intelligence services who ran double agents against the DO for much of the Cold War.[29]

Despite the DO's dismal, if not derelict, performances over the years, it has escaped a sustained and rigorous examination of its operations and business practices. The DO bureaucracy is set in its ways and operates in a rut created during the Cold War, and many of the organization's benefactors want to avoid the political atmosphere of the Church Committee Hearings in the 1970s. Indeed, the DO has no vested interest in instigating controversial reforms in the absence of strong White House and Congressionally imposed investigations capable of fathoming the origins of DO systemic HUMINT collection failures. Nor is it likely to chart a course of reforms that amount to anything more than window dressing to appease the House and Senate oversight committees that approve the CIA's budget. The DO is even resistant to reform should the DCI be so inclined, and opts to wait out politically appointed DCIs and let any reform agendas wither on the vine after their departures from agency headquarters. Although George Tenet has had a longer tenure at the CIA than most of his predecessors as DCI, it is hard to discern any revolutionary changes he has implemented in the DO. In light of the poor showing in Iraq and failure to warn of the conspiracy of September 11, the often heard CIA refrain and defense to its critics that "agency successes cannot be made public" is wearing thin.

The DO still clings to stealing secrets from states that want to hide them from American policy makers as its *raison d'être.* But by looking back over the battlegrounds littered with CIA intelligence failures—in all of which the lack of accurate and reliable HUMINT reports was a decisive contributing factor—it is clear that the DO has performed dismally against its own core mission requirements. It is long past time for outside bodies, whether in the White House or Congress, to investigate the origins of the DO's failures and to set the organization right because American policy makers and citizens deserve better than the DO has produced.

## *The Requisite Analytic Talent*

The CIA produced shoddy strategic intelligence in the period prior to the cataclysmic events of September 11. CIA analysis of the terrorist target was shallow, in no small measure because "many analysts were inexperienced, unqualified, under-trained, and without access to critical information" the Joint House-Senate Committee assessed.[30] As David Ignatius, who has keen judgment regarding the intelligence world, writes: "The DI analysts work hard, but their product is too often mediocre. . . . America's intelligence analysts should be a match for the best college faculty in the world. They're far from that now, and life outside the CIA cocoon might do them some good."[31]

The lack of analytic expertise in the CIA undoubtedly also contributed to the series of intelligence failures surrounding the prewar assessment of

Iraq's WMD programs. It is no wonder, then, that British intelligence—with a close history of sharing intelligence widely and deeply with American intelligence—produced more balanced and nuanced prewar assessments of Iraq's WMD programs. To be sure, British intelligence suffered from inadequate assessments of Iraq's programs, but its judgments were closer to the mark than the CIA's assessments contained in the October 2002 NIE. For example, in 1995 British intelligence made this accurate judgment: "We assess that [Iraq] may also have hidden some specialized equipment and stocks of precursor chemicals but it is unlikely they have a covert stockpile of weapons or agent in any significant quantity; Hussein Kamil claims there are no remaining stockpiles of agents."[32] More generally, British intelligence in 1998 assessed that "UNSCOM and the IAEA have succeeded in destroying or controlling the vast majority of Saddam's 1991 weapons of mass destruction (WMD) capability."[33] What accounts for the discrepancy between British and American intelligence estimates based on shared raw intelligence information must be the superior quality of British intelligence analysts. The United States would be well advised to seek some pointers from its British comrade in arms on how to recruit, train, and retain solid analysts for working the demanding WMD problem.

The CIA needs to improve its batting stance by changing the mind-sets and talent set of CIA managers, who in turn need to cultivate a new generation of strategic analysts. CIA managers on the analytic side of the business are poorly suited intellectually to oversee strategic analysis, the interface between the realms of politics and military affairs. CIA managers most often come up the analytic ranks as political and economic analysts who are unfamiliar with even a basic knowledge of military affairs. Only a small percentage of agency managers have backgrounds in military affairs, a longstanding agency tradition. The author vividly recalls being counseled by veteran agency military analysts who got their analytic spurs in the heated agency disputes with military intelligence and policy makers over the course of the Vietnam War. They advised in the late 1980s that specializing in military analysis was no place to stay for any analyst who aspired to advance to senior agency ranks.

What was true in the 1980s about the CIA's neglect of strategic and military analysis is true today. The military expertise honed in the Cold War against the Soviet Union and Warsaw Pact suffered significant attrition caused by neglect and retirements after the Berlin Wall collapsed. Agency managers saw no need to replenish their ranks because they labored under a mistaken philosophical worldview that military affairs would not be relevant to the post–Cold War world. During the 1990s, the small cadre of military analysts was a beleaguered band of misfits stretched few and far between analyzing conflicts in the Balkans and the Persian Gulf. Agency management judged that conducting military analysis was a tertiary function; they invested more substantially in political and economic analyses, even while policy makers most often judged the quality of these analyses to be mixed.

All the while, agency managers increasingly allowed military analyst workloads to be driven by tasking from the military services and the Pentagon, while forcing military analysts to devote less attention to more significant customers of strategic analysis in the White House. By the end of the 1990s, most CIA military analysts judged that their collective manpower was so limited by answering the daily deluge of questions, especially from the tactical and operationally oriented military, they were unable to look over the horizon to examine longer-range warning challenges for civilian policy makers, who are not always well served by intelligence analysis coming from operationally oriented military service interests, as Pearl Harbor and Vietnam showed. Reforming the CIA's managerial ranks and pushing aside the mind-sets that have hampered the CIA's ability to conduct strategic warning will require sunshine from the White House and Capitol Hill, as well as the new DNI, because the CIA mind-set that has permeated managerial ranks is doggedly resistant.

The CIA has now launched an unprecedented advertising campaign on its Web site and in major newspapers and periodicals looking for analysts familiar with military affairs and the Middle East and South Asia. But now is, in many respects, too late; American policy makers need expert analysis today in the heat of battle, not several years from now when new recruits begin to emerge as competent military analysts. As a stopgap measure, agency managers are press-ganging analysts to work on Afghanistan, Iraq, and Al Qaeda. That management philosophy, though, is akin to lemmings jumping from a cliff. Siphoning analysts from other issues will further erode the CIA's already poor ability to conduct strategic warning for policy makers of armed conflicts *before* they have broken out.

The CIA has systemically failed to recruit, train, and retain top analytic talent in WMD fields. The CIA was the most vocal within an intelligence community arguing that Saddam's nuclear weapons program was robustly reconstituting. That argument stemmed from one lone CIA analyst in the Office of Weapons Intelligence, Proliferation, and Arms Control whose qualifications to make those judgments were questionable.[34] And even perhaps more damning is that no manager in the CIA's excessively heavy management layers had the expertise needed to critically question that analyst's faulty case.

Policy makers dealing with WMD proliferation know to turn for technical expertise to the National Laboratories run by the Department of Energy, not to the CIA. The dirty little secret in the intelligence community is that the National Labs go out of their way to recruit and train highly qualified Ph.D.s in a variety of disciplines, while CIA management culturally discriminates against Ph.D.s in its hallways. The CIA has even lost a fair share of its best analysts to the National Labs where the working environment is much better than the environs at Langley. As former Assistant Secretary of Defense Ashton Carter has recommended: "The intelligence community needs to increase the size and technical training of its workforce. Because intelligence agencies have difficulty recruiting and training top talent with more lucrative prospects in private industry, they need to forge better links with the outside scientific community so that advice and insight are 'on call.'"[35]

## Bureaucratic Additions, but Not Fixes

The good news is that the executive and legislative branches and the American public have woken up to the fact that American intelligence is in desperate need of major reforms. That belated recognition, unfortunately, came only on the heels of the two most stunning and catastrophic failures in intelligence since Pearl Harbor: 9/11 and the prewar assessments of Iraq's weapons of mass destruction programs. The bad news is that the newly undertaken reforms of the intelligence community do not take aim at the root causes of our intelligence failures—the pathetically poor HUMINT and shoddy analysis primarily produced by the CIA.

Well-meaning intelligence reform advocates including members of Congress and families of 9/11 victims mistakenly fixed their sights on measures recommended by the 9/11 Commission, most notably the creation of the new position, director of national intelligence. The establishment of the DNI, however, unconstructively adds to the already bureaucratically bloated intelligence community. Analysts and HUMINT operators at the CIA had roughly eight layers of bureaucracy separating them from their boss, the director of central intelligence (DCI). The DNI will add at least another layer between the DCI and the president to further separate intelligence officers and their products from their most important customer. Such excessive bureaucratic layering will likely make the CIA even more sluggish in serving the commander in chief and is a far cry from the flatter organizations in the significantly more nimble information-technology firms in the private sector. And if the CIA is to keep pace with American adversaries, it needs to be organized less like an IBM of the 1960s and more like the Intel of today.

The 9/11 Commission's recommendations to form new intelligence centers responsible for terrorism and proliferation is another mistaken target for reforms. The creation of these centers bloats the intelligence community's bureaucracy and does nothing to increase competency in HUMINT collection or analysis. In fact, such fusion centers already exist within the CIA, are staffed by intelligence professionals from the entire intelligence community (not just the CIA), and benefit from the synergetic effects of analysts working hand-in-hand with operations officers. These centers have to improve their future performances to be sure, but the 9/11 Commission failed to argue why the creation of new centers under a DNI—which will have more removed working relations with the CIA's HUMINT collectors—will perform any better. Even if all the 9/11 Commission's recommendations had been in place on September 10, 2001, it would be hard to see how the intelligence community would have averted its failures.

Would-be reformers need to retrain their sights from extremities such as the DNI to the vital targets of HUMINT collection and analysis. Post-9/11 investigations, including the Joint House-Senate Investigation, the Senate Select Committee's investigation, and the 9/11 Commission, all found profound shortcomings in HUMINT collection and the quality of intelligence analysis. The last report in particular shied away from a menu of deep, thoughtful recommendations on how best to improve these performances from the grassroots up at the CIA. Instead, it defaulted to the easier task of making superficial recommendations for changing bureaucratic wiring diagrams and calling the work done.

## Fixing the Fundamentals of Spying and Analysis

To the military's and Defense Department's credit there is much discussion, debate, and movement toward a transformation in military affairs. In contrast, there is little debate and discussion, much less implementation, in a parallel effort to transform the business practices of the CIA and American intelligence. This disconnect is significant in that wielding the military instrument in preemptive or preventive action in the future will in large measure depend on high-quality strategic intelligence to inform civilian policy makers as well as to guide military planning for execution of strikes against WMD-armed adversaries.

The CIA's leadership had determined that spying in Saddam's Iraq was simply too hard, a conclusion that reflects the mistaken institutional and cultural bias for running agents in place. Such an objective is a failed business practice and legacy of the Cold War. DO officers are loath to admit it, but spies, by and large, are not seduced by case officers to commit treason—they volunteer. As former CIA case officer Reuel Marc Gerecht rightly observes, the best spies are not seduced; they volunteer their services to the United States.[36] In the cases of volunteers or "walk-ins" from hard targets like Iraq, the DO would be far better off trying to get them to defect for debriefings rather than trying to turn them around and go back into their countries to report in place.

The CIA needs to concentrate on vetting walk-ins who volunteer intelligence to CIA case officers overseas while encouraging and facilitating defections from the intelligence targets that matter most to American interests. The CIA today, for example, should be sparing no expense or effort to encourage defections from Iran, whether from members of the Revolutionary Guard or technicians and scientists from Iran's suspected nuclear weapons program for intelligence, as well as to disrupt Tehran's suspected nuclear weapons aspirations. Although defections offer a one-time snapshot of clandestine activities, one snapshot is better than none. And if given a critical mass of reporting from dozens of defectors from Iran, CIA analysts probably could paste together a clearer picture of Iran's nuclear weapons program than the CIA has today.

Many commentators and observers take a lesson learned from the CIA's Iraq intelligence failure that defectors cannot be trusted. But neither can paid or "controlled" human assets on the CIA's payroll be trusted. They too have a vested interest in telling the United States what they think Washington wants to hear if they are to keep those CIA paychecks coming. What CIA analysts need—much like investigative reporters—is a wide variety of reporting sources that can be cross-checked against other sources to fathom ground truth.

The CIA needs to sever its umbilical cord to U.S. official facilities overseas as its primary infrastructure for HUMINT collection. SSCI found that human operations "against a closed society like Iraq prior to Operation Iraqi Freedom were hobbled by the Intelligence Community's dependence on having an official U.S. presence in-country to mount clandestine HUMINT collection efforts." To be sure, case officers based under diplomatic cover in these facilities will be critical to receiving and debriefing walk-ins and defectors, as well as for liaison with host intelligence services. The United States, however, needs more robust HUMINT collection means against hard targets such as China, Iran, North Korea, Pakistan, Russia, Saudi Arabia, and Syria where the United States has no official diplomatic presence or where the counterintelligence environment is extremely tight for case officers under official cover to effectively operate.

The CIA needs to substantially bolster the use of nonofficial cover officers who have no connections to the American diplomatic infrastructure. Such officers can melt into areas rich in hard target HUMINT collection opportunities, such as the Muslim expatriate communities in Europe that are hotbeds for Al Qaeda recruitment, indoctrination, and logistics or in the Chinese expatriate communities in Asia. CIA nonofficial cover officers in the past have suffered from neglect at the hands of old school DO managers who dominate the CIA's bureaucratic power structure. As intelligence expert James Bamford has observed, such cover officers are forced to operate under the authority of the DO's regional offices, which traditionally look down upon their nonofficial counterparts.[37] One of the first orders of business for the DNI is to ensure that the DCI gives nonofficial cover officers their own separate chain of command that runs directly into the DCI's suite. The infusion of competition between a rejuvenated and independent nonofficial cover officers program could be constructively managed by the DCI to put a "fire" under the backsides of the recalcitrant, risk adverse, and old school DO officers. As it stands today, the old school DO is a monopoly that needs to be deregulated under the DCI's authority.

As for the CIA's strategic analysis shortcomings, the DNI needs to ensure that the agency recruits, trains, and rewards top analytic talent in WMD fields. President Bush's order to the CIA to increase by 50 percent the number of operations officers and analysts—paying special attention to target terrorism and WMD proliferation—represents a surge in hiring and a slight acknowledgment of this problem.[38] But merely throwing a larger quantity of analysts at the problem will not yield qualitatively better intelligence. Unless training and expertise expectations are higher and CIA management nurtures them, the CIA will merely have even younger and more inexperienced analysts than before 9/11.

As it stands today, case officers and analysts are ill equipped and not empowered by CIA managers to undertake the critical task of expertly monitoring WMD proliferation. DO case officers are still trained to troll for potential human source recruits in the diplomatic cocktail party circuit. But foreign diplomats are not likely to be very valuable intelligence sources because more often than not they are not privy to—or are completely ignorant about—information regarding the strategic intent, purposes, or status of their countries' WMD programs. And CIA analysts are force-fed a data stream from classified sources by their managers, an intellectual feeding tube that deprives analysts of opportunities to fully exploit the explosion in publicly available literature relevant to WMD proliferation.

Analysts would do much better in monitoring WMD if they had the charge, responsibility, and freedom to track down open source leads and have discussions inside American and foreign scientific and WMD-related communities much like any investigative reporter does working for a major news organization. Alas, the security bubble in which analysts work is simply too restrictive, if not oppressive, for such business practices. If the United States is to improve its intelligence performance against WMD, it needs to substantially lessen the defensive counterintelligence and security crouch it has assumed for too long. CIA case officers and analysts need to take substantially more risks to run effective offenses against an array of increasingly sophisticated and dangerous WMD aspirants and possessors that threaten American national security interests.

What is now required from both the DNI and the CIA is a cadre of strong, enlightened, aggressive, and innovative leaders who recognize the profound need to change the business cultures and practices of collection and analysis. And like changing the course of the USS *Ronald Reagan,* that process will require a steely determination to counter deep bureaucratic rot as well as a steady and firm hand at the helm for a period of time beyond the usual year and a half for an average presidential appointee. DCI Porter Goss may indeed be working along these lines in his shake-up of the CIA's senior management. The story line in the media is that Goss is politicizing the agency with the retirements of numerous senior officers from the analytic and operational sides of the agency, a line that is no doubt fueled by leaks to the press by unnamed senior officials who are encrusted in the status quo bureaucracy.

It is imperative for the American public to put these senior departures in context. If a *Fortune 500* company were to have performed as poorly as the CIA has done in the past several years, shareholders would rightly be demanding the mass dismissal of the company's senior management. In this light, the retirements of a handful—out of hundreds of senior CIA officials—could hardly be characterized as a "blood bath," and those few should not satisfy

American shareholders, that is, taxpayers. Judging from my own seventeen-year career at the CIA, the mid-level ranks of the agency—which do not leak opinions to the press out of fear of being fired—would welcome more retirements from the staid senior CIA rungs to clear the stale and stagnant air at Langley and to get on with the essential business of stealing secrets and figuring out the next moves of America's adversaries.

The gauntlet has been thrown down; the DNI and the CIA will have to boldly move to redress profound intelligence collection and analysis shortcomings if American intelligence is to be able to give the commander in chief the quality of intelligence that is needed in a high-stakes international security environment littered with nation-states armed with WMD and Islamic extremists eager to lay their hands on them. The challenge for the DNI, the director of the CIA, and the next generation of intelligence officers is to prove Clausewitz's dismal assessment of intelligence wrong. Regrettably, the wisdom of Clausewitz is a rare commodity today, and no more so than in the CIA and the American intelligence community.

# Notes

1. Carl von Clausewitz, *On War,* Michael Howard and Peter Paret, eds. and trans. (Princeton, N.J.: Princeton University Press, 1989), 117.

2. George W. Bush, *The National Security Strategy of the United States of America* (Washington, D.C.: The White House, September 2002), 15. Available at http://www.whitehouse.gov/nsc/nss.pdf. In strategic discourse, a preemptive strike traditionally has referred to attacking an adversary in anticipation of an impending, imminent war while preventive strikes generally refer to destroying a potential adversary's nascent capabilities to stop them from growing into a significant threat over time. In the contemporary debate, preemptive and preventive strikes are commonly used interchangeably.

3. See Richard L. Russell, "Military Retaliation for Terrorism: The 1998 Cruise Missile Strikes Against al-Qaeda in Afghanistan and Sudan," Pew Case Study (Washington, D.C.: Institute for the Study of Diplomacy, 2002).

4. Jason D. Ellis and Geoffrey D. Kiefer, *Combating Proliferation: Strategic Intelligence and Security Policy* (Baltimore, Md.: The Johns Hopkins University Press, 2004), 62.

5. "Key Judgments," Iraq's Continuing Programs for Weapons of Mass Destruction, National Intelligence Estimate, October 2002. Available at http:// www.fas.org/ irp/cia/product/iraq-wmd.html.

6. Bob Woodward, *Plan of Attack* (New York: Simon & Schuster, 2004), 249.

7. "Secretary Powell at the UN: Iraq's Failure to Disarm," Bureau of Public Affairs, U.S. Department of State. Available at http://www.state.gov/p/nea/disarm/.

8. For an excellent examination of the turning point from hiding, preserving and rebuilding WMD capabilities to weathering economic hardships until UN sanctions would be lifted, see Barton Gellman, "Iraq's Arsenal of Ambitions," *Washington Post*, 7 January 2004.

9. Special Advisor to the DCI on Iraq's WMD, *Comprehensive Report*, Volume I, 30 September 2004.

10. Bob Drogin, "Spy Work in Iraq Riddled by Failures," *Los Angeles Times*, 17 June 2004, A1.

11. Senate Select Committee on Intelligence, "Conclusions," *Report on the U.S. Intelligence Community's Prewar Intelligence Assessments on Iraq* (Washington, D.C.: U.S. Senate, 9 July 2004), 14.

12. See Richard L. Russell, "CIA's Strategic Intelligence in Iraq," *Political Science Quarterly* 117, no. 2 (Summer 2002), 201.

13. Walter Pincus, "Spy Agencies Faulted for Missing Indian Tests," *Washington Post*, 3 June 1998, A18.

14. David E. Sanger and William J. Broad, "Pakistan's Nuclear Earnings," *New York Times*, 16 March 2004.

15. William J. Broad and David E. Sanger, "The Bomb Merchant," *New York Times*, 26 December 2004, A1.

16. For an examination of the strategic rationales for nuclear weapons in Saudi Arabia, see Richard L. Russell, "Saudi Nukes: A Looming Intelligence Failure," *Washington Times*, 5 January 2004.

17. Douglas Franz and William C. Rempel, "New Find in a Nuclear World," *Los Angeles Times*, 28 November 2004, A1.

18. David E. Sanger and William J. Broad, "From Rogue Nuclear Programs, Web of Trails Leads to Pakistan," *New York Times*, 4 January 2004, A1.

19. Douglas Frantz, "Iran Moving Methodically Toward Nuclear Capability," *Los Angeles Times*, 21 October 2004, A1. For more discussion of the massive scope and sophistication of Iran's uranium enrichment program, see Joby Warrick and Glenn Kessler, "Iran's Nuclear Program Speeds Ahead," *Washington Post*, 10 March 2003, A1. For Iran's strategic rationale for a nuclear weapons program, see Richard L. Russell, "Iran in Iraq's Shadow: Dealing with Tehran's Nuclear Weapons Bid," *Parameters* XXXIV, no. 3 (Autumn 2004), 32–34.

20. David E. Sanger, "Pakistan Found to Aid Iran Nuclear Efforts," *New York Times*, 2 September 2004.

21. Sanger and Broad, "From Rogue Nuclear Programs."

22. Jonathan D. Pollack, "The United States, North Korea, and the End of the Agreed Framework," *Naval War College Review* LVI, no. 3 (Summer 2003), 13.

23. Douglas Jehl, "Bush's Arms Intelligence Panel Works in Secret," *New York Times,* 6 December 2004.

24. Phillip van Niekerk, "South Africa Had Six A-Bombs," *Washington Post,* 25 March 1993. For analyses of South Africa's nuclear weapons program and its decision to abandon it, see Peter Liberman, "The Rise and Fall of the South African Bomb," *International Security* 26, no. 2 (Fall 2001), 45–86; and J. W. de Villiers, Roger Jardine, and Mitchell Reiss, "Why South Africa Gave Up the Bomb," *Foreign Affairs* 72, no. 5 (November/December), 98–109.

25. On the funding and personnel strength of the intelligence community, see Greg Miller, "Iraq Envoy to be Chief of Intelligence," *Los Angeles Times,* 18 February 2005, A1.

26. Woodward, *Plan of Attack,* 299.

27. Senate Select Committee on Intelligence, "Conclusions," *Report on the U.S. Intelligence Community's Prewar Intelligence Assessments on Iraq* (Washington, D.C.: U.S. Senate, 9 July 2004), 24. For more in-depth examinations of CIA's human collection and analytic shortcomings, see Richard L. Russell, "Spies Like Them," *National Interest* 77 (Fall 2004), 59–62; and Richard L. Russell, "Intelligence Failures: The Wrong Model for the War on Terror," *Policy Review* 123 (February & March 2004), 61–72.

28. Findings of the Final Report of the Joint Inquiry Investigating the Attacks of September 11, 2001 (Washington, D.C.: U.S. Government, 10 December 2002). The report is available on the House Permanent Select Committee on Intelligence's website at http://intelligence.house.gov/CaseStudies.aspx?Section=11.

29. For a revealing and damning discussion of CIA's poor human intelligence performances including during the Cold War and today, see John Diamond, "CIA's Spy Network Thin," *USA Today,* 22 September 2004, A13.

30. Findings of the Final Report of the Joint Inquiry Investigating the Attacks of September 11, 2001.

31. David Ignatius, "Spying: Time to Think Outside the Box," *Washington Post,* 29 August 2004, B7.

32. Report of a Committee of Privy Counsellors, *Review of Intelligence on Weapons of Mass Destruction* (Norwich, U.K.: Her Majesty's Stationery Office, 14 July 2004), 47.

33. Ibid., 49.

34. For an examination how one analyst of questionable authority in CIA dominated the analytic assessment that Iraq was actively reconstituting its nuclear weapons program, see David Barstow, "How the White House Embraced Disputed Arms Intelligence," *New York Times,* 3 October 2004.

35. Ashton B. Carter, "How to Counter WMD," *Foreign Affairs* 83, no. 5 (September/October 2004), 85.

36. Reuel Marc Gerecht, "The Sorry State of the CIA," *The Weekly Standard*, 19 July 2004.

37. James Bamford, *A Pretext for War: 9/11, Iraq, and the Abuse of America's Intelligence Agencies* (New York: Doubleday, 2004), 200.

38. Walter Pincus, "Bush Orders the CIA to Hire More Spies," *Washington Post*, 24 November 2004, A4.

# TWENTY-SIX

# Making the Case

## *Defense Counterintelligence as a Strategic Asset*

Anthony D. Mc Ivor and Roy L. Reed Jr.

### Rethinking Defense Counterintelligence

The one lesson that leadership of the Defense Department's counterintelligence (Defense CI) community has taken to heart in recent years came from a former deputy director at the Defense Intelligence Agency (DIA) in the early 1990s. After an exhaustive review of a proposal to move responsibility for Defense CI operations to the DIA, he observed that "more support, respect, and confidence in the endeavors of these [CI] people would help the Defense CI mission more than any restructure initiative." The observation is as true today as it was when written.

Intelligence reform proposals have for the last year been circulating far and wide, under many flags. In the near future, more remodeling blueprints will be offered; the plans of others will come undone as the new director for national intelligence establishes his authority; and because the United States remains at war with terrorists abroad and at home as well as insurgents in Iraq, some political points will indubitably be scored. But for all the storm and thunder in the public arena, one important development will continue to be overlooked: major reform in the practice and processes, if not always the structure, of American intelligence has been well under way for some time.

The views expressed herein are the authors' alone. They do not reflect the policies or positions of the U.S. government, nor those of the organizations where the authors are based.

Nowhere is this more evident than in Defense CI. The focus is on moving counterintelligence *back* to the center of the projection of national power and influence in response to evolving and emerging conflicts and threats.

In a very short time, the practice of counterintelligence has experienced profound, and we suspect, enduring changes. In part, this is a change in perception. But in greater measure, it is a change in expectations, both those internal to the discipline and those imposed externally. For Defense CI, meeting the challenges posed by those changes opens the door to opportunity. From the definition of the mission to the professionalization of the counterintelligence corps, Defense CI has been rethinking its ability to bring CI capabilities to bear in support of the Department of Defense (DoD)'s military missions, to protect America's technology, and to contribute to the broader intelligence community requirements as well.

This head-to-toe reevaluation is being encouraged at the highest levels of government and throughout the community. As the pace of critical reviews of intelligence practices intensified in the years after the Cold War, the language was not often flattering. An influential 1997 study, chaired by Lt. Gen. William E. Odom, U.S. Army (Ret.), concluded that the problems and issues confronting counterintelligence "present a disturbing picture of CI as a fragmented, poorly coordinated, and often amateurish discipline within the intelligence community." In subsequent remarks at the Hudson Institute, General Odom identified counterintelligence reform as "the most urgent and in his view the most seriously needed."[1] While he was referring to counterintelligence writ large, we would argue that General Odom's remarks were very close to the mark for Defense CI in particular—at that time.

The Defense CI community, which absorbed the "two antagonistic cultures" analysis in the House Permanent Select Committee on Intelligence (HPSCI ) staff study of the previous year (*IC21*),[2] was already in motion. Responses to several of the key points made by the Odom study were under way in Defense CI components by the turn of the decade. The specific recommendations for a national-level CI school and for a multidisciplinary approach to CI collection were soon to be realized by a new Defense CI organization. In the process, the argument for a fresh appreciation of counterintelligence as a strategic asset was being built. The following paragraphs will examine how this happened. The focus here is on Defense CI today.

With change, some might say transformation, reshaping the country's entire national security posture, why is a review of counterintelligence particularly relevant? Skeptics will wonder whether there isn't more smoke than fire.

Looking beyond the crisp new tactical CI role on the battlefield in Iraq and some innovative data management approaches to the global war on terror, old hands will ask whether the Defense CI's traditional "silent war" isn't still the greater threat to American security. They will rightly insist that the human dimension to the traditional clandestine or covert threats met by counterintelligence capabilities has scarcely changed. Those are valid points. We concede them, but we also wish to set the argument in a broader context.

We will argue that while the traditional missions have not disappeared, the traditional methods of accomplishing them are no longer sufficient. More importantly, CI is being tasked with new roles, in new environments, that will change business practices—in some cases, beyond recognition. In short, counterintelligence is and ought to be at the forefront of the general movement of intelligence from the periphery to the center of U.S. national security policy. Much of the professional excitement in the field today comes from recognition among practitioners that the discipline is being reshaped to its core. And in the process, new opportunities to raise the profile and the field of play for counterintelligence professionals are present as never before. A quick survey of four leading CI issues in the community provides ample illustration.

## Integration

Integration is foremost among those issues. Defense CI components are undergoing an unprecedented effort to standardize and synchronize their activities. Structure, policies, and practices are all on the table. Efforts to improve information sharing run the full gamut of intelligence functions, from CI functions like investigations and operations to foreign intelligence functions, such as human intelligence (HUMINT), signals intelligence (SIGINT), measurement and signature intelligence (MASINT), and back again. Aggressive development of a secure system of systems is changing the information flow and the expectations of CI consumers. Defense CI is closing on the goal of integrating functions and technology at the point of operational execution. As that happens, the insertion of CI into the full spectrum of military operations offers new tactical and strategic tools for combatant commanders.

## Strategic Analysis

Historically, in a Defense CI perspective strategic analysis has not been a strongpoint, within the DoD or elsewhere. A long-term analytic capability to identify, neutralize, or exploit "unknown, unseen, and unexpected threats" was cited as an aspiration in the Defense Annual Report to the President as recently as 2002. Since then, extraordinary strides in interagency collaboration have changed the equation. Today, the issues surrounding analysis have shifted fundamentally. Services and the combatant commands, together with the intelligence, security, and law communities, are engaged in joint horizontal risk assessments that were unimaginable in the 1990s. This has raised the performance bar. Strategic analysis is now beginning to drive orchestrated Defense counterintelligence activity directed toward the strategic goals, objectives, and outcomes of the Department of Defense. Enhanced capabilities, fueled by shared information, create the common operational pictures vital to a commander's ability to assign campaign priorities and targeting.

### Professional and Organizational Development

The third area that reveals the profound changes taking shape in the Defense CI discipline is professional and organizational development. Efforts under way today to professionalize the CI corps are giving counterintelligence analysts and operators access to new skill sets and career paths. By redefining technical disciplines and creating general road maps for them, with the requisite qualification and training benchmarks, Defense CI is making a significant new investment in that mission-essential weapon—the human being. Not surprisingly, these efforts place renewed emphasis on language fluency and foreign thought worlds in order to remedy occasionally dismal levels of "intercultural effectiveness."[3] This activity goes hand-in-hand with better outreach to new partners in allied communities.

In combination, these initiatives are developing a new breed of CI professional with a greater range of technical skills, better communications and relationships with interagency counterparts, and sufficient prospects for advancement to stay in the CI career field. The classes now emerging from the recently established Joint Counterintelligence Training Academy in Maryland bear witness to these changes. The men and women passing through the academy are better positioned than ever to produce what Undersecretary of Defense for Intelligence Dr. Stephen Cambone has called "exquisite intelligence," the knowledge of our adversaries' secrets without them knowing that we know.

### Innovative Use of Technology

Innovative uses of technology as a critical enabler are the fourth bellwether of new trends in the community. Creative applications of forensic technology have grown exponentially in the past several years. To leverage them, Defense CI is building a common IT framework that will synchronize resources and share information in ways that were inconceivable a generation ago. Secure communications that bring operators and analysts into real-time collaboration at the tactical, operational, and strategic levels are changing traditional workflows and relationships. New approaches to collection, enhanced analytical programs (such as nonobvious relationships), and the next generation of credibility assessment tools all point to new ways of conducting CI's business.

The leading elements in this changed environment began to emerge long before the sea change in expectations and practices that occurred in the wake of 9/11. A leading harbinger of that environment was the small group of CI visionaries from across the government known collectively as CI-21. Their goal was to refocus national counterintelligence to better deal with novel and emerging twenty-first-century clandestine or covert threats. Within Defense, this effort transformed Defense counterintelligence thinking and resulted in the creation of the Joint Counterintelligence Assessment Group (JCAG), the precursor to the present Defense Counterintelligence Field Activity.

## A New Environment

*"Knowledge is of two kinds. We know a subject ourselves, or we know where we can find information upon it."*

Samuel Johnson

The original charter of the JCAG organization was to develop more effective horizontal risk assessments for high technology programs. A proof of concept, the JCAG was established by the deputy secretary of defense as part of the stand-up of a new Defense counterintelligence effort, the Defense Joint CI Program (DJCIP), within the DoD's Joint Military Intelligence Program. The DJCIP was explicitly created to fund counterintelligence capabilities within the military departments. It was also to support the protection of critical Defense assets in concert with the counterintelligence missions identified in a new Presidential Decision Directive on U.S. Counterintelligence, the latter itself a result of the CI-21 effort. The JCAG soon became a focal point in the Defense-wide piece of the DJCIP.

The pioneering work done at JCAG initially concentrated on leveraging new and emerging technologies to conduct net assessments of risks to critical Defense assets, primarily in research and technology. JCAG's mission was to focus on horizontal, integrated risk assessments that ran across organizational boundaries. The JCAG concept and resulting program capitalized on the earlier work done by the U.S. Army's Land Information Warfare Activity and came to reside in the Information Dominance Center there.

As JCAG activity stood up, it soon began to blur and then to bridge the borders between traditional information technology and analysis functions. By bringing the two fields into direct collaboration, old-staff-versus-line barriers were broken, unleashing a new source of creative energy. This process is now beginning to include the operators as well. As information technologists, analysts, and collectors began to anticipate and reckon with the needs and the limitations of each other's disciplines, the work tempo increased. The idea that significantly more timely and more actionable products could reach the desks of Defense planners and decision makers took root.

The JCAG actually began a process of long-term intellectual capital building through technical test and experimentation. Some intelligence organizations commit up to a third of their analytic effort to this type of production; for Defense CI this was a new venture. The process borrowed from such concepts as augmenting human intellect and collective IQ to not only execute faster and smarter but also to understand how the core capabilities could continuously improve their processes and their performance. Most importantly, the change management process itself came under scrutiny. The early ideas—specifically those on collaborative knowledge and bootstrapping the improvement process—of Dr. Douglas Engelbart and other knowledge management researchers at SRI International, and elsewhere, were instrumental to the success of these initiatives.[4]

Adoption of standard knowledge management protocols began to make tacit CI knowledge explicit and extendible. Along the way, analysts started to think more like investigators, technologists more like collectors. To build upon these changes, JCAG explored new business structures, engaged the private sector in partnerships, and borrowed ideas from successful commercial enterprises such as Amazon.com. New skills, more in tune with library science than surveillance, were enlisted. At the water cooler, discussions of complexity and chaos theory mixed with assessments of new communications technology.

While JCAG continued to use technology to open doors to new processes, two events had an inexorable impact on the original JCAG concept. First, the terrorist attacks of 9/11 redirected the JCAG to immediate problems related to the global war on terrorism. For JCAG, the focus became support for domestic threat analysis for the protection of DoD personnel and installations in the continental United States. This evolved into a strong commitment of resources for the stand-up of a new combatant command, USNORTHCOM.

Second, the Defense Department was preparing for full implementation of the new Presidential Decision Directive 75, which originated in an interagency study undertaken by the CIA, the FBI, and the DoD. Beginning in June 1999, this joint effort eventually produced a blueprint titled "U.S. Counterintelligence Effectiveness for the 21st Century," or simply CI-21— the moniker embraced by its early advocates. The shortened name stuck. The president signed PDD-75 in December 2000.

The CI-21 draft had been extensively coordinated over a period of many months. In the legislative branch, both the HPSCI and the Senate Select Committee on Intelligence provided review and comment. Wide circulation assured that the study included a consensus view of the problems facing counterintelligence—across the community. As Roy Jonkers pointed out at the time, CI-21 was erected on the earlier National Counterintelligence Center concept and gave the latter an "integrated interdepartmental coordination."[5] CI-21 established a framework for reorganizing national counterintelligence.

Traditionally organized around the CI functions—investigations, operations, collection, analysis, production, and functional services—Defense CI was to be reorganized around the four CI missions called out in PDD-75. Those core missions were identified as: (1) support to force protection, (2) support to research and technology protection, (3) support to critical infrastructure protection, and (4) countering clandestine and covert threats. The Defense CI community was put on notice that daily practices and organization were to be aligned to emphasize these missions. The primary instrument for that realignment was to be a new counterintelligence field activity.

In February 2002 those changes were institutionalized with the formal establishment of the Counterintelligence Field Activity (CIFA). The text of the implementation directive makes very clear that the DoD fully supported the National Counterintelligence Program and the National Counterintelligence Executive created by PDD-75. CIFA became the single coordination focal point for the department's "CI Policy implementation, Defense-wide CI resource and budget planning and for DOD CI implementation liaison with the NCIX staff."[6] With CIFA, a new strategic vision for Defense CI was born.

The ills identified in the CI-21 study were immediately recognizable to anyone with intelligence community experience. They included

- inadequate coordination between policy and counterintelligence
- incomplete cooperation and information sharing

- lack of strategic counterintelligence threat analysis
- a reactive rather than proactive focus
- failure to exploit new technologies
- lack of a national plan for integration of information and analysis
- inadequately prepared workforce and insufficient resources
- absence of a national advocate for resources and policies
- inadequate coordination with the private sector—the evident and enduring popularity of Ray Semko's security briefs notwithstanding

Remedy of these shortcomings was the primary objective of CI-21's interagency study team. Perhaps most importantly, by signaling a consensus, one that included the DoD, on the most critical issues at that juncture, it gives us a means to measure subsequent activity.

Why is all of this important to the state of Defense CI today? The answer lies in the DoD's response to these developments. While the 107th Congress held hearings on the Robert Hannsen espionage case, reviewed systemic security issues at the FBI (as documented in the Webster Commission's report in March 2002), and examined the effectiveness of the counterintelligence function at the Department of Energy's Office of Counterintelligence and the National Nuclear Security Administration's Office of Defense Nuclear Counterintelligence, the Defense Department quietly prepared substantive changes in both counterintelligence policy and organization, in close alignment with the objectives of PDD-75. The new field activity was given a charter for change.

## CIFA: A New Strategic Direction

The Defense Department established CIFA to deal with the need for an increasingly integrated Defense CI system that could orchestrate strategic CI activity across its multiple components—Agencies, Commands and Departments—some with their own CI stovepipe and others exposed by gaps in continuous CI coverage. Whether enshrined in history, culture, or technology investments, barriers to close cooperation and information sharing were going to fall. Standing up the new organization was the plainest possible statement that for Defense CI, business as usual was over. In fact, the DoD had argued earlier that, among U.S. intelligence priorities, a premium on the importance of strategic counterintelligence was warranted.

Establishing a separate defense component for counterintelligence had additional benefits. First, it ensured that, as is the case with every defense component, the secretary of defense himself had an organization to call on to ensure that his Title 10 responsibilities related to routine counterintelligence were met, as needed. Second, it allowed the current defense secretary, who has a good understanding of the importance of counterintelligence, direct visibility and access to an organization leveraging the importance of CI as a strategic asset. Third, it gave Congress better visibility and insight into defense-wide CI matters since there would be a defense component with counterintelligence as its principal mission.

Fourth, in a way consistent with the thesis behind the Presidential Decision Directive on counterintelligence, it established a mechanism for Defense to implement a new national model for counterintelligence. The point is significant. While the FBI and the CIA are routinely seen as the big national players in counterintelligence matters, the lion's share of the equities, by far, is under the mantle of Defense.

Finally, the notion of a separate component within the DoD for counterintelligence gave the department an acceptable structure to "bridge" law enforcement information into the all source intelligence analysis venue. Again, this is not a trivial benefit. A distinct organizational construct for counterintelligence for defense-wide strategic matters is consistent with how Title 50, the National Security Act, distinguishes between the content of counterintelligence as opposed to foreign intelligence. Post–9/11, this is particularly relevant to the DoD's ability to conduct meaningful analysis of foreign threats to Defense Department installations and facilities *within U.S. borders,* without aggravating the cultural, perceptual, and political issues related to the nation's historic aversion to such domestic activity by U.S. foreign intelligence entities (such as the CIA).

Because law enforcement is an unavoidable aspect of CI, which by its definition in Executive Order 12333 deals with issues that are violations of the U.S. Criminal Code (espionage, sabotage, international terrorism), the DoD has recognized the need for a defense-wide counterintelligence organization to have some, even if limited, law enforcement authority. To that end, the department concluded that any Defense component addressing defense-wide strategic CI issues needed to be a predominately counterintelligence rather than foreign intelligence organization. While this helps not only to ensure that appropriate safeguards are in place but also to address matters of perception, the solution remains problematic.

Counterintelligence as a strategic capability does not stop at America's borders. The constant concern for the civil rights of citizens renders the CI–law enforcement relationship one of continual oversight and judicial discretion. In the end, however, being an organization with a singular core competency allows CIFA to bridge successfully law enforcement, intelligence, operational, and security issues from a truly global perspective—and to do so in a manner consistent with U.S. law and legal custom. It is a defining influence on the conduct of CIFA's missions.

The four mission areas derived from PDD-75 and the Defense CI strategic and performance goals being developed under the guidance of the undersecretary of defense for intelligence locate CIFA's work on familiar ground. Whether supporting the missions of the services and combatant commanders, influencing adversaries and competitors, maintaining DoD advantages, or avoiding strategic surprise, these are becoming defining vectors for Defense CI. The difference comes out in closer examination of CIFA's role in the CI community. As a coordinator linking the Office of the Secretary of Defense (OSD) with the National Counterintelligence Executive (NCIX) and the larger community, as a developer building new tools for analysis and collection, and as an integrator of a common operational picture and the system to deploy it in CI campaigns, CIFA is changing the ways Defense counterintelligence gets its business done.

CIFA is working with OSD and intelligence community leadership to bring into play new perspectives on CI capabilities. By engaging those leaders, CIFA is opening opportunities for CI professionals to contribute to the evolving national security framework. These discussions are challenging long-held assumptions about the interplay of counterintelligence, law enforcement, and foreign intelligence. In the present international environment, where joint and coalition approaches are becoming standard procedure, the need for this sort of fresh thinking is urgent.

## Emerging Defense CI Strategy Moves to the Field

The CI community–oriented activity undertaken at CIFA is beginning to make a difference. But the real story is still being written in the trenches. Execution is the province of the service and agency CI corps. Defense leadership across administrations have cited service and defense agency CI investigations, operations, and related activities as being critical to successful operational outcomes in both Gulf Wars. Counterintelligence analysts in the services and the DIA have become critical components in bringing the necessary range of points of view to the table in the indications and warnings venue. Since the terrorist attacks on Khobar Towers and the USS *Cole,* all of the DoD's counterintelligence assets, across all defense components, have emerged as crucial contributors to the DoD's force protection and combating terrorism programs—at home and abroad.

To cite just one example, the Air Force Office of Special Investigations (AFOSI) program Eagle Eyes and its companion reporting vehicle, the Threat and Local Observation Notice (TALON), are among the most recognized antiterrorism tools in the Defense Department. In 2002, the department designated TALON as the DoD standard for reporting suspicious activity. The program highlights processes for rapid follow-up investigations and information sharing with other echelons of command and other law enforcement agencies. Eagle Eyes is effective because it elicits the support not just of military personnel, but the entire community associated with an installation: civilian workers, family members, contractors, off-base merchants, community organizations, neighborhoods, and civilian news media. Joint task forces that include the FBI and state police organizations are leveraging the program. The communications technology that drives the Eagle Eyes/TALON system is facilitating new forms of interagency collaboration.

Today, CI operators, analysts, and support personnel are embracing change. Their accomplishments—and the methods of accomplishment—speak volumes. Again, seen through the prism of our initial four CI issues, the cumulative beneficial effects of improved integration, stronger tools for analysis, professionalization opportunities, and new technology are becoming visible across Defense CI. This is good news, but is it enough? A quick look at some persistent limitations and shortfalls among those same issues suggests the answer may still be "not yet."

## Perennial CI Challenges

The benefits of integration seem beyond question. Better planning, shared information, and synchronization of efforts can all deliver solid payoffs. But is the huge investment in collaboration really worth the risk and the effort? Recent surveys conducted by Helsinki-based knowledge management professor Karl-Erik Sveiby suggest that we don't really know. In Sveiby's words, the time and money spent enhancing collaboration is still "a leap of faith."[7] Worse, according to Sveiby, is that almost all we do know is anecdotal evidence provided by consultants, managers, and IT vendors—all of whom have a vested interest in the message.

Many of the same knowledge management or intellectual capital questions are applicable to evaluating recent changes in the conduct of strategic analysis in CI. The danger posed by inherited assumptions remains a perennial problem. Automating and distributing preconceived ideas hardly enhances analysis. Another lingering question is, "What is the true yield?" Is it measurable in any meaningful way? If, as Sveiby and others have suggested, knowledge management and intellectual capital practices are today still in their infancy, the road ahead will be long indeed. Referring to the private sector, Sveiby identified the "corporate culture of competition, hoarding, and power" as the single greatest impediment to effective knowledge management strategies.[8] In some respects, he could quite easily have been writing about the CI community.

On the surface, a positive view of CI education, professionalization, and career enhancement efforts is widely shared. But intelligence scholar Ernest May's observation about the need to understand tribes in government much as anthropologists strive to understand tribal communities elsewhere still has the ring of truth.[9] It is easy to make light of cultural differences among institutions, branches, and agencies, but we do so at considerable risk.

Bringing adult learning and systems integration together, MIT's Michael Schrage maintains that "it is human behavior issues that are fundamentally both the opportunities and the obstacles in managing learning and innovation in organizations." In a capsule summation that could have been written for Defense CI, Schrage continues, "It is not a matter of wiring everybody . . . it is a much harder task for adults to open up to new learning, new behaviors, new design approaches and new ways of modeling things."[10] Advance claims for the power of an integrated, synchronized system of systems may for the moment be just that: plausible aspirations. This is a useful corrective to bear in mind because advances in knowledge management and information technology do loom large in the plans for Defense CI.

CI strategists are certainly counting on technology as a critical enabler. But they are not so seduced as not to be mindful of historian Michael Howard's prescient admonition: "The light provided by our knowledge of technological capabilities and our capacity for strategic analysis is so dazzling as to be almost hypnotic, . . . but it is those shadowing regions of human understanding based on our knowledge of social development, cultural diversity and patterns of human behavior that we have to look for answers."[11] Moreover, the very nature of counterintelligence may be sufficient armor against claims that any particular technical breakthrough will "change warfare forever." After all, CI practitioners lived through the advent of the stirrup, the longbow, the telegraph, the railroad, the airplane, the bomb, and the computer, to select just a few watershed innovations, and still confront adversaries that behave in oddly familiar ways.

Counterintelligence is principally concerned with human beings—their social structures, networks, personal motivations, and cultural drivers. The effectiveness of Defense CI, especially at the strategic level, depends on multidisciplinary and multifunctional assessments, with a heavy emphasis on behavior. Where technology can make the difference lies in such efforts as social network analysis and perhaps the emerging discipline of multi-INT fusion. Defense CI is now exploiting models and methods of proof from the behavioral sciences and drawing upon the academic community to meet the demand for esoteric knowledge.

These reality checks to our vision of changed CI practices are at the same time reinforced and modified by some of the early lessons from Operation Enduring Freedom in Afghanistan, Operation Iraqi Freedom, and the continuing global campaign against terrorist organizations. A high-profile CI role in support of tactical operations, significant innovations in document exploitation capabilities in the field, and new methods of force protection reflect successful Defense CI changes. While there have been some disappointments, delayed programs and the ones that "got away," and the persistent difficulty in connecting a "tactical to strategic" continuum for the sharing of information at all operational levels, the balance seems to favor the new direction.

All the same, it would seem prudent to bear in mind the recent lessons from Madrid. Apart from the train atrocity at Atocha, it was the spectacular deployment of SMS technology—Howard Rheingold's "smart mobs" in full cry—that swung the elections and toppled the Aznar government.[12] People are communicating, moving, coordinating, and acting in ways we have yet to reckon with. For us or against us, new technology continues to be a CI challenge across all of our leading-edge CI issues. For all our prowess, "owning the street" remains as difficult as ever. We still need the basics.

James M. Olson, former director of the CIA's Counterintelligence Center, believes that "the basic rules of CI are immutable and should be scrupulously followed." Olson is increasingly concerned that principles fundamental to good CI practices are not being followed as carefully as they should be. Drawing on his own experience, Olson offers his "Ten Commandments," making "no claim that they are inspired or even definitive."[13] We believe they are both street-smart and wise. Collectively, they touch the heart of every recent effort to transform CI practices. Looking ahead, we would do well to keep them in mind.

## Defense CI Tomorrow: Moving to the Offensive

Historically, practitioners, or strategists, and decision makers have not always viewed counterintelligence as a strategic asset. We would further argue that, as an instrument for achieving national security objectives today, Defense counterintelligence can be most relevant at the strategic end of the tactical-to-strategic continuum. Failure to recognize and understand how to employ the strategic potential of CI removes an operational capability from our arsenal, without cause. This limits U.S. ability to neutralize, deter, or combat terrorism, espionage, sabotage, and other covert or clandestine elements in the twenty-first-century threat environment.

Despite the constraints imposed by Title X, and the contemporary demands placed on service assets, these limitations are *not* inevitable. In fact, drawing upon that work, there are several venues where Defense counterintelligence can be deployed immediately as a strategic asset—and in the process transform counterintelligence itself. Perhaps foremost among these are strategic direction, the notion that strategic counterintelligence in and of itself can be used to effect a strategic goal or objective (effects-based operations) and the synchronization of operational activities.[14]

For example, one could consider the strategic advantages that counterintelligence capabilities could bring to the DoD's renewed interest in effects-base operations. The department's former lead on transformation, Vice Adm. Arthur K. Cebrowski, U.S. Navy (Ret.), speaks of the need to move from bomb damage assessment to effects-based assessments. Cebrowski's office is prototyping defense capabilities that are based on "knowledge-enabled warfare and demand-centered intelligence."[15] All of this generates brilliant copy for Washington briefings. To get a better measure of its staying power, we need to look at how these initiatives are received—and acted upon—in the field.

The notion that counterintelligence is an integral component of modern military operations is not new. In considerable measure, such a vision brings the discipline full circle back to the stature it attained during World War II. In his posthumously published account, *FORTITUDE: The D-Day Deception Campaign,* Roger Hesketh suggests there "is only one method which combines the qualities of precision, certainty and speed necessary for the conduct of strategic deception at long range and over an extended period, and that is the double-cross agent." He further argues that by destroying the German spy system in England, the British Security Service "laid the foundation for all that FORTITUDE achieved."[16] The present focus on effects-based operations suggests that CI is today equally, if not more, important as a vital part of twenty-first-century military operations.

The work on effects-based operations under way at Joint Forces Command, specifically where it addresses intelligence, suggests that firm leadership commitment, informed decision making, and strong investment can indeed create capabilities that leverage counterintelligence as a strategic asset. In support of strategic direction, effects-based operations offer great promise as "a process for obtaining a desired strategic outcome or effect on the enemy through the synergistic and cumulative application of the full range of military and non-military capabilities at all levels of conflict."[17] An "effect" in this context is the physical, functional, or psychological outcome, event, or consequence that results from an action.

Planning fully orchestrated strategically focused counterintelligence activities requires both clarity in desired outcomes and considerable analytic skill sets, many of which are not normally associated with Defense CI. Defense CI begins with deep knowledge of the adversary who is viewed as operating in a complex, adaptive organization, certainly within a complex adaptive system, set against the context of U.S. capabilities. In these operations, the synchronization of counterintelligence activities is not focused principally on the collection and production of what is typically defined as counterintelligence content in Title 50 or Executive Order 12333.

Because counterintelligence is concerned principally with human beings—their social structures, networks, motivations, cultural drivers, etc.—success, especially at the strategic level, depends on multidisciplinary and multifunctional assessments with a heavy emphasis on behavior. The ability to forecast how the human sources of particular threats will respond to actions taken, particularly when integrated with other operational activity, is critical. In the conduct of campaigns, synchronization of counterintelligence input at all levels (tactical through strategic) is now imperative. In fact, current DoD efforts to remodel defense intelligence are taking stock of the need for both the comprehensive integration of counterintelligence into effects-based operations and the notion of "operationalization" of foreign intelligence to better support the department's strategic goals and objectives as well as the operations of the combatant commands. If counterinsurgencies are indeed intelligence wars in their fundamentals, then U.S. forces have rarely needed these capabilities more urgently than they do today.

Fresh thinking on the contributions of Defense CI can be found in the doctrinal guidance provided in *Joint and National Intelligence Support to Military Operations*.[18] It is no longer beneficial for a military department's CI component to operate solely in "vertical" support of a service's Title X needs. Defense counterintelligence must continue to move to a center-based system with centralized synchronization of CI actions to assure adequate understanding and sufficient planning for the *consequences* of all such activity. Counterintelligence, viewed as a strategic asset, is a tool with enormous untapped potential both as a change agent and an additional source of predictive capability.

Many intelligence observers, we among them, have called for a more proactive stance for Defense CI. Performing the traditional functions (collections, dissemination, investigation) faster and better remains absolutely essential. But could the products of those labors not also be leveraged to more directly support the objectives of larger campaigns and strategies? The great deception that preceded D-Day, as we have seen, is one often-cited instance of such operational success. The diversion of German attention from Normandy suggests the potential for a larger strategic role on basis of understanding of/by CI practitioners. Even while the campaigns of 1943–44 enjoyed singular circumstantial advantages, their very scope and imagination makes the point for bold thinking. But where can counterintelligence engage to similar effect today?

One answer may lie in rethinking the fields of play. We could begin with the recognition that brilliant execution solely within the Defense Department may be of limited utility in the final analysis. If counterintelligence is not fully synchronized with interagency efforts in support of national strategic goals/objectives—for example, in combination with efforts at State, Commerce, Energy, etc.—we may well miss the main chance. In one example, a stronger nexus among counterintelligence, law enforcement, and information operations could support offensive efforts through contributions to effects-based operations. The same reasoning applies to the potential gains from greater cooperation with the business community. On both counts, this takes us back to consideration of some of the novel ideas, such as governing by networks, mentioned above.

In short, the introduction of new ideas, combined with a rapid succession of changes, has created a rare opportunity. If we seize it, we can constructively pose the most fundamental questions about our discipline and its place in the department and the community we serve and serve with. Choosing the right questions will enable us to think anew about counterintelligence—in each of its dimensions. At this juncture, we cannot afford to squander the prospect.

## A New Assessment of CI?

So, is the field of Defense CI really becoming a new game? We have argued that the preponderance of the evidence suggests that it is. The changes set in motion at the end of the 1990s are a work in progress. Their potential can be seen, even as it is not yet fully realized. The "more support, respect, and confidence" prescription is making a difference. Not all barriers have fallen. Still, the direction is clear. If the trend continues, it will position Defense CI to play a larger and more proactive role in national security. Indeed, it now appears that counterintelligence is poised to be recognized by defense planners and decision makers in a distinctly new light.

This is not an idle or a narcissistic question. It matters because it shapes far-ranging decisions: who we recruit, how we train, what we buy, how we fight . . . in short, how we "weaponize" counterintelligence and move it more to the center of the projection of national power and influence. After an extended decade of relative neglect, which persisted despite the undeniable successes logged by Defense CI service components, the profession is moving, however tentatively, closer to an operational intelligence capability that is increasingly at the center of the projection of U.S. national power and influence. At the same time, the CI discipline and its leadership face abiding challenges particular to counterintelligence. These suggest a need to treat counterintelligence and foreign intelligence as equally important strategic assets, which focus on distinctly different content, if not outcomes.

We can expect that foreign intelligence services and other organizations will be relentless in their quest to diminish U.S. industrial advantages, and that their tradecraft will evolve in unsuspected ways. New technology will enliven our adversaries' toolboxes every bit as much as our own. The threats posed by insiders, long-term visitors, and unwitting enablers will certainly grow. And the very definition of "adversary" will be open to reinterpretation as globalization pressures and coalition responses to them increase. An agile, reinvigorated Defense CI is already turning these challenges into opportunities.

The effort to recast Defense counterintelligence for the future cannot stop here. Many essential goals remain works in progress. In its report, the

Commission on the Intelligence Capabilities of the United States Regarding Weapons of Mass Destruction noted one critical area to address: "the absence of a systematic and integrated technical counterintelligence capability." The report goes on to observe that counterintelligence has largely been exclusively devoted to the human intelligence (HUMINT) efforts of foreign intelligence services while leaving other organizations (for example, the National Security Agency) to protect the U.S. information infrastructure.[19] While the report does not elaborate, for Defense counterintelligence to become integrated fully into military operations will require an end to the myopic view that it is "counter-HUMINT." Moreover, an acceptance of the twenty-first-century need for a full range of CI capabilities . . . counter-SIGINT, counter-Imagery, even counter-MASINT . . . is now overdue.

In chapters 10 and 11 of the commission's report, they further suggest a modest reorganization of the FBI, which appears at first glance to move national counterintelligence in the direction of the British MI 5 model. The more recent revival of high-level debate over domestic intelligence suggests that the MI 5 question is far from resolved.[20] The pros and cons of the MI 5 model take us too far afield here. We would note, however, that today the Defense Department contains the only truly global U.S. counterintelligence capability for which U.S. geographic borders do not create seams that can be exploited by a potential competitor or adversary. As they cross over U.S. borders, DoD counterintelligence components are required to coordinate separately with the CIA (when overseas) and the FBI (when inside the United States). As the discussion on moving U.S. counterintelligence to an MI 5 model unfolds, it seems to us that history and current experiences related to the war on terrorism and Operation Iraqi Freedom would suggest a center of gravity in the DoD, which coincidentally has the single largest equity in the success of U.S. counterintelligence in the twenty-first century.[21]

Defense counterintelligence is opening a new chapter; before it concludes, many time-honored CI activities will be done quite differently than in the past. New mission support requirements will stretch tomorrow's operators and analysts. Whether tracking spies in the backstreets of Minneapolis or Mumbai, protecting the department's most sensitive technologies, or supporting the combatant commanders in the field, Department of Defense counterintelligence is a strategic capability, a weapon, and a profession that is rethinking its very raison d'être to meet the complex threats of the twenty-first century.

# Notes

1. Bernard E. Victory, ed., *Modernizing Intelligence: Structure and Change for the 21st Century.* National Institute for Public Policy, January 2002. Study was originally published in September 1997 and subsequently became the basis for General Odom's book *Fixing Intelligence* (New Haven, Conn.: Yale University Press, 2003). Quote cited appears in 2002 edition. Gen Odom's remarks at the Hudson Institute were carried by C-SPAN on May 18, 2003.

2. See "Intelligence and Law Enforcement" Part XIII in *IC21: The Intelligence Community in the 21st Century.* HPSCI Staff Study, 104th Congress. (Washington, D.C.: GPO, 1996).

3. See J. W. Barnett, "Insight into Foreign Thoughtworlds for National Security Decision Makers," Institute for Defense Analysis, Document D-2665, January 2004. We are still far from fulfilling Barnett's recommendations, but the issue is acknowledged and some steps are being taken.

4. See Douglas C. Englebart, "Toward High-Performance Organizations: A Strategic Role for Groupware." Available at www.bootstrap.org, accessed on February 23, 2004. Engelbart, a pioneer in online computing, e-mail, collaborative tools and "dynamic knowledge repositories," has long argued that the highest payoff comes from efforts to improve the improvement process itself.

5. The National Counterintelligence Center itself an innovation owing to an earlier opportunity to rethink counterintelligence practices and procedures—PDD 24 issued following disclosure of the Aldrich Ames intelligence disaster in 1994.

6. See full text in DODD 5105.67.

7. Dr. Karl-Erik Sveiby, "Why Collaborate." Available at www.sveiby.com/articles/ccs.pdf, accessed on February 20, 2004.

8. Simon Lelic, "The Knowledge: Karl-Erik Sveiby." Available at www. kmmagazine.com, posted September 2, 2002, accessed on February 20, 2004.

9. Ernest R. May, "Studying and Teaching Intelligence," *Studies in Intelligence,* Vol. 38, No. 5, 1995. Given as the keynote address to the Symposium for Teaching Intelligence, sponsored by the CIA's Center for the Study of Intelligence in October 1993.

10. Marcia Conner, "Our Shared Playground: An Interview with Michael Schrage." Available at www.linezine.com, accessed on December 19, 2003.

11. Michael Howard, "The Future of Deterrence," *RUSI Journal,* Vol. 10, June, 1986.

12. Howard Rheingold, *Smart Mobs: The Next Social Revolution* (Philadelphia: Basic Books, 2002). For the role of SMS technology in the Spanish elections, see Maria Valerio, "Miles de personas protestan en toda España." Available at www.elmundo.es, posted on March 15, 2004. For a related perspective, outside the

scope of this article, see Glenn Harlan Reynolds, "The Blogs of War" in *The National Interest*, No. 75, Spring, 2004.

13. Olson's "Ten Commandments" are as follows: Be offensive; Honor your professionals; Own the street; Know your history; Do not ignore analysis; Do not be parochial; Train your people; Do not be shoved aside; Do not stay too long; and Never give up. Many of these principles are echoed, in distinct contexts, by other authors in this anthology.

14. These ideas were more fully developed in a DOD CI Whitepaper prepared for Andrew Marshall in late 2003.

15. Greta Wodele, "Pentagon transformation chief seeks helping hand from Congress," *GovExec Daily Brief*, February 26, 2004; and Lisa Troshinsky, "Transformation Chief Cites Progress, But Says Barriers Remain," *Aerospace Daily*, February 27, 2004.

16. Roger Hasketh, *FORTITUDE, The D-Day Deception Campaign* (New York: The Overlook Press, 2000), p. 352. Hesketh, who orchestrated the operation, wrote this account as an internal British counterintelligence report after the war.

17. USJFCOM, *Effects-Based Operations White Paper*, Version 1, Norfolk, 2001.

18. See treatment of counterintelligence roles in Joint Publication 2-01, issued in October 2004.

19. See the "Report of Commission on the Intelligence Capabilities of the United States Regarding Weapons of Mass Destruction," March 31, 2005, chapter 11.

20. See "9/11 Panel Questions Domestic Intelligence" in *Washington Post*, June 6, 2005.

21. At this writing, the 2005 Base Realignment and Closure Commission (BRAC) recommendation to consolidate CIFA and the Defense Security Service (DSS) into a new DoD Counterintelligence and Security Agency is under examination. The recommendation would co-locate the new agency with the Military Department investigation agencies (the Navy's NCIS, the Air Force OSI, and Army's CID) at new quarters in Quantico, Virginia. In addition to the value of having the Joint CI training academy at the new location and the proximity of the FBI training facilities there, the long-term potential for operational synergy appears considerable.

# TWENTY-SEVEN

# Does Intelligence Have a Future Tense?

### WESLEY K. WARK

Canadian journalist and military historian Gwynne Dyer recently published a polemic with an ominous (and clever) title: *Future: Tense: The Coming World Order.*[1] Intelligence has the same problem with a future tense. We need to guess at the future and to gauge the tensions that will confront the practice of intelligence.[2] The best way to read the future of intelligence is in the tea leaves of its past, and in the lineaments of its contemporary crisis. Notions of the value and role of intelligence have been so badly rocked by the tragedy of 9/11 and the failures of the Iraq War that we are in danger of forgetting or misreading the significance of intelligence's past. The crisis of confidence and direction that faces intelligence seems so vast and unprecedented that it is easily assumed to be insurmountable.

History does not guarantee survival. A crisis of confidence can sink the ship. But the argument of this paper is that so long as we understand the history of intelligence, we will find good reason for maintaining its survival into the future. Intelligence as a form of statecraft will survive its current crisis of confidence, but not without change and scars. Intelligence the world over will undoubtedly face a tense future and is in the midst of revolutionary transformation.

The greatest challenge facing twenty-first-century intelligence communities is that of adapting to a fundamentally new *operating* environment. Adaptive sounds evolutionary, but this doesn't convey the scale of the challenge. To begin to get a fix on the challenge, we have to invoke history and understand the origins of modern intelligence services. Those origins are relatively recent, although popular mythology might suggest otherwise.

In his magisterial account *Intelligence Power in Peace and War,* the British writer and former intelligence officer Michael Herman argues that modern intelligence was a Victorian innovation.[3] Herman is right to root the distinction between modern and premodern intelligence in signs of bureaucratic permanence, as opposed to ad-hoc arrangements that depended on princely whim. But his historical chronology needs some fine-tuning. In fact, modern intelligence services really took shape only in the decade before World War I. The upsurge of European great power antagonisms in the years before 1914 was the seedbed for the expansion of intelligence practice and the birthplace of yet another form of modernity.[4] We need to understand the birth of modern intelligence in order to be able to plot its initial operating environment and to understand the kinds of changes that have occurred in the past one hundred years.

Prewar great power rivalries were marked by arms races, concerns over technological innovation in weaponry, fear of surprise attack, general anxiety over the maintenance of great power status in a world defined by crude social Darwinist measurements of power, and worries about domestic stability and popular support. This was a world of seagoing behemoths (known as dreadnoughts), of undersea creatures (known as submarines), and of the first generation of flying machines. War suddenly had two new dimensions: undersea and aerial. The traditional dimensions of war on land and on the waves were suddenly more lethal; this was the world of precision artillery, machine guns, more lethal rifles, and—at sea—of fast, heavily armored and heavily gunned battleships.

Although we like to assume that the twenty-first century is uniquely characterized by a communications revolution, the world a hundred years ago was the world of the telegraph and Marconi's wireless, of expanding mass literacy, mass culture, and mass participation in politics, often surging around support for socialist and nationalist doctrines. The world of pre-1914 was a world ripe for stealing secrets and guarding secrets. It was a world of national insecurity, centered, of course, on Europe.[5] National insecurity bred an appetite, which proved permanent, for intelligence.

The operating environment in which modern-day intelligence was born matched a new appetite for acquiring secrets and protecting secrets unequally against a very modest capability to perform either task. Intelligence services of pre-1914 vintage were simply not equipped to make any serious contribution to the political and military problems of the day. They were all hobbled by lack of resources, lack of talent, lack of experience, and peripheral status in the decision-making apparatus of the day. They were all overwhelmed, as the essays in Ernest May's classic collection, *Knowing One's Enemies,* show so well, by the scale of the challenges they confronted.[6]

But if the birth of modern intelligence marked also the birth of modern intelligence failure as a theme, there were some compensatory features. One was that intelligence began, early on, to acquire its trademark mystique. This was aided, in part, by the eccentricities of an early generation of spy chiefs, among whom the most notable was the first head of the British secret service, Mansfield Cumming.[7] It was also shaped by a fictional landscape rapidly being invoked by the first generation of spy fiction novelists. These novelists, among them Erskine Childers, William le Queux, and E. Phillips Oppenheim—hugely popular in their day but now forgotten—had to struggle just to imagine espionage and embed the word in popular consciousness. They had also to solve a Victorian cultural dilemma, namely, how to craft fictional gentlemen's spies in an age which assumed that spying and the gentlemanly virtues were quite incompatible.[8] The gulf was bridged by appeals to patriotism and the acknowledged virtues of game playing (as in Kipling's exhortation to "play up and play the game").

The operating environment of pre-1914 days was thus marked by complex intelligence challenges, extraordinary incapacity, and the beginnings of popular cachet. Intelligence services functioned, barely, in only a handful of European great powers—they had yet to be exported to the global community. Their focus was primarily military; political, economic, social, and cultural issues were, at this point, far beyond their ken. Modern intelligence was born before 1914, but only just.

Since 1914, intelligence services have won permanence and power in a global system. Successive historical challenges have pushed the expansion of intelligence practice outwards in huge concentric waves. World War I saw the mobilization of technological resources, especially in communications intelligence, also known as signals intelligence (SIGINT), and the adaptation of the Wright brothers' invention to the collection of aerial photographic intelligence, later to be christened IMINT. The troubled interwar years brought the challenges of reading ideological systems (first Bolshevism, then Fascism) and introduced the idea of economic intelligence. World War II bred, especially through the triumphs of Ultra and the sophistication of deception operations, the revolutionary notion that intelligence could be a war winner.

The Cold War and the Damocles sword of nuclear weapons introduced a demand for early warning, matched with a need for precise intelligence on the strategic arsenals and doctrine of the major powers. The call for 'no atomic Pearl Harbors' was a fearsome charge to level on the shoulders of Western intelligence communities. On top of this, the effort to parlay the experience of World War II behind-the-lines operations into a new doctrine of covert operations that might serve as a surrogate for superpower conflict brought a new prominence and role for intelligence. Beginning with the so-called golden age of covert operations in the 1950s, intelligence was acquiring an instrumental status alongside the more traditional elements of diplomatic and military power. The end of the Cold War hastened the assumption that, in addition to now-traditional and ongoing tasks related to the assessment of foreign state power and practice, intelligence services could tackle a wide range of transnational and subnational threats, including international crime, drug trafficking, illegal people migrations, ecological threats, and, of course, terrorism.[9]

Even the most superficial of historical sketches, like the one above, suggest just how much the operating environment of intelligence has changed since 1914. Intelligence services operate in a true global environment and are no longer restricted to a handful of European powers. They tackle an immense range of international and national security issues, and do so with technological, organizational, and human resources that weren't even a gleam in the eye of the founding generation of intelligence officers. They have been thrust from the uttermost periphery of government to the inner circles of decision making. In short, intelligence power has come to be a reality.

Intelligence power does not mean intelligence performance, however. The rise of intelligence power in the twentieth century has to be plotted against a mixed record of success and failure, with points of failure dominant. Richard Betts's argument that intelligence failures are inevitable looks persuasive when set against the historical record.[10] A paradox beckons. How can an institution, in this case intelligence, have risen to power and permanence based on a record marked by failure? The answer illustrates the elements that continue to link intelligence's modern birth to its current and future conditions.

Modern intelligence services were created as a result of a growing appetite for secrets, for information of all kinds. That appetite on the part of states has not abated, nor is it likely to abate in the future. Spy novelist E. Phillips Oppenheim (mentioned earlier), used to say of spy fiction that, "so long as the world lasts its secret international history will suggest the most fascinating of all material for the writing of fiction."[11] What goes for the fictional realm goes as well for the factual. The international system provides a flood of "fascinating" stories for intelligence services to puzzle over. Some of these "stories" come in the shape of secrets; some represent what Greg Treverton distinguishes as "mysteries," ultimately less knowable.[12]

Appetite for information persists, and is potentially gargantuan. Intelligence services struggle to meet that appetite. The nature of the struggle has changed, but the original impulse that gave birth to modern intelligence is unchanged.

More paradoxically, the original gap that existed between the demands placed on nascent intelligence services at the dawn of modernity and their capacity to deliver also persists. We can call this the "expectations gap." In the beginning the gap was simple—a gulf separated the complexity of the issues that faced European intelligence services before 1914, as war approached, from their capacity to report in meaningful ways. Today the expectations gap operates in a fundamentally changed environment. Complex questions persist, but the capacity of intelligence services to respond to them has been transformed by power, technology, and accumulated experience. Why, then, an expectations gap? The answer lies, in part, in an expectation of intelligence perfectibility that has long dogged its work, particularly since prediction entered the clandestine portfolio.

Perfectibility has been the attendant curse of the growth of intelligence power; it has also been the attendant curse of the increasing role that technology and technical systems for intelligence collection and data management have come to play. Intelligence perfectibility is, of course, a chimera. It obscures a necessary focus on the pursuit of reform and improvement; it hampers a clear-eyed appreciation of the limits of what intelligence can do. Too great expectations of intelligence, rooted in facile notions of progress and technological fixes, can also stimulate their opposite—disenchantment, low prestige, and negative expectations. Both extremes are dangerous; both are commonplace as well.[13]

The history of intelligence has also imparted, along its hundred-year course, some misleading messages. The two most profound have to do with constructions of monopoly and secrecy. When modern intelligence practice first became institutionalized, two implicit assumptions reigned. One was that governments would have a monopoly over the provision of intelligence. The other assumption was that intelligence services would have some kind of internal monopoly over the collection, assessment, and dissemination of intelligence within government. The assumption of an internal monopoly has always remained a wish, rather than a reality, in most government systems.

Intelligence has always been bedeviled by bureaucratic rivalries; by definitional imprecision over what constitutes intelligence as opposed, say, to military or foreign policy information; by competing models that favor centralized or decentralized bureaucratic structures; and by the ever-present tendency on the part of decision makers, especially strong in authoritarian governments, to consider themselves their own best intelligence advisers.

The future of intelligence will continue to witness unresolvable disputes over all these features of the proper organization of intelligence within government. The future performance of intelligence will undoubtedly be affected by the degree to which individual state systems manage to find solutions to organizational issues.[14] But if internal, bureaucratic monopoly has always been contested, something more radical has happened to the broader claim for government monopoly over the practice.

The notion that only states could practice intelligence was akin to the claim that only states had a legitimate monopoly over the conduct of war. Both notions have been challenged by developments in global politics and global infrastructure since the turn of the twentieth century. The difference is that while states still stake a strong claim to a monopoly of organized violence, the claim to a monopoly of intelligence has been considerably undermined.

The assault has come from a number of directions. Competitive economic pressures drove one direction. Governments had practiced forms of economic intelligence since the institution of blockade measures and economic warfare during World War I and, notably, in peacetime with the British experiment conducted by the Industrial Intelligence Centre during the 1930s to monitor the Nazi economy.[15] This effort had always been focused on state assessments of threatening economic developments on the part of competitor or enemy states. The rise of the global economy and of multinational enterprises brought a new player into the game. Corporations with global interests found themselves in need of risk assessments and market knowledge of overseas trading areas. Such forms of intelligence gathering and assessment were considered proprietary and could not be shared or reaped from access to government information.

Demands on larger corporations to stay competitive provided the stimulus for their own appetite for intelligence, with a wide range of objects. Corporate intelligence needs might be focused on competitors, marketplaces, or even consumers. Demands for corporate security, whether of ideas, products, communications, or databases have provided another strong stimulus to the growth of business intelligence practices. The parallel development of economic intelligence by the private sector to serve distinct private sector needs has been one of the great changes to affect the modern practice of intelligence. But because private-sector intelligence and government practice have been so isolated from each other (barring the competition for scarce professionals), students of state intelligence have paid too little attention to the implications of the spread of intelligence into the world of business. Nor have students of business, and notably among the faculty of the business schools, been particularly alert to the experience and lessons of state practice.[16]

How might the rise of business intelligence affect the future of intelligence itself? The main impact will depend on the extent to which shared global and domestic security concerns and interests compel convergence. It remains to be seen to what extent government intelligence services might engage in economic intelligence in order to benefit identified "national" enterprises.[17] There is no clear-cut trend that can yet be identified. Some states have been implicated in the use of their intelligence services for economic espionage; others (notably the CIA) have proclaimed their intention to eschew such practice. Whatever the cumulative picture, full convergence from such an angle seems unlikely.

Nor is a strong push likely from the private sector. While there will be strong pressure to continue to identify self-generated business intelligence as proprietary, a zone of potential convergence does exist with regard to privacy legislation, protection of critical infrastructure, security alerts at home and abroad and, most radically, the development of an interface between government and private sector global risk assessment. To develop this convergence will require mutual understanding of the benefits, and the creation of new entities to handle the flow of information across the walls of government and business secrecy. These are tasks for the future.

The transformation wrought by the global communications revolution has also had a huge impact on intelligence. Not only has it transformed information flows and database management, but it has also established "open source" intelligence as the principal well of information from which so-called secret services must drink.[18] The paradox of secret intelligence systems relying on open source information is a marker of the change that has come over the practice of intelligence. But the ascendancy of open source information has also opened the door to the privatization of aspects of the old state monopoly of intelligence.

A host of private-sector analytical agencies have arisen that engage in for-fee assessments of current and future global security challenges. Some of these agencies are building new business on old foundations, for example, Jane's Information Group or the RAND Corporation. Others, such as George Friedman's STRATFOR, and the UK-based Oxford Analytica, are essentially new competitors seeking to create a new marketplace for ideas.[19] Competition is the essence of what these services offer and what their future impact might be. Competition cuts into the intelligence monopoly at several angles.

In a basic sense, government services have to match the quality and spread of their reporting against what private-sector concerns can supply on a contract or subscription basis. In theory, the result is a better product. Competition from private-sector agencies can also introduce a much-needed element of "contestability" into the government process, although for the most part—because of secrecy concerns—this remains a one-way street. But for contestability to work, government services have to be willing to treat private-sector intelligence providers as equal partners, equal in talent and analytical acuity. This requires willingness to relinquish some part of the monopoly of wisdom, which is such a strong and closeted feature of the government culture of intelligence.

To the extent that private-sector intelligence services offer their analyses in the public domain, they add their voice to the "competitive" world of media news, becoming one more source of an alternate public understanding of events.

All these forms of competition from private-sector companies alter to some degree the operating environment for government intelligence communities. The more radical possibility that the future holds concerns outsourcing of intelligence. This possibility has already emerged in limited form in both intelligence collection (think, for example, of commercial satellite systems) and assessment. The barriers to greater outsourcing reside largely in intangible issues of trust and control. Of the two, trust depends first on private-sector companies acquiring their own permanence in a volatile marketplace, and subsequently on the creation of an extended track record with government intelligence communities.

Control may be a more elusive and difficult issue. The relationship between government decision makers and their intelligence providers can be difficult and delicate. Intelligence communities have to operate so as to be both independent in judgment and also relevant to the concerns and agenda of those in power, thus loyal servants. The political calculations involved in this relationship for decision makers and intelligence officials within government alike might be extremely difficult to replicate in a relationship between an outsourced private-sector intelligence system and government. The limits to the privatization of intelligence will most likely stem from the desire on the part of governments to keep intelligence close. The appetite for information that originally spawned intelligence was a political appetite, informed by a desire to possess secrets. Possession will continue to shape the relationship between decision makers and intelligence, even in an increasingly privatized system.

The question of who does intelligence in a twenty-first-century world has a clear answer. The practice has spread outward, beyond government, into the business community and into a world of private-sector intelligence providers. Within the body of the nation-state, monopoly has been replaced already by quasi-monopoly and will erode further but not, I predict, completely.

If we look beyond the boundaries of the nation-state we can see a similar process of expansion at work. At the dawn of intelligence modernity, the practice of espionage was the preserve of a handful of European great powers. One hundred years later, the practice is ubiquitous among states, no longer a perquisite of Europe or of great power status. But as the structure of international society has grown more complex and less rooted in national sovereignty, a universe of NGOs and international organizations has emerged, each with informational needs that will sooner or later be defined as intelligence.[20]

International aid agencies, the private-sector development community, and global charities all must have the capacity to generate their own forms of knowledge about the international environment in which they operate. This requirement is driven by the needs of operational effectiveness, security, and independence. The last factor rules out any real convergence between NGO intelligence and government intelligence. NGO "intelligence" will be spared the onerous task of reinventing the wheel only by the mobility of intelligence professionals freely crossing boundaries between public- and private-sector careers, and by the ability of private-sector intelligence providers to be of (affordable) service.

Another dimension of the internationalization of intelligence stems from the post-1945 effort to erect forms of international governance, especially the United Nations. The UN'S role in conflict resolution, peacekeeping, arms control, counterproliferation, international law, and development has generated a massive requirement for intelligence. But the requirement has not been an easy one to meet. Cultural and procedural rigidities at the UN have banished the "I" word itself from the UN vocabulary. The UN has, by tradition, operated without its own intelligence capacity and through reliance on the provision of information by member states. The system can easily break down in times of crisis. As unsatisfactory as the traditional system might be, it is unlikely to be altered in the future. The supranational pretensions of the UN will remain held in check by the reality of national sovereignty, and that reality will remain particularly strong in the field of intelligence. To the extent that an antiglobalization backlash remains a potent force into the twenty-first century, it will also act as a brake on the enlargement of the powers of the UN.[21]

Where we might look for substantial change is not in the system of UN intelligence itself but in the nature of its use. The UN's trump card of international legitimacy, if sustained, might encourage member states to accept, and ultimately take for granted, much better conditions of intelligence sharing with the world body.[22] Intelligence sharing within the auspices of UN operations awaits its own modernity. It requires that same condition of permanence that distinguishes the modern from the premodern world of intelligence.

While it seems doubtful that any radical future awaits intelligence when it comes to its role in supranational organizations, the perspective shifts when we consider the prospects of regional systems and alliance structures. While full-fledged regional integrations such as the European Union, and trading partnerships such as NAFTA (North American Free Trade Agreement) have struggled in their different ways with developing a framework for intelligence sharing, the greatest potential for change seems to exist within intelligence alliances. Bilateral and multilateral intelligence sharing relationships are, by now, a common feature of the international system, though they are usually shrouded in considerable secrecy.

At the pinnacle of all such endeavors stands the UKUSA alliance that links the United Kingdom, the United States, Canada, Australia, and New Zealand in an intelligence-sharing partnership. UKUSA has a long history (dating back, in effect, to the period of World War II) and benefits from an extraordinary degree of intimacy rooted in that history.[23] It is an alliance founded upon sharing the burden of intelligence collection and exchanging the intelligence product. Its success to date reflects the power of its historical roots: the fact that it is insulated, for the most part, from political controversies between its member states, and that it is an acknowledgment of the tremendous challenges of a global intelligence watch.[24] Intelligence power is, and always has been, very unequally shared among the UKUSA members. Its superpower member, the United States, in effect, sustains the alliance. But the fact that the United States continues to see value in a multilateral intelligence partnership is recognition of the reality that even a superpower cannot go it alone when it comes to global intelligence requirements.

The UKUSA alliance and similar multilateral efforts that may emerge in the future have great, but unrealized potential to transform intelligence. UKUSA remains largely a partnership of individual state agencies, especially in the SIGINT field, rather than a fully integrated and collective effort. Yet the failure of UKUSA to provide an added shield to prevent the 9/11 attacks, and the inability of UKUSA to assist in formulating accurate assessments of Iraq's weapons of mass destruction (WMD) program in the run-up to the 2003 campaign, suggest that UKUSA may be at a watershed. Its preeminence as an intelligence alliance rooted in World War II and Cold War targeting may decline if it is unable to make a significant contribution in confronting such new transnational threats as terrorism and WMD proliferation. Alternatively, UKUSA may prove the wave of the future if it is able to create an environment for collective intelligence tasking and assessment, and to provide a forum for multilateral intelligence contestability. To reach this point of evolution, the UKUSA alliance would have to overcome its partnership roots and impediments; in effect, it would have to be a system in which nation-state intelligence monopolies were discarded in favor of a collective effort. The obstacles to achieving such a transformed UKUSA are large, not least in the political distrust of loss of control over intelligence and in the very unequal power distribution within the alliance. But if UKUSA cannot transform itself in this way, it is hard to imagine any other multilateral or international structure that is likely to do better.

Ranging over the set of challenges to the traditional monopoly of nation-state intelligence services that I have delineated, we can conclude that the greatest impact has been made by privatization, whether in the form of the rise of business intelligence or the emergence of private-sector intelligence services. Globalization, on the other hand, while sustaining a wealth of nongovernmental actors in the international arena, and nourishing hopes for progress toward world governance and norms, has not yet had a transformative impact. The global communications revolution has fed developments and has worked its massive contribution through both privatization and globalization.

The future of intelligence will continue to rest on two foundations present at the birth of intelligence modernity. Informational appetite will continue to be the stimulus for the maintenance and growth of intelligence systems within and without nation-states. Informational appetite linked to national security concerns and sovereignty doctrine will continue to dictate that the intelligence services of nation-states will remain at the center of any global system. Complete monopoly has been, and will continue to be, eroded. But the ruling notion of intelligence as a form of statecraft will remain.

But informational appetites, plus sovereignty, are only part of the equation of intelligence modernity and only partly a guide to what comes next. A third element, present at the creation, was secrecy. The modern origins of intelligence services were rooted in a desire to obtain the secrets of foreign powers, to uncover secret conspiracies and threats within the domestic body politic, and to perform these tasks, at home and abroad, under cover of secrecy. Secrecy was, from the beginning, linked to a desire to maintain deniability. Pursuing secrets by secret means was a reflection of democratic (and authoritarian) permissibility at the turn of the twentieth century. Much has changed, but the change has also been of relatively recent vintage.

For the first seventy years of intelligence modernity, from the turn of the century until the 1970s, the veil of secrecy that surrounded intelligence operations was penetrated only occasionally by spy fiction writers, enterprising journalists, and varieties of rogue and semiofficial memoirists. None of these sources posed a fundamental challenge to the doctrine of official secrecy. It was only in the last quarter of the twentieth century, with the creation, spearheaded by the United States, of machinery of public accountability and of the right of access to government records that the doctrine of official secrecy began to unravel.[25] It has done so at a rapid pace since that time, but a pace differentiated according to national cultures.

The accelerating erosion of official secrecy had beneficial effects. It stimulated greater public knowledge, made possible academic scholarship, and provided for more consistent and more informed media coverage. It also, of course, allowed for a public check on abuses of power by intelligence communities.[26]

These developments altered traditional doctrines of secrecy, but they did not end secrecy itself. The core concept of intelligence as dealing with secrets in ways that needed protection from public gaze remained intact. Sources and methods were considered sacrosanct. So was the notion that intelligence services provided privileged assessments under conditions of confidentiality to government clients (and no one else).

In an operating environment still marked by secrecy and what we can call client privilege, the utility of intelligence was traditionally restricted to direct or diffuse influence on government decision making. There have been some notable exceptions to this general rule. The use made by British signal intelligence of the decrypted Zimmermann telegram in early 1917 had a decided impact on U.S. public opinion about the war and on the U.S. decision to enter that conflict. But this effect was an unplanned byproduct of the effort on the part of the British to induce the government of Woodrow Wilson to abandon its neutrality policy.[27]

During the tense standoff that marked the Cuban Missile Crisis in October 1962, the U.S. government decided to publicize at the UN its U-2 photoreconnaissance images of Soviet missile sites.[28] The intent was to influence global public opinion and to add pressure to the American insistence on a Soviet withdrawal. But the Zimmermann telegram and Cuban Missile Crisis episodes were aberrations. They did not alter the general understanding that intelligence served internal government decision making, not public diplomacy; nor was intelligence an instrument for winning public approval.

There are contemporary signs of a seismic shift in the doctrine of intelligence secrecy. These signs emerge from the run-up to the invasion of Iraq in March 2003. The two leading Coalition partners, the United Kingdom and the United States, both took the unprecedented step of utilizing secret intelligence assessments in public in an effort to win domestic and international support for an attack on Iraq. In the British case, the government decided to have the Joint Intelligence Committee prepare a public version of an intelligence assessment on Iraqi WMD.[29]

The United States followed suit with the release of a version of an assessment prepared by the National Intelligence Council on the same topic.[30] This effort to use intelligence to sway public opinion culminated in the speech delivered by Secretary of State Colin Powell to the UN Security Council in February 2003, on the eve of the invasion. Powell's presentation focused on selected evidence of Iraq's illegal weapons program. A global audience watched with rapt attention as the secretary of state unveiled images of suspect Iraqi installations and listened with equal attentiveness to recordings of intercepted Iraqi military communications.

It remains unclear whether the public use of intelligence in the case of Iraq marks the beginning of a fundamentally new understanding of the uses of intelligence. The question mark is present because the experiment in 2002–3 proved to be such a disaster. The intelligence base on which American and British policy was founded proved faulty to a remarkable degree. Public exposure of the intelligence assessments allowed for highly public refutation.[31]

The public use of intelligence did not seem to dramatically sway opinion in either the United States or the United Kingdom; in the latter, massive antiwar protests dogged the British policy. Certainly, Secretary Powell's presentation at the UN was not enough to secure a second resolution authorizing the use of force against Iraq. The gambit thus failed.

It may well be that governments will prove much more cautious in the future about using intelligence to try to influence political support for controversial actions. It may even be that governments will feel compelled to retreat altogether from the idea and will return to the traditional understanding of intelligence as a secret source of knowledge.[32]

But the public use of intelligence in the case of the Iraq War reveals just how tempting intelligence secrets can be. Feeling confident of their intelligence assessments and, it must be surmised, confident of the necessity of a war option, both the U.S. and the UK governments felt the public use of intelligence was in their best interest. The Iraq case does not appear to be unique, and the temptation to shape public opinion through dramatic intelligence revelations will remain.

Even the acknowledgment of the reality of temptation is enough to alter the operating environment in which intelligence services must act in the twenty-first century. The public use of intelligence remains a heretical proposition for many professionals within intelligence communities and will no doubt be strongly resisted in the future. But the failure of the Iraq experiment should not blind us to the possible benefits of public intelligence.

Shaping public opinion through the selective and edited release of intelligence material may be designed as politically manipulative. But it also gives publics a capacity to interrogate the informational basis on which a government might choose to act, in ways impossible under older doctrines of secrecy. Public intelligence can thus play a significant role in the democratic process. Moreover, public intelligence can serve as an important instrument of quality control, forcing the intelligence assessment process to be as rigorous and informed as possible. The more that intelligence has to operate in the public domain, the more that its analysis is subject to public scrutiny, arguably, and the better its product must be.

But some cautionary notes are necessary. Public intelligence may offer benefits for both intelligence communities and the publics they (indirectly) serve. But public intelligence could also stifle efforts to build a creative, risk-taking intelligence environment. Safe and tepid analysis might be the ultimate outcome of a public process, at least until a new working relationship among government, intelligence communities, and the public could be engineered. The practice of using intelligence to shape public opinion might also have deleterious effects on the relationship between intelligence communities and government leaders. The more accepted the practice becomes, the greater the degree of politicization that might occur in the formulation of key intelligence judgments. Government leaders would have a great deal at stake in terms of the intelligence they offered to their publics, and the desire to ensure that it conformed to policy prescriptions would be strong.

We find in the prospect of public intelligence a nexus of competing interests, benefits, and costs. The overriding factor will be temptation. Yielding to temptation may be wise so long as a number of conditions are met. I would include in these:

- The independent power of intelligence communities to determine what product can be made public;
- The recognition by political leaders that intelligence can be an important facet of public education but must be used reasonably and in conditions where maximum confidence in the intelligence judgments exists;
- Public intelligence should be considered an instrument of public diplomacy only when the circumstances seem serious enough to warrant such use; and
- Publics themselves are sufficiently aware of the capacities and limitations of intelligence to exercise independent critical judgment about the assessments on offer.

Wherever the future takes us, public intelligence will be a phenomenon worth watching. The temptations and difficulties involved reflect one final facet of the operating environment that confronts intelligence services. From the beginning, modern intelligence has relied on a mystique of clandestinity.[33] This cachet, sometimes cultivated, sometimes the product of wholly independent popular culture forces, is an important contributor to intelligence legitimacy. Intelligence services depend to a degree on their cachet to stimulate recruitment of talent, to win public confidence and support, and to garner the trust and confidence of decision makers.

The greatest single force behind the mystique of intelligence has been the popularity of fictional narratives of espionage. These narratives have typically invested espionage and its operatives with dramatic powers to uncover plots, catch villains, and save civilization.[34] Fictional spy narratives are a strong bastion, perhaps the last surviving one, of the great man theory of history. Their history mirrors the history of twentieth-century intelligence. Spy fiction and modern intelligence were born at the exact same time and have remained an intertwined phenomenon in the public consciousness ever since.[35] There is a curious dependency here. Spy fiction relies for its cachet on the imagined powers of real-life spy services. Spy services, in turn, rely for public support on the imaginings of spy fiction writers. Should either lose its luster, its counterpart must falter.

The current crisis of confidence over intelligence, stimulated by the failures of 9/11 and the Iraq war, provides an opening challenge for twenty-first-century spy narratives. It is too early to say how they will cope with this challenge of representation. But it may well be that the solution of the crisis will be found in fictional representations. We have already seen the emergence of the 9/11 novel as a mainstream literary phenomenon.[36] We will sooner or later see the emergence of the 9/11 spy novel, whose redemptive task will be to save civilization (and intelligence with it) from terrorists and terror.

Crises of confidence are destined to become a permanent part of the new operating environment of modern intelligence. They are a natural consequence of the unappeasable appetite for information that sustains intelligence, unreal expectations, the myths that have accumulated around intelligence, the march of privatization, and the erosion of secrecy.

Intelligence has a future, rooted both in its past and in the new. That future will be tense as new forces grind against old. The key to the survival of intelligence power will be a capacity for creative adaptation. Intelligence practitioners of the future will have to find ways to tame information flows, inject realism into expectations, profit sensibly from mythology, exploit privatization for all its worth, and learn to live with a greater public identity, greater interaction, and lessened secrecy. This may be a tall order, but is it any less tall than the task that confronted intelligence practitioners a century ago, namely, to invent intelligence itself? Modern intelligence is here to stay—to marvel at its birth and at its survival.

# Notes

1. Gwynne Dyer, *Future: Tense: The Coming World Order* (Toronto: McClelland and Stewart, 2004). I admire the title, though not the substance, of this account.

2. I have previously tried my hand at this in Wesley K. Wark, ed., "Twenty-First Century Intelligence," special issue of *Intelligence and National Security,* 18, no. 4 (Winter 2003).

3. Michael Herman, *Intelligence Power in Peace and War* (New York: Cambridge University Press, 1996), p. 15.

4. The best account of the origins of a modern spy service is to be found in Christopher Andrew's history of British intelligence, *Secret Service: the Masking of the British Intelligence Community* (London: Heinemann, 1985), ch. 2.

5. The literature on European insecurity before 1914 is immense, but see the following synoptic studies: Donald Kagan, *On the Origins of War and the Preservation of Peace* (New York: Doubleday, 1995), ch. 2; and Paul Kennedy, *The Rise and Fall of the Great Powers* (New York: Random House, 1987), ch. 5.

6. Ernest R. May, ed., *Knowing One's Enemies: Intelligence Assessment before the Two World Wars* (Princeton, N.J.: Princeton University Press, 1984).

7. Mansfield Cumming and the world of early twentieth century British intelligence is wonderfully evoked in Alan Judd, *The Quest for C: Mansfield Cumming and the Founding of the Secret Service* (London: Harper Collins, 1995). The author had access to Cumming's diaries, still held by the British Secret Intelligence Service.

8. See the essays in Wesley K. Wark, ed., *Spy Fiction, Spy Films and Real Intelligence* (London: Frank Cass, 1991).

9. International histories of twentieth century intelligence remain rare, but one satisfactory account is provided by Jeffrey Richelson, *A Century of Spies: Intelligence in the Twentieth Century* (New York: Oxford University Press, 1995).

10. Richard K. Betts, "Analysis, War and Decision: Why Intelligence Failures are Inevitable," *World Politics,* Vol. 31, No. 1 (Oct. 1978), 61–89.

11. Quoted in David Stafford, *The Silent Game: The Real World of Imaginary Spies* (Toronto: Lester and Orpen Dennys, 1988), p. 37.

12. Gregory F. Treverton, *Reshaping National Intelligence for an Age of Information,* RAND Studies in Policy Analysis Series (New York: Cambridge University Press, 2001).

13. A brief, but insightful account of some of these problems can be found in Mark M. Lowenthal, *Intelligence: From Secrets to Policy,* (Washington, D.C.: CQ Press, 2000), especially ch. 9.

14. Walter Laqueur surveyed some of the problems involved in improving intelligence performance in *A World of Secrets: The Uses and Limits of Intelligence* (New York: Basic Books, 1985), ch. 11, "The Future of Intelligence."

15. For an account of the pioneering work of the Industrial Intelligence Centre, see Wesley K. Wark, *The Ultimate Enemy: British Intelligence and Nazi Germany 1933–1939* (Ithaca, N.Y.: Cornell University Press, 1985), ch. 7.

16. The literature that examines the evolution of business intelligence or its links to state practice is scanty. For one account of the problems that would beset any U.S. effort at state-sponsored economic espionage, see Randall M. Fort, "Economic Espionage," in Roy Godson et al., eds., *US Intelligence at the Crossroads: Agendas for Reform* (Washington, D.C.: Brassey's, 1995), 181–96. On the uses of economic intelligence,

see Philip Zelikow, "American Intelligence and the World Economy," in *In From the Cold: The Report of the Twentieth Century Fund Task Force on the Future of US Intelligence* (New York: Twentieth Century Fund Press, 1996).

17. For an account of the impact of foreign economic espionage on the United States, see John J. Fialka, *War by Other Means: Economic Espionage in America* (New York: Norton, 1997).

18. See the useful essays collected in Roger George and Robert Kline, eds., *Intelligence and the National Security Strategist: Enduring Issues and Challenges* (Washington, D.C.: Sherman Kent Center for Intelligence Studies, 2004), Part VI, "The Open Source Revolution."

19. Stratfor can be found on the web at www.stratfor.com. George Friedman is the author of *America's Secret War: Inside the Hidden Worldwide Struggle between America and its Enemies* (New York: Doubleday, 2004). Oxford Analytica can be found at www.oxon.com

20. For a survey of developments in the global system, see Ian Clark, *Globalization and Fragmentation: International Relations in the Twentieth Century* (New York: Oxford University Press, 1997)

21. UN Secretary General Kofi Annan proposed a major overhaul of the UN in his report *In Larger Freedom: Towards Development, Security and Human Rights for All*, 21 March 2005. See also the UN High-Level Panel on Threats, Challenges and Change, "A More Secure World: Our Shared Responsibility," December 2004, online at www.un.org/secureworld

22. For some interesting reflections on the problem of using intelligence for UN arms control monitoring, see the memoirs of Hans Blix, *Disarming Iraq* (New York: Pantheon, 2004).

23. For a study of the origins of UKUSA see Bradley F. Smith, *The Ultra-Magic Deals, and the Most Secret Special Relationship, 1940–1946* (Novato, Calif.: Presidio Press, 1993). For a broader history, see Jeffrey Richelson and Desmond Ball, *The Ties that Bind: Intelligence Cooperation between the UKUSA Countries* (London: Allan and Unwin, 1985).

24. Martin Rudner surveys British approaches to SIGINT alliances in "Britain Betwixt and Between: UK SIGINT Alliance Strategy's Transatlantic and European Connections," *Intelligence and National Security,* 19, no. 4 (Winter 2004), 571–609.

25. For an account of the role played by the US Congress in these developments, see Loch Johnson, *A Season of Inquiry* (Lexington, Ky.: University of Kentucky Press, 1985).

26. For a wise study, see Daniel Patrick Moynihan, *Secrecy: The American Experience* (New Haven, Conn.: Yale University Press, 1998).

27. The classic account remains Barbara Tuchman, *The Zimmermann Telegram* (New York: Ballantine Books, 1985, first ed. 1958). The best account of the role of British intelligence in the affair is Patrick Beesly, *Room 40: British Naval Intelligence 1914–1918* (New York: Harcourt, Brace, Jovanovich, 1982).

28. For an assessment of the role played by intelligence services in the international dimensions of the Cuban Missile Crisis, see the essays collected by James Bight and David Welch in a special issue of *Intelligence and National Security,* Vol. 13, No. 3 (Autumn 1998).

29. UK Joint Intelligence Committee, "Iraq's Weapons of Mass Destruction," 24 September 2002, online at www.pm.gov.uk

30. U.S. National Intelligence estimate, October 2202, "Iraq's Continuing Programs for Weapons of Mass Destruction," online at www.cia.gov/cia/reports/iraq_wmd/iraq_oct_2002.htm

31. UK intelligence assessments were the subject of a critical public inquiry chaired by Lord Butler, "Review of Intelligence on Weapons of Mass Destruction," July 2004, online at www.butlerreview.org.uk. The flawed U.S. assessments were dissected by the US Senate, Select Committee on Intelligence, "Report of the U.S. Intelligence Community's Prewar Intelligence Assessments on Iraq," July 2004, online at http://intelligence.senate.gov

32. Kenneth Pollack suggests a ten-year freeze on the public release of intelligence assessments in his insightful essay, "Spies, Lies and Weapons: What Went Wrong," *The Atlantic Monthly,* Vol. 293, No. 1 (Jan–Feb. 2004), pp. 79–92.

33. For a partial study, see John G. Cawelti and Bruce A. Rosenberg, *The Spy Story* (Chicago: University of Chicago Press, 1987), ch. 1, "The Appeal of Clandestinity."

34. See Wesley Wark, ed., *Spy Fiction, Spy Films and Real Intelligence* (London: Frank Cass, 1991) and Wesley Wark, "The Spy Thriller," in Robin Winks, ed., *Mystery and Suspense Writers,* vol. 2 (New York: Scribner's, 1998).

35. For an account of this intertwining, see Fred Hitz, *The Great Game: the Myth and Reality of Espionage* (New York: Knopf, 2004).

36. Notably, Ian McEwan's superb novel, *Saturday* (New York: Knopf, 2005).

# TWENTY-EIGHT

# Intelligence Transformation Past and Future

## The Evolution of War and U.S. Intelligence

### MICHAEL WARNER

Readers of the Washington press in late 2004 could have been forgiven for believing that the city had become the battleground of two titans grappling to dominate a vast federal structure. The stakes were high, the rhetoric sharp, and the combat bruising. In this case, the contestants were not, it seemed, the rival presidential nominees, but the secretary of defense and the director of central intelligence, supposedly locked in an epic duel for control of the nation's intelligence community. At least that was how reporters depicted the events leading to the passage of the Intelligence Reform and Terrorism Prevention Act of 2004—the biggest overhaul of the U.S. intelligence community in six decades.

As with much of the reporting in that political season, however, the impression one gained from the news media was incomplete. Like watching a boxing match on the deck of a ship at sea, the competition was riveting, but many other things were happening out of sight of the spectators. The 2004 act was indeed a landmark for the intelligence community, although for reasons quite different from what many observers and even participants in the debates over its drafts imagined. While it did shift increments of power over intelligence assets from the Department of Defense (DoD) toward a new, national chief of intelligence (the director of national intelligence, or DNI), the act will likely be remembered more for the effect it had on the more mundane details of domestic intelligence. Simply put, the Intelligence Reform and Terrorism Prevention Act created something unprecedented in American history: a single officer to bridge the historic divide between foreign and domestic intelligence.

The creation of such a post was timely because warfare is changing. Indeed, the act came three years late; an effective DNI (as the authors of the act argued) might have prevented the terror attacks of September 11, 2001. At any rate, the United States and the world are scrambling to adjust to a mode of warfare undreamt of by the great theorist Carl von Clausewitz—one in which human beings make themselves into weapons for the joy of killing and the promise of paradise, where enemies do not need to control tracts of land on which to marshal the smaller and smaller items they need to mount horrific attacks, and where violence is not politics "by other means" but a nonnegotiable end in itself. The White House recognized this threat and articulated the United States' changing outlook in its *National Security Strategy* in September 2002, but the process of shifting traditional perspectives, and institutions, will take years to complete.

A new grand strategy is emerging by fits and starts, and with it a need for a new model for the U.S. intelligence community. We are in a betwixt and between world, still dominated by state-actors who play by familiar rules and deploy their militaries in time-honored (and deadly) ways. In some respects, the U.S. military and the intelligence community are like the British Army in 1913. Most of their daily work is patrolling and peacekeeping, punctuated with combat against outlaws and insurgents in faraway places, but they have to be ready to contend with more powerful (if more conventional) opponents as well. Reform of the intelligence community's portion of that common enterprise began only with the passage of the Intelligence Reform and Terrorism Prevention Act. What follows are some notes that can help us understand how U.S. intelligence must adjust as the future unfolds and the nature of war does—or does not—transform itself into something that Clausewitz would not have recognized.

## Origins Have Consequences

Let us posit for the sake of argument that U.S. intelligence developed four main emphases or functions over the last century. These emerged in conjunction with America's growing involvement in events in the Eastern Hemisphere, as the nation's grand strategy—its implicit and explicit balancing of threats, capabilities, and objectives—shifted after 1914 from commercial neutrality to active global engagement, and as science spawned devastating new weapons that could reach across the ocean moats that had largely isolated the United States from Old World quarrels. These changes in strategy and technology prompted the creation of new intelligence disciplines and organizations.

The four main missions for American intelligence were homeland defense, clandestine activities abroad, support to military operations, and support to the president and his lieutenants. These missions emerged gradually, not replacing but supplementing one another.

- The first of the intelligence community's missions—homeland defense— can be traced to World War I and the need for the federal government to fight German saboteurs in the United States as well as German forces in Europe. Permanent counterintelligence and signals intelligence establishments eventually emerged from the ad hoc measures deployed in 1917.
- The second mission met President Franklin Roosevelt's need for an intelligence capability that could work secretly abroad to implement American policies, quietly influencing nations and events in ways favorable to the United States. This capability was warehoused just after the end of World War II for future assignment to the still-to-be-determined peacetime intelligence structure.
- The third major mission was to exploit sophisticated new techniques to support commanders at the strategic and "operational" levels in the active theaters of war. The volumes of high-quality intelligence needed to beat the Axis were vast, and several intelligence disciplines had to be created or mastered to provide senior commanders the answers they required.
- Finally, President Harry Truman created the fourth major mission for American intelligence: a permanent capability to support national (as opposed to departmental) decision making. This evolved under Presidents Truman and Eisenhower into a strategic estimative function, a current intelligence brief, and a watch office vital for situational awareness—and the analysts to staff all three. It was partly to build this support capability that President Truman created the post of Director of Central Intelligence (DCI) in January 1946.

The first of these missions was forced upon the nation to protect itself from foreign manipulation; the latter three evolved because the United States had suddenly changed its outlook on the world and its grand strategy. The United States had hitherto relied on its navy and its vast latent power to deter hypothetical foreign threats, but President Roosevelt and Congress committed the United States to a forward defense posture almost overnight in the summer of 1940, when Hitler seemed poised to control all of Europe. Roosevelt argued that global economic interdependence and the destructiveness of modern weapons had made the Old World's wars into matters of weight for the United States. America, he warned, could not sit idle while aggressors consolidated bases from which to strike at its vital interests. As a consequence of President Roosevelt's strategic turn, America's intelligence organizations were haltingly reorganized, retooled, and rechartered to serve the new needs of policy makers and commanders.

All of this change laid the foundation for the intelligence establishment that America needed in order to serve a new grand strategy for the Cold War. President Roosevelt had sketched such a strategy but left to his successors the job of building the American and international institutions to contain the new threat posed by Communism and an expansionist Soviet Union. The National Security Act of 1947 is another of FDR's posthumous monuments; it marked the ascension of the new strategy and guided the reform of the foreign and military structures (and their intelligence components) to provide for a new, global role. It is still the charter document of America's national security establishment.

The National Security Act that President Truman signed in 1947 essentially codified the four main missions for American intelligence that had been fashioned over the previous thirty years. Curiously, however, the act by no means unified the nation's intelligence efforts. President Truman and the 80th Congress insisted on greater coordination only to a point.[1] A new, statutory DCI would now hold sway over two of the four main missions: he would ensure that clandestine activities overseas were coordinated and de-conflicted, and his staff would prepare intelligence for the president, relieving the chief executive of having to serve as his own intelligence analyst. The other two basic missions of American intelligence were explicitly withheld from the DCI.

The intelligence structure codified by the National Security Act was such an improvement over the wartime setup that the act merits the credit it is generally accorded for enhancing U.S. intelligence capabilities. That credit should not, however, obscure the many less-than-optimal compromises that had to be written into the legislation to ensure its passage.

The act made the DCI first among equals among a set of competing intelligence barons, most of them based in the military. The reason behind this confederation scheme was a lack of faith in the untried idea of a "central" director. The Federal Bureau of Investigation (FBI) and the armed services weighed their chances and made convincing arguments for autonomy. Having spent years building up capabilities on their own, at times incurring opposition from other departments (especially the Department of State), the services and the bureau were not about to cede their hard-won competencies to some intelligence czar outside their control. Congressmen, in turn, feared an "American Gestapo" conducting law enforcement and intelligence work at home and abroad. The National Security Act thus drew a bright line between these realms and left the "internal security" mission (and virtually everything associated with domestic intelligence) with J. Edgar Hoover's FBI, which had recently proven that it could defeat Axis agents in America. At the same time, generals and admirals convinced President Truman that "every department required its own intelligence"; that is, they could not rely on a civilian agency to meet their intelligence needs.[2] Thus, the act left the military free to provide for its own "departmental" intelligence.

The DCI thus had no law enforcement powers or domestic intelligence mission, and would have little role in producing intelligence for military commanders. The FBI went its own way on internal security matters, barely cooperating with the DCI and his new Central Intelligence Agency (CIA). The armed services tried but often failed to provide their own "departmental" intelligence. Key capabilities they had constructed in wartime had been demobilized by 1947, and at any rate, the new job of monitoring the Soviet Union proved too big for the services even when they began pooling their efforts in the 1950s and 1960s. The problems that the act created, ignored, or worsened would take decades to solve.

This is why one can read the history of American intelligence between 1947 and 2004 as a long series of remedies for certain problems locked in place by the National Security Act. A few modest organizational reforms were implemented, mainly in improving the White House's oversight of the community, and in giving the secretary of defense his own "national" intelligence agencies.[3] The military and the CIA gradually learned how to collect and analyze overhead imagery, at least at the "national" level—tactical and operational-level commanders had mediocre imagery support well into the 1990s. The more sweeping ideas to be proposed, however, foundered on the immemorial constitutional separation of powers and inevitable organizational rivalries, and sometimes on the clashing personalities of the leaders involved. The National Security Act had created an institutional system in which change required consensus among key players in the executive and legislative branches, while several of these actors on their own could veto proposed improvements that would benefit the system as a whole.

Such problems should not be overblown. During and after the Cold War, the intelligence community did a credible job of monitoring and dealing with foreign entities—usually states—that wanted to gain or maintain control over territory and were not willing to commit suicide for this end. The structural weaknesses that the National Security Act had rigidified could be papered over. The FBI had done passably well against the Soviet espionage threat, which had no domestic support infrastructure after the American Communist Party was all but driven underground in the early 1950s. Intelligence support to commanders helped preserve the U.S. military from disaster in Korea and Vietnam, but was never tested by a war with the Soviet Union. By the late 1990s, however, signs of increasing strain had become worrisome even outside the community, as the digital revolution forced a wholesale review and recapitalization of intelligence capabilities.

The terror attacks of September 11 forced Washington to contemplate a broader transformation. The last decade has seen considerable strain on the institutions fashioned to win World War II and the Cold War. Conventional counterintelligence doctrine was impotent against Al Qaeda, an enemy that burrowed into the American population and exploited the seam between intelligence and law enforcement. The growing need to provide real-time intelligence support to military deployments swamped the organic capabilities of the DoD and taxed the national systems based in Washington. Both of these structural weaknesses could be traced to the compromises embodied in the National Security Act. More important, however, was the uncertainty introduced into the nation's larger strategic thinking. The strategy that Franklin Roosevelt had bequeathed was at its root Westphalian, based on the assumption that nation-states are the key actors in the world and thus should be the primary concern of U.S. foreign policy.[4] Al Qaeda showed, however, that it does not take a powerful nation-state to visit mass destruction on America, and that resentments against globalization will be violently directed at the United States.

Historians in future decades will refer to the rush of lawmaking and strategy writing since 2001 as one of those pivotal periods in American history that come only two or three times a century. A new national strategy is being worked out in legislation and practice, through White House pronouncements, statutes such as the USA Patriot Act and the Homeland Security Act, judicial rulings, and command decisions in various theaters of war. It may take several more years before the full scope of the threats and opportunities to the United States and the civilized world are clearly defined, and just as many years or more before the new grand strategy of the nation is fully articulated. A shift in national strategy must, eventually, dictate a corresponding shift in the missions and resources provided to a nation's commanders, its diplomats, and its intelligence officers. That lesser shift, however, is already under way.

## The Intelligence Reform and Terrorism Prevention Act of 2004

This roundabout historical survey is prologue to understanding the changes that emerged from the debates over the Intelligence Reform and Terrorism Prevention Act. It is crucial to note for the record that this new law is for the most part a series of amendments to the National Security Act of 1947, which remains the dominant paradigm for the nation's intelligence effort.[5] The new act thus represents an attempt at integrating those functions that were originally codified in the National Security Act, and at adjusting the nation's institutions to the emerging national security strategy.

The place to begin is to notice that the Intelligence Reform Act has done away with the position of DCI, splitting that job into two functions: head of the intelligence community, and director of the CIA. The former officer—the DNI—has been empowered to lead an intelligence enterprise that is "more unified, coordinated, and effective" than ever before, as President George W. Bush stated in signing the act on 17 December 2004. The DNI is the principal intelligence adviser to the president, with the responsibility of ensuring the quality of intelligence provided to the nation's leaders.[6] The DNI will also have more clout to manage the intelligence system—and enough responsibility for its performance to be held accountable for what happens on his or her watch as head of the community.

Recent DCIs tried to perform these duties, but they lacked the tools that have now been handed to the DNI by the recent act. When it came to tasking intelligence assets, DCIs could approve collection requirements and determine collection priorities; DNIs, in contrast, can establish objectives and priorities, determine requirements, and manage and direct the actual tasking.[7] Furthermore, the DNI should be more effective than any DCI in determining and monitoring intelligence community budgets, plans, policies, acquisitions, and personnel. The act, for instance, decrees that funds for national intelligence activities will still pass through the comptrollers of their respective departments but now will do so at the "exclusive direction" of the DNI.[8] In addition, the DNI can transfer up to $150 million a year per agency between different national intelligence programs (if the White House and Congress concur).[9] Powers like these amount to real influence in Washington.

The act also gave the DNI authorities that are not strictly analogous to any that could be found in the former DCI's portfolio. That is why it would be a mistake to regard the DNI as merely a stronger DCI. The DNI will be just what his title implies—the officer in charge of providing "national intelligence," which the act defines as any intelligence, "regardless of the source," that pertains to two or more agencies and which involves a threat to the United States or its interests—or involves "any other matter bearing on the United States national or homeland security."[10] This expansive new definition is the key to grasping the difference between the DNI and the old DCI, and the news media's comparative silence on this development during the debates over the act's passage is partly why the reporting in late 2004 was often misleading.

One can observe the practical difference between the new DNI and the old DCI by contrasting the new state of affairs with the way in which the Truman administration and the 80th Congress in 1947 had made domestic and foreign intelligence into separate realms. The DNI is the principal intelligence adviser "to the President, the National Security Council, *and the Homeland Security Council*" (emphasis added).[11] Indeed, the DNI was made the overseer of the National Intelligence Program (the funding stream for national intelligence activities), which was formerly titled the National Foreign Intelligence Program. The change in names was intended by the authors of the act precisely to signal the DNI's responsibility for foreign *and* domestic intelligence.[12] The act also gave the DNI considerable powers to integrate the community's data management enterprise architecture "to ensure maximum availability of access to intelligence information."[13]

Congress preserved traces of the old foreign-domestic seam in the new act. The FBI remains, in effect, the primary domestic intelligence agency, although it is never called that. The CIA is still banned from "internal security" functions, having "no police, subpoena, or law enforcement powers."[14] Although this language is not applied to the DNI or his or her office, the DNI must bow to the intelligence officers of the Department of Homeland Security (DHS) in the direct dissemination of intelligence to state, local, and private officials.[15] The act also implies that the DNI's National Counterterrorism Center (NCTC) shall concentrate on foreign threats, but then Congress, in practically its next sentence, said that the NCTC (and by inference the DNI) may "receive intelligence pertaining exclusively to domestic counterterrorism from any Federal, State, or local government or other source" if the president directs.[16] Implementing all the new authorities will require the

DNI to undertake a delicate calibration of competing bureaucratic interests and political concerns, but there can be no doubt that the 2004 act created something new in making the DNI a *de facto* coordinator of domestic as well as foreign intelligence. Indeed, it is not implausible that the DNI might come to be viewed as an honest broker between the rival interests of the FBI and DHS—as the one official who can keep both working together.

How then will the DNI work in the domestic realm? The act gave him or her no formal agency to conduct this business, but it did create a mechanism that DNIs might adapt for the role. Official Washington in 2004 was smitten with the idea of deploying interagency "centers" against some of the harder foreign intelligence targets, such as terrorism and the international proliferation of weapons of mass destruction.[17] Several of the intelligence reform bills introduced in the spring and summer of that year proposed centers in one form or another, and the 9/11 Commission recommended them (particularly a national counterterrorism center) in its best-selling report. President Bush soon provided for centers and even created one—the NCTC—in his Executive Orders 13354 and 13355.[18] The Intelligence Reform Act brought this process to closure, establishing the NCTC in statute and authorizing the DNI to create more centers as needed to "address intelligence priorities, including, but not limited to, regional issues."[19]

Intelligence centers mean many things to many people, but the rough idea is to focus the community's dispersed systems and disciplines on particular and difficult problems so as to gain greater insight or capability. They have existed since DCI William Casey's founding of the counterterrorism center in 1986, but in their original form, they were operationally focused and dominated by the CIA, with a few representatives of other agencies on staff to give them a light "community" tint.[20] The most important center for the DNI, at least at the outset, will be the NCTC, mainly because it already exists, has a statutory foundation, and has sponsors on both ends of Pennsylvania Avenue. The NCTC can receive domestic terrorism information, as noted above, but its unique role in "strategic operational planning" of activities across the entire federal government makes it an unlikely model for other centers under the DNI. Fortunately for him or her, however, the act specifies no one design for centers, and this leaves the DNI with the latitude to establish (or dismantle) them when and how he or she deems appropriate.[21] Thus, centers can truly be many things, as the DNI wills, as long as he can garner cooperation—and staffers—from the other community agencies. It seems likely that he or she would, for the act empowers the DNI to transfer in up to a hundred staffers from other agencies in the first year of a center's existence.[22] With that power in his or her pocket, a DNI will have a persuasive case in asking for detailees to staff small and temporary offices to share sensitive data on topics of critical national interest. Indeed, the DNI's mandate to create centers gives him an opportunity to experiment in covering politically loaded domestic intelligence topics; DNIs might well come to use centers more often on the domestic side than the foreign.

Finally, Congress insisted that all this be done with scrupulous attention to the rights and privileges of Americans. One of the act's principal authors, Sen. Joseph Lieberman (D-CT), conceded on the floor of the Senate that waging the war on terrorism "has required that the federal government take steps that consolidate government authority and increase the government's presence in our lives."[23] This necessary increase, however, needed to be closely monitored. The act ordered the DNI to appoint a Civil Liberties Protection Officer who works for him alone to "ensure that the protection of civil liberties and privacy" is incorporated in the policies and procedures of his organization.[24] Indeed, the act also (at the urging of the 9/11 Commission) imposed upon the White House an advisory Privacy and Civil Liberties Oversight Board, noting that the government's enhanced powers for prosecuting the war on terrorism called, in turn, for "an enhanced system of checks and balances."[25]

In summary, the act makes the DNI the government's principal intelligence adviser and head of the community for both foreign and domestic intelligence matters. The old foreign-domestic divide from 1947 has been bridged, but it still exists. What remains to be seen is which is more important: the bridge or the divide.

## Intelligence Reform and the Future of War

Reform of the intelligence community did not end with the passage of the recent Intelligence Reform and Terrorism Prevention Act; that was but the beginning of the work. The act could be implemented in ways that frustrate the purposes of its authors, and even result in a less-coordinated and less-capable intelligence system. We have already seen that the original National Security Act of 1947 embodied problematic compromises and was not well implemented, at least at the outset. Translating the 2004 act into practice will have to be done with due regard to the constitutional separation of powers, the divisions of labor in the executive branch and of oversight in Congress, and the inevitable bureaucratic inertia of the established agencies. The next few years will see new and old players realign in patterns that they are likely to hold for decades, as it is not likely that Congress will soon pass another intelligence bill as sweeping as this one.

The intelligence community will settle into new patterns while it adjusts to a world in which ubiquitous computers, networked global markets, and new, lighter, and smaller materials are shrinking time and space dramatically. In the twentieth century, intelligence was transformed from a hobby of generals and princes into a profession that harnessed the electromagnetic spectrum to serve the ancient crafts of espionage and codebreaking. U.S. intelligence organized itself according to the physical properties—and even the wavelengths—of the media it exploited. Now, however, intelligence is undergoing a second revolution. The old is not being swept away, but it is being made with cheaper, lighter, and much smaller components. Global communications and miniaturization have given to individuals certain powers that only states could exploit in the past, while states themselves are gaining awesome powers of personal intrusion and social control.

Whether nation-states will still be the dominant actors on the world scene at the close of the twenty-first century hence is not yet certain.

The Intelligence Reform Act allows the community to bend—and presumably not break—under the pressure of future developments. We cannot know how the act will actually be implemented, but we can say some things with reasonable certainty. The act gives the new DNI the ability to address some of the structural weaknesses left in place by the original National Security Act. To see how well he or she is doing in implementing the new act, watch for answers to these questions over the next few years:

- Does the DNI foster greater collaboration between signals intelligence, law enforcement, and the domestic intelligence staffs between them?
- Does the DNI advance the DoD's quest to build organic intelligence support for military operations, impede that effort, or sit on the fence?
- Does the DNI honor the sensitivities of Americans and their elected representatives with regard to civil liberties; that is, does he or she understand that the act's warrant to develop the discipline of domestic intelligence in the United States will be revoked if it is somehow seen to be politicized?

Answers to these questions—and many others—will emerge in the fullness of time, but it remains to be seen whether those answers will be for good or for ill, how soon they develop, and whether they do so according to plan or simply by default.

## Signals Intelligence and Law Enforcement

Intelligence themes in the United States have always been accompanied by a *continuo* interplay between counterintelligence and signals intelligence. The original interagency synergy in America was the cooperation of U.S. Army codebreakers, Bureau of Investigation agents, and local law enforcement to thwart German saboteurs in World War I. It is no coincidence that the most valuable intelligence—the decrypted communiqués of the enemy—winds up in the service of the most pressing national need—the imperative to thwart foreign agents in the homeland. Something similar will surely happen again, even if its details remain obscure to outside observers for a generation or more. The digital revolution is changing the means of communication and the targets of investigation, but it cannot annul the mutual dependence of policy makers, cryptanalysts, and investigators. Their changing interplay will determine a great deal about the institutional and operational context in which the rest of the U.S. intelligence establishment does its business.

The new act creates the possibility to expand cooperation between cryptology and domestic intelligence. Electronic snooping on Americans has long been a sensitive topic, tightly fenced by legal restrictions and implicitly deterred by the perils of media exposure and congressional ire. The Patriot Act and other changes made it easier for federal law enforcement to eavesdrop on private communications and to share the resulting intelligence with national security agencies, without, however, enlisting the full services of the National Security Agency—America's primary signals intelligence organization. The Intelligence Reform Act gives the DNI the authority to create centers that can do that, albeit in relatively small and experimental organizations. Thus, the DNI may be by default the best-positioned official in Washington to promote cooperation in this field.

## Growing Defense Intelligence

The deepest worry for some members of Congress crafting the provisions of the new act was that voiced by Rep. Duncan Hunter (R-CA), chairman of the House Armed Services Committee: Will the DNI neglect the needs of battlefield commanders for national intelligence support? This remains a serious concern, despite the obligations that the act imposed on the DNI.[26] Equally serious would be an inversion of this scenario; some authors of the act worried that the Pentagon could subvert DNI authorities. In their thinking, a strong secretary of defense (and history suggests there is rarely any other kind) could manipulate the intelligence assets housed in the DoD so that the DNI could not properly task and coordinate them—and the national intelligence system thereby loses the benefits of their efforts.[27] No statute implements itself; the act's insistence on mutual defense and national intelligence support will have to be carefully fostered and tended if it is to produce a truly unified military and intelligence effort.

The urgency of doing so is growing for two reasons. First, terrorists still need bases where they can arm, train, and equip, but they seem to be less and less dependent on the control of actual territory as coalition military operations uproot them from their sanctuaries. The insurgency continued in Iraq, for instance, even after the U.S. Marines took Falluja in late 2004. It is conceivable that in some not-so-distant future, a global terrorist movement will be virtually independent of physical bases—a free-floating virus in the midst of civilization.

Second, the United States, despite its current military predominance, has not repealed the laws of military science or human nature; sooner or later it will fight a "peer competitor," even if one of a local and temporary sort. The American military has to be ready for such a contingency. In other words, the nation's military and intelligence establishments have to know how to fight two utterly different kinds of war: one conventional and familiar to students of Clausewitz; the other radically new.

To succeed, the United States military has to get better at providing intelligence for itself. Intelligence support from agencies in Washington can help, but no national-level system can be as responsive to a commander's rapidly evolving needs for tactical and theater-level systems under his control. Success in war requires (among other things) unity of effort. Such unity has always been difficult for U.S. intelligence—divided as it is among agencies, missions, and disciplines—to provide and nurture. Both intelligence officers and commanders are still learning how to integrate the intelligence and operations functions on higher-level military staffs. The act provides a stronger framework for innovation and cooperation between the secretary of defense and the DNI in providing this support, but it will ultimately be up to the former and his or her undersecretary for defense for intelligence to do this job in the Pentagon.

### Guarding Civil Liberties

Clausewitz famously reminded us that war is "nothing but the continuation of policy with other means."[28] His dictum has stood the test of time, but is it still true for *all* of the adversaries that the United States and so many nations now confront? Clausewitz's rule certainly holds for states or entities that wish to become states; in other words, for leaders who control swaths of territory, and thus have a goal—a policy—that lends itself to negotiation or even to tests of strength. Disputes over land, or things based on land, can always be settled by compromise or conquest. But what if the resentments over globalization and modernity breed a free-floating, "land-less" terror threat that opposes civilization as such? Then the front that our enemy chooses would lie in our own cities, where violence would be pursued for its sake, with the only "policy" being the cosmic goal of promoting fear and forcing governments to impose restrictions that "unmask" the modern state's supposedly tyrannical nature (and thus breed more resistance and violence).

Some argue that this is exactly the threat that the West faces. Others are unconvinced. As a result, American society is now in the midst of a deepening and necessary debate over when and how, if ever, the threat of terrorism reaches the point that citizens, aliens, and travelers (both in physical and cyber realms) forfeit the presumption of innocence. Both the executive and legislative branches are reexamining the boundaries between "foreign" and "domestic," and among "intelligence," "law enforcement," and "security." The new Intelligence Reform Act allows the DNI to organize domestic intelligence in the United States, but the political reality is such that this mandate will hold good only as long as he or she works in a way that defends rather than infringes on American liberties—or "we lose the values that we are struggling to defend," in the words of the 9/11 Commission.[29]

## Conclusion

The British commentator Alistair Cooke once described the three principles that had animated the framing of the American Constitution and then sustained the Great Republic as "compromise, compromise, compromise."[30] President Truman and the 80th Congress relearned these principles in drafting the National Security Act; their compromises established the enduring strengths and weaknesses of the nation's intelligence community. In late 2004, another president and Congress appended a new layer of compromises—in the form of the Intelligence Reform and Terrorism Prevention Act—to add flexibility and accountability to the system that had been codified in 1947. The question is whether, in light of the shifting nature of warfare and the headlong rush of technology, the new set of compromises has done enough to empower an intelligence community to combat terrorists who want to destroy American cities.

The Intelligence Reform Act does not guarantee success. Consequently, watching the reform of U.S. intelligence over the next few years will take strong nerves as well as sharp eyes. There will be more excitement of the sort of blow-by-blow narration offered by the news media during the debates over the 2004 act. Most of the real business of reform will likely take place belowdecks, as it were. Will some new "Peace of Westphalia" set the outlines for the global system of the future? Will international terrorism prove to be a passing danger or a permanent shift in the nature of war, and if the latter, will some new Clause-witz divine and explain the changes in the military art? The new act gives the U.S. government, and the intelligence community, an opportunity to adjust (and perhaps to excel at supporting) the emerging national security strategy while these larger historical questions are being resolved.

## Notes

1. I discuss the debates and compromises over the intelligence provisions of the National Security Act at length in *Central Intelligence: Origins and Evolution* (Washington, D.C.: Central Intelligence Agency, 2001), pp. 2–7.

2. Harry S. Truman, *Memoirs,* Volume II, *Years of Trial and Hope* (Garden City, N.Y., Doubleday, 1956), p. 57.

3. These were, in order of creation, the National Security Agency, the Defense Intelligence Agency, and the National Reconnaissance Office. The National Photographic Interpretation Center, a joint CIA-DoD entity, became part of the Secretary of Defense's National Imagery and Mapping Agency in 1997 (it is now the National Geospatial-Intelligence Agency).

4. The Peace of Westphalia in 1648 ended the Thirty Years War and is widely credited by scholars of international relations with marking the close of the lingering diplomatic and military power of feudal entities and the dominance of nation-states in Continental (and ultimately world) affairs.

5. It is also worth repeating for the record that the inspiration and impetus behind that Act came primarily from the findings of the independent panel to study the events of 11 September 2001, the National Commission on Terrorist Attacks Upon the United States (or the 9/11 Commission for short).

6. See Section 1011 of S.2845, the Intelligence Reform and Terrorism Prevention Act, amending the National Security Act at Sections 102(a) and 102A(a); see also Section 1019 of S.2845.

7. Contrast Section 103(c)3 of the National Security Act, as amended, with the change to that language mandated by Section 1011 of the Intelligence Reform and Terrorism Prevention Act amending Section 102A(f)1(A) of the National Security Act.

8. Section 1011 of S.2845, the Intelligence Reform and Terrorism Prevention Act, amending Section 102A(c)5 of the National Security Act.

9. Ibid., amending Section 102A(d)5(A).

10. Section 1012 of the Intelligence Reform and Terrorism Prevention Act, amending the National Security Act at 50 USC 401a. Interestingly, the amendments to the National Security Act did not adjust that Act's note that the term *intelligence* "includes foreign intelligence and counterintelligence" (see 50 USC 401a). It is possible to read the amended Act as now saying that the DNI's purview extends only to "intelligence" matters, which supposedly must by definition include some foreign component. Such a reading, however, is not likely to be endorsed by Congress, or by the National Security Act, which further on speaks of "domestic counterterrorism intelligence" in terms of the NCTC. That reference would seem to foreclose any argument that the DNI's writ extends only to matters related to foreign intelligence; see Section 1021 of the Intelligence Reform and Terrorism Prevention Act amending the National Security Act at Section 119(e).

11. Section 1011 of the Intelligence Reform and Terrorism Prevention Act, amending Section 102(b)2 of the National Security Act.

12. Sen. Susan Collins (R-ME), a principal sponsor of the Act, noted on the floor of the Senate that "The renaming of the [National Foreign Intelligence] Program signifies that the national security threats of the 21st Century straddle the foreign/domestic divide and that our Intelligence Community must have capabilities that cross this seam." This remark and many other comments on the bill are included in *Congressional Record* (Senate), Vol. 150, No. 139, 8 December 2004, p. 11968; this is available online at www.gpo.gov/gpoaccess.

13. Section 1011 of the Intelligence Reform and Terrorism Prevention Act, amending Section 102A[g] of the National Security Act. Almost before the new Act was signed, Congress amended Section 103G of the National Security Act to give the DNI a Community-wide Chief Information Officer with sweeping new powers; see Section 303 of the Intelligence Authorization Act for Fiscal Year 2005.

14. Section 1011 of the Intelligence Reform and Terrorism Prevention Act, amending Section 104A(d) of the National Security Act.

15. Ibid., see Section 102A(f)1(B).

16. Section 1021 of the Intelligence Reform and Terrorism Prevention Act, amending Sections 119(d & e) of the National Security Act.

17. For an example of the thinking behind the center idea, see Larry C. Kindsvater, "The Need to Reorganize the Intelligence Community," *Studies in Intelligence* 47 (2003). Mr. Kindsvater became Deputy Director of Central Intelligence for Community Management not long after this article was published.

18. President Bush established the NCTC in his Executive Order 13354, "National Counterterrorism Center," on 27 August 2004. His Executive Order

13355, "Strengthened Management of the Intelligence Community," empowered the DCI to establish other centers; it was signed the same day.

19. Sections 1011 and 1023 of the Intelligence Reform and Terrorism Prevention Act, amending Sections 102A(f)2 and 119B(a), respectively, of the National Security Act. The NCTC will work for the DNI to serve as the nation's primary organization for analysis of foreign-sponsored terrorism and to plan and assign counter-terrorism roles and responsibilities to the departments and agencies to perform under their own unique authorities. See Section 1021 of the Intelligence Reform and Terrorism Prevention Act of 2004 amending the National Security Act of 1947.

20. National Commission on Terrorist Attacks Upon the United States (the 9/11 Commission), *The 9/11 Commission Report,* (Washington, D.C.: Government Printing Office, 2004), pp. 92, 342.

21. Section 1023 of the Intelligence Reform and Terrorism Prevention Act, amending Section 119B of the National Security Act. The Act new specifies that a National Counterproliferation Center be established along the same lines as the NCTC by the end of 2006, but it also gives the President the option to waive this requirement; see Section 119A(c).

22. Section 1011 of the Intelligence Reform and Terrorism Prevention Act, amending Section 102A(e)1 of the National Security Act.

23. *Congressional Record* (Senate), p. 11974.

24. Ibid., amending Section 103D(a).

25. Section 1061 of the Intelligence Reform and Terrorism Prevention Act. The commission's recommendation can be read in the *The 9/11 Commission Report,* p. 395.

26. See, for instance, Section 1018 of the Intelligence Reform and Terrorism Prevention Act, as well as its Section 1011 amending Section 102A(a) of the National Security Act.

27. For an example of this concern, see Milton Kady II, "Intelligence Bill Conferees Push for Agreement on Budget Authority," *Congressional Quarterly* on-line edition, 9 November 2004.

28. Carl von Clausewitz, *On War,* Michael Howard and Peter Paret, editors and translators (Princeton, N.J.: Princeton University Press, 1976), p. 69.

29. *The 9/11 Commission Report,* p. 395.

30. Alistair Cooke, *Alistair Cooke's America* (New York: Knopf, 1973), p. 144.

# TWENTY-NINE

# Refocusing Intelligence

## The Art of Analysis

KEITH J. MASBACK WITH SEAN TYTLER

You cannot depend on your eyes when your imagination is out of focus.

Mark Twain

## Introduction

In their treatment of military innovation during the period between World War I and World War II, Williamson Murray and Allan Millett point to the German military establishment, particularly its officer corps, as the model for reflective, self-investigative analysis. Fearing no reprisals, and holding nothing back, the Germans embarked on a decade-long process of identifying what they had done wrong, which doctrines were flawed, and which innovative systems or tactics were needed to improve battlefield performance. As a lesson for turning defeat into an opportunity for self-improvement, there may be none better.[1]

Today, the United States finds itself with a similar opportunity to fundamentally rethink its intelligence enterprise, assuming it is willing to be as open about its modern intelligence failures as postwar Germany was about its military failures. It has been nearly impossible the last few years to open a newspaper and not see an accounting of the latest "intelligence failure" or missed opportunity. The 9/11 Commission and the Intelligence Reform and

---

The views expressed in this chapter are the author's alone and do not necessarily reflect the views of the National Geospatial-Intelligence Agency, the Department of Defense, or the United States Government.

Terrorist Prevention Act have identified the failures of policies, capabilities, management, and imagination in most of the seminal events of the post–Cold War era.[2] Not surprisingly, the recommended solutions being debated in Congress and the media tend to focus on the first three while ignoring the fourth—failures of imagination.

While the intelligence enterprise cannot indoctrinate imagination into its analysts, it can create an environment that best positions its people to apply the critical thinking necessary to reduce uncertainty. If this nation is to successfully anticipate emerging threats, our leadership must deemphasize structural reform and rely less on technology. Instead, the intelligence enterprise must focus on strengthening analytics through a mission-focused concept of operations, long-term, multidisciplinary education of our people, and a work environment that empowers our analysts to do the most with the data at hand.

Whether transnational or local, traditional or unconventional, the wars we face in the coming decades will present our nation with multiple challenges. Today our focus, and rightly so, is on missions like counter-proliferation, counterinsurgency, counterterrorism, and homeland security. Tomorrow it may be these, it may be traditional warfare, or it may be something entirely unforeseen. Since none of us can confidently predict the future, the intelligence enterprise must be positioned to recognize evolving threats before they emerge and muster its assets before a crisis is imminent.[3]

Intelligence reorganization, information technology, information sharing, terrorism task forces, streamlined oversight—all of these hot-button issues address important elements of the problem, but they do not get to the heart of the problem. Too often, we rely on organizational restructuring and technology to address our shortcomings, thinking it possible to organize or buy our way to improved performance. At the center of the intelligence enterprise lies not its structure or its tools, but its people, the analysts and users who ultimately are responsible for understanding our adversaries and providing decision makers with actionable intelligence.

This chapter will first describe the changing intelligence landscape and the challenges the intelligence enterprise faces in the coming years, including more dynamic and unpredictable adversaries and new and voluminous sources of data. Next, we will examine three ways to address these challenges—a new concept of operations, an updated educational philosophy, and a deeper focus on how analysts turn data into information—that, taken together, will strengthen the intelligence enterprise's ability to solve the complex and dynamic problems facing today's and tomorrow's operators and decision makers.

## The Changing Challenge

Beauty is in the eye of the beholder; the same can be said of the "revolution in military affairs" (RMA). Some swear by it, pointing to amazing advances in information technology and network-centric warfare.[4] Others recognize these advances as a shift not in military affairs themselves, but in performance, seeing more change in the efficiency of combat operations than its key functions.[5] Too much time has already been spent discussing the truth or fiction behind the RMA and its relevance to operations today. What is important to recognize is the enormous pressure being placed on the intelligence enterprise by the growing demands for information by a wide variety of customers.

Before going further, it is important to define what this chapter refers to as the intelligence enterprise. The intelligence enterprise comprises the full scope of people, processes, and technologies resident across many agencies and organizations that together produce information for a wide range of consumers. It is purposefully broad in definition, encompassing more than such traditional elements of the intelligence community as the Central Intelligence Agency or Defense Intelligence Agency, and more than such traditional consumers of intelligence as national policy makers and the military. The intelligence enterprise includes consumers, such as squad leaders of military units, first responders, border patrol agents, and police officers, and sources, such as blogs, open source information, and foreign intelligence. While such a broad definition might seem impractical, it is important that any changes or improvements be made with not just the traditional intelligence community in mind. Our shortfalls extend beyond this traditional community, and our solutions must be broader as well.

National decision makers, military operators, and homeland security officials—among the primary customers for intelligence products—need better information derived from multiple and varied sources, and they need it sooner than ever before because of the dynamic and elusive nature of many of today's—and most likely tomorrow's—adversaries.[6] This situation raises two new challenges for the intelligence enterprise to address.

### First Challenge: Facing New Adversaries
### with Our Old Intelligence Enterprise

In an enterprise designed and built to counter the slow, often predictable actions of a peer competitor, adapting to a more dynamic, disjointed, and agile adversary can present daunting challenges. Major weapon systems and the intelligence assets developed to provide them information were designed to operate in a Cold War–style environment, not during a global war on terrorism.[7] Information becomes dated and irrelevant more rapidly than ever before. As a result, all echelons, or levels, of the intelligence enterprise need to become more agile and responsive to the time lines and product types required by the full range of customers, from forward-deployed military operators to emergency first responders to the president, and everyone in between. These different consumers should not be treated as a single, monolithic entity; each has different tendencies, reporting requirements, and time lines that have several implications for the speed, openness, and specificity of intelligence products.

In an age of fleeting targets and short information half-lives, intelligence is shifting from an enabling staff function to a core mission for many military and homeland security operations. Unfortunately, cultural and technological divides still exist across the enterprise that preclude a complete merger of forward-deployed operators, emergency first responders and national agency analysts into a single, unified enterprise focused on information responsiveness and relevance at both the strategic and tactical levels.

For military users, the persistent myth that the intelligence enterprise and its component agencies are too slow and too focused on broader, more strategic problems has for the most part been proven false; many intelligence professionals and organizations have proven their responsiveness to tactical demands in Operation Enduring Freedom and Operation Iraqi Freedom. There remain, however, some technological and cultural divides, with operators often unable to access information and products generated by the intelligence enterprise. This may be due to bandwidth restrictions, formatting differences, multilevel security concerns, policy issues, or the simple fact that many users do not know the types of intelligence actually available to them. The problems are even more pronounced for homeland security, where legal constraints and the growing pains associated with a new executive department subsuming a wide variety of missions cloud the relationship between the intelligence enterprise and homeland security officials. In this case, both cultural and technological barriers remain. These barriers must be overcome before we can legitimately describe the enterprise as responsive and adaptable enough to meet emerging threats.

## Second Challenge: Absorbing an Increasing Volume and Variety of Data

An influx of data from new and varied sources threatens to overwhelm the intelligence enterprise and the customers it serves. A variety of data sources are being designed to gather more information on a wider range of targets of interest more often, in the belief that collecting more data will result in more actionable and relevant information. Put simply, the enterprise does not need more data; it needs better information. This is not to imply that volume is necessarily a bad thing; it certainly can be an ally if we have the analytic capabilities necessary to separate the signal from the noise and to discern needles in the haystack. However, with today's emphasis on enhanced information collection and sharing, analysts will soon face a bow wave of data that risks frustrating and overwhelming them. Such an emphasis increases the likelihood that we will overlook key pieces of information critical to national security.

This focus on collection and reliance on new technologies as a panacea for our ills is flawed in principle. The 9/11 Commission recognized this in part through its insistence on information sharing as a key tenet of enterprise improvement. But the problem requires more than merely information sharing, not least because such sharing could add to the deluge of data. The key to overcoming this deluge of data rests in creating an environment that emphasizes the process of problem solving as an almost scientific exercise: understanding the nature of the problem, developing hypotheses, collecting data to test (and potentially disprove) those hypotheses, deciding between relevant and irrelevant pieces of evidence, refining hypotheses based on these new data, and developing and defending an argument, complete with possible counterarguments.[8] Such a process organizes the problem into manageable chunks that serve as guides through what otherwise would be an unwieldy volume of data.

In designing an intelligence enterprise to handle a growing volume and variety of data, we cannot assume a static and predictable analytic mission where one size fits all; users sometimes want access to raw data to analyze themselves, and sometimes they want conclusions derived by others. In those cases where users need more than just raw data, getting to actionable intelligence and ensuring that it is both relevant and accurate requires that analysts understand the complete picture—the story behind the facts.[9] By focusing on the analysts—on the operational concepts that guide them, on the educational program that nurtures them, and on their ability to interact with data in new and intuitive ways—the intelligence enterprise will regain its edge and begin acting more like an information-age provider of knowledge than an industrial-age manufacturer of product.

## Focusing Our Efforts:
## Addressing the Changing Challenge

The fundamental objective of tomorrow's analyst remains the same as today's: to reduce uncertainty by providing the predictive and anticipatory intelligence an operator or decision maker needs to stay ahead of our adversaries, while recognizing the gaps in data and the uncertainties in conclusions. This is true whether we refer to the cycle as tasking-processing-exploitation-dissemination (TPED), tasking-processing-posting-using (TPPU), or some other acronym.[10] This does not mean that the intelligence enterprise—specifically, how it goes about executing these principles—is not ripe for a fresh look. This section introduces three critical components to updating the enterprise into a more efficient, relevant, and holistic knowledge producer.

### First Component: Rethinking the Concept of Operations

The intelligence enterprise is beginning to understand that as an entity charged with solving problems, it needs a new concept of operations that switches the focus from assets and tools to the problems themselves—from what systems are available to help, to what information is needed to do the job. Concepts of operation are critical components of any organization, providing the procedures under which people perform their jobs and execute their missions. Much more than philosophical treatises, concepts of operation provide the guiding principles under which key acquisition and operational decisions are made. Two critical actions are necessary to properly redefine the operational concepts guiding the intelligence enterprise: instituting problem-centric analysis, and instilling a culture of stewardship of intelligence assets rather than ownership.

### Instituting Problem-Centric Analysis

A key component of the evolving concept of operations for the intelligence enterprise is problem-centric analysis. Too often consumers of intelligence—those asking the questions—color their information requests based on the data sources they believe are available or under their span of control. Users must focus more on the problem at hand—what do I need to know?; why do I need to know it?; and when do I need to know it?—than on individual data sources and the intelligence disciplines they are most comfortable handling. This is the heart of problem-centric analysis.

Under this concept of operations, a consumer communicates the problem—specifically, the information she needs to do her job—and the time frame within which she needs an answer to the intelligence enterprise in much the same way we request information from the Internet today. Within the enterprise, an information-brokering function breaks the problem into manageable information queries (what information is needed to address the problem), tracks down existing data, and tasks the collection of any necessary but unavailable data. The information broker is not a fundamentally new concept; today, for example, collection managers perform this function for their local intelligence needs. However, these human brokers cannot know the full range of information available to them across the entire intelligence enterprise; they are able to use only the information and assets of which they are aware. Tomorrow's brokers will be virtual, with connectivity to information throughout the enterprise that today's individual collection managers cannot dream of accessing.

This concept of operations, while it will take time and patience to implement, will help to more efficiently task and use limited resources while making more information sources available to consumers. If a consumer asks only for an image of a location, he may miss more valuable snippets of information that other technical or human sources can provide, simply because he does not have access or insight into these sources. Enterprise-wide information brokering will free the consumer from his own constraints and improve the chances that relevant information can be brought to bear on any given problem.

This problem-centric concept of operations has several implications. First, such a concept will position the enterprise to break through the long-standing segregation between strategic and tactical intelligence problems, which often have vastly different time lines and call for vastly different products. Too often, different elements within the community handle these different analytic missions. A forward-deployed team at an operations center might deal well with time-critical response, whereas the traditional organizations within the intelligence community might excel at long-term forensic research. Knowledge derived by one intelligence echelon often fails to make it to the other, resulting in the continued segregation of the enterprise with potential consequences for future operations. Perhaps more disturbing, the "tyranny of the present" often precludes our ability to focus on the future: analysts working long-term research issues are often assigned to support current operations, leaving our long-term trend analysis woefully understaffed.

The enterprise must offer analysis and support in varying levels from "triage to post-op," to borrow a phrase from the medical field. In some cases, military operators or emergency first responders need immediate intelligence on targets "just around the bend." Old information in existing databases is of little utility for real-time, dynamic operations. In other cases, military leaders and national decision makers need more in-depth analysis, where quality and depth are more important than speed. Old information is essential to such forensic analysis, providing activity trends and possibly predicting future activity.

Instituting a problem-centric concept of operations that focuses the attention and resources of the entire enterprise on any type of problem will break this strategic/tactical segregation by providing consumers, through information brokering, access to data and information throughout the enterprise they would otherwise not have received. Such streamlined information retrieval and tasking functions will free more people to do actual analysis, thus mitigating the high-demand, low-supply problem characteristic of today's intelligence analysts. This concept will be successful only if we recognize that consumers need to be collectors as well as users within this enterprise. We have already seen the line between collector and consumer blurring, as squad leaders and commanders in Iraq and Afghanistan often rely on information that they and their teams locally collect and assess. These operators must be able to provide this hard-earned local knowledge to the enterprise and make it available to other analysts working other issues.

Second, this concept of operations implies that rarely will a single analyst, or even a single intelligence discipline like imagery or signals intelligence, be able to provide the complete story. Enabling collaboration across the enterprise—across echelons and across disciplines—must be a priority. For near-term intelligence problems, the organizing principle will include virtual teams of analysts who coalesce based upon the needs of the issue and, once the issue is addressed, dissolve as quickly as they formed. A team of analysts with both technical and cultural knowledge is necessary. For long-term, strategic issues, teams of analysts will have to be more permanent. The power of collaboration and brainstorming among specialists in different areas and regions will ensure that the full range of possible scenarios is considered, thus reducing uncertainty and mitigating against surprise.

The intelligence enterprise must converge data sources and intelligence disciplines into a single, integrated analytic component that has access and insight into all current and historical information. Analysts must be empowered with the tools to make it happen: personnel databases that indicate who has expertise in what areas; chat rooms that enable near-instantaneous online communication; and connectivity to experts outside the traditional intelligence community—even those without security clearances—who can provide unique and relevant insight into an adversary or an event.

Today's intelligence enterprise has already made great strides in converging and integrating disparate components. The National Geospatial-Intelligence Agency (NGA) has recognized the inefficiencies associated with stovepiped intelligence capabilities, and is currently working to transition to a converged analysis architecture for all geospatial-intelligence. The National Security Agency and NGA have jointly created small experimental pockets of multi-intelligence analysts, with plans to increase the size and scope of these pockets until they become standard across the community. Other agencies and intelligence components have taken similar steps.

Third, asking users to focus more on their needs and less on the intelligence enterprise's methods does not eliminate the need for educated consumers of intelligence.[11] Too many users and consumers remain uneducated on the abilities of the intelligence enterprise they rely upon for information. Moving toward a problem-centric concept of operations where users communicate their needs does not mean they no longer ought to have an appreciation for what intelligence can and cannot do for them. It is incumbent upon the enterprise to constantly educate consumers on the types of products that can be generated, and it is equally incumbent on the consumers to understand these products.

Such education will facilitate and improve cross-community conversation. Consumers will be better able to formulate a sensible, actionable operations plan if they understand and have confidence in the art of the possible. Of course, as new problems and threats arise, users should never be afraid to request, even demand, information that may push the enterprise to rethink its own limits.

## Instilling a Culture of Stewardship rather than Ownership

To truly unify the enterprise across echelons, users and consumers must trust that they will have access to the intelligence necessary to respond to time-sensitive needs. If they do not, users will continue to hold their own organic assets in reserve. Given the wide variety of data sources, opinions, and experience resident across the intelligence enterprise, we do ourselves a disservice by not making the entire breadth and depth of the intelligence enterprise available against any problem for any user.

To help address this issue of stovepiped assets, the enterprise needs to examine the information requests and allocate appropriate resources to meet these requests, regardless of who controls or "owns" these resources. Such integrated tasking and allocation authority places assets—collection systems or analysts—owned by agencies or services under the stewardship of the intelligence enterprise. Without the ability to optimize resource allocation and ensure that appropriate assets are put against the most pressing problems, the enterprise will continue to function as a collection of organizational stovepipes, each relying on its own organic assets.

Many consumers will initially be quite uncomfortable with an arrangement that places intelligence assets under the stewardship of others. Such a change should not—and probably could not—be forced. Instead, the enterprise must work to gain the trust and confidence of the users over a period of time, providing them with information they would not have discovered through local assets alone, and assuring them that time-sensitive tasks can be managed and executed with no loss of effectiveness.

This section has discussed a concept of operations that, if enacted enterprise wide, would integrate intelligence assets at all echelons and foster cross-community collaboration much more closely than any changes to an organizational chart could do. Such a concept of operations is but one part of the solution. A key recommendation of the recent 9/11 Commission and the Intelligence Reform and Terrorist Prevention Act—information sharing across the community—requires more than new ways of facing problems and new methods of efficiently using available collection and analysis assets.[12] Information sharing is more than integrating stovepipes, updating security networks, and creating a culture of need to share versus need to know; it is as much art as technology. Analyst education and information interaction also need to be addressed. The next two sections will look at these issues and propose some guiding principles for how to put the analyst back in analysis.

### Second Component: Implementing Innovative Education Programs

In a recent speech, Bill Gates, chairman of Microsoft Corporation, lamented the state of high school education in America today and called for a redesigned curriculum that better reflects the needs of tomorrow rather than the needs of yesterday. According to Mr. Gates, while a number of high school students are well prepared by specialized high school programs geared toward university preparation, "Today, only one-third of our students graduate from high school ready for college, work and citizenship."[13] In the same way, while pockets of incredibly effective education can be found within the intelligence enterprise, on the whole we poorly prepare our analysts, and the consumers, to take maximum advantage of the capabilities available to address and solve problems.

Besides new concepts of operation, the analytic community needs to review its training and education system with an eye toward augmenting the traditional curriculum with more advanced programs that stress problem solving and cross-community collaboration as much as technical proficiency in one or more intelligence disciplines. Education programs should focus intensely on two areas: teaching and reinforcing problem-solving skills, and overcoming intellectual or cultural bias in intelligence reporting.

## Equating Analysis with Investigation: Teaching Problem Solving

For the purposes of this chapter, it is important to recognize the key difference between training and education for the intelligence enterprise. Training creates the mental toolkit necessary to analyze different forms of data—be they different forms of imagery or different types of signals—and pull out relevant tidbits of information. Such training is often provided to analysts as new data sources emerge that they must be prepared to exploit. Education, however, must be far more dynamic. Education prepares the analyst to know which tools and techniques can be applied to a problem and how to deconstruct a problem and form (and test) hypotheses. Education in the art of intelligence focuses on problem-solving skills and forensic investigation, while training focuses on techniques of the tradecraft and data manipulation.

Simply knowing how to run a fingerprint database or trace a license plate number does not help law enforcement detectives solve a crime and catch a criminal; education and experience teach them the tricks of the trade, some overt and some subtle, that help them link seemingly disparate pieces of data into a compelling mosaic. As they learn more and more about a given case, detectives identify the information that would help them close any remaining gaps, and determine the likely sources for this information. This iterative process of collection and analysis, of linking facts and identifying unknowns, must be the focus of education for analysts within the intelligence enterprise. Education programs must cultivate in our analysts the mental agility to face new challenges, either through confidence in their own abilities or through knowledge of the expertise and abilities of other analysts within the community.

Today's introductory training, which mostly involves class work and lecture, teaches our analysts how to use the tools and techniques at their disposal and how to read and exploit data; specialized courses later in an analyst's career, or uneven on-the-job training, slowly educate them in the nuances of forensic investigation. A focus on collaboration and problem solving requires an education process that transcends the lecture hall. Education and training should be executed in a laboratory-style environment, with cutting-edge techniques being applied to real-world problems. In recent years, elements of the intelligence enterprise have placed seasoned analysts in small experimental environments and given them free reign to try new and different techniques to discover information, often with unexpected and largely beneficial results. Such an environment ought not be the exclusive domain of veterans; rather, it should become the norm across the enterprise, where rookies and seasoned pros share insights and investigate imaginative problem-solving techniques.

In addition, educating our analysts as part of a cross-disciplinary team does not mean the intelligence enterprise should sacrifice depth of expertise for breadth of familiarity. On the contrary, it will remain vital that intelligence analysts maintain a depth of expertise in one or more disciplines. With a variety of new sources on the horizon, our community will need experts trained in the many and subtle nuances of each source, able to constantly enhance the tradecraft associated with their respective disciplines.

## Understanding Our Limitations: Educating to Mitigate Bias

Even with perfect and insightful intelligence, it may be too difficult to overcome the biases of operators or decision makers, or perhaps even the biases of the analysts themselves. All too often, limited information exists on targets of interest. Left to their own devices, operators or decision makers may take this limited information and apply their own biased, often simplistic assumptions of how a typical adversary would act. In his book on causes of war, Greg Cashman illustrates this tendency through the work of Ole Hosti on John Foster Dulles. According to Hosti, Dulles had an ingrained image of the Soviet Union that colored his perception of intelligence—if the intelligence fit his view, he would accept it; if it did not, he would reject it as unreliable.[14]

Today's operators and decision makers face the same tendency, and it is up to intelligence to provide not only specific information but also, where desired by the consumer, the context that tells a story relevant to the consumer. Education that stresses investigative skills, tests assumptions, and informs our analysts with a complete range of cultural, economic, and social understanding will best position them to overcome their biases, and strengthen their argument as they seek to mitigate the biases of their customers.

A critical enabler to mitigating analyst and user bias is cultural awareness. Operators and decision makers must understand our adversaries—their biases, cultural beliefs, and image of the United States—in order to truly understand their motivations and intentions.[15] The theoretical framework under which we see and characterize our adversaries often dictates our conclusions about their intentions. Today the intelligence enterprise relies on its regional experts to provide the contextual framework under which many conclusions are reached. Given the fleeting nature of activities and the demand for rapid, relevant information, analysts and consumers beyond these regional experts need a broader understanding of the mind-set and cultural framework under which any given adversary operates. Since the lines between collector and analyst continue to merge, such cultural knowledge must reside within the consumer community as well as the intelligence enterprise.

Of course, our best resources for understanding regional tendencies and cultural trends are our coalition partners. The importance of coalition allies for future operations—military, intelligence, diplomatic, and economic—cannot be understated. Integrating the expertise of our partners into the intelligence enterprise can only strengthen our knowledge base and reduce the uncertainty in any underlying assumptions analysts make about intentions or tendencies.

Nothing promises that a new concept of operations or an innovative education program can *completely* eliminate human error or the tendency to assume the worst-case scenario that could be justified by the available intelligence.[16] No perfect solution exists for fixing the analytic problems evident today. By educating analysts and users alike, and providing the information-handling concept of operations that focuses attention on problems, not assets, we can instantiate a culture that encourages collaboration and strengthens the overall ability of the analytic enterprise, in much the same way free and open exchange of experimental results and procedures strengthens the civil scientific community.

### Third Component: Understanding Human-Information Interaction

The third pillar in strengthening analytics is human-information interaction. All the innovative concepts of operation and ingenious educational programs will fail to refocus the intelligence enterprise unless the work environment is conducive to innovative, imaginative, and well-grounded analysis. Only an analyst can provide the context to data that translates an image into an order of battle, or a signal into an actionable warning. The enterprise must empower its analysts to combine technical data with historical knowledge, area analysis, and cultural tendencies. Only through development of new and advanced ways for analysts to interact with data (and other analysts) can the enterprise be best positioned to assist its customers; these new ways can be organized under the umbrella of human-systems effectiveness.

Human-systems effectiveness is an integrative approach that optimizes intelligence analysis through the application of advanced engineering, scientific research, and organizational management practices to promote the most effective and efficient interaction with the analytic environment.[17] It recognizes that knowledge is created when analysts interact with data and information to piece together a puzzle. Interaction with data—arguably the key to the entire process—is an intensely personal experience that depends on the insights, preferences, and experience of the analyst. With an influx of new data sources, increasing collaboration between analysts and operators from across the intelligence and operational spectrum, and a multitude of problems ranging from mundane to crisis and traditional to unconventional, how humans deal with data is critically important. Future efforts geared toward implementing human-systems effectiveness should focus on the human side (understanding and accounting for generational differences in the intelligence workforce), the systems side (rethinking the goals and objectives of the research and development), and the all-important interplay between the two.

## The Human Side: Understanding and Accounting for Generational Differences

The intelligence enterprise is composed of a diverse workforce crossing multiple generations, each of which has a unique view on how humans interact with data and the proper work environment within which that interaction should occur. While the breakdowns are subjective, most analysts classify the generations as: Veterans (born between 1922 and 1945); Boomers (born between 1946 and 1964); Generation X (born between 1965 and 1980); and Generation Y, or Millenials (born between 1981 and today).[18] At the risk of painting each generation with too broad a brush (and recognizing that it is not about age but attitude), more seasoned analysts generally prefer quiet, dimly lit rooms where collaboration is done in face-to-face meetings or via telephone and email. Newer analysts generally like their environment bright, loud, and "busy," with whiteboards, teleconferencing, digital music, chat rooms, and open source news coverage.

Neither approach is right nor wrong, but both must be taken into account. Mixing dissimilar generations in work teams creates differences in work management expectations (meeting times, standard work hours) and habits that demand new, flexible policies that recognize the strengths and weaknesses of each analyst. Such policies are necessary to retain skilled senior analysts for their experience and to welcome new analysts for their creativity and flexibility. "One size fits all" cannot be a guiding principle of the future workplace (or education programs and acquisition initiatives for that matter).

## The Systems Side: Re-thinking Research and Development Requirements

In an enterprise designed to provide knowledge and dependent upon its analysts for success, there is surprisingly little emphasis on cognitive science or human factors engineering. Fortunately, industry and academia are filling the void. Smart agents to aid analysts by executing routine tasks are being developed based on enhanced knowledge of how different senses are linked to information portrayal. Research into virtual environments that integrate multiple senses is ongoing and includes immersive and multisensory three-dimensional displays. Such an immersive capability will offer analysts a work environment that allows them to more fully understand and manipulate information. This is more than simply visualization of information (eyesight); it is perceptualization of information (using multiple senses).

Immersive work environments do not offer a panacea for the failures in imagination over the past few years. The science and art that go into designing and developing these environments do not guarantee that new and better information will automatically flow from the analyst. In some cases, one can imagine that such work environments might cloud the issue, "hiding" the nugget of information that a consumer really needs within a sea of equally valuable, but perhaps less relevant, data. However, the potential rewards of empowering analysts with more than a computer monitor and graphics program seem logically to vastly outweigh the potential growing pains of multisensory, immersive work environments.

When developing new tools and techniques for analysis, the typical approach has been to design to requirements such as bandwidth, processing speed, and storage capacity and cost, with little or no consideration of a user's interface desires. While government acquisition officials love nothing more than cookie-cutter designs—one workstation replicated many times—this approach to acquisition ignores the unique needs of individual analysts and cannot evolve to face the demands of a dynamic and multifaceted workforce. The intelligence enterprise needs individually customizable work environments, facilities that can accommodate collaborative brainstorming sessions as well as library-style research, portals that create a workable environment within which virtual teams of analysts can share information, and an ongoing operations and maintenance process that identifies potential areas for improvement and the potentially changing tendencies of new hires.

## Conclusion

This chapter has described the challenges being faced by today's intelligence enterprise and advanced three discrete concepts beyond information technology and information sharing to face these challenges. By altering our concepts of operation to include problem-centric analysis and collaboration, updating our education programs to train new generations of analysts in investigative analysis as well as data manipulation, and adopting human-systems effectiveness as a critical enabler that best positions analysts and users to perceive and understand data, the intelligence enterprise can make intelligence far more relevant and actionable to users across the political and military spectrum. In building a path to achieve this environment, this essay has focused on both philosophical and tangible changes necessary to grow and mature the type of intelligence analysis vital to the nation's security. This is not a short-term effort. Cultures, biases, training programs, and acquisition processes simply do not change overnight.

This essay began with an instructive lesson from the interwar period, and it will end with one. During the period between the two world wars, few in the U.S. defense establishment recognized the potential impact of the airplane to change the nature of warfare. But some, such as Billy Mitchell, had the imagination and foresight to see airpower not for what it was but for what it could be. Unable to force the change on a military establishment comfortable with existing strategies for land and sea warfare, the pioneers of American military aviation began by slowly, if zealously, changing the mind-set of military leaders. New training and education programs focused on airpower; clearly defined promotional paths for air corps officers were created; involvement in exercises and war games was advanced; and the occasional change agent like Billy Mitchell challenged the existing mind-set.[19]

Intelligence analysis must evolve in the same way as airpower, with dynamic and flexible concepts of operation, an ongoing education and career growth program, and technological solutions that enhance the enterprise. Still, as with airpower, the intelligence enterprise will be unable to overcome failures in imagination without the right people—without an analytic workforce and leadership that imagines new possibilities, demands the tools and environment to do their job, and prods the establishment along.

## Notes

1. Murray R. Williamson and Allan R. Millett, *Military Innovation in the Interwar Period* (Cambridge: Cambridge University Press, 1998).

2. General information from *The 9/11 Commission Report* and *The Intelligence Reform and Terrorist Prevention Act of 2004,* p. 339.

3. This point was illuminated clearly by reading Audrey Kurth Cronin, "Rethinking Sovereignty: American Strategy in the Age of Terrorism," *Survival* Vol. 44, No. 2 (Summer 2002), p. 129.

4. For example, see William A. Owens, *Lifting the Fog of War* (New York: Farrar Straus Giroux, 2000).

5. Martin Van Creveld, *Technology and War: From 2000 B.C. to the Present* (London: Macmillan, 1989), p. 314.

6. Discussion of emerging threats taken from *The National Security Strategy of the United States,* September 2002.

7. This comment is based on the work of Van Creveld, *Technology and War,* p. 303. Van Creveld speaks about yesterday's forces being prepared to fight mechanical forces, not today's smaller, less technologically advanced enemies.

8. Though many sources exist for defining the "scientific method," an interesting one for our purposes is the analytical process defined for students of political science in Graham Allison and Philip Zelikow, *Essence of Decision: Explaining the Cuban Missile Crisis* (Reading, Mass.: Longman: 1999), pp. 4–7.

9. For a discussion on making information and intelligence more relevant to decision-makers and operators, see the quote by George Ball on comprehending the Cuban Missile Crisis in Alexander L. George, *Bridging the Gap* (Washington D.C.: United States Institute of Peace Press, 1993), p. 8: "We were presented . . . with an equation of compound variables and multiple unknowns. No one has yet devised a computer that will digest such raw data as was available to us and promptly print out a recommended course of action. . . ." This was spoken some years ago, but remains true today, even with advances in information technology.

10. These acronyms are used within the intelligence community to describe the functions necessary to collect data, analyze it, and get it out to the customers.

11. With thanks to New York Clothier Syms, Inc., and its motto, "An educated consumer is our best customer."

12. *The 9/11 Commission Report,* pp. 416–17; also look at the Information Sharing Environment (ISE) initiative in the *Intelligence Reform and Terrorist Prevention Act of 2004*

13. Excerpts from speech to the National Education Summit on High Schools, February 26, 2005, as quoted in the *Washington Post,* March 6, 2005, p. B02.

14. Greg Cashman, *What Causes War?* (Boston: Lexington Books, 1993), p. 55

15. For a general overview of educational needs, to include knowledge of foreign cultures, see Robert H. Scales, "Studying the Art of War" in *Washington Times,* February 17, 2005. For a discussion of the need to understand how an adversary views the United States, see George, *Bridging the Gap,* p. 129.

16. Eugene B. Skolnikoff, *The Elusive Transformation* (Princeton, N.J.: Princeton University Press, 1993), p. 60.

17. The working definition of human systems effectiveness provided here is from a white paper prepared by the author and William Morgan titled "Human Systems Effectiveness."

18. *Generation Diversity,* National Technology Alliance Report #TR-011-052804-081, November 24, 2004.

19. Williamson and Millett, *Military Innovation in the Interwar Period.*

APPENDIX

# U.S. Principles
# of Joint Operations

## 1. Objective

- The purpose of the objective is to direct every military operation toward a clearly defined, decisive, and attainable objective.
- The purpose of military operations is to achieve the military objectives that support the overall political goals of the conflict. This frequently involves the destruction of the enemy armed forces' capabilities and their will to fight. The objective of joint operations not involving this destruction might be more difficult to define; nonetheless, it too must be clear from the beginning. Objectives must directly, quickly, and economically contribute to the purpose of the operation. Each operation must contribute to strategic objectives. Joint force commanders should avoid actions that do not contribute to achieving the objective(s).
- Additionally, changes to the military objectives may occur because political and military leaders gain a better understanding of the situation, or it may occur because the situation itself changes. The joint force commander should anticipate these shifts in political goals that necessitate a change in military objectives. The changes may be very subtle, but if not made, attainment of the military objectives may no longer support the political goals, legitimacy may be undermined, and force security may be compromised.

---

From the Department of Defense, Joint Publication 3-0 (revised first draft), *Doctrine for Joint Operations,* Appendix A. Special thanks to Dr. Antulio Echevarria, SSI, of the Army War College for providing the text from the source document in the following format.

## 2. Offensive

- The purpose of an offensive action is to seize, retain, and exploit initiative.
- Offensive action is the most effective and decisive way to attain a clearly defined objective.
- Offensive actions are the means by which a military force seizes and holds the initiative while maintaining freedom of action. The importance of offensive action is fundamentally true across all levels of war.
- Commanders adopt the defensive only as a temporary expedient and must seek every opportunity to seize or re-seize the initiative. An offensive spirit must therefore be inherent in the conduct of all operations.

## 3. Mass

- The purpose of mass is to concentrate the effects of combat power at the most advantageous place and time to achieve decisive results.
- To achieve mass is to synchronize and/or integrate appropriate joint force capabilities where they will have decisive effect in a short period of time. Mass often must be sustained to have the desired effect. Massing effects, rather than concentrating forces, can enable even numerically inferior forces to achieve decisive results and minimize human losses and waste of resources.

## 4. Economy of Force

- The purpose of the economy of force is to allocate minimum essential combat power to secondary efforts.
- Economy of force is the judicious employment and distribution of forces. It is the measured allocation of available combat power to such tasks as limited attacks, defense, delays, deception, or even retrograde operations in order to achieve mass elsewhere at the decisive point and time.

## 5. Maneuver

- The purpose of maneuver is to place the enemy in a position of disadvantage through the flexible application of combat power.
- Maneuver is the movement of forces in relation to the enemy to secure or retain positional advantage, usually in order to deliver—or threaten

delivery of—the direct and indirect fires of the maneuvering force. Effective maneuver keeps the enemy off balance and thus also protects the friendly force.

- It contributes materially in exploiting successes, preserving freedom of action, and reducing vulnerability by continually posing new problems for the enemy.
- At all levels of war, successful maneuver requires not only fire and movement but also agility and versatility of thought, plans, operations, and organizations. It requires designating and then, if necessary, shifting the main effort and applying the principles of mass and economy of force.

## 6. Unity of Command

- The purpose of unity of command is to ensure unity of effort under one responsible commander for every objective.
- Unity of command means that all forces operate under a single commander with the requisite authority to direct all forces employed in pursuit of a common purpose. Unity of effort, however, requires coordination and cooperation among all forces toward a commonly recognized objective, although they are not necessarily part of the same command structure. During multinational and interagency operations, unity of command may not be possible, but the requirement for unity of effort becomes paramount. Unity of effort—coordination through cooperation and common interests—is an essential complement to unity of command.

## 7. Security

- The purpose of security is to never permit the enemy to acquire unexpected advantage.
- Security enhances freedom of action by reducing friendly vulnerability to hostile acts, influence, or surprise. Security results from the measures taken by commanders to protect their forces. Staff planning and an understanding of enemy strategy, tactics, and doctrine will enhance security. Risk is inherent in military operations. Application of this principle includes prudent risk management, not undue caution. Protecting the force increases friendly combat power and preserves freedom of action.

## 8. Surprise

- The purpose of surprise is to strike at a time or place or in a manner for which the enemy is unprepared.
- Surprise can help the commander shift combat power and thus achieve success well out of proportion to the effort expended. Factors contributing to surprise include speed in decision-making, information sharing, and force movement; effective intelligence; deception; application of unexpected combat power; operations security; and variations in tactics and methods of operation.

## 9. Simplicity

- The purpose of simplicity is to prepare clear, uncomplicated plans and concise orders to ensure thorough understanding.
- Simplicity contributes to successful operations. Simple plans and clear, concise orders minimize misunderstanding and confusion. When other factors are equal, the simplest plan is preferable. Simplicity in plans allows better understanding and execution planning at all echelons. Simplicity and clarity of expression greatly facilitate mission execution in the stress, fatigue, and other complexities of modern combat and are especially critical to success in combined operations.

## 10. Restraint

- A single act could cause significant military and political consequences; therefore, judicious use of force is necessary. Restraint requires the careful and disciplined balancing of the need for security, the conduct of military operations, and the desired end state. An example would be the exposure of intelligence gathering activities (i.e., interrogation of detainees and prisoners of war) that could have significant political and military repercussions and therefore should be conducted with sound judgment. Excessive force antagonizes those parties involved, thereby damaging the legitimacy of the opposing party.
- Commanders at all levels must take proactive steps to ensure their personnel are properly trained, i.e., know and understand the rules of engagement (ROE) and are quickly informed of changes. Failure to understand and comply with established ROE can result in fratricide,

mission failure, and/or national embarrassment. ROE in some operations may be more restrictive, detailed, and sensitive to political concerns than during combat, consistent always with the inherent right of self-defense. ROE should be unclassified, if possible, and widely disseminated. Restraint is best achieved when ROE issued at the beginning of an operation address most anticipated situations that may arise. ROE should be consistently reviewed and revised as necessary. Additionally, ROE should be carefully scrutinized to ensure the lives and health of military personnel involved in joint operations are not necessarily endangered. In multinational operations, use of force may be dictated by coalition or allied force ROE. Commanders at all levels must take proactive steps to ensure an understanding of ROE and how to influence changes to them. Since the domestic law of some nations may be more restrictive concerning the use of force than permitted under coalition or allied force ROE, commanders must be aware of national restrictions imposed on force participants.

## 11. Perseverance

- Prepare for the measured, protracted application of military capability in support of strategic aims.
- Some operations may require years to achieve the termination criteria. The underlying causes of the crisis may be elusive, making it difficult to achieve decisive resolution. The patient, resolute, and persistent pursuit of national goals and objectives is often a requirement for success. This will frequently involve diplomatic, economic, and informational measures to supplement military efforts.

## 12. Legitimacy

- Legitimacy is based on the legality, morality, and rightness of the actions undertaken as well as the will of the U.S. public to support the actions. The purpose of legitimacy is to develop and maintain the will necessary to achieve the desired end state. Legitimacy is frequently a decisive element. The perception of legitimacy by the U.S. public is strengthened if there are obvious national or humanitarian interests at stake, and if there is assurance that American lives are not being needlessly or carelessly placed at risk. Other interested audiences may include the foreign nations, civil populations in the operational area, and the participating forces. Information operations should be used prudently to enhance both domestic and international perceptions of the legitimacy of an operation.

- Committed forces must sustain the legitimacy of the operation and of the host government, where applicable. Security actions must be balanced with legitimacy concerns. All actions must be considered in the light of potentially competing factions where appropriate. Legitimacy may depend on adherence to objectives agreed to by the international community, ensuring the action is appropriate to the situation, and fairness in dealing with various factions. Restricting the use of force, restructuring the types of forces employed, and ensuring the disciplined conduct of the forces involved may reinforce it.

- Another aspect of this principle is the legitimacy bestowed upon a local government through the perception of the populace it governs. Humanitarian and civil military operations help develop a sense of legitimacy for the supported government. During operations where a host government does not exist, extreme caution should be used when dealing with individuals and organizations and to avoid inadvertently legitimizing them.

# LIST OF CONTRIBUTORS

## Dr. John B. Alexander

For more than a decade, Dr. John Alexander has been a leading advocate for the development of nonlethal weapons. He entered the U.S. Army as a private in 1956 and rose through the ranks to sergeant first class, attended OCS, and was a colonel of Infantry in 1988 when he retired. During his varied career, he held many key positions in special operations, intelligence, and research and development. From 1966 through early 1969 he commanded Special Forces "A" Teams in Vietnam and Thailand.

Joining Los Alamos National Laboratory he organized and chaired the six major conferences on nonlethal weapons, served as a U.S. delegate to four NATO studies on the topic, and was a member of the first Council on Foreign Relations study that led to creation of the Joint Non-Lethal Weapons Directorate. He wrote many of the seminal articles on nonlethal weapons and was a member of the National Research Council Committee for Assessment of Non-Lethal Weapons Science and Technology.

Currently he is a senior fellow with the Joint Special Operations University. His books include *Future War* and *Winning the War*, St. Martin's Press.

## Ms. Deborah G. Barger

Deborah G. Barger is a senior executive in the newly established Office of the Director of National Intelligence. She spent a year at RAND on a fellowship in 2002–3, and used that time to research and develop her ideas on how to bring about a revolution in intelligence affairs (RIA). Her independent views do not necessarily reflect those of the U.S. government.

557

## Dr. James Jay Carafano

Dr. James Jay Carafano is a senior fellow for Defense and Homeland Security at The Kathryn and Shelby Cullom Davis Institute for International Studies, The Heritage Foundation. A historian and educator, he served as an assistant professor of history at the U.S. Military Academy, director of military studies at the Army's Center of Military History, and taught at Georgetown University, Mount Saint Mary's College, the U.S. Naval War College, and the National Defense University.

Dr. Carafano has authored *Waltzing into the Cold War: The Struggle for Occupied Austria* (Texas A & M, 2002) and *After D-Day: Operation Cobra and the Normandy Breakout (The Art of War)* (Lynne Rienner, 2000), a Military Book Club selection, and coauthored *Winning the Long War: Lessons from the Cold War for Defeating Terrorism and Preserving Freedom* (Heritage Press, 2005) and *Homeland Security: A Complete Guide to Understanding, Preventing, and Surviving Terrorism* (McGraw-Hill, 2005).

Dr. Carafano joined Heritage after serving as a senior fellow at the Center for Strategic and Budgetary Assessments, a Washington Policy Institute, where he refined his defense analysis skills after a twenty-five-year U.S. Army career. Before retiring as a lieutenant colonel, he served as executive editor of *Joint Force Quarterly*, the Defense Department's premier professional journal.

## Mr. Craig Cohen

Craig Cohen is a researcher in the Post-Conflict Reconstruction Project at the Center for Strategic and International Studies (2004– ). Mr. Cohen has experience working for the United Nations and nongovernmental organizations in Rwanda (1996–97), Malawi (1997 and 1999), Azerbaijan (1998), and the former Yugoslavia (1996), as well as experience at the headquarters of the UN High Commissioner for Refugees in Geneva (1998) and the International Center for Transitional Justice in New York (2002). He holds a master's degree from the Fletcher School of Law and Diplomacy at Tufts University (2004), and a bachelor's from Duke University.

## Ms. Bathsheba N. Crocker, J.D.

Bathsheba Crocker is a fellow in the International Security Program and co-director of the Post-Conflict Reconstruction Project at the Center for Strategic and International Studies (CSIS) in Washington D.C. From 2002–3, she was a Council on Foreign Relations International Affairs Fellow, working on post-conflict reconstruction issues at CSIS. Ms. Crocker has served as an assistant professor at the Johns Hopkins School of Advanced International Studies and George Washington University's Elliott School of International Affairs. She was a member of a CSIS-led reconstruction assessment team that went to Iraq in July 2003 at the request of the U.S. Department of Defense, and she coauthored numerous CSIS reports on post-conflict Iraq as well as a report on planning for the post-conflict phase in Sudan.

She has published several journal articles on post-conflict Iraq, chapters on Iraq, Kosovo, and Sierra Leone, which appear in *Winning the Peace: An American Strategy for Post-Conflict Reconstruction* (CSIS Press, 2004), and numerous op-eds on Iraq and Sudan. Before joining CSIS, Ms. Crocker worked as an attorney in the Legal Adviser's Office at the U.S. Department of State, where she focused on foreign assistance, appropriations law, and economic sanctions issues. Prior to that, she served as the deputy U.S. special representative for the Southeast Europe Initiative in Rome, Italy, working on economic reconstruction in the Balkans. She served as the executive assistant to the deputy national security adviser at the White House from 1999 to 2000. She received a B.A. from Stanford University, a J.D. from Harvard Law School, and an M.A. in law and Diplomacy from the Fletcher School of Law and Diplomacy at Tufts University.

## Lt. Gen. James M. Dubik

Lt. Gen. James M. Dubik is currently the commanding general of the U.S. Army's 1st Corps. Formerly, he was the commanding general of the 25th Infantry Division, deputy commanding general for transformation of the U.S. Army's Training and Doctrine Command, and the assistant division commander of the 1st Cavalry Division. He also has extensive experience in the 82nd Airborne Division, the 10th Mountain Division, and the 1st and 2nd Ranger Battalions. He commanded U.S. and multinational forces in northern Haiti during the intervention of 1994–95 and was the deputy commanding general of Multinational Division (North) in Bosnia in 1999. General Dubik has a B.A. in philosophy from Gannon University, an M.A. in philosophy from Johns Hopkins University, and an M.A. in military arts and science from the U.S. Army's Command and General Staff College. He also held a Seminar XXI Fellowship with MIT, attended the executive program for national and international security at Harvard University's John F. Kennedy School of Government, and attended the National Security Leadership Course at the Maxwell School of Citizenship and Public Affairs at Syracuse University.

## Dr. Antulio J. Echevarria II

Dr. Antulio Echevarria is the director of research at the U.S. Army War College in Carlisle Barracks, Pennsylvania. A retired army officer, he graduated from the U.S. Military Academy in 1981 and held a variety of command and staff assignments in Germany and the United States. He is a graduate of the U.S. Army's Command and General Staff College and the U.S. Army War College. He holds M.A. and Ph.D. degrees in history from Princeton University, where his study of Clausewitz and German military thinking was facilitated by Peter Paret. His book, *After Clausewitz: German Military Thinkers before the Great War*, was published by the University Press of Kansas (2001).

Dr. Echevarria has also written extensively on military-historical themes, publishing articles in a number of scholarly and professional military journals, including *The Journal of Strategic Studies, The Journal of Military History, War in History, War & Society, Naval War College Review, Parameters, Joint Force Quarterly, Royal United Services Institute, Military Review, Airpower Journal, Marine Corps Gazette,* and *Military History Quarterly.* He is currently engaged in writing two book-length manuscripts: one on "Imagining Wars to Come," which examines how military professionals and amateurs imagined future war before 1914; and one that offers a reinterpretation of Clausewitz's *On War.*

## Col. John Ewers

Col. John Ewers is the commandant of the Marine Corps Fellow at CSIS for the 2004–5 academic year. A Washington, D.C., native, he was commissioned in 1984 and was certified as a U.S. Marine judge advocate in 1986. As a junior officer, Colonel Ewers served as a prosecutor, defense counsel, special assistant U.S. attorney, deputy SJA, military justice officer, and a recruit training series commander. From 1996 to 1999 he was a military judge on the Navy–Marine Corps Trial Judiciary, presiding over more than five hundred special and general courts-martial.

From 2000 to 2002, Colonel Ewers was the officer in charge, Legal Service Support Section, 1st FSSG—the busiest military justice section in the armed forces—and was responsible for developing the legal service support plan for I Marine Expeditionary Force in Operation Iraqi Freedom. In 2002, he was assigned as staff judge advocate, 1st Marine Division, where his responsibilities included working law of war, rules of engagement, and other operational law issues during OIF (1). His most recent assignment before reporting to CSIS was as commanding officer, Third Recruit Training Battalion, San Diego.

Colonel Ewers received a B.A. in philosophy/political science from the University of Delaware in 1981, a J.D., cum laude, from Georgetown University Law Center in 1985, and a master of laws as an honor graduate of the Judge Advocate General School, U.S. Army in 1994. He is also a distinguished graduate of the U.S. Marine Corps Command and Staff College.

## Brig. Gen. David A. Fastabend

Brig. Gen. David A. Fastabend currently serves as the deputy director/chief of staff, Futures Center, United States Army Training and Doctrine Command, Fort Monroe, Virginia, and is on the promotion list to major general. His previous assignment was director, Concepts Development and Experimentation. General Fastabend is an engineer officer who has served in command and staff positions throughout the army. His command assignments include: command of B Company, 10th Combat Engineer Battalion in Germany; command of 13th Engineer Battalion, Fort Ord, California; command of 555th Engineer Group, Fort Lewis, Washington, and commanding general, United States Army Engineer Division, Northwestern, Portland, Oregon.

General Fastabend has also served in several key operations, plans, and training staff assignments, including plans officer, G-3, 9th Infantry Division, Fort Lewis, Washington; FM 1090-5 Writing Team, Command and General Staff College, Fort Leavenworth, Kansas; speechwriter to CINPAC, PACOM, Camp Smith, Hawaii; and executive officer to the vice chief of staff, Army, Washington, DC. A 1974 graduate of the United States Military Academy, General Fastabend also holds a master's of science in civil engineering from MIT and a master's in military arts and science.

## Adm. Sir Ian Forbes

Adm. Sir Ian Forbes joined the Royal Navy in 1965 and retired in 2004. He commanded four ships (including *Invincible*) and was engaged in crisis operations off the Falklands, Bosnia, in the Gulf, and off Kosovo. As a rear admiral, he was chief of staff in Sarajevo in the organization charged with rebuilding Bosnia. He commanded the United Kingdom Maritime Battle Group in the Gulf and the Adriatic, working closely with U.S. forces. Following this, he was the Royal Navy's surface fleet commander. His final appointment as a four star admiral was in the United States, in the NATO headquarters in Norfolk, Virginia, as deputy SACLANT and deputy SACT. Admiral Forbes's operational and politico-military experience during ten years at the senior command level in multinational contexts is extensive. He commanded in every rank he held during this time in combined and joint operations. In his final tour, he was the driving force in establishing the Allied Command Transformation, the new NATO Strategic Command responsible for transforming the NATO Alliance. A Royal College of Defence Studies graduate, he was awarded Commander of the British Empire in 1994, and the Queen's Commendation for Valuable Service in 1996. He was made a Knight Commander of the Bath in 2003. In the same year, Admiral Forbes was honored as the first recipient of the NATO Meritorious Medal for his work in transforming the alliance.

## Dr. Colin S. Gray

Colin S. Gray is professor of international politics and strategic studies at the University of Reading, England. He worked at the International Institute for Strategic Studies (London), and at Hudson Institute (Croton-on-Hudson, New York), before founding a defense-oriented think tank—the National Institute for Public Policy—in the Washington area. Dr. Gray served for five years in the Reagan administration on the President's General Advisory Committee on Arms Control and Disarmament. He has served as an adviser both to the United States and the British governments (he has dual citizenship). His government work has included studies on nuclear strategy, arms control policy, maritime strategy, space strategy, and the use of special forces. He has written nineteen books, most recently *Another Bloody Century: Future Warfare* (Weidenfeld and Nicolson, 2005) and *The Sheriff: America's Defense of the New World Order* (University Press of Kentucky, 2004). He is a graduate of the universities of Manchester and Oxford.

## Dr. Grant T. Hammond

Grant T. Hammond is deputy director of the Center for Strategy and Technology (CSAT) and professor of strategy and international security at the Air

War College (AWC), Maxwell AFB, Alabama. He holds an A.B. from Harvard in history and an M.A. and Ph. D. in international relations from Johns Hopkins University's Paul H. Nitze School of Advanced International Studies. Dr. Hammond was a major participant in two chief of staff of the Air Force (CSAF)–commissioned yearlong studies—SPACECAST 2020 in 1993–94 and AF 2025 in 1995–96, for which he was twice awarded the Meritorious Civilian Service Award.

He is the author of three books: *Countertrade, Offsets and Barter in International Political Economy* (London: Pinter Publishers, 1990, paperback 1993); *Plowshares into Swords: Arms Races in International Politics, 1840–1991* (University of South Carolina Press, 1993), and *The Mind of War: John Boyd and American Security* (Smithsonian Press, 2001, paperback edition, 2004). He has published in a number of journals and contributed numerous book chapters as well. He is currently working on a book titled *The Revolution in Security Affairs: The Transformation of War and Politics in the 21st Century.*

## Col. Thomas X. Hammes

A graduate of the U.S. Naval Academy and a career Marine, Thomas X. Hammes spent most of his thirty years on active duty in infantry and intelligence assignments in the operating forces, successfully avoiding both joint and headquarters duties.

## Lt. Col. Frank G. Hoffman

Frank G. Hoffman is a national security affairs analyst with more than twenty-five years of policy and operational experience. He is currently a research fellow at the Center for Emerging Threats and Opportunities at the Marine Corps Combat Development Command, Quantico, Virginia. Prior to this position, Colonel Hoffman was appointed by the secretary of defense to the staff of the U.S. Commission on National Security/21st Century (Hart-Rudman Commission). He was the principal analyst for the commission's recommendations, which led to the establishment of the Department of Homeland Security.

Previously, Colonel Hoffman was the national security analyst and director, Marine Strategic Studies Group, at the Marine Corps Combat Development Command in Quantico. In addition to participating in several defense science boards and contributing to a number of government reports, Colonel Hoffman has authored *Decisive Force: The New American Way of War* (Praeger,

1996). He has published more than a hundred articles and reviews on national security strategy, defense economics, and military history. He holds degrees from the Wharton Business School, George Mason University, and the U.S. Naval War College.

## Dr. Mary H. Kaldor

Mary Kaldor is school professor and director of the Centre for the Study of Global Governance, London School of Economics and Political Science. Prior to coming to the LSE, she worked at Sussex University as the Jean Monnet Reader in Contemporary European Studies. Her published works include, among others, *The Imaginary War: Understanding the East-West Conflict* (Blackwell 1990) and *New and Old Wars: Organised Violence in a Global Era* (1999, Polity Press). She is coeditor and coauthor of the annual *Global Civil Society* yearbook, which is published by Oxford University Press and Sage Publications. Her latest book, *Global Civil Society: An Answer to War,* was published by Polity Press in 2003.

Mary Kaldor was the convenor of the Study Group on European Security Capabilities, established at the request of Javier Solana. Their report, "A Human Security Doctrine for Europe," was launched in September 2004.

## Lt. Col. Robert R. Leonhard

Lt. Col. Robert R. Leonhard, U.S. Army (Ret.), served twenty-four years in the army as an infantry officer, war planner, and combat developer. He earned a B.A. in history from Columbus College, Georgia; an M.S. in international relations from Troy State University; and he is completing his doctorate in U.S. nineteenth century history at West Virginia University. His army schooling included Command and General Staff College, the School of Advanced Military Studies, and French Commando School. He is currently on the senior staff of the Johns Hopkins University Applied Physics Laboratory, working national security issues. He is the author of numerous articles and three books on modern warfare, including his latest, *The Principles of War for the Information Age.*

## Mr. Keith J. Masback

Mr. Masback is the deputy director, Office of Strategic Transformation, at the National Geospatial-Intelligence Agency. Mr. Masback directs efforts to shape the agency's future and develop visionary concepts to strengthen the nation's geospatial-intelligence enterprise. Mr. Masback began his career in the U.S. Army as an infantry officer with the Berlin Brigade, Germany. He also served in command and staff positions at XVIII Airborne Corps, Fort Bragg, North Carolina. As an army civilian, he led the Army Intelligence Master Plan and served as the director of Intelligence, Surveillance, and Reconnaissance Integration.

Mr. Masback was born and raised in White Plains, New York. He holds a bachelor's degree in political science from Gettysburg College. He completed the Post-Graduate Intelligence Program at the Joint Military Intelligence College, Defense Intelligence Agency. He is also a United States Soccer Federation–certified referee.

## Dr. Anthony D. Mc Ivor

Dr. Anthony D. Mc Ivor is vice president for national security studies at ManTech-Gray Hawk Systems, Inc., in Alexandria, Virginia. Dr. Mc Ivor writes on defense and security issues with present research focusing on intelligence, modern theories of conflict, and the future of warfare. His joint article with Vice Adm. John G. Morgan on "Rethinking the Principles of War" for *Proceedings* (October 2003) was the catalyst for this book project. Dr. Mc Ivor serves as the project coordinator. Concurrently the editor of the *American Intelligence Journal,* he has recently written several articles on defense counterintelligence. A historian of modern Europe and the Mediterranean, he is now preparing a history of U.S. counterintelligence since the First Gulf War. Dr. Mc Ivor earned a Ph.D. in European economic history from the University of California, San Diego.

## Dr. Steven Metz

Dr. Steven Metz is chairman of the Regional Strategy and Planning Department and research professor of national security affairs at the U.S. Army War College Strategic Studies Institute. He has been with SSI since 1993. Dr. Metz has also been on the faculty of the Air War College, the U.S. Army Command and General Staff College, and several universities. He has been an adviser to political organizations, campaigns, and commissions; served on many national security policy task forces; testified in both houses of Congress; and spoken on military and security issues around the world. He is the author of more than ninety publications on world politics, national security policy, and military strategy. He is currently writing a book on the insurgency in Iraq. Dr. Metz holds a B.A. in philosophy and an M.A. in international studies from the University of South Carolina, and a Ph.D. in political science from Johns Hopkins University.

## Dr. William M. Nolte

William M. Nolte is chancellor of the National Intelligence University system and assistant deputy director of national intelligence for education and training. In these capacities, he coordinates on behalf of the director of national intelligence (DNI) the education and training activities of the U.S. intelligence community. In previous assignments, he has served as deputy assistant director of central intelligence for analysis, director of training at the National Security Agency, and head of congressional liaison for NSA. He has served two tours at the National Intelligence Council.

On behalf of the DNI, Dr. Nolte manages the Intelligence Fellows program. He serves on the board of *Studies in Intelligence* and is an adjunct member of the faculty at the Joint Military Intelligence College. In the 2005–6 academic year, he will join the faculty of the University of Maryland College Park as a research professor in the school of public policy. He holds an A.B. from La Salle University and a Ph.D. from the University of Maryland, both in history.

## Lt. Col. Ralph Peters

Ralph Peters is a retired U.S. Army officer and the author of the recent book *New Glory: Expanding America's Global Supremacy,* which recommends extensive changes in our approach to warfare and national strategy. He has published nineteen other books, as well as several hundred essays and columns. With experience in sixty countries, he continues to travel extensively in the developing world.

## Mr. Roy L. Reed Jr.

Mr. Roy L. Reed Jr. is the former director for strategy and transformation at the Department of Defense's Counterintelligence Field Activity (CIFA). His military and civil service careers included coordinating with community leaders from the armed services, defense agencies, the intelligence community, and the Office of the Secretary of Defense to develop, implement, or recommend DoD counterintelligence strategic plans, policies, and performance management programs. Highlights of his government career include special assistant, Security and Investigative Activities, OSD; director, Strategic Planning, HQ AFOSI; chief, Security Operations, HQ, AFOSI; and multiple overseas tours including Central America and the Mediterranean. He is currently president of Kingfisher Systems, LLC, in Alexandria, Virginia, and consults for the Institute for Physical Sciences in McLean, Virginia.

## Maj. Gen. Paulette M. Risher

Maj. Gen. Paulette Risher serves as the president of United States Special Operations Command's Joint Special Operations University and as the director of the Center for Knowledge and Futures (J7/J9). Her areas of specialization are civil-military operations and organizational development. As a U.S. Army officer, General Risher has more than thirty-two years of active and reserve service, and in her most recent civilian position, she has served as a psychologist with the U.S. Air Force. General Risher holds a bachelor's degree in psychology from Arizona State University; an M.A. in psychology (organizational development) from the University of West Florida; and is completing an educational specialist (EdS) degree at the University of West Florida. General Risher has been selected to serve as the deputy commander for Mobilization and Reserve Affairs for USSOCOM.

## Dr. Richard L. Russell

Richard L. Russell is a professor of national security affairs at the National Defense University. He also is an adjunct associate professor for the Security Studies Program and a research associate for the Institute for the Study of Diplomacy at Georgetown University. Russell previously served as a political-military analyst at the Central Intelligence Agency. He is the author of *Strategic Contest: Weapons Proliferation and War in the Greater Middle East* (New York and London: Routledge, 2005).

## Dr. Robert H. Scales Jr.

Dr. Robert Scales is currently president of Colgen, Inc, a consulting firm specializing in issues relating to land power, war gaming and strategic leadership. Prior to joining the private sector, Dr. Scales served more than thirty years in the U.S. Army, retiring as a major general. He commanded two units in Vietnam and was awarded the Silver Star for action during the battles around Dong Ap Bia (Hamburger Hill) during the summer of 1969. Subsequently, he served in command and staff positions in the United States, Germany, and Korea, and ended his military career as commandant of the United States Army War College. In 1995 he created the Army After Next program, which was the army's first attempt to build a strategic game and operational concept for future land warfare. He is the author of two books on military history: *Certain Victory,* the official account of the army in the Gulf War, and *Firepower in Limited War,* a history of the evolution of firepower doctrine since the end of the Korean War. He has also written two books on the theory of warfare: *Future Warfare,* a strategic anthology on America's wars to come, and *Yellow Smoke: The Future of Land Warfare for America's Military.* His latest work, *The Iraq War: A Military History,* was written with Williamson Murray. Dr. Scales is the senior military analyst for the BBC, National Public Radio, and Fox News Network. He is a graduate of West Point and earned his Ph.D. in history from Duke University.

## Dr. Anna Simons

Anna Simons is an associate professor of defense analysis at the Naval Postgraduate School. Prior to teaching at NPS she taught in the anthropology department at UCLA. She holds a Ph.D. in social anthropology from Harvard University (1992) and an A.B. from Harvard College (1979).

She is the author of *Networks of Dissolution: Somalia Undone* (Westview Press, 1995) and *The Company They Keep: Life Inside the U.S. Army Special Forces* (The Free Press, 1997). Her published articles include, among others, "The Death of Conquest" (*The National Interest*); "U.S. Special Operations Forces and the War on Terrorism" (with David Tucker, *Small Wars & Insurgencies*); and "War: Back to the Future" (*Annual Review of Anthropology*). Trained as an anthropologist and an Africanist, her work examines ties that bind members of groups together as well as the rifts that drive groups apart, both in this country and abroad.

## Dr. Jon T. Sumida

Dr. Jon Sumida is the author of *In Defence of Naval Supremacy: Finance, Technology, and British Naval Policy, 1889–1914,* (1989) and *Inventing Grand*

*Strategy and Teaching Command: The Classic Works of Alfred Thayer Mahan Reconsidered* (1997) and editor of *The Pollen Papers: The Privately Circulated Printed Works of Arthur Hungerford Pollen 1901–1916* (1984). He has published twenty-five articles, three of which have won the Moncado prize from the Society for Military History. Sumida is an associate professor at the University of Maryland, College Park. He was a distinguished visiting professor in the Department of Military Strategy and Operations at the National War College (2000), Major General Matthew C. Horner Chair of Military Theory at the U.S. Marine Corps University (2004–6), and chair of the Department of the Army Historical Advisory Committee (2003–5). Dr. Sumida has received post-doctoral fellowships from Churchill College, Cambridge University (1983), the Woodrow Wilson International Center for Scholars (1986/1995–96), and the John Simon Guggenheim Foundation (1990–91).

## Dr. Robert R. Tomes

Robert R. Tomes is senior adviser to the technical executive at the National Geospatial-Intelligence Agency, a director of the Anna Sobol Levy Foundation, and a member of the Council for Emerging National Security Affairs (CENSA). His recent publications include articles in defense studies, small wars and insurgencies, armed forces and society, and CENSA's forthcoming *The Faces of Intelligence Reform: Perspectives on Direction and Form.* He has a B.A. in history and political science from the University of Iowa, an M.A. in political science from Iowa State University, and was an Anna Sobol Levy Fellow of Middle East studies at the Hebrew University of Jerusalem. He also holds a Ph.D. in government and politics from the University of Maryland, College Park. He is member of the Rethinking the Principles of War steering group.

## Mr. Sean Tytler

Sean Tytler is an Aerospace Engineer with SRA International, supporting the National Geospatial-Intelligence Agency. Since graduating from the Massachusetts Institute of Technology in 1998, he has provided analysis and insight into a wide range of intelligence collection and analytical capabilities for the United States Air Force, Office of Secretary of Defense, and NGA. He received a Master's degree in National Security Studies from Georgetown University in 2004 and resides in Falls Church, Virginia.

# Dr. Harlan K. Ullman

Dr. Harlan Ullman works in the worlds of both policy and business. In the former, he is an author, columnist, and television regular on foreign, defense, and national security matters. His latest book, *Finishing Business: Ten Steps to Defeat Global Terror* (Naval Institute Press, 2004), was named one of the ten best policy books of 2004. His columns appear in the *Washington Times* and he is a contributor for Fox, BBC, and Al Jazeera.

A U.S. Naval Academy graduate, he was skipper of both swift boats (completing in excess of 150 missions and patrols in Vietnam) and destroyers, and served as professor of military strategy at the National War College. He is senior adviser at the Center for Strategic and International Studies and chairman of CNAC's senior seminar (Center for Naval Analyses).

Dr. Ullman chairs two private company boards. His next books are *Shock and Awe Ten Years Later—New, Old and Changing Faces of War* and *Shattered Dreams—How Culture and Ideology are Destroying America's Relevance Abroad and Society at Home,* both due out in late 2006.

# Dr. Milan Vego

Dr. Milan Vego served for twelve years as a naval officer in the former Yugoslav Navy and three years as 2nd mate (deck) on board former West German merchant ships. Subsequently, he served as adjunct professor, the George Washington University (1984); visiting fellow, Center for Naval Analyses (CNA), Alexandria, Virginia (1985–87); research fellow, Foreign Military Studies Office (formerly Soviet Army Studies Office, U.S. Combined Arms Center, Fort Leavenworth, Kansas (1987–89); adjunct professor at the Defense Intelligence College, DIA (1985–91); and at the Wargaming and Simulation Center, National Defense University (1988–91). Since August 1991, he has served as professor of operations, JMO Department, the U.S. Naval War College.

Dr. Vego has published five books: *Soviet Navy Today* (Arms & Armour Press, 1986); *Soviet Naval Tactics* (Naval Institute Press, 1992); *Austro-Hungarian Naval Policy, 1904–1914* (Frank Cass Publishers, 1996); *Naval Strategy and Operations in Narrow Seas* (Frank Cass Publishers, 1999; 2nd revised and enlarged edition, 2003); and *Operational Warfare* (Naval War College, 2000). The book *Battle for Leyte, September–December 1944: An Operational Analysis* will be published by the Naval Institute Press in November

2005. Dr. Vego has written two monographs: *Yugoslavia and Soviet Policy of Force in the Mediterranean* (Center for Naval Analysis, 1981) and *Recce-Strike Complexes in Soviet Theory and Practice* (Soviet Army Studies Office, U.S. Army Combined Arms Center, 1990). He has also published around 250 articles/essays in various professional journals.

Dr. Vego holds a B.A. in modern history, University of Belgrad; an M.A. in modern history, University of Belgrade; a Master Mariner's License (1973); and a Ph.D. in European history from the George Washington University.

## Dr. Michael Vlahos

Michael Vlahos is principal staff in the National Security Analysis Department (NSAD) at Johns Hopkins University's Applied Physics Laboratory. He was director of the State Department's Center for the Study of Foreign Affairs from 1988–91, and director of security studies at Johns Hopkins University's School of Advanced International Studies from 1981–88. Dr. Vlahos has worked with anthropologists and Islamic studies specialists to develop a culture-area concept to help the defense world better understand and respond operationally to the changing environment of the Muslim world.

This concept appears in his two recent monographs—*Terror's Mask: Insurgency within Islam* (2002), and *Culture's Mask: War and Change after Iraq* (2004)—and the paper "Two Enemies: Non-State Actors and Change in the Muslim World." Dr. Vlahos earned his doctorate in history and strategic studies from the Fletcher School of Law and Diplomacy at Tufts University and is a 1973 graduate of Yale College. In addition to eight books and monographs, Dr. Vlahos has published four-score articles in, among other periodicals, *Foreign Affairs, Washington Quarterly, The Times Literary Supplement, Foreign Policy, National Review,* and *Rolling Stone.*

## Dr. Wesley K. Wark

Dr. Wesley Wark is a professor at the University of Toronto's Munk Centre for International Studies. He teaches both graduate and undergraduate courses on intelligence, terrorism, and security.

He has published numerous books and articles since his first book, *British Intelligence and Nazi Germany,* appeared in 1985. He is the author, most recently, of the edited collection *Twenty-First Century Intelligence* (Routledge, 2005). He was the guest editor and contributor for a special issue of the *International Journal,* published by the Canadian Institute of International Affairs, on the subject of "Security in an Age of Terrorism" (Winter 2004/05). He is preparing for publication an official history of the Canadian intelligence community during the Cold War. Professor Wark's next book project will be a study of Canada and the war on terror, due out in 2006.

Professor Wark serves as elected president of the Canadian Association for Security and Intelligence Studies.

## Dr. Michael Warner

Dr. Michael Warner is a Central Intelligence Agency historian currently serving in the Office of the Director of National Intelligence. He has written and lectured extensively on the origins and evolution of U.S. intelligence. The views expressed in this essay are his own and do not reflect official positions or views of the CIA or any other U.S. government entity.